Birkhäuser

Progress in Nonlinear Differential Equations and Their Applications
Volume 83

For further volumes:
http://www.springer.com/series/4889

Gyula Csató • Bernard Dacorogna
Olivier Kneuss

The Pullback Equation for Differential Forms

Birkhäuser

Gyula Csató
Lausanne, Switzerland

Bernard Dacorogna
Lausanne, Switzerland

Olivier Kneuss
Lausanne, Switzerland

ISBN 978-0-8176-8312-2 e-ISBN 978-0-8176-8313-9
DOI 10.1007/978-0-8176-8313-9
Springer New York Dordrecht Heidelberg London

Library of Congress Control Number: 2011941790

Mathematics Subject Classification: 15A75, 35FXX, 58AXX

Printed on acid-free paper

Springer is part of Springer Science+Business Media (www.birkhauser-science.com)

Preface

In the present book we study the pullback equation for differential forms

$$\varphi^*(g) = f,$$

namely, given two differential k-forms f and g we want to discuss the equivalence of such forms. This turns out to be a system of nonlinear first-order partial differential equations in the unknown map φ.

The problem that we study here is a particular case of the equivalence of tensors which has received considerable attention. However, the pullback equation for differential forms has quite different features than those for symmetric tensors, such as Riemannian metrics, which has also been studied a great deal. In more physical terms, the problem of equivalence of forms can also be seen as a problem of mass transportation.

This is an important problem in geometry and in analysis. It has been extensively studied, in the cases $k = 2$ and $k = n$, but much less when $3 \leq k \leq n-1$. The problem considered here of finding normal forms (Darboux theorem, Pfaff normal form) is a fundamental question in symplectic and contact geometry. With respect to the literature in geometry, the main emphasis of the book is on regularity and boundary conditions. Indeed, special attention has been given to getting optimal regularity; this is a particularly delicate point and requires estimates for elliptic equations and fine properties of Hölder spaces.

In the case of volume forms (i.e., $k = n$), our problem is clearly related to the widely studied subject of optimal mass transportation. However, our analysis is not in this framework. As stated before, the two main points of our analysis are that we provide optimal regularity in Hölder spaces and, at the same time, we are able to handle boundary conditions.

Our book will hopefully appeal to both geometers and analysts. In order to make the subject more easily attractive for the analysts, we have reduced as much as possible the notations of geometry. For example, we have restricted our attention to domains in $\mathbb{R}^n$, but it goes without saying that all results generalize to manifolds with or without boundary.

In Part I we gather some basic facts about exterior and differential forms that are used throughout Parts II and IV. Most of the results are standard, but they are presented so that the reader may be able to grasp the main results of the subject without getting too involved with the terminology and concepts of differential geometry.

Part II presents the classical Hodge decomposition following the proof of Morrey, but with some variants, notably in our way of deriving the Gaffney inequality. We also give applications to several versions of the Poincaré lemma that are constantly used in the other parts of the book. Part II can be of interest independently of the main subject of the book.

Part III discusses the case $k = n$. We have tried in this part to make it, as much as possible, independent of the machinery of differential forms. Indeed, Part III can essentially be read with no reference to the other parts of the work, except for the properties of Hölder spaces presented in Part V.

Part IV deals with the general case. Emphasis in this part is given to the symplectic case $k = 2$. We also briefly deal with the simpler cases $k = 0, 1, n-1$. The case $3 \leq k \leq n-2$ is much harder and we are able to obtain results only for forms having a special structure. The difficulty is already at the algebraic level.

In Part V we gather several basic properties of Hölder spaces that are used extensively throughout the book. Due to the nonlinearity of the pullback equation, Hölder spaces are much better adapted than Sobolev spaces. The literature on Hölder spaces is considerably smaller than the one on Sobolev spaces. Moreover, the results presented here cannot be found solely in a single reference. We hope that this part will be useful to mathematicians well beyond those who are primarily interested in the pullback equation.

Acknowledgments Several results of Part IV find their origins from joint works and discussions with S. Bandyopadhyay. During the preparation of the manuscript, we have benefited from many helpful comments by P. Bousquet, G. Cupini, W. Gangbo, N. Kamran, T. Ratiu, K.D. Semmler, D. Serre and D. Ye. The discussions with M. Troyanov have been particularly fruitful. We also thank H. Brézis for accepting, with enthusiasm, our book in the Birkhäuser series that he edits.

The research of the third author has been, in part, subsidized by a grant of the Fonds National Suisse de la Recherche Scientifique.

Lausanne

CSATÓ Gyula
DACOROGNA Bernard
KNEUSS Olivier

Contents

Chapter 1
Introduction

1.1 Statement of the Problem

The aim of this book is the study of the *pullback equation*

$$\varphi^* (g) = f. \tag{1.1}$$

More precisely, we want to find a map $\varphi : \mathbb{R}^n \to \mathbb{R}^n$; preferably we want this map to be a diffeomorphism that satisfies the above equation, where f and g are differential k-forms, $0 \leq k \leq n$. Most of the time we will require these two forms to be closed. Before going further, let us examine the exact meaning of (1.1). We write

$$g(x) = \sum_{1 \leq i_1 < \cdots < i_k \leq n} g_{i_1 \cdots i_k}(x)\, dx^{i_1} \wedge \cdots \wedge dx^{i_k}$$

and similarly for f. The meaning of (1.1) is that

$$\sum_{1 \leq i_1 < \cdots < i_k \leq n} g_{i_1 \cdots i_k} \circ \varphi \, d\varphi^{i_1} \wedge \cdots \wedge d\varphi^{i_k} = \sum_{1 \leq i_1 < \cdots < i_k \leq n} f_{i_1 \cdots i_k} dx^{i_1} \wedge \cdots \wedge dx^{i_k},$$

where

$$d\varphi^i = \sum_{j=1}^{n} \frac{\partial \varphi^i}{\partial x_j} dx^j.$$

This turns out to be a *nonlinear* (if $2 \leq k \leq n$) homogeneous of degree k (in the derivatives) first-order system of $\binom{n}{k}$ partial differential equations. Let us see the form that the equation takes when $k = 0, 1, 2, n$.

Case: $k = 0$. Equation (1.1) reads as

$$g(\varphi(x)) = f(x)$$

while

$$dg = 0 \Leftrightarrow \operatorname{grad} g = 0.$$

We will be, only marginally, interested in this elementary case, which is trivial for closed forms. In any case, (1.1) is *not*, when $k = 0$, a differential equation.

Case: $k = 1$. The form g, and analogously for f, can be written as

$$g(x) = \sum_{i=1}^{n} g_i(x)\, dx^i.$$

Equation (1.1) then becomes

$$\sum_{i=1}^{n} g_i(\varphi(x))\, d\varphi^i = \sum_{i=1}^{n} f_i(x)\, dx^i$$

while

$$dg = 0 \Leftrightarrow \operatorname{curl} g = 0 \Leftrightarrow \frac{\partial g_i}{\partial x_j} - \frac{\partial g_j}{\partial x_i} = 0, \quad 1 \le i < j \le n.$$

Writing

$$d\varphi^i = \sum_{j=1}^{n} \frac{\partial \varphi^i}{\partial x_j} dx^j$$

and substituting into the equation, we find that (1.1) is equivalent to

$$\sum_{j=1}^{n} g_j(\varphi(x)) \frac{\partial \varphi^j}{\partial x_i}(x) = f_i(x), \quad 1 \le i \le n.$$

This is a system of $\binom{n}{1} = n$ first-order *linear* (in the first derivatives) partial differential equations.

Case: $k = 2$. The form g, and analogously for f, can be written as

$$g = \sum_{1 \le i < j \le n} g_{ij}(x)\, dx^i \wedge dx^j$$

while

$$dg = 0 \Leftrightarrow \frac{\partial g_{ij}}{\partial x_k} - \frac{\partial g_{ik}}{\partial x_j} + \frac{\partial g_{jk}}{\partial x_i} = 0, \quad 1 \le i < j < k \le n.$$

The equation $\varphi^*(g) = f$ becomes

$$\sum_{1 \le p < q \le n} g_{pq}(\varphi(x))\, d\varphi^p \wedge d\varphi^q = \sum_{1 \le i < j \le n} f_{ij}(x)\, dx^i \wedge dx^j.$$

We get, as before, that (1.1) is equivalent, for every $1 \le i < j \le n$, to

$$\sum_{1 \le p < q \le n} g_{pq}(\varphi(x)) \left(\frac{\partial \varphi^p}{\partial x_i} \frac{\partial \varphi^q}{\partial x_j} - \frac{\partial \varphi^p}{\partial x_j} \frac{\partial \varphi^q}{\partial x_i} \right)(x) = f_{ij}(x),$$

which is a *nonlinear* homogeneous of degree 2 (in the derivatives) system of $\binom{n}{2} = \frac{n(n-1)}{2}$ first-order partial differential equations.

Case: $k=n$. In this case we always have $df=dg=0$. By abuse of notations, if we identify volume forms and functions, we get that the equation $\varphi^*(g)=f$ becomes

$$g(\varphi(x))\det\nabla\varphi(x)=f(x).$$

It is then a nonlinear homogeneous of degree n (in the derivatives) first-order partial differential equation.smallskip

The main questions that we will discuss are the following.

1) *Local existence.* This is the easiest question. We will handle fairly completely the case of closed 2-forms, which is the case of the Darboux theorem. The cases of 1 and $(n-1)$-forms as well as the case of n-forms will also be dealt with. It will turn out that the case $3\le k\le n-2$ is much more difficult and we will be able to handle only closed k-forms with special structure.
2) *Global existence.* This is a much more difficult problem. We will obtain results in the case of volume forms and of closed 2-forms.
3) *Regularity.* A special emphasis will be given on getting sharp regularity results. For this reason we will have to work with Hölder spaces $C^{r,\alpha}$, $0<\alpha<1$, not with spaces C^r. Apart from the linear problems considered in Part II, we will not deal with Sobolev spaces. In the present context the reason is that Hölder spaces form an algebra contrary to Sobolev spaces (with low exponents).

1.2 Exterior and Differential Forms

In Chapter 2 we have gathered some algebraic results about exterior forms that are used throughout the book.

1.2.1 Definitions and Basic Properties of Exterior Forms

Let $1\le k\le n$ be an integer. An exterior k-form will be denoted by

$$f=\sum_{1\le i_1<\cdots<i_k\le n} f_{i_1\cdots i_k}e^{i_1}\wedge\cdots\wedge e^{i_k}.$$

The set of exterior k-forms over $\mathbb{R}^n$ is a vector space and is denoted $\Lambda^k(\mathbb{R}^n)$ and its dimension is

$$\dim(\Lambda^k(\mathbb{R}^n))=\tbinom{n}{k}.$$

If $k=0$, we set

$$\Lambda^0(\mathbb{R}^n)=\mathbb{R}.$$

By abuse of notations, we will, when convenient and in order not to burden the notations, identify k-forms with vectors in $\mathbb{R}^{\binom{n}{k}}$.

(i) The *exterior product* of $f \in \Lambda^k(\mathbb{R}^n)$ with $g \in \Lambda^l(\mathbb{R}^n)$, denoted by $f \wedge g$, is defined as usual (cf. Definition 2.2) and it belongs to $\Lambda^{k+l}(\mathbb{R}^n)$. The *scalar product* between two k-forms f and g is denoted by

$$\langle g; f \rangle = \sum_{1 \leq i_1 < \cdots < i_k \leq n} g_{i_1 \cdots i_k} f_{i_1 \cdots i_k}.$$

The *Hodge star operator* (cf. Definition 2.9) associates to $f \in \Lambda^k(\mathbb{R}^n)$ a form $(*f) \in \Lambda^{n-k}(\mathbb{R}^n)$. We define (cf. Definition 2.11) the *interior product* of $f \in \Lambda^k(\mathbb{R}^n)$ with $g \in \Lambda^l(\mathbb{R}^n)$ by

$$g \lrcorner f = (-1)^{n(k-l)} * (g \wedge (*f)).$$

These definitions are linked through the following elementary facts (cf. Proposition 2.16). For every $f \in \Lambda^k(\mathbb{R}^n)$, $g \in \Lambda^{k+1}(\mathbb{R}^n)$ and $h \in \Lambda^1(\mathbb{R}^n)$,

$$|h|^2 f = h \lrcorner (h \wedge f) + h \wedge (h \lrcorner f),$$
$$\langle h \wedge f; g \rangle = \langle f; h \lrcorner g \rangle.$$

(ii) Let $A \in \mathbb{R}^{n \times n}$ be a matrix and let $f \in \Lambda^k(\mathbb{R}^n)$ be given by

$$f = \sum_{1 \leq i_1 < \cdots < i_k \leq n} f_{i_1 \cdots i_k} e^{i_1} \wedge \cdots \wedge e^{i_k}.$$

We define (cf. Definition 2.17) the *pullback of f by A*, denoted $A^*(f)$, by

$$A^*(f) = \sum_{1 \leq i_1 < \cdots < i_k \leq n} f_{i_1 \cdots i_k} A^{i_1} \wedge \cdots \wedge A^{i_k} \in \Lambda^k(\mathbb{R}^n),$$

where A^j is the jth row of A and is identified by

$$A^j = \sum_{k=1}^{n} A_k^j e^k \in \Lambda^1(\mathbb{R}^n).$$

If $k = 0$, we then let

$$A^*(f) = f.$$

The present definition is consistent with the one given at the beginning of the chapter; just set $\varphi(x) = Ax$ in (1.1).

(iii) We next define the notion of *rank* (also called rank of order 1 in Chapter 2) of $f \in \Lambda^k(\mathbb{R}^n)$. We first associate to the linear map

$$g \in \Lambda^1(\mathbb{R}^n) \to g \lrcorner f \in \Lambda^{k-1}(\mathbb{R}^n)$$

a matrix $\overline{f} \in \mathbb{R}^{\binom{n}{k-1} \times n}$ such that, by abuse of notations,

$$g \lrcorner f = \overline{f} g \quad \text{for every } g \in \Lambda^1(\mathbb{R}^n).$$

In this case, we have

$$g \lrcorner f$$
$$= \sum_{1 \leq j_1 < \cdots < j_{k-1} \leq n} \left(\sum_{\gamma=1}^{k} (-1)^{\gamma-1} \sum_{j_{\gamma-1} < i < j_\gamma} f_{j_1 \cdots j_{\gamma-1} i j_\gamma \cdots j_{k-1}} g_i \right) e^{j_1} \wedge \cdots \wedge e^{j_{k-1}}.$$

More explicitly, using the lexicographical order for the columns (index below) and the rows (index above) of the matrix $\overline{f}$, we have

$$(\overline{f})_i^{j_1 \cdots j_{k-1}} = f_{i j_1 \cdots j_{k-1}}$$

for $1 \leq i \leq n$ and $1 \leq j_1 < \cdots < j_{k-1} \leq n$. The rank of the k-form f is then the rank of the $\binom{n}{k-1} \times n$ matrix $\overline{f}$ (or similarly the rank of the map $g \to g \lrcorner f$). We then write (in Chapter 2, we write $\operatorname{rank}_1[f]$, but in the remaining part of the book we write only $\operatorname{rank}[f]$)

$$\operatorname{rank}[f] = \operatorname{rank}(\overline{f}).$$

Note that only when $k = 2$ or $k = n$, the matrix $\overline{f}$ is a square matrix. We will get our best results precisely in these cases and when the matrix $\overline{f}$ is invertible.

We then have the following elementary result (cf. Proposition 2.37).

Proposition 1.1. *Let* $f \in \Lambda^k(\mathbb{R}^n)$, $f \neq 0$.

(i) If $k = 1$, *then the rank of* f *is always* 1.

(ii) If $k = 2$, *then the rank of* f *is even. The forms*

$$\omega_m = \sum_{i=1}^{m} e^{2i-1} \wedge e^{2i}$$

are such that $\operatorname{rank}[\omega_m] = 2m$. *Moreover,* $\operatorname{rank}[f] = 2m$ *if and only if*

$$f^m \neq 0 \quad \text{and} \quad f^{m+1} = 0,$$

where $f^m = \underbrace{f \wedge \cdots \wedge f}_{m \text{ times}}$.

(iii) If $3 \leq k \leq n$, *then*

$$\operatorname{rank}[f] \in \{k, k+2, \ldots, n\}$$

and any of the values in $\{k, k+2, \ldots, n\}$ *can be achieved by the rank of a k-form. In particular, if* $k = n-1$, *then* $\operatorname{rank}[f] = n-1$, *whereas if* $k = n$, *then* $\operatorname{rank}[f] = n$.

Remark 1.2 (cf. Propositions 2.24 and 2.33). The rank is an invariant for the pullback equation. More precisely, if there exists $A \in \mathrm{GL}(n)$ (i.e., A is an invertible $n \times n$ matrix) such that

$$A^*(g) = f,$$

then

$$\operatorname{rank}[g] = \operatorname{rank}[f].$$

Conversely, when $k = 1, 2, n-1, n,$ if $\operatorname{rank}[g] = \operatorname{rank}[f]$, then there exists $A \in \mathrm{GL}(n)$ such that

$$A^*(g) = f.$$

However, the converse is not true, in general, if $3 \le k \le n-2$. For example (cf. Example 2.36), when $k = 3$, the forms

$$f = e^1 \wedge e^2 \wedge e^3 + e^4 \wedge e^5 \wedge e^6,$$
$$g = e^1 \wedge e^2 \wedge e^3 + e^1 \wedge e^4 \wedge e^5 + e^2 \wedge e^4 \wedge e^6 + e^3 \wedge e^5 \wedge e^6$$

have both rank $= 6$, but there is no $A \in \mathrm{GL}(6)$ so that

$$A^*(g) = f.$$

Similarly and more strikingly (cf. Example 2.35), when $k = 4$ and

$$f = e^1 \wedge e^2 \wedge e^3 \wedge e^4 + e^1 \wedge e^2 \wedge e^5 \wedge e^6 + e^3 \wedge e^4 \wedge e^5 \wedge e^6,$$

there is no $A \in \mathrm{GL}(6)$ such that

$$A^*(f) = -f.$$

1.2.2 Divisibility

We then discuss the notion of *divisibility* for exterior forms. Given two integers $1 \le l \le k \le n$, a k-form f and a l-form g, we want to know if we can find a $(k-l)$-form u so that

$$f = g \wedge u.$$

This is an important question in the theory of Grassmann algebras. A well-known result is the so called Cartan lemma (cf. Theorem 2.42).

Theorem 1.3 (Cartan lemma). *Let $1 \le k \le n$ and $f \in \Lambda^k(\mathbb{R}^n)$ with $f \ne 0$. Let $1 \le l \le k$ and $g_1, \ldots, g_l \in \Lambda^1(\mathbb{R}^n)$ be such that*

$$g_1 \wedge \cdots \wedge g_l \ne 0.$$

Then there exists $u \in \Lambda^{k-l}(\mathbb{R}^n)$ verifying

$$f = g_1 \wedge \cdots \wedge g_l \wedge u$$

if and only if

$$f \wedge g_1 = \cdots = f \wedge g_l = 0.$$

Remark 1.4. In the same spirit, the following facts can easily be proved (cf. Proposition 2.43):

(i) The form $f \in \Lambda^k(\mathbb{R}^n)$ is totally divisible, meaning that there exist $f_1, \cdots, f_k \in \Lambda^1(\mathbb{R}^n)$ such that

$$f = f_1 \wedge \cdots \wedge f_k$$

if and only if

$$\operatorname{rank}[f] = k.$$

(ii) If k is odd and if $f \in \Lambda^k(\mathbb{R}^n)$ with $\operatorname{rank}[f] = k+2$, then there exist $u \in \Lambda^1(\mathbb{R}^n)$ and $g \in \Lambda^{k-1}(\mathbb{R}^n)$ such that

$$f = g \wedge u.$$

Our main result (cf. Theorem 2.45 for a more general statement) will be the following theorem obtained by Dacorogna–Kneuss [31]. It generalizes the Cartan lemma.

Theorem 1.5. *Let $0 \leq l \leq k \leq n$ be integers. Let $g \in \Lambda^l(\mathbb{R}^n)$ and $f \in \Lambda^k(\mathbb{R}^n)$. The following statements are then equivalent:*

(i) There exists $u \in \Lambda^{k-l}(\mathbb{R}^n)$ verifying

$$f = g \wedge u.$$

(ii) For every $h \in \Lambda^{n-k}(\mathbb{R}^n)$, the following implication holds:

$$[h \wedge g = 0] \quad \Rightarrow \quad [h \wedge f = 0].$$

1.2.3 Differential Forms

In Chapter 3 we have gathered the main notations concerning differential forms.

Definition 1.6. Let $\Omega \subset \mathbb{R}^n$ be open and $f \in C^1\left(\Omega; \Lambda^k\right)$, namely

$$f = \sum_{1 \leq i_1 < \cdots < i_k \leq n} f_{i_1 \cdots i_k} dx^{i_1} \wedge \cdots \wedge dx^{i_k}.$$

(i) The *exterior derivative* of f denoted df belongs to $C^0\left(\Omega; \Lambda^{k+1}\right)$ and is defined by

$$df = \sum_{1 \leq i_1 < \cdots < i_k \leq n} \sum_{m=1}^{n} \frac{\partial f_{i_1 \cdots i_k}}{\partial x_m} dx^m \wedge dx^{i_1} \wedge \cdots \wedge dx^{i_k}.$$

If $k = n$, then $df = 0$.

(ii) The *interior derivative* or *codifferential* of f denoted δf belongs to $C^0(\Omega;\Lambda^{k-1})$ and is defined by

$$\delta f = (-1)^{n(k-1)} * (d(*f)).$$

Remark 1.7. (i) If $k = 0$, then the operator d can be identified with the gradient operator, while $\delta f = 0$ for any f.

(ii) If $k = 1$, then the operator d can be identified with the curl operator and the operator δ is the divergence operator.

We next gather some well-known properties of the operators d and δ (cf. Theorems 3.5 and 3.7).

Theorem 1.8. *Let $f \in C^2(\Omega;\Lambda^k)$. Then*

$$ddf = 0, \quad \delta\delta f = 0 \quad \textit{and} \quad d\delta f + \delta d f = \Delta f.$$

We also need the following definition. In the sequel we will denote the exterior unit normal of $\partial\Omega$ by ν.

Definition 1.9. The *tangential component* of a k-form f on $\partial\Omega$ is the $(k+1)$-form

$$\nu \wedge f \in \Lambda^{k+1}.$$

The *normal component* of a k-form f on $\partial\Omega$ is the $(k-1)$-form

$$\nu \lrcorner f \in \Lambda^{k-1}.$$

We easily deduce the following properties (cf. Theorem 3.23).

Proposition 1.10. *Let $0 \le k \le n$ and $f \in C^1(\overline{\Omega};\Lambda^k)$; then*

$$\begin{aligned} \nu \wedge f = 0 \text{ on } \partial\Omega \;&\Rightarrow\; \nu \wedge df = 0 \text{ on } \partial\Omega, \\ \nu \lrcorner f = 0 \text{ on } \partial\Omega \;&\Rightarrow\; \nu \lrcorner \delta f = 0 \text{ on } \partial\Omega. \end{aligned}$$

We will constantly use the integration by parts formula (cf. Theorem 3.28).

Theorem 1.11. *Let $1 \le k \le n$, $f \in C^1(\overline{\Omega};\Lambda^{k-1})$ and $g \in C^1(\overline{\Omega};\Lambda^k)$. Then*

$$\int_\Omega \langle df; g\rangle + \int_\Omega \langle f; \delta g\rangle = \int_{\partial\Omega} \langle \nu \wedge f; g\rangle = \int_{\partial\Omega} \langle f; \nu \lrcorner g\rangle.$$

We will adopt the following notations.

Notation 1.12. *Let $\Omega \subset \mathbb{R}^n$ be open, $r \ge 0$ be an integer and $0 \le \alpha \le 1 \le p \le \infty$. Spaces with vanishing tangential or normal component will be denoted in the following way:*

$$C_T^{r,\alpha}(\overline{\Omega};\Lambda^k) = \{f \in C^{r,\alpha}(\overline{\Omega};\Lambda^k) : \nu \wedge f = 0 \quad \text{on } \partial\Omega\},$$
$$C_N^{r,\alpha}(\overline{\Omega};\Lambda^k) = \{f \in C^{r,\alpha}(\overline{\Omega};\Lambda^k) : \nu \lrcorner f = 0 \quad \text{on } \partial\Omega\},$$
$$W_T^{r+1,p}(\Omega;\Lambda^k) = \{f \in W^{r+1,p}(\Omega;\Lambda^k) : \nu \wedge f = 0 \quad \text{on } \partial\Omega\},$$
$$W_N^{r+1,p}(\Omega;\Lambda^k) = \{f \in W^{r+1,p}(\Omega;\Lambda^k) : \nu \lrcorner f = 0 \quad \text{on } \partial\Omega\}.$$

The different sets of harmonic fields *will be denoted by*

$$\mathcal{H}(\Omega;\Lambda^k) = \{f \in W^{1,2}(\Omega;\Lambda^k) : df = 0 \text{ and } \delta f = 0 \text{ in } \Omega\},$$
$$\mathcal{H}_T(\Omega;\Lambda^k) = \{f \in \mathcal{H}(\Omega;\Lambda^k) : \nu \wedge f = 0 \quad \text{on } \partial\Omega\},$$
$$\mathcal{H}_N(\Omega;\Lambda^k) = \{f \in \mathcal{H}(\Omega;\Lambda^k) : \nu \lrcorner f = 0 \quad \text{on } \partial\Omega\}.$$

We now list (cf. Section 6.1) some properties of the harmonic fields.

Theorem 1.13. *Let $\Omega \subset \mathbb{R}^n$ be an open set. Then*

$$\mathcal{H}(\Omega;\Lambda^k) \subset C^\infty(\Omega;\Lambda^k).$$

Moreover if Ω is bounded and smooth, then the next statements are valid.

(i) The following inclusion holds:

$$\mathcal{H}_T(\Omega;\Lambda^k) \cup \mathcal{H}_N(\Omega;\Lambda^k) \subset C^\infty(\overline{\Omega};\Lambda^k).$$

Furthermore, if $r \geq 0$ is an integer and $0 \leq \alpha \leq 1$, then there exists $C = C(r,\Omega)$ such that for every $\omega \in \mathcal{H}_T(\Omega;\Lambda^k) \cup \mathcal{H}_N(\Omega;\Lambda^k)$,

$$\|\omega\|_{W^{r,2}} \leq C\|\omega\|_{L^2} \quad \text{and} \quad \|\omega\|_{C^{r,\alpha}} \leq C\|\omega\|_{C^0}.$$

(ii) The spaces $\mathcal{H}_T(\Omega;\Lambda^k)$ and $\mathcal{H}_N(\Omega;\Lambda^k)$ are finite dimensional and closed in $L^2(\Omega;\Lambda^k)$.

(iii) Furthermore, if Ω is contractible (cf. Definition 6.1), then

$$\mathcal{H}_T(\Omega;\Lambda^k) = \{0\} \quad \text{if } 0 \leq k \leq n-1,$$
$$\mathcal{H}_N(\Omega;\Lambda^k) = \{0\} \quad \text{if } 1 \leq k \leq n.$$

(iv) If $k = 0$ or $k = n$ and $h \in \mathcal{H}(\Omega;\Lambda^k)$, then h is constant on each connected component of Ω. In particular, $\mathcal{H}_T(\Omega;\Lambda^0) = \{0\}$ and $\mathcal{H}_N(\Omega;\Lambda^n) = \{0\}$.

Remark 1.14. If $k = 1$ and assuming that Ω is smooth, then the sets $\mathcal{H}_T$ and $\mathcal{H}_N$ can be rewritten, as usual by abuse of notations, as

$$\mathcal{H}_T(\Omega;\Lambda^1) = \left\{ f \in C^\infty(\overline{\Omega};\mathbb{R}^n) : \begin{bmatrix} \operatorname{curl} f = 0 \text{ and } \operatorname{div} f = 0 \\ f_i\nu_j - f_j\nu_i = 0,\ \forall 1 \leq i < j \leq n \end{bmatrix} \right\},$$
$$\mathcal{H}_N(\Omega;\Lambda^1) = \left\{ f \in C^\infty(\overline{\Omega};\mathbb{R}^n) : \begin{bmatrix} \operatorname{curl} f = 0 \text{ and } \operatorname{div} f = 0 \\ \sum_{i=1}^n f_i\nu_i = 0 \end{bmatrix} \right\}.$$

Moreover, if Ω is simply connected, then

$$\mathscr{H}_T\left(\Omega;\Lambda^1\right)=\mathscr{H}_N\left(\Omega;\Lambda^1\right)=\{0\}.$$

1.3 Hodge–Morrey Decomposition and Poincaré Lemma

1.3.1 A General Identity and Gaffney Inequality

In the proof of Morrey of the Hodge decomposition, one of the key points to get compactness is the following inequality (cf. Theorem 5.16).

Theorem 1.15 (Gaffney inequality). *Let $\Omega\subset\mathbb{R}^n$ be a bounded open smooth set. Then there exists a constant $C=C(\Omega)>0$ such that*

$$\|\omega\|_{W^{1,2}}^2\leq C\left(\|d\omega\|_{L^2}^2+\|\delta\omega\|_{L^2}^2+\|\omega\|_{L^2}^2\right)$$

for every $\omega\in W_T^{1,2}(\Omega;\Lambda^k)\cup W_N^{1,2}(\Omega;\Lambda^k)$.

Remark 1.16. When $k=1$, the inequality says, identifying 1-forms with vector fields,

$$\|\omega\|_{W^{1,2}}^2\leq C\left(\|\operatorname{curl}\omega\|_{L^2}^2+\|\operatorname{div}\omega\|_{L^2}^2+\|\omega\|_{L^2}^2\right)$$

for every $\omega\in W^{1,2}(\Omega;\mathbb{R}^n)$ satisfying either one of the following two conditions:

$$\nu\wedge\omega=0\ \Leftrightarrow\ \omega_i\nu_j-\omega_j\nu_i=0,\ \forall\, 1\leq i<j\leq n,$$
$$\nu\lrcorner\,\omega=\langle\nu;\omega\rangle=\sum_{i=1}^n\omega_i\nu_i=0.$$

The inequality, as stated above, has been proved by Morrey [76, 77], generalizing results of Gaffney [44, 45]. We will prove in Section 5.3 the inequality appealing to a very general identity (see Theorem 5.7) proved by Csató and Dacorogna [24].

Theorem 1.17 (A general identity). *Let $0\leq k\leq n$ and let $\Omega\subset\mathbb{R}^n$ be a bounded open smooth set and with exterior unit normal ν. Then every $\alpha,\beta\in C^1(\overline{\Omega};\Lambda^k)$ satisfy the equation*

$$\begin{aligned}&\int_\Omega\left(\langle d\alpha;d\beta\rangle+\langle\delta\alpha;\delta\beta\rangle-\langle\nabla\alpha;\nabla\beta\rangle\right)\\&\quad=-\int_{\partial\Omega}\left(\langle\nu\wedge d(\nu\lrcorner\,\alpha);\nu\wedge\beta\rangle+\langle\nu\lrcorner\,\delta(\nu\wedge\alpha);\nu\lrcorner\,\beta\rangle\right)\\&\qquad+\int_{\partial\Omega}\left(\langle L^\nu(\nu\wedge\alpha);\nu\wedge\beta\rangle+\langle K^\nu(\nu\lrcorner\,\alpha);\nu\lrcorner\,\beta\rangle\right).\end{aligned}$$

The operators L^ν and K^ν (cf. Definition 5.1) can be seen as matrices acting on $(k+1)$-forms and $(k-1)$-forms respectively (identifying, as usual, a k-form with a $\binom{n}{k}$ vector). They depend only on the geometry of Ω and on the degree k of the form. They can easily be calculated explicitly for general k-forms and, when Ω is a ball of radius R (cf. Corollary 5.9), it turns out that

$$L^{\nu}(\nu \wedge \omega) = \frac{k}{R}\,\nu \wedge \omega \quad \text{and} \quad K^{\nu}(\nu \lrcorner \omega) = \frac{n-k}{R}\,\nu \lrcorner \omega$$

and, thus,

$$\langle L^{\nu}(\nu \wedge \omega); \nu \wedge \omega \rangle = \frac{k}{R}\,|\nu \wedge \omega|^2 \quad \text{and} \quad \langle K^{\nu}(\nu \lrcorner \omega); \nu \lrcorner \omega \rangle = \frac{n-k}{R}\,|\nu \lrcorner \omega|^2.$$

In the case of a 1-form and for general open sets Ω (cf. Proposition 5.11), it can be shown that K^{ν} is a scalar and it is a multiple of κ, the mean curvature of the hypersurface $\partial\Omega$, namely

$$K^{\nu} = (n-1)\,\kappa.$$

Summarizing the results for a 1-form ω in $\mathbb{R}^n$ (cf. Corollary 5.12) with vanishing tangential component (i.e., $\nu \wedge \omega = 0$ on $\partial\Omega$), we have

$$\int_{\Omega} \left(|\operatorname{curl} \omega|^2 + |\operatorname{div} \omega|^2 - |\nabla \omega|^2 \right) = (n-1) \int_{\partial\Omega} \kappa\, [\langle \nu; \omega \rangle]^2,$$

where κ is the mean curvature of the hypersurface $\partial\Omega$ and $\langle .;. \rangle$ denotes the scalar product in $\mathbb{R}^n$.

1.3.2 The Hodge–Morrey Decomposition

We now turn to the celebrated Hodge–Morrey decomposition (cf. Theorem 6.9).

Theorem 1.18 (Hodge–Morrey decomposition). *Let $\Omega \subset \mathbb{R}^n$ be a bounded open smooth set. Let $0 \le k \le n$ and $f \in L^2(\Omega; \Lambda^k)$. Then there exist*

$$\alpha \in W_T^{1,2}(\Omega; \Lambda^{k-1}), \quad \beta \in W_T^{1,2}(\Omega; \Lambda^{k+1}),$$
$$h \in \mathcal{H}_T(\Omega; \Lambda^k) \quad \text{and} \quad \omega \in W_T^{2,2}(\Omega; \Lambda^k)$$

such that, in Ω,

$$f = d\alpha + \delta\beta + h, \quad \alpha = \delta\omega \quad \text{and} \quad \beta = d\omega.$$

Remark 1.19. (i) We have quoted only one of the three decompositions (cf. Theorem 6.9 for details). Another one, completely similar, is by replacing T by N and the other one mixing both T and N.

(ii) If $k \le n-1$ and if Ω is contractible, then $h = 0$.

(iii) If $k = 0$, then the theorem reads as

$$f = \delta\beta = \delta d\omega = \Delta\omega \text{ in } \Omega \quad \text{with} \quad \omega = 0 \text{ on } \partial\Omega.$$

(iv) When $k = 1$ and $n = 3$, the decomposition reads as follows. Let ν be the exterior unit normal. For any $f \in L^2(\Omega;\mathbb{R}^3)$, there exist

$$\omega \in W^{2,2}\left(\Omega;\mathbb{R}^3\right) \quad \text{with } \omega_i\nu_j - \omega_j\nu_i = 0 \text{ on } \partial\Omega,\ \forall 1 \le i < j \le 3$$

$$\alpha \in W_0^{1,2}\left(\Omega\right) \quad \text{and} \quad \alpha = \operatorname{div}\omega,$$

$$\beta \in W^{1,2}\left(\Omega;\mathbb{R}^3\right) \quad \text{with } \beta = -\operatorname{curl}\omega \quad \text{and} \quad \langle\nu;\beta\rangle = 0 \text{ on } \partial\Omega$$

$$h \in \left\{ h \in C^\infty\left(\overline{\Omega};\mathbb{R}^3\right) : \left[\begin{array}{c} \operatorname{curl} h = 0 \text{ and } \operatorname{div} h = 0 \\ h_i\nu_j - h_j\nu_i = 0,\ \forall 1 \le i < j \le 3 \end{array} \right. \right\}$$

such that

$$f = \operatorname{grad}\alpha + \operatorname{curl}\beta + h \text{ in } \Omega.$$

Furthermore, if Ω is simply connected, then $h = 0$.

(v) If f is more regular than in L^2, then α, β and ω are in the corresponding class of regularity (cf. Theorem 6.12). More precisely if, for example, $r \ge 0$ is an integer, $0 < q < 1$ and $f \in C^{r,q}(\overline{\Omega};\Lambda^k)$, then

$$\alpha \in C^{r+1,q}\left(\overline{\Omega};\Lambda^{k-1}\right), \quad \beta \in C^{r+1,q}\left(\overline{\Omega};\Lambda^{k+1}\right) \quad \text{and} \quad \omega \in C^{r+2,q}\left(\overline{\Omega};\Lambda^k\right).$$

(vi) The proof of Morrey (cf. Theorem 6.7) uses the direct methods of the calculus of variations. One minimizes

$$D_f(\omega) = \int_\Omega \left(\frac{1}{2}|d\omega|^2 + \frac{1}{2}|\delta\omega|^2 + \langle f;\omega\rangle \right)$$

in an appropriate space, Gaffney inequality giving the coercivity of the integral.

1.3.3 *First-Order Systems of Cauchy–Riemann Type*

It turns out that the Hodge–Morrey decomposition is in fact equivalent (cf. Proposition 7.9) to solving the first-order system

$$\begin{cases} d\omega = f \quad \text{and} \quad \delta\omega = g & \text{in } \Omega, \\ \nu \wedge \omega = \nu \wedge \omega_0 & \text{on } \partial\Omega \end{cases}$$

or the similar one,

$$\begin{cases} d\omega = f \quad \text{and} \quad \delta\omega = g & \text{in } \Omega, \\ \nu \lrcorner \omega = \nu \lrcorner \omega_0 & \text{on } \partial\Omega. \end{cases}$$

Both systems are discussed in Theorems 7.2 and 7.4. We here state a simplified version of the first one.

Theorem 1.20. *Let $r \geq 0$ and $1 \leq k \leq n-2$ be integers, $0 < q < 1$ and $\Omega \subset \mathbb{R}^n$ be a bounded contractible open smooth set and with exterior unit normal ν. Let $g \in C^{r,q}\left(\overline{\Omega};\Lambda^{k-1}\right)$ and $f \in C^{r,q}\left(\overline{\Omega};\Lambda^{k+1}\right)$ be such that*

$$\delta g = 0 \text{ in } \Omega, \quad df = 0 \text{ in } \Omega \quad \text{and} \quad \nu \wedge f = 0 \text{ on } \partial\Omega.$$

Then there exists $\omega \in C^{r+1,q}\left(\overline{\Omega};\Lambda^k\right)$, such that

$$\begin{cases} d\omega = f \quad \text{and} \quad \delta\omega = g & \text{in } \Omega, \\ \nu \wedge \omega = 0 & \text{on } \partial\Omega. \end{cases}$$

Remark 1.21. (i) It turns out that the sufficient conditions are also necessary (cf. Theorems 7.2 and 7.4).

(ii) When $k = n-1$, the result is valid provided

$$\int_\Omega f = 0.$$

Note that in this case the conditions $df = 0$ and $\nu \wedge f = 0$ are automatically fulfilled.

(iii) Completely analogous results are given in Theorems 7.2 and 7.4 for Sobolev spaces.

(iv) If Ω is not contractible, then additional necessary conditions have to be added.

(v) When $k = 1$ and $n = 3$, the theorem reads as follows. Let $\Omega \subset \mathbb{R}^3$ be a bounded contractible smooth open set, $g \in C^{r,q}\left(\overline{\Omega}\right)$ and $f \in C^{r,q}\left(\overline{\Omega};\mathbb{R}^3\right)$ be such that

$$\operatorname{div} f = 0 \text{ in } \Omega \quad \text{and} \quad \langle f;\nu\rangle = 0 \text{ on } \partial\Omega.$$

Then there exists $\omega \in C^{r+1,q}\left(\overline{\Omega};\mathbb{R}^3\right)$ such that

$$\begin{cases} \operatorname{curl}\omega = f \quad \text{and} \quad \operatorname{div}\omega = g & \text{in } \Omega, \\ \omega_i\nu_j - \omega_j\nu_i = 0 \ \forall 1 \leq i < j \leq 3 & \text{on } \partial\Omega. \end{cases}$$

1.3.4 Poincaré Lemma

We start with the classical Poincaré lemma (cf. Theorem 8.1).

Theorem 1.22 (Poincaré lemma). *Let $r \geq 1$ and $0 \leq k \leq n-1$ be integers and $\Omega \subset \mathbb{R}^n$ be an open contractible set. Let $g \in C^r\left(\Omega;\Lambda^{k+1}\right)$ with $dg = 0$ in Ω. Then there exists $G \in C^r\left(\Omega;\Lambda^k\right)$ such that*

$$dG = g \quad \text{in } \Omega.$$

With the help of the Hodge–Morrey decomposition, the result can be improved (cf. Theorem 8.3) in two directions. First, one can consider general sets Ω, not only contractible sets. Moreover, one can get sharp regularity in Hölder and in Sobolev spaces. We quote here only the case of Hölder spaces. We also give the theorem with the d operator. Analogous results are also valid for the δ operator; see Theorem 8.4.

Theorem 1.23. *Let $r \geq 0$ and $0 \leq k \leq n-1$ be integers, $0 < \alpha < 1$ and $\Omega \subset \mathbb{R}^n$ be a bounded open smooth set. Let $f : \overline{\Omega} \to \Lambda^{k+1}$. The following statements are equivalent:*

(i) Let $f \in C^{r,\alpha}\left(\overline{\Omega};\Lambda^{k+1}\right)$ be such that

$$df = 0 \text{ in } \Omega \quad \text{and} \quad \int_\Omega \langle f;\psi\rangle = 0 \text{ for every } \psi \in \mathcal{H}_N\left(\Omega;\Lambda^{k+1}\right).$$

(ii) There exists $\omega \in C^{r+1,\alpha}\left(\overline{\Omega};\Lambda^k\right)$ such that

$$d\omega = f \quad \text{in } \Omega.$$

Remark 1.24. (i) When $k = n-1$, there is no restriction on the solvability of $d\omega = f$.

(ii) Recall that if Ω is contractible and $0 \leq k \leq n-1$, then

$$\mathcal{H}_N\left(\Omega;\Lambda^{k+1}\right) = \{0\}.$$

We finally consider the boundary value problems

$$\begin{cases} d\omega = f & \text{in } \Omega, \\ \omega = \omega_0 & \text{on } \partial\Omega \end{cases} \quad \text{and} \quad \begin{cases} \delta\omega = g & \text{in } \Omega, \\ \omega = \omega_0 & \text{on } \partial\Omega. \end{cases}$$

We give a result for the first one and for $\omega_0 = 0$ (cf. Theorem 8.16 for general ω_0), but a similar one (cf. Theorem 8.18) exists for the second problem. We only discuss the case of Hölder spaces, but the result is also valid in Sobolev spaces (see Theorems 8.16 and 8.18 for details).

Theorem 1.25. *Let $r \geq 0$ and $0 \leq k \leq n-1$ be integers, $0 < \alpha < 1$ and $\Omega \subset \mathbb{R}^n$ be a bounded open smooth set and with exterior unit normal ν. Then the following statements are equivalent:*

(i) Let $f \in C^{r,\alpha}\left(\overline{\Omega};\Lambda^{k+1}\right)$ satisfy

$$df = 0 \text{ in } \Omega, \quad \nu \wedge f = 0 \text{ on } \partial\Omega,$$

and, for every $\chi \in \mathcal{H}_T\left(\Omega;\Lambda^{k+1}\right)$,

$$\int_\Omega \langle f;\chi\rangle = 0.$$

(ii) There exists $\omega \in C^{r+1,\alpha}\left(\overline{\Omega};\Lambda^k\right)$ *such that*

$$\begin{cases} d\omega = f & \textit{in } \Omega, \\ \omega = 0 & \textit{on } \partial\Omega. \end{cases}$$

1.4 The Case of Volume Forms

1.4.1 Statement of the Problem

In Part III, we will discuss the following problem. Given Ω a bounded open set in $\mathbb{R}^n$ and $f,g : \mathbb{R}^n \to \mathbb{R}$, we want to find $\varphi : \overline{\Omega} \to \mathbb{R}^n$ verifying

$$\begin{cases} g(\varphi(x)) \det \nabla\varphi(x) = f(x) & x \in \Omega, \\ \varphi(x) = x & x \in \partial\Omega. \end{cases} \tag{1.2}$$

Writing the functions f and g as volume forms through the straightforward identification

$$g = g(x)dx^1 \wedge \cdots \wedge dx^n \quad \text{and} \quad f = f(x)dx^1 \wedge \cdots \wedge dx^n,$$

problem (1.2) can be written as

$$\begin{cases} \varphi^*(g) = f & \text{in } \Omega, \\ \varphi = \text{id} & \text{on } \partial\Omega, \end{cases}$$

where $\varphi^*(g)$ is the pullback of g by φ.

The following preliminary remarks are in order.

(i) The case $n = 1$ is completely elementary and is discussed in Section 1.4.2.

(ii) When $n \geq 2$, the equation in (1.2) is a nonlinear first-order *partial differential equation* homogeneous of degree n in the derivatives. It is *underdetermined*, in the sense that we have n unknowns (the components of φ) and only one equation. Related to this observation, we have that if there exists a solution to our problem, then there are infinitely many ones. Indeed, for example, if $n = 2$, Ω is the unit ball and $f = g = 1$, the maps φ_m (written in polar and in Cartesian coordinates) defined by

$$\varphi_m(x) = \varphi_m(x_1,x_2) = \begin{pmatrix} r\cos\left(\theta + 2m\pi r^2\right) \\ r\sin\left(\theta + 2m\pi r^2\right) \end{pmatrix}$$
$$= \begin{pmatrix} x_1 \cos\left(2m\pi\left(x_1^2 + x_2^2\right)\right) - x_2 \sin\left(2m\pi\left(x_1^2 + x_2^2\right)\right) \\ x_2 \cos\left(2m\pi\left(x_1^2 + x_2^2\right)\right) + x_1 \sin\left(2m\pi\left(x_1^2 + x_2^2\right)\right) \end{pmatrix}$$

satisfy (1.2) for every $m \in \mathbb{Z}$.

(iii) An integration by parts, or, what amounts to the same thing, an elementary topological degree argument (see (19.3)), immediately gives the *necessary condition* (independently of the fact that φ is a diffeomorphism or not and of the fact that $\varphi(\Omega)$ contains strictly or not Ω)

$$\int_\Omega f = \int_\Omega g. \tag{1.3}$$

In most of our analysis, it will turn out that this condition is also sufficient.

(iv) We will always assume that $g > 0$. If g is not strictly positive, then hypotheses other than (1.3) are necessary; for example, f cannot be strictly positive. Indeed if, for example, $f \equiv 1$ and g is allowed to vanish even at a single point, then no C^1 solution of our problem exists (cf. Proposition 11.6). However, in a very special case (cf. Lemma 11.21), we will deal with functions f and g that *both* change sign.

(v) We will, however, allow f to change sign, but the analysis is very different if $f > 0$ or if f vanishes, even at a single point, let alone if it becomes negative. The first problem will be discussed in Chapter 10, whereas the second one will be dealt with in Chapter 11. One of the main differences is that in the first case, any solution of (1.2) is necessarily a diffeomorphism (cf. Theorem 19.12), whereas this is never true in the second case.

(vi) It is easy to see (cf. Corollary 19.4) that any solution of (1.2) satisfies

$$\varphi(\Omega) \supset \Omega \quad \text{and} \quad \varphi(\overline{\Omega}) \supset \overline{\Omega}. \tag{1.4}$$

If $f > 0$, we have, since φ is a diffeomorphism, that (cf. Theorem 19.12)

$$\varphi(\Omega) = \Omega \quad \text{and} \quad \varphi(\overline{\Omega}) = \overline{\Omega}.$$

If this is not the case, then, in general, the inclusions can be strict. We will discuss in Chapter 11 this matter in details.

(vii) Problem (1.2) admits a *weak formulation*. Indeed, if φ is a diffeomorphism, we can write (cf. Theorem 19.7) the equation $g(\varphi)\det\nabla\varphi = f$ as

$$\int_{\varphi(E)} g = \int_E f \quad \text{for every open set } E \subset \Omega$$

or, equivalently,

$$\int_\Omega g\,\zeta\left(\varphi^{-1}\right) = \int_\Omega f\,\zeta \quad \text{for every } \zeta \in C_0^\infty(\Omega).$$

We observe that both new writings make sense if φ is only a homeomorphism.

(viii) The problem can be seen as a question of *mass transportation*. Indeed, we want to transport the mass distribution g to the mass distribution f without moving the points of the boundary of Ω. In this context, the equation is usually written as

$$\int_E g = \int_{\varphi^{-1}(E)} f \quad \text{for every open set } E \subset \Omega.$$

The problem of *optimal* mass transportation has received considerable attention. We should point out that our analysis is not in this framework. The two main strong points of our analysis are that we are able to find smooth solutions, sometimes with the optimal regularity and to deal with fixed boundary data.

1.4.2 The One-Dimensional Case

As already stated, the case $n = 1$ is completely elementary (cf. Proposition 11.4), but it exhibits some striking differences with the case $n \geq 2$. However, it may shed some light on some issues that we will discuss in the higher-dimensional case. Let $\Omega = (a,b)$,

$$F(x) = \int_a^x f(t)\,dt \quad \text{and} \quad G(x) = \int_a^x g(t)\,dt.$$

Then problem (1.2) becomes

$$\begin{cases} G(\varphi(x)) = F(x) & \text{if } x \in (a,b), \\ \varphi(a) = a \quad \text{and} & \varphi(b) = b. \end{cases}$$

If G is invertible and this happens if, for example, $g > 0$ and if

$$F([a,b]) \subset G(\mathbb{R}), \tag{1.5}$$

and this happens if, for example, $g \geq g_0 > 0$, then the problem has the solution

$$\varphi(x) = G^{-1}(F(x)).$$

The necessary condition (1.3)

$$\int_a^b f = \int_a^b g$$

ensures that

$$\varphi(a) = a \quad \text{and} \quad \varphi(b) = b.$$

This very elementary analysis leads to the following conclusions:

1) Contrary to the case $n \geq 2$, the necessary condition (1.3) is not sufficient. We need the extra condition (1.5); see Proposition 11.4 for details.

2) The problem has a *unique* solution, contrary to the case $n \geq 2$.

3) If f and g are in the space C^r, then the solution φ is in C^{r+1}.

4) If $f>0$, then φ is a diffeomorphism from $[a,b]$ onto itself.

5) If f is allowed to change sign, then, in general,

$$[a,b] \underset{\neq}{\subset} \varphi([a,b]).$$

For example, this always happens if $f(a)<0$ or $f(b)<0$.

1.4.3 The Case $f \cdot g > 0$

In Chapter 10 we will study problem (1.2) when $f \cdot g > 0$. It will be seen that (1.3) is sufficient to solve (1.2) and that any solution is in fact a diffeomorphism from $\overline{\Omega}$ to $\overline{\Omega}$ (see Theorem 19.12). This last observation implies, in particular, a symmetry in f and g and allows us to restrict ourselves, without loss of generality, to the case $g \equiv 1$. Our main result (cf. Theorem 10.3) will be the following.

Theorem 1.26 (Dacorogna–Moser theorem). *Let $r \geq 0$ be an integer and $0<\alpha<1$. Let $\Omega \subset \mathbb{R}^n$ be a bounded connected open smooth set. Then the two following statements are equivalent:*

(i) The function $f \in C^{r,\alpha}(\overline{\Omega})$, $f>0$ in $\overline{\Omega}$ and satisfies

$$\int_\Omega f = \operatorname{meas} \Omega.$$

(ii) There exists $\varphi \in \operatorname{Diff}^{r+1,\alpha}(\overline{\Omega};\overline{\Omega})$ satisfying

$$\begin{cases} \det \nabla \varphi(x) = f(x) & x \in \Omega, \\ \varphi(x) = x & x \in \partial\Omega. \end{cases}$$

Furthermore, if $c>0$ is such that

$$\left\| \frac{1}{f} \right\|_{C^0}, \quad \|f\|_{C^{0,\alpha}} \leq c,$$

then there exists a constant $C = C(c,r,\alpha,\Omega) > 0$ such that

$$\|\varphi - \mathrm{id}\|_{C^{r+1,\alpha}} \leq C \|f-1\|_{C^{r,\alpha}}.$$

The study of this problem originated in the seminal work of Moser [78]. The above optimal theorem was obtained by Dacorogna and Moser [33]. Burago and Kleiner [19] and Mc Mullen [73], independently, proved that the result is false if $r=\alpha=0$, suggesting that the gain of regularity is to be expected only when $0<\alpha<1$.

In Section 10.5 (cf. Theorem 10.11), we present a different approach proposed by Dacorogna and Moser [33] to solve our problem. This method is constructive and does not use the regularity of elliptic differential operators; in this sense, it is more elementary. The drawback is that it does not provide any gain of regularity, which is the strong point of the above theorem. However, the advantage is that it is much more flexible. For example, if we assume in (1.2) that

$$\operatorname{supp}(f-g)\subset\Omega,$$

then we will be able to find φ such that

$$\operatorname{supp}(\varphi-\mathrm{id})\subset\Omega.$$

This type of result, unreachable by the method of elliptic partial differential equations, will turn out to be crucial in Chapter 11.

1.4.4 The Case with No Sign Hypothesis on f

In Chapter 11, we discuss the case where the function f is allowed to change sign and we will follow Cupini, Dacorogna and Kneuss [25]. As already pointed out, we will however (apart from a very special case) assume that $g>0$. In fact, contrary to the case $f\cdot g>0$, the problem is no longer symmetric in f and g.

We start by observing that if f vanishes even at a single point, then the solution φ cannot be a diffeomorphism, although it can be a homeomorphism. In any case, if f is negative somewhere, it can never be a homeomorphism (see Proposition 19.14). Furthermore, if f is negative in some parts of the boundary, then any solution φ must go out of the domain (see Proposition 11.3); more precisely,

$$\overline{\Omega}\underset{\neq}{\subset}\varphi(\overline{\Omega}).$$

A special case of our theorem (cf. Theorem 11.1) is the following.

Theorem 1.27. *Let $n\geq 2$ and $r\geq 1$ be integers. Let $B_1\subset\mathbb{R}^n$ be the open unit ball. Let $f\in C^r(\overline{B}_1)$ be such that*

$$\int_{B_1} f=\operatorname{meas}B_1\,.$$

Then there exists $\varphi\in C^r(\overline{B}_1;\mathbb{R}^n)$ satisfying

$$\begin{cases}\det\nabla\varphi(x)=f(x) & x\in B_1,\\ \varphi(x)=x & x\in\partial B_1\,.\end{cases}$$

Furthermore, the following conclusions also hold:

(i) If either $f>0$ on ∂B_1 or $f\geq 0$ in $\overline{B}_1$, then φ can be chosen so that

$$\varphi(\overline{B}_1)=\overline{B}_1\,.$$

(ii) If $f\geq 0$ in $\overline{B}_1$ and $f^{-1}(0)\cap B_1$ is countable, then φ can be chosen as a homeomorphism from $\overline{B}_1$ onto $\overline{B}_1$.

1.5 The Case $0 \leq k \leq n-1$

Having dealt with the case $k = n$, we now discuss the equation

$$\varphi^*(g) = f$$

when $0 \leq k \leq n-1$. The cases $k = 0, 1, n-1$ are the simplest ones. The most important results of Part IV are for the case $k = 2$, where we obtain not only a local result but also a global one; we, moreover, obtain sharp regularity results for both cases. The case $3 \leq k \leq n-2$ is considerably harder, even at the algebraic level and we will be able to obtain results only for forms having a special structure.

We first point out the following necessary conditions (cf. Proposition 17.1).

Proposition 1.28. *Let $\Omega \subset \mathbb{R}^n$ be a bounded open smooth set and $\varphi \in \operatorname{Diff}^1\left(\overline{\Omega}; \varphi\left(\overline{\Omega}\right)\right)$. Let $1 \leq k \leq n$, $f \in C^1\left(\overline{\Omega}; \Lambda^k\right)$ and $g \in C^1\left(\varphi\left(\overline{\Omega}\right); \Lambda^k\right)$ be such that*

$$\varphi^*(g) = f \text{ in } \Omega.$$

(i) For every $x \in \Omega$,

$$\operatorname{rank}\left[g\left(\varphi(x)\right)\right] = \operatorname{rank}\left[f(x)\right] \quad \text{and} \quad \operatorname{rank}\left[dg\left(\varphi(x)\right)\right] = \operatorname{rank}\left[df(x)\right].$$

In particular,

$$dg = 0 \text{ in } \varphi(\Omega) \Leftrightarrow df = 0 \text{ in } \Omega.$$

(ii) If $\varphi(x) = x$ for $x \in \partial\Omega$, then

$$\nu \wedge f = \nu \wedge g \text{ on } \partial\Omega,$$

where ν is the exterior unit normal to Ω.

If we drop the condition that φ is a diffeomorphism, then the rank is, in general, not conserved. We have already seen such a phenomenon when $k = n$ in Theorem 1.27.

1.5.1 The Flow Method

One of the simplest and most elegant tools that we will use for the pullback equation is Theorem 12.7 and it was first established by Moser in [78], who, however, dealt only with manifolds without boundary. Its main drawback is that it does not provide the expected gain in regularity.

Theorem 1.29. *Let $r \geq 1$ and $0 \leq k \leq n$ be integers, $0 \leq \alpha \leq 1$, $T > 0$ and $\Omega \subset \mathbb{R}^n$ be a bounded open Lipschitz set. Let*

$$u \in C^{r,\alpha}\left([0,T] \times \overline{\Omega}; \mathbb{R}^n\right),\ u = u(t,x) = u_t(x),$$
$$f \in C^{r,\alpha}\left([0,T] \times \overline{\Omega}; \Lambda^k\right),\ f = f(t,x) = f_t(x)$$

be such that for every $t \in [0,T]$,

$$u_t = 0 \text{ on } \partial\Omega, \quad df_t = 0 \text{ in } \Omega \quad \text{and} \quad d(u_t \lrcorner f_t) = -\frac{d}{dt} f_t \text{ in } \Omega.$$

Then for every $t \in [0,T]$, the solution φ_t of

$$\begin{cases} \dfrac{d}{dt}\varphi_t = u_t \circ \varphi_t, & 0 \le t \le T \\ \varphi_0 = \mathrm{id} \end{cases}$$

belongs to $\mathrm{Diff}^{r,\alpha}\left(\overline{\Omega}; \overline{\Omega}\right)$, *satisfies $\varphi_t = \mathrm{id}$ on $\partial\Omega$ and*

$$\varphi_t^*(f_t) = f_0 \quad \text{in} \quad \Omega.$$

1.5.2 The Cases k = 0 and k = 1

We start with the case $k = 0$, which is particularly elementary. We have for example the following local result (cf. Theorem 13.1). For a global result, see Theorem 13.2.

Theorem 1.30. *Let $r \ge 1$ be an integer, $x_0 \in \mathbb{R}^n$ and $f, g \in C^r$ in a neighborhood of x_0 and such that $f(x_0) = g(x_0)$,*

$$\nabla f(x_0) \ne 0 \quad \text{and} \quad \nabla g(x_0) \ne 0.$$

Then there exist a neighborhood U of x_0 and $\varphi \in \mathrm{Diff}^r(U; \varphi(U))$ such that

$$\varphi^*(g) = f \text{ in } U \quad \text{and} \quad \varphi(x_0) = x_0.$$

The results for $k = 0$ extend in a straightforward way to the case of closed 1-forms (cf. Corollaries 13.3 and 13.5).

We now give a theorem (cf. Theorems 13.8 and 13.10) for nonclosed 1-forms. It can be considered as the 1-form version of the Darboux theorem. We will see below that it is equivalent to the Darboux theorem for closed 2-forms.

Theorem 1.31. *Let $2 \le 2m \le n$ be integers, $x_0 \in \mathbb{R}^n$ and ω be a C^∞ 1-form such that $\omega(x_0) \ne 0$ and*

$$\operatorname{rank}[d\omega] = 2m \quad \text{in a neighborhood of } x_0.$$

Then there exist an open set U and

$$\varphi \in \mathrm{Diff}^\infty(U; \varphi(U))$$

such that $\varphi(U)$ *is a neighborhood of* x_0 *and*

$$\varphi^*(\omega) = \begin{cases} \sum_{i=1}^{m} x_{2i-1} dx^{2i} & \text{if } \omega \wedge (d\omega)^m = 0 \text{ in a neighborhood of } x_0 \\ \sum_{i=1}^{m} x_{2i-1} dx^{2i} + dx^{2m+1} & \text{if } \omega \wedge (d\omega)^m \neq 0 \text{ in a neighborhood of } x_0 . \end{cases}$$

Remark 1.32. (i) In the theorem, we have adopted the notation

$$(d\omega)^m = \underbrace{d\omega \wedge \cdots \wedge d\omega}_{m \text{ times}} .$$

(ii) Note that if $n = 2m$, then $\omega \wedge (d\omega)^m \equiv 0$.

1.5.3 The Case $k = 2$

Our best results besides the ones for volume forms are in the case $k = 2$.

We start with two *local* results. The first one is the celebrated *Darboux theorem*, but as stated it is due to Bandyopadhyay and Dacorogna [8] (cf. Theorem 14.1). The difference between the following theorem and all of the classical ones is in terms of regularity of the diffeomorphism. We provide the optimal possible regularity in Hölder spaces; the other ones give only that if $\omega \in C^{r,\alpha}$, then $\varphi \in C^{r,\alpha}$.

Theorem 1.33. *Let* $r \geq 0$ *and* $n = 2m \geq 4$ *be integers. Let* $0 < \alpha < 1$ *and* $x_0 \in \mathbb{R}^n$. *Let* ω_m *be the* standard symplectic form *of* $\operatorname{rank}[\omega_m] = 2m = n$,

$$\omega_m = \sum_{i=1}^{m} dx^{2i-1} \wedge dx^{2i}.$$

Let ω *be a 2-form. The two following statements are then equivalent:*

(i) The 2-form ω *is closed, is in* $C^{r,\alpha}$ *in a neighborhood of* x_0 *and verifies*

$$\operatorname{rank}[\omega(x_0)] = n.$$

(ii) There exist a neighborhood U *of* x_0 *and* $\varphi \in \operatorname{Diff}^{r+1,\alpha}(U; \varphi(U))$ *such that*

$$\varphi^*(\omega_m) = \omega \text{ in } U \quad \text{and} \quad \varphi(x_0) = x_0 .$$

One possible proof of the theorem could be to use Theorem 1.31 with $n = 2m$ (cf. Remark 13.7 for details). We, however, will go the other way around and prove Theorem 1.31 using Theorem 1.33.

We next discuss the case of forms of lower rank. This is also well known in the literature. However, our theorem (cf. Theorem 14.3, proved in [9] by Bandyopadhyay, Dacorogna and Kneuss) provides, as the previous theorem, one class higher degree of regularity than the other results. Indeed, in all other theorems it is proved that if $\omega \in C^{r,\alpha}$, then $\varphi \in C^{r-1,\alpha}$. It may appear that the theorem below is still not optimal, since it only shows that $\varphi \in C^{r,\alpha}$ when $\omega \in C^{r,\alpha}$. However, since there are some missing variables, it is probably the best possible regularity.

Theorem 1.34. *Let $n \geq 3$, $r,m \geq 1$ be integers and $0 < \alpha < 1$. Let $x_0 \in \mathbb{R}^n$ and ω_m be the standard symplectic form of* $\operatorname{rank}[\omega_m] = 2m < n$, *namely*

$$\omega_m = \sum_{i=1}^{m} dx^{2i-1} \wedge dx^{2i}.$$

Let ω be a $C^{r,\alpha}$ closed 2-form such that

$$\operatorname{rank}[\omega] = 2m \text{ in a neighborhood of } x_0.$$

Then there exist a neighborhood U of x_0 and $\varphi \in \operatorname{Diff}^{r,\alpha}(U;\varphi(U))$ such that

$$\varphi^*(\omega_m) = \omega \text{ in } U \quad \text{and} \quad \varphi(x_0) = x_0.$$

We now turn to a *global* result (cf. Theorem 14.5). It has been obtained under slightly more restrictive hypotheses by Bandyopadhyay and Dacorogna [8] and as stated by Dacorogna and Kneuss [32]. The theorem provides the first global result on manifolds with boundary. It is also nearly optimal.

Theorem 1.35. *Let $n > 2$ be even and $\Omega \subset \mathbb{R}^n$ be a bounded open smooth set with exterior unit normal ν. Let $0 < \alpha < 1$ and $r \geq 1$ be an integer. Let $f,g \in C^{r,\alpha}(\overline{\Omega};\Lambda^2)$ satisfying $df = dg = 0$ in Ω,*

$$\nu \wedge f, \nu \wedge g \in C^{r+1,\alpha}\left(\partial\Omega;\Lambda^3\right), \quad \nu \wedge f = \nu \wedge g \text{ on } \partial\Omega,$$

$$\int_\Omega \langle f;\psi\rangle = \int_\Omega \langle g;\psi\rangle \quad \text{for every } \psi \in \mathscr{H}_T\left(\Omega;\Lambda^2\right) \tag{1.6}$$

and, for every $t \in [0,1]$,

$$\operatorname{rank}[tg + (1-t)f] = n \quad \text{in } \overline{\Omega}.$$

Then there exists $\varphi \in \operatorname{Diff}^{r+1,\alpha}\left(\overline{\Omega};\overline{\Omega}\right)$ such that

$$\begin{cases} \varphi^*(g) = f & \text{in } \Omega, \\ \varphi = \operatorname{id} & \text{on } \partial\Omega. \end{cases}$$

Remark 1.36. (i) In a similar way, we can consider a general homotopy f_t with $f_0 = f$, $f_1 = g$, provided

$$df_t = 0, \quad \nu \wedge f_t = \nu \wedge f_0 \text{ on } \partial\Omega \quad \text{and} \quad \operatorname{rank}[f_t] = n \text{ in } \overline{\Omega},$$

$$\int_\Omega \langle f_t;\psi\rangle = \int_\Omega \langle f_0;\psi\rangle \quad \text{for every } \psi \in \mathscr{H}_T\left(\Omega;\Lambda^2\right).$$

(ii) If Ω is contractible, then $\mathcal{H}_T(\Omega;\Lambda^2) = \{0\}$ and, therefore, (1.6) is automatically satisfied.

1.5.4 The Case $3 \leq k \leq n-1$

The presentation in Chapter 15 follows closely the results of Bandyopadhyay, Dacorogna and Kneuss [9]. We start with the case $k = n-1$. We have as a consequence of Theorems 15.3 and 15.5 the following result.

Theorem 1.37. *Let $x_0 \in \mathbb{R}^n$ and f be a $(n-1)$-form such that $f \in C^\infty$ in a neighborhood of x_0 and $f(x_0) \neq 0$. Then there exist a neighborhood U of x_0 and*

$$\varphi \in \operatorname{Diff}^\infty(U;\varphi(U))$$

such that φ satisfies one of the two following equations in U:

(i) If $df = 0$ in a neighborhood of x_0, then

$$f = \nabla\varphi^1 \wedge \cdots \wedge \nabla\varphi^{n-1} = \varphi^*\left(dx^1 \wedge \cdots \wedge dx^{n-1}\right).$$

(ii) If $df(x_0) \neq 0$, then

$$f = \varphi^n\left(\nabla\varphi^1 \wedge \cdots \wedge \nabla\varphi^{n-1}\right) = \varphi^*\left(x_n\, dx^1 \wedge \cdots \wedge dx^{n-1}\right).$$

Remark 1.38. (i) The present theorem, when $df = 0$, is a consequence of Theorem 15.1, which is valid for k-forms of rank k.

(ii) With our usual abuse of notations, identifying a $(n-1)$-form with a vector field and observing that the d operator can then be essentially identified with the divergence operator, we can rewrite the theorem as follows (cf. Corollaries 15.4 and 15.7). For any C^∞ vector field f such that $f(x_0) \neq 0$, there exist an open set U and

$$\varphi \in \operatorname{Diff}^\infty(U;\varphi(U))$$

such that $\varphi(U)$ is a neighborhood of x_0 and

$$f = \begin{cases} *\left(\nabla\varphi^1 \wedge \cdots \wedge \nabla\varphi^{n-1}\right) & \text{if } \operatorname{div} f = 0 \\ *\left(\varphi^n\left(\nabla\varphi^1 \wedge \cdots \wedge \nabla\varphi^{n-1}\right)\right) & \text{if } \operatorname{div} f \neq 0, \end{cases}$$

where $*$ denotes the Hodge $*$ operator.

We now turn to the case $3 \leq k \leq n-2$, which is, as already said, much more difficult. This is so already at the algebraic level, since there are no known canonical forms. Additionally, even when the algebraic setting is simple, the analytical situation is more complicated than in the cases $k = 0,1,2,n-1,n$ (see Proposition 15.14 for such an example). The only cases that we will be able to study in Chapter 15 are those that are combinations of 1 and 2-forms that we can handle separately.

For 1-forms, we easily obtain local (cf. Proposition 15.8) as well as global results (cf. Proposition 15.10). We now give a simple theorem (a more general statement can be found in Theorem 15.15) that deals with 3-forms obtained by product of a 1-form and a 2-form (in the same spirit, Theorem 15.12 allows to deal with some k-forms that are product of 1 and 2-forms).

Theorem 1.39. *Let $n = 2m \geq 4$ be integers, $x_0 \in \mathbb{R}^n$ and f be a C^∞ symplectic (i.e., closed and with* $\operatorname{rank}[f] = n$*) 2-form and a be a nonzero closed C^∞ 1-form. Then there exist a neighborhood U of x_0 and $\varphi \in \operatorname{Diff}^\infty(U;\varphi(U))$ such that $\varphi(x_0) = x_0$ and*

$$\varphi^*(\omega_m) = f \quad \text{and} \quad \varphi^*(dx^n) = a \quad \text{in } U,$$

where

$$\omega_m = \sum_{i=1}^{m} dx^{2i-1} \wedge dx^{2i}.$$

In particular, if

$$G = \left[\sum_{i=1}^{m-1} dx^{2i-1} \wedge dx^{2i}\right] \wedge dx^n = \omega_m \wedge dx^n,$$

then

$$\varphi^*(G) = f \wedge a \quad \text{in } U.$$

1.6 Hölder Spaces

Throughout the book we have used very fine properties of Hölder continuous functions. Most of the results discussed in Chapter 16 are "standard," but they are scattered in the literature. There does not exist such a huge literature as the one for Sobolev spaces. Some of the best references are Fefferman [42], Gilbarg and Trudinger [49] and Hörmander [55].

1.6.1 Definition and Extension of Hölder Functions

We give here the definition of Hölder continuous functions (cf. Definition 16.2).

Definition 1.40. Let $\Omega \subset \mathbb{R}^n$ be a bounded open set, $f : \overline{\Omega} \to \mathbb{R}$ and $0 < \alpha \leq 1$. Let

$$[f]_{C^{0,\alpha}(\overline{\Omega})} = \sup_{\substack{x,y\in\overline{\Omega}\\ x\neq y}} \left\{ \frac{|f(x) - f(y)|}{|x-y|^\alpha} \right\}.$$

(i) The set $C^{0,\alpha}\left(\overline{\Omega}\right)$ is the set of $f \in C^0\left(\overline{\Omega}\right)$ so that

$$\|f\|_{C^{0,\alpha}(\overline{\Omega})} = \|f\|_{C^0(\overline{\Omega})} + [f]_{C^{0,\alpha}(\overline{\Omega})} < \infty,$$

where

$$\|f\|_{C^0(\overline{\Omega})} = \sup_{x \in \Omega} \{|f(x)|\}.$$

If there is no ambiguity, we drop the dependence on the set $\overline{\Omega}$ and write simply

$$\|f\|_{C^{0,\alpha}} = \|f\|_{C^0} + [f]_{C^{0,\alpha}} .$$

(ii) If $r \geq 1$ is an integer, then the set $C^{r,\alpha}\left(\overline{\Omega}\right)$ is the set of functions $f \in C^r\left(\overline{\Omega}\right)$ so that

$$[\nabla^r f]_{C^{0,\alpha}(\overline{\Omega})} < \infty.$$

We equip $C^{r,\alpha}\left(\overline{\Omega}\right)$ with the following norm:

$$\|f\|_{C^{r,\alpha}(\overline{\Omega})} = \|f\|_{C^r(\overline{\Omega})} + [\nabla^r f]_{C^{0,\alpha}(\overline{\Omega})},$$

where

$$\|f\|_{C^r(\overline{\Omega})} = \sum_{m=0}^{r} \|\nabla^m f\|_{C^0(\overline{\Omega})} .$$

Remark 1.41. (i) $C^{r,\alpha}\left(\overline{\Omega}\right)$ with its norm $\|\cdot\|_{C^{r,\alpha}}$ is a Banach space.

(ii) If $\alpha = 0$, we set

$$\|f\|_{C^{r,0}} = \|f\|_{C^r} .$$

(iii) If we assume that Ω is bounded and Lipschitz, then the norms

$$\|f\|_{C^{r,\alpha}} = \sum_{m=0}^{r} \|\nabla^m f\|_{C^{0,\alpha}}$$

and

$$\|f\|_{C^{r,\alpha}} = \begin{cases} \|f\|_{C^0} + [\nabla^r f]_{C^{0,\alpha}} & \text{if } 0 < \alpha \leq 1 \\ \|f\|_{C^0} + \|\nabla^r f\|_{C^0} & \text{if } \alpha = 0. \end{cases}$$

are equivalent to the one defined above. We should, however, point out that these norms are, in general, not equivalent for very wild sets.

(iv) When $\alpha = 1$, we note that $C^{0,1}\left(\overline{\Omega}\right)$ is in fact the set of *Lipschitz continuous* and bounded functions.

The following result (cf. Theorem 16.11) is a remarkable extension result due to Calderon [20] and Stein [92].

Theorem 1.42. *Let $\Omega \subset \mathbb{R}^n$ be a bounded open Lipschitz set. Then there exists a continuous linear extension operator*

$$E : C^{r,\alpha}\left(\overline{\Omega}\right) \to C_0^{r,\alpha}\left(\mathbb{R}^n\right)$$

for any integer $r \geq 0$ and any $0 \leq \alpha \leq 1$. More precisely, there exists a constant $C = C\left(r, \Omega\right) > 0$ such that for every $f \in C^{r,\alpha}\left(\overline{\Omega}\right)$,

$$E(f)|_{\overline{\Omega}} = f, \quad \operatorname{supp}[E(f)] \text{ is compact,}$$
$$\|E(f)\|_{C^{r,\alpha}(\mathbb{R}^n)} \leq C\|f\|_{C^{r,\alpha}(\overline{\Omega})}.$$

Remark 1.43. The extension is universal, in the sense that the same extension also leads to

$$\|E(f)\|_{C^{s,\beta}(\mathbb{R}^n)} \leq C\|f\|_{C^{s,\beta}(\overline{\Omega})}$$

for any integer s and any $0 \leq \beta \leq 1$, with, of course, $C = C(s,\Omega)$ as far as $f \in C^{s,\beta}(\overline{\Omega})$. The same extension is also valid for Sobolev spaces.

1.6.2 Interpolation, Product, Composition and Inverse

We now state the interpolation theorem (cf. Theorem 16.26) that plays an essential role in our analysis.

Theorem 1.44. *Let $\Omega \subset \mathbb{R}^n$ be a bounded open Lipschitz set, $s \geq r \geq t \geq 0$ be integers and $0 \leq \alpha, \beta, \gamma \leq 1$ with*

$$t + \gamma \leq r + \alpha \leq s + \beta.$$

Let $\lambda \in [0,1]$ be such that

$$r + \alpha = \lambda(s+\beta) + (1-\lambda)(t+\gamma).$$

Then there exists a constant $C = C(s,\Omega) > 0$ such that

$$\|f\|_{C^{r,\alpha}} \leq C\|f\|^{\lambda}_{C^{s,\beta}}\|f\|^{1-\lambda}_{C^{t,\gamma}}.$$

As a byproduct of the interpolation theorem, we get the following result (cf. Theorem 16.28).

Theorem 1.45. *Let $\Omega \subset \mathbb{R}^n$ be a bounded open Lipschitz set, $r \geq 0$ an integer and $0 \leq \alpha \leq 1$. Then there exists a constant $C = C(r,\Omega) > 0$ such that*

$$\|fg\|_{C^{r,\alpha}} \leq C(\|f\|_{C^{r,\alpha}}\|g\|_{C^0} + \|f\|_{C^0}\|g\|_{C^{r,\alpha}}).$$

The next theorem (cf. Theorem 16.31) will also be intensively used.

Theorem 1.46. *Let $\Omega \subset \mathbb{R}^n$, $O \subset \mathbb{R}^m$ be bounded open Lipschitz sets, $r \geq 0$ an integer and $0 \leq \alpha \leq 1$. Let $g \in C^{r,\alpha}(\overline{O})$ and $f \in C^{r,\alpha}(\overline{\Omega};\overline{O}) \cap C^1(\overline{\Omega};\overline{O})$. Then*

$$\|g \circ f\|_{C^{0,\alpha}(\overline{\Omega})} \leq \|g\|_{C^{0,\alpha}(\overline{O})}\|f\|^{\alpha}_{C^1(\overline{\Omega})} + \|g\|_{C^0(\overline{O})},$$

whereas if $r \geq 1$, there exists a constant $C = C(r, \Omega, O) > 0$ such that

$$\|g \circ f\|_{C^{r,\alpha}(\overline{\Omega})} \leq C \left[\|g\|_{C^{r,\alpha}(\overline{O})} \|f\|_{C^1(\overline{\Omega})}^{r+\alpha} + \|g\|_{C^1(\overline{O})} \|f\|_{C^{r,\alpha}(\overline{\Omega})} + \|g\|_{C^0(\overline{O})} \right].$$

We easily deduce, from the previous results, an estimate on the inverse (cf. Theorem 16.32).

Theorem 1.47. *Let $\Omega, O \subset \mathbb{R}^n$ be bounded open Lipschitz sets, $r \geq 1$ an integer and $0 \leq \alpha \leq 1$. Let $c > 0$. Let $f \in C^{r,\alpha}\left(\overline{\Omega}; \overline{O}\right)$ and $g \in C^{r,\alpha}\left(\overline{O}; \overline{\Omega}\right)$ be such that*

$$g \circ f = \mathrm{id} \quad \textit{and} \quad \|g\|_{C^1(\overline{O})}, \|f\|_{C^1(\overline{\Omega})} \leq c.$$

Then there exists a constant $C = C(c, r, \Omega, O) > 0$ such that

$$\|f\|_{C^{r,\alpha}(\overline{\Omega})} \leq C \|g\|_{C^{r,\alpha}(\overline{O})}.$$

1.6.3 Smoothing Operator

The next theorem (cf. Theorem 16.43) is about smoothing C^r or $C^{r,\alpha}$ functions. We should draw the attention that in order to get the conclusions of the theorem, one proceeds, as usual, by convolution. However, we have to choose the kernel very carefully.

Theorem 1.48. *Let $\Omega \subset \mathbb{R}^n$ be a bounded open Lipschitz set. Let $s \geq r \geq t \geq 0$ be integers and $0 \leq \alpha, \beta, \gamma \leq 1$ be such that*

$$t + \gamma \leq r + \alpha \leq s + \beta.$$

Let $f \in C^{r,\alpha}\left(\overline{\Omega}\right)$. Then, for every $0 < \varepsilon \leq 1$, there exist a constant $C = C(s, \Omega) > 0$ and $f_\varepsilon \in C^\infty\left(\overline{\Omega}\right)$ such that

$$\|f_\varepsilon\|_{C^{s,\beta}} \leq \frac{C}{\varepsilon^{(s+\beta)-(r+\alpha)}} \|f\|_{C^{r,\alpha}},$$
$$\|f - f_\varepsilon\|_{C^{t,\gamma}} \leq C \varepsilon^{(r+\alpha)-(t+\gamma)} \|f\|_{C^{r,\alpha}}.$$

We also need to approximate closed forms in $C^{r,\alpha}\left(\overline{\Omega}; \Lambda^k\right)$ by smooth closed forms in a precise way (cf. Theorem 16.49).

Theorem 1.49. *Let $\Omega \subset \mathbb{R}^n$ be a bounded open smooth set and ν be the exterior unit normal. Let $s \geq r \geq t \geq 0$ with $s \geq 1$ and $1 \leq k \leq n-1$ be integers. Let $0 < \alpha, \beta, \gamma < 1$ be such that*

$$t + \gamma \leq r + \alpha \leq s + \beta.$$

Let $g \in C^{r,\alpha}\left(\overline{\Omega};\Lambda^k\right)$ *with*

$$dg = 0 \text{ in } \Omega \quad \text{and} \quad \nu \wedge g \in C^{s,\beta}\left(\partial\Omega;\Lambda^{k+1}\right).$$

Then for every $\varepsilon \in (0,1]$, *there exist* $g_\varepsilon \in C^\infty\left(\Omega;\Lambda^k\right) \cap C^{s,\beta}\left(\overline{\Omega};\Lambda^k\right)$ *and a constant* $C = C(s,\alpha,\beta,\gamma,\Omega) > 0$ *such that*

$$dg_\varepsilon = 0 \text{ in } \Omega, \quad \nu \wedge g_\varepsilon = \nu \wedge g \text{ on } \partial\Omega,$$

$$\int_\Omega \langle g_\varepsilon;\psi\rangle = \int_\Omega \langle g;\psi\rangle \quad \text{for every } \psi \in \mathscr{H}_T\left(\Omega;\Lambda^k\right),$$

$$\|g_\varepsilon\|_{C^{s,\beta}(\overline{\Omega})} \leq \frac{C}{\varepsilon^{(s+\beta)-(r+\alpha)}}\|g\|_{C^{r,\alpha}(\overline{\Omega})} + C\|\nu\wedge g\|_{C^{s,\beta}(\partial\Omega)},$$

$$\|g_\varepsilon - g\|_{C^{t,\gamma}(\overline{\Omega})} \leq C\varepsilon^{(r+\alpha)-(t+\gamma)}\|g\|_{C^{r,\alpha}(\overline{\Omega})}.$$

Remark 1.50. We recall that if Ω is contractible and since $1 \leq k \leq n-1$, then

$$\mathscr{H}_T\left(\Omega;\Lambda^k\right) = \{0\}.$$

Part I
Exterior and Differential Forms

Chapter 2
Exterior Forms and the Notion of Divisibility

The present chapter is divided into three parts.

In Section 2.1, we recall the definitions and basic properties of exterior forms. All notions introduced there are standard and, therefore, our presentation will be very brief. We refer for further developments to the classic books on the subject—for example, Bourbaki [15], Bryant, Chern, Gardner, Goldschmidt and Griffiths [18], Godbillon [51], Godement [52], Greub [54], or Lang [67]. In what follows we will only consider the finite vector space $\mathbb{R}^n$, $n \geq 1$, over $\mathbb{R}$. However, we can obviously replace $\mathbb{R}^n$ by any finite n-dimensional vector space over a field K of characteristic 0.

In Section 2.2, we introduce the different notions of rank and corank for exterior forms. The material presented here is new and has been introduced by Dacorogna and Kneuss [31]. However, the notion of rank (in our terminology below, rank of order 1) of an exterior form is standard for 2-forms (see, e.g., Abraham, Marsden and Ratiu [1], Bryant et al. [18], Godbillon [51], Mc Duff and Salamon [72], Postnikov [82], Sternberg [93], or Taylor [96]) and also exists, although less standard and sometimes expressed in a different but equivalent way, for general k-forms; see, for example, Bandyopadhyay, Dacorogna and Kneuss [9], Godbillon [51], Marcus [74], Martinet [71], or Sternberg [93]. All of the other notions of rank and corank of an exterior form are new. The importance of these notions will be clear in our study of the pullback equation; they are indeed invariant under pullback (see Proposition 2.33 and Proposition 17.1).

In Section 2.3, we discuss the central result of this chapter; it concerns the notion of divisibility of an exterior form. More precisely, given $f \in \Lambda^k(\mathbb{R}^n)$ and $g \in \Lambda^l(\mathbb{R}^n)$ with $0 \leq l \leq k$, we want to find $u \in \Lambda^{k-l}(\mathbb{R}^n)$ such that

$$f = g \wedge u.$$

We will give here a result due to Dacorogna and Kneuss [31] (cf. Theorem 2.45) which generalizes the celebrated Cartan lemma (cf. Theorem 2.42). This lemma asserts that if g has the additional structure

$$g = g_1 \wedge \cdots \wedge g_l \neq 0,$$

where $g_1,\dots,g_l \in \Lambda^1(\mathbb{R}^n)$, then a necessary and sufficient condition for finding such a $u \in \Lambda^{k-l}(\mathbb{R}^n)$ is that

$$f \wedge g_1 = \cdots = f \wedge g_l = 0.$$

Our theorem will rely on the notions of annihilators introduced in Section 2.2. This question of divisibility leads in a natural way to the notion of prime exterior forms (cf. Definition 2.39 and Corollary 2.49).

2.1 Definitions

2.1.1 Exterior Forms and Exterior Product

Definition 2.1 (Exterior form). Let $k \geq 1$ be an integer. An *exterior k-form over* $\mathbb{R}^n$ is a map

$$f : \underbrace{\mathbb{R}^n \times \cdots \times \mathbb{R}^n}_{k \text{ times}} \to \mathbb{R}$$

such that

(i) f is linear is each variable,

(ii) for every $X_1,\dots,X_k \in \mathbb{R}^n$ and for every $\sigma \in \mathrm{Sym}(k)$,

$$f(X_{\sigma(1)},\dots,X_{\sigma(k)}) = \mathrm{sign}(\sigma) f(X_1,\dots,X_k),$$

where $\mathrm{Sym}(k)$ is the set of the permutations of $\{1,\dots,k\}$ and $\mathrm{sign}(\sigma)$ denotes the sign of the permutation σ.

We denote by $\Lambda^k(\mathbb{R}^n)$ the set of exterior k-forms over $\mathbb{R}^n$. If $k = 0$, we set

$$\Lambda^0(\mathbb{R}^n) = \mathbb{R}.$$

We have that

$$\Lambda^k(\mathbb{R}^n) = \{0\} \text{ if } k > n.$$

Definition 2.2 (Exterior product). Let $f \in \Lambda^k(\mathbb{R}^n)$ and $g \in \Lambda^l(\mathbb{R}^n)$. The *exterior product of f with g*, denoted by $f \wedge g$, belongs to $\Lambda^{k+l}(\mathbb{R}^n)$ and is defined by

$$\begin{aligned}(f \wedge g)&(X_1,\dots,X_{k+l}) \\ &= \sum_{\sigma \in \mathrm{Sym}(k,l)} \mathrm{sign}(\sigma) f(X_{\sigma(1)},\dots,X_{\sigma(k)}) g(X_{\sigma(k+1)},\dots,X_{\sigma(k+l)}),\end{aligned}$$

where

$$\mathrm{Sym}(k,l) = \{\sigma \in \mathrm{Sym}(k+l) : \sigma(1) < \cdots < \sigma(k); \sigma(k+1) < \cdots < \sigma(k+l)\}.$$

If $k = 0$ (i.e., $f \in \Lambda^0(\mathbb{R}^n) = \mathbb{R}$), we define

$$f \wedge g = g \wedge f = fg.$$

Theorem 2.3 (Properties of the exterior product). *Let $f \in \Lambda^k(\mathbb{R}^n)$, $g \in \Lambda^l(\mathbb{R}^n)$ and $h \in \Lambda^p(\mathbb{R}^n)$. The exterior product is bilinear and the following properties are verified:*

$$(f \wedge g) \wedge h = f \wedge (g \wedge h),$$
$$f \wedge g = (-1)^{kl} g \wedge f.$$

Moreover, if $f_1, \ldots, f_k \in \Lambda^1(\mathbb{R}^n)$ and $X_1, \ldots, X_k \in \mathbb{R}^n$, then

$$f_1 \wedge \cdots \wedge f_k(X_1, \ldots, X_k) = \det\left[f_i(X_j)\right]_{1 \le i,j \le k}. \tag{2.1}$$

In particular, the family $\{f_1, \ldots, f_k\}$ is linearly independent if and only if

$$f_1 \wedge \cdots \wedge f_k \neq 0.$$

Let $E_1, \ldots, E_n$ be the canonical basis of $\mathbb{R}^n$ and let e^i, $1 \le i \le n$, be its dual basis, which means that

$$e^i(E_j) = \delta_{ij};$$

where $\delta_{ij} = 1$ if $i = j$ and $\delta_{ij} = 0$ otherwise.

Proposition 2.4. *Let $1 \le k \le n$. The set*

$$\{e^{i_1} \wedge \cdots \wedge e^{i_k}\}, \quad 1 \le i_1 < \cdots < i_k \le n,$$

is a basis of $\Lambda^k(\mathbb{R}^n)$; in particular,

$$\dim(\Lambda^k(\mathbb{R}^n)) = \binom{n}{k}.$$

Therefore, any $f \in \Lambda^k(\mathbb{R}^n)$ can be written as

$$f = \sum_{1 \le i_1 < \cdots < i_k \le n} f_{i_1 \cdots i_k} e^{i_1} \wedge \cdots \wedge e^{i_k}$$

for some unique $f_{i_1 \cdots i_k} \in \mathbb{R}$. Moreover, the coefficients of f can be recovered by the formula

$$f_{i_1 \cdots i_k} = f(E_{i_1}, \ldots, E_{i_k}). \tag{2.2}$$

Notation 2.5. *(i) We will denote the set of strictly increasing tuples of length k by*

$$\mathscr{T}_k = \{I = (i_1, \ldots, i_k) \in \mathbb{N}^k,\ 1 \le i_1 < \cdots < i_k \le n\}.$$

In this way, for $I \in \mathscr{T}_k$ we write

$$e^I = e^{i_1} \wedge \cdots \wedge e^{i_k},$$

and for $f \in \Lambda^k(\mathbb{R}^n)$, *we write*

$$f_I = f_{i_1 \cdots i_k}.$$

With these notations, we have

$$f = \sum_{1 \le i_1 < \cdots < i_k \le n} f_{i_1 \cdots i_k} e^{i_1} \wedge \cdots \wedge e^{i_k} = \sum_{I \in \mathcal{T}_k} f_I e^I. \tag{2.3}$$

(ii) When in an index we write $\hat{\imath}$, *this means that* i *is omitted. For example,*

$$f_{1 \cdots \hat{4} \cdots k} = f_{1235 \cdots k}.$$

(iii) Sometimes it will be more convenient to assign meaning to $f_{i_1 \cdots i_k}$ *for any* k*-index* $1 \le i_1, \ldots, i_k \le n$, *by letting*

$$f_{i_1 \cdots i_k} = \begin{cases} \operatorname{sign}(\sigma) f_{i_{\sigma(1)} \cdots i_{\sigma(k)}} & \text{if } i_{\sigma(1)} < \cdots < i_{\sigma(k)} \\ 0 & \text{if two indices coincide} \end{cases}$$

for $\sigma \in \operatorname{Sym}(k)$.

Proposition 2.6 (Formula for the exterior product). *Let* $f \in \Lambda^k(\mathbb{R}^n)$ *and* $g \in \Lambda^l(\mathbb{R}^n)$. *The following formula holds true:*

$$\begin{aligned} & f \wedge g \\ & = \sum_{j_1 < \cdots < j_{l+k}} \left(\sum_{i_1 < \cdots < i_l} (f \wedge e^{i_1} \wedge \cdots \wedge e^{i_l})_{j_1 \cdots j_{l+k}} \left(g_{i_1 \cdots i_l} \right) \right) e^{j_1} \wedge \cdots \wedge e^{j_{l+k}}. \end{aligned}$$

In particular, when $k = 1$ *(i.e.,* $f \in \Lambda^1(\mathbb{R}^n)$*), the formula reads as*

$$f \wedge g = \sum_{1 \le j_1 < \cdots < j_{l+1} \le n} \left(\sum_{\gamma=1}^{l+1} (-1)^{\gamma-1} f_{j_\gamma} g_{j_1 \cdots j_{\gamma-1} j_{\gamma+1} \cdots j_{l+1}} \right) e^{j_1} \wedge \cdots \wedge e^{j_{l+1}}.$$

2.1.2 Scalar Product, Hodge Star Operator and Interior Product

We now introduce the notions of scalar product, Hodge star operator and interior product. We also state some basic properties of these operators.

Definition 2.7 (Scalar product). Let $f, g \in \Lambda^k(\mathbb{R}^n)$. We define the *scalar product* of f with g as

$$\langle f; g \rangle = \sum_{1 \le i_1 < \cdots < i_k \le n} f_{i_1 \cdots i_k} g_{i_1 \cdots i_k} = \sum_{I \in \mathcal{T}_k} f_I g_I$$

and we let

$$|f|^2 = \langle f; f \rangle.$$

Remark 2.8. From (2.2) it follows that

$$\langle f;g\rangle = \sum_{1\le i_1<\ldots<i_k\le n} f(E_{i_1},\ldots,E_{i_k})g(E_{i_1},\ldots,E_{i_k}).$$

In fact, we could have taken any orthonormal basis $\{Y_1,\ldots,Y_n\}$ of $\mathbb{R}^n$ and defined

$$\langle f;g\rangle = \sum_{1\le i_1<\cdots<i_k\le n} f(Y_{i_1},\ldots,Y_{i_k})g(Y_{i_1},\ldots,Y_{i_k}).$$

The independence of the chosen orthonormal basis follows from Remark 2.18 and the fact that $\langle f;g\rangle = \langle A^*f;A^*g\rangle$ if $A\in O(n)$ (cf. Proposition 2.19).

Definition 2.9 (Hodge duality). The *Hodge star operator* is the linear operator

$$*:\Lambda^k(\mathbb{R}^n)\to\Lambda^{n-k}(\mathbb{R}^n)$$

defined by

$$f\wedge g = \langle *f;g\rangle\, e^1\wedge\cdots\wedge e^n$$

for every $g\in\Lambda^{n-k}(\mathbb{R}^n)$.

Theorem 2.10 (Properties of the Hodge star operator). *Let $0\le k\le n$. Then*

$$*(e^1\wedge\cdots\wedge e^n)=1 \quad \textit{and} \quad *1=e^1\wedge\cdots\wedge e^n.$$

Moreover, for every $f,g\in\Lambda^k(\mathbb{R}^n)$,

$$f\wedge(*g)=\langle f;g\rangle\, e^1\wedge\cdots\wedge e^n \quad \textit{and} \quad *(*f)=(-1)^{k(n-k)}f.$$

Finally, let $I\in\mathscr{T}_k$, $I^c=\{1,\ldots,n\}\backslash I\in\mathscr{T}_{n-k}$ and let r be such that

$$e^I\wedge e^{I^c}=(-1)^r\, e^1\wedge\cdots\wedge e^n.$$

Then

$$*(e^I)=(-1)^r e^{I^c}.$$

This last statement holds also true if the elements of I and I^c are not ordered increasingly.

We next turn to the definition of the interior product.

Definition 2.11 (Interior product). Let $0\le k,l\le n$, $f\in\Lambda^k(\mathbb{R}^n)$ and $g\in\Lambda^l(\mathbb{R}^n)$. We define the *interior product* of f with g by

$$g\lrcorner f=(-1)^{n(k-l)}*(g\wedge(*f)).$$

Proposition 2.12 (Formula for the interior product). *Let $0\le k,l\le n$, $f\in\Lambda^k(\mathbb{R}^n)$ and $g\in\Lambda^l(\mathbb{R}^n)$. The following formulas then hold.*

(i) If $l \leq k$, then $g \lrcorner f \in \Lambda^{k-l}(\mathbb{R}^n)$ and

$$g \lrcorner f$$
$$= (-1)^{k(k-l)} \sum_{1 \leq j_1 < \cdots < j_{k-l} \leq n} \left(\sum_{1 \leq i_1 < \cdots < i_l \leq n} f_{i_1 \cdots i_l j_1 \cdots j_{k-l}} g_{i_1 \cdots i_l} \right) e^{j_1} \wedge \cdots \wedge e^{j_{k-l}}.$$

In particular, when $l = 1$ (i.e., $g \in \Lambda^1(\mathbb{R}^n)$), the formula reads as

$$g \lrcorner f = \sum_{1 \leq j_1 < \cdots < j_{k-1} \leq n} \left(\sum_{\gamma=1}^{k} (-1)^{\gamma-1} \sum_{j_{\gamma-1} < i < j_\gamma} f_{j_1 \cdots j_{\gamma-1} i j_\gamma \cdots j_{k-1}} g_i \right) e^{j_1} \wedge \cdots \wedge e^{j_{k-1}},$$

where if $\gamma = 1$, it is understood that $f_{j_1 \cdots j_{\gamma-1} i j_\gamma \cdots j_{k-1}} = f_{i j_1 \cdots j_{k-1}}$ and, similarly, when $\gamma = k$, it is understood that $f_{j_1 \cdots j_{\gamma-1} i j_\gamma \cdots j_{k-1}} = f_{j_1 \cdots j_{k-1} i}$.

(ii) If $l = k$, then

$$f \lrcorner g = g \lrcorner f = \langle f; g \rangle.$$

(iii) If $l > k$, then

$$g \lrcorner f = 0.$$

Remark 2.13. When $g = e^I = e^{i_1} \wedge \cdots \wedge e^{i_l}$ and $f = e^J = e^{j_1} \wedge \cdots \wedge e^{j_k}$, the proposition leads to

$$e^I \lrcorner e^J = \begin{cases} 0 & \text{if } I \not\subset J \\ (-1)^{k(k-l)+r} e^{J \setminus I} & \text{if } I \subset J, \end{cases}$$

where $r \in \{0, 1\}$ is given by

$$e^I \wedge e^{J \setminus I} = (-1)^r e^J.$$

Proposition 2.14 (Properties of the interior product). *Let $0 \leq k, l \leq n$, $f \in \Lambda^k(\mathbb{R}^n)$ and $g \in \Lambda^l(\mathbb{R}^n)$. The following properties are then satisfied:*

(i) The interior product is linear in both arguments.

(ii) The coefficients of f can be recovered by the formula

$$\begin{aligned} f_{i_1 \cdots i_k} &= \langle f; e^{i_1} \wedge \cdots \wedge e^{i_k} \rangle = f \lrcorner \left(e^{i_1} \wedge \cdots \wedge e^{i_k} \right) \\ &= * \left[f \wedge \left(* \left(e^{i_1} \wedge \cdots \wedge e^{i_k} \right) \right) \right]. \end{aligned} \tag{2.4}$$

(iii) When $l = 1$, writing

$$g = \sum_{i=1}^{n} g_i e^i \quad \text{and} \quad G = \sum_{i=1}^{n} G^i E_i,$$

where $G^i = g_i$, then for every $X_1, \ldots, X_{k-1} \in \mathbb{R}^n$,

$$(g \lrcorner f)(X_1, \ldots, X_{k-1}) = f(G, X_1, \ldots, X_{k-1}).$$

Remark 2.15. Sometimes in the literature (see, e.g., [96]), one finds a different definition for the interior product of a k-form with a vector; namely let $G \in \mathbb{R}^n$ and $f \in \Lambda^k(\mathbb{R}^n)$; then $i_G(f) \in \Lambda^{k-1}(\mathbb{R}^n)$ is defined by

$$i_G(f)(X_1, \dots, X_{k-1}) = f(G, X_1, \dots, X_{k-1}).$$

Identifying $G = \sum_{i=1}^n G^i E_i$ with $g = \sum_{i=1}^n g_i e^i$ ($G^i = g_i$), statement (iii) in Proposition 2.14 tells us that the two definitions coincide; that is,

$$i_G(f) = g \lrcorner f.$$

Proposition 2.16. *Let $f \in \Lambda^k(\mathbb{R}^n)$, $g \in \Lambda^l(\mathbb{R}^n)$ and $h \in \Lambda^p(\mathbb{R}^n)$. Then*

$$(h \wedge g) \lrcorner f = (-1)^{k+l} h \lrcorner (g \lrcorner f).$$

Furthermore, if $p = k + l$, then

$$\langle f \wedge g; h \rangle = (-1)^{l(k+1)} \langle g; f \lrcorner h \rangle = (-1)^k \langle f; g \lrcorner h \rangle.$$

Let $w \in \Lambda^1(\mathbb{R}^n)$; then

$$\begin{aligned} w \lrcorner (f \wedge g) &= (w \lrcorner f) \wedge g + (-1)^{kl} (w \lrcorner g) \wedge f \\ &= (w \lrcorner f) \wedge g + (-1)^k f \wedge (w \lrcorner g). \end{aligned}$$

In particular, if k is even and m is an integer and letting $f^m = \underbrace{f \wedge \cdots \wedge f}_{m \text{ times}}$, then

$$w \lrcorner f^{m+1} = (m+1) \left[(w \lrcorner f) \wedge f^m \right]. \tag{2.5}$$

If $v, w \in \Lambda^1(\mathbb{R}^n)$, then

$$w \lrcorner (v \wedge f) + v \wedge (w \lrcorner f) = \langle w; v \rangle f \tag{2.6}$$

and, thus,

$$w \lrcorner (w \wedge f) + w \wedge (w \lrcorner f) = |w|^2 f, \tag{2.7}$$

$$|w|^4 |f|^2 = |w \lrcorner (w \wedge f)|^2 + |w \wedge (w \lrcorner f)|^2 = |w|^2 \left(|w \wedge f|^2 + |w \lrcorner f|^2 \right).$$

2.1.3 Pullback and Dimension Reduction

We start with the following definition and properties.

Definition 2.17 (Pullback). Let $A \in \mathbb{R}^{n \times m}$ be a $(n \times m)$-matrix and $f \in \Lambda^k(\mathbb{R}^n)$ be given by

$$f = \sum_{1 \le i_1 < \cdots < i_k \le n} f_{i_1 \cdots i_k} e^{i_1} \wedge \cdots \wedge e^{i_k}.$$

We define the *pullback of f by A,* denoted by $A^*(f)$ and belonging to $\Lambda^k(\mathbb{R}^m)$, through

$$A^*(f) = \sum_{1\leq i_1<\cdots<i_k\leq n} f_{i_1\cdots i_k} A^{i_1} \wedge \cdots \wedge A^{i_k},$$

where A^j is the jth row of A and is identified by

$$A^j = \sum_{k=1}^{m} A^j_k e^k \in \Lambda^1(\mathbb{R}^m).$$

If $k = 0$, we then let

$$A^*(f) = f.$$

Remark 2.18. There is an equivalent definition of the pullback, namely

$$A^*(f)(X_1,\ldots,X_k) = f(A\cdot X_1,\ldots,A\cdot X_k),$$

for every $X_1,\ldots,X_k \in \mathbb{R}^m$.

Proposition 2.19 (Properties of the pullback). *Let $f \in \Lambda^k(\mathbb{R}^n)$, $g \in \Lambda^l(\mathbb{R}^n)$, $A \in \mathbb{R}^{n\times m}$ and $B \in \mathbb{R}^{m\times p}$. Then*

$$(AB)^*(f) = B^*(A^*(f)),$$

$$A^*(f\wedge g) = A^*(f)\wedge A^*(g). \tag{2.8}$$

If $k = l$, then

$$A^*(f+g) = A^*(f) + A^*(g).$$

If $n = m$, then

$$A^*(e^1\wedge\cdots\wedge e^n) = \det(A)\, e^1\wedge\cdots\wedge e^n.$$

If $A \in \mathrm{GL}(n)$, then

$$A^*(g\lrcorner f) = ((A^{-1})^t)^*(g))\lrcorner A^*(f), \tag{2.9}$$

$$A^*(*f) = \det(A)\left[*\left(((A^{-1})^t)^*(f)\right)\right]. \tag{2.10}$$

If $A \in O(n)$, then

$$A^*(g\lrcorner f) = A^*(g)\lrcorner A^*(f),$$

$$A^*(*f) = \det(A)\left[*(A^*(f))\right], \tag{2.11}$$

and if $k = l$,

$$\langle A^*(f);A^*(g)\rangle = \langle f;g\rangle.$$

Definition 2.20. Let $f \in \Lambda^k(\mathbb{R}^n)$. We define, for $k \geq 1$,

$$\Lambda^1_f = \{u \in \Lambda^1(\mathbb{R}^n) : \exists g \in \Lambda^{k-1}(\mathbb{R}^n) \quad \text{with} \quad g\lrcorner f = u\}.$$

Remark 2.21. (i) Let $f \in \Lambda^k(\mathbb{R}^n)$ and $A \in \mathrm{GL}(n)$. Using (2.9), we immediately deduce that

$$A^*(\Lambda_f^1) = \Lambda_{A^*(f)}^1 . \tag{2.12}$$

(ii) It will be seen that if $f \neq 0$, then

$$\dim(\Lambda_f^1) = \operatorname{rank}_1[f] \in \{k, k+2, \ldots, n\}$$

(see Definition 2.28 for the definition of $\operatorname{rank}_1[f]$, Proposition 2.37(i) for the equivalence and Proposition 2.37(ii) for the range of values of $\operatorname{rank}_1[f]$).

The following lemma is very useful for reducing dimension. Below we give a purely algebraic proof; later (cf. Theorem 4.5) we will give two analytical proofs; one of them being based on the Frobenius theorem (cf. Theorem 4.2).

Theorem 2.22 (Dimension reduction). *Let $1 \leq k \leq n$ and $f \in \Lambda^k(\mathbb{R}^n)$ with $f \neq 0$. Let $\{a^1, \ldots, a^l\}$ be a basis of Λ_f^1. Then there exist $\widetilde{f}_{i_1 \cdots i_k} \in \mathbb{R}$, $1 \leq i_1 < \cdots < i_k \leq l$, such that*

$$f = \sum_{1 \leq i_1 < \cdots < i_k \leq l} \widetilde{f}_{i_1 \cdots i_k} a^{i_1} \wedge \cdots \wedge a^{i_k}.$$

In particular, there exists $A \in \mathrm{GL}(n)$ such that

$$A^*(f) = \sum_{1 \leq i_1 < \cdots < i_k \leq l} \widetilde{f}_{i_1 \cdots i_k} e^{i_1} \wedge \cdots \wedge e^{i_k}$$

and

$$\Lambda_{A^*(f)}^1 = \operatorname{span}\left\{e^1, \ldots, e^l\right\}.$$

Remark 2.23. (i) Looking at $\Lambda^k(\mathbb{R}^l)$ as a subset of $\Lambda^k(\mathbb{R}^n)$, by abuse of notations the theorem implies that

$$A^*(f) \in \Lambda^k(\mathbb{R}^l).$$

(ii) The above theorem (cf. Remark 2.21(ii)) therefore tells us that any $f \in \Lambda^k(\mathbb{R}^n)$ with $\operatorname{rank}_1[f] = l$ can be seen (up to a pullback) as a k-form over $\mathbb{R}^l$. Moreover, if $m \geq n$ and, by abuse of notations, we consider $f \in \Lambda^k(\mathbb{R}^m)$, we see that Λ_f^1 is independent of whether we see f as an element of $\Lambda^k(\mathbb{R}^n)$ or $\Lambda^k(\mathbb{R}^m)$.

Proof. Step 1. Let $\{a^1, \ldots, a^l\}$ be a basis of Λ_f^1 and we complete it as a basis of $\mathbb{R}^n$, namely $\{a^1, \ldots, a^n\}$. Let $B \in \mathrm{GL}(n)$ be the matrix whose ith row is a^i, $1 \leq i \leq n$. Finally, we define

$$A = B^{-1}.$$

Note that, by definition,

$$A^*(a^i) = e^i \quad \text{and} \quad B^*(e^i) = a^i, \quad 1 \leq i \leq n. \tag{2.13}$$

Assume (cf. Step 2) that we can find some $\widetilde{f}_{i_1 \cdots i_k} \in \mathbb{R}$ such that

$$A^*(f) = \sum_{1 \leq i_1 < \cdots < i_k \leq l} \widetilde{f}_{i_1 \cdots i_k} e^{i_1} \wedge \cdots \wedge e^{i_k} \tag{2.14}$$

(i.e., $A^*(f) \in \Lambda^k(\mathbb{R}^l)$). We will then have the result since, using (2.13), we get

$$\begin{aligned} f = B^*(A^*(f)) &= B^*\left(\sum_{1\le i_1<\cdots<i_k\le l} \widetilde{f}_{i_1\cdots i_k} e^{i_1}\wedge\cdots\wedge e^{i_k}\right) \\ &= \sum_{1\le i_1<\cdots<i_k\le l} \widetilde{f}_{i_1\cdots i_k} a^{i_1}\wedge\cdots\wedge a^{i_k}. \end{aligned}$$

Step 2. We finally show (2.14). Writing

$$A^*(f) = \sum_{1\le i_1<\cdots<i_k\le n} \widetilde{f}_{i_1\cdots i_k} e^{i_1}\wedge\cdots\wedge e^{i_k},$$

we see that (2.14) is equivalent to proving

$$\widetilde{f}_{i_1\cdots i_k} = 0 \quad \text{for every } i_k > l.$$

Let $1 \le i_1 < \cdots < i_k \le n$. Using the properties of the interior product we have that

$$(e^{i_1}\wedge\cdots\wedge e^{i_{k-1}}) \lrcorner A^*(f) = \sum_{j=1}^{n} \pm \widetilde{f}_{i_1\cdots i_{k-1}j} e^j.$$

Appealing to (2.12) and to (2.13), we deduce that

$$\Lambda^1_{A^*(f)} = \operatorname{span}\{e^1,\ldots,e^l\}.$$

Combining this with the definition of $\Lambda^1_{A^*(f)}$, we must have

$$\widetilde{f}_{i_1\cdots i_{k-1}j} = 0, \quad \text{for every } j > l,$$

which shows the assertion. □

2.1.4 Canonical Forms for 1, 2, $(n-2)$ and $(n-1)$-Forms

We now prove that when $k = 1, 2, (n-2), (n-1)$, it is possible by a linear transformation to pull back any form to a canonical one. When $3 \le k \le n-3$, no standard canonical form is known. Statements (i), (ii) and (iv) of Proposition 2.24 will be proved in the more general analytical context (see Proposition 15.8 for statement (i), Theorems 14.1 and 14.3 for the case $k = 2$ and Theorem 15.3 for the case $k = n-1$). It is not presently known if statement (iii) for $(n-2)$-forms can be extended to the analytical setting.

Proposition 2.24. *The following four statements hold true:*

(i) Let $1 \le k \le n$ and $f_1,\ldots,f_k \in \Lambda^1(\mathbb{R}^n)$ be such that

$$f_1\wedge\cdots\wedge f_k \neq 0.$$

Then there exists $A \in \mathrm{GL}(n)$ *such that*

$$A^*(f_i) = e^i \text{ for every } 1 \leq i \leq k.$$

In particular, for every $f \in \Lambda^1(\mathbb{R}^n)$ *with* $f \neq 0$, *there exists* $A \in \mathrm{GL}(n)$ *such that*

$$A^*(f) = e^1.$$

(ii) Let $f \in \Lambda^2(\mathbb{R}^n)$ *with* $f \neq 0$. *Let* m *be the integer such that*

$$f^m \neq 0 \quad \text{and} \quad f^{m+1} = 0.$$

Then there exists $A \in \mathrm{GL}(n)$ *such that*

$$A^*(f) = \omega_m = \sum_{i=1}^{m} e^{2i-1} \wedge e^{2i}.$$

Moreover, if $g, h \in \Lambda^1(\mathbb{R}^n)$ *with* $g, h \neq 0$, *then, in addition to* $A^*(f) = \omega_m$, *the following can be ensured:*

$$A^*(g) = h$$

provided

$$\{g \in \Lambda_f^1 \text{ and } h \in \Lambda_{\omega_m}^1\} \quad \text{or} \quad \{g \notin \Lambda_f^1 \text{ and } h \notin \Lambda_{\omega_m}^1\}.$$

(iii) Let $f \in \Lambda^{n-2}(\mathbb{R}^n)$ *with* $f \neq 0$. *Let* m *be the integer such that*

$$(*f)^m \neq 0 \quad \text{and} \quad (*f)^{m+1} = 0.$$

Then there exists $A \in \mathrm{GL}(n)$ *such that*

$$A^*(f) = \begin{cases} *(\omega_m) & \text{if } n > 2m \text{ or if } \{n = 2m \text{ and } m \text{ even}\} \\ \operatorname{sign}[(*f)^m]\,[*(\omega_m)] & \text{if } n = 2m \text{ and } m \text{ odd.} \end{cases}$$

Moreover, when $n = 2m$ *and* m *is odd, there exists no* $A \in \mathrm{GL}(n)$ *such that*

$$A^*(*\omega_m) = -[*(\omega_m)].$$

(iv) Let $f \in \Lambda^{n-1}(\mathbb{R}^n)$ *with* $f \neq 0$; *then there exists* $A \in \mathrm{GL}(n)$ *such that*

$$A^*(f) = e^1 \wedge \cdots \wedge e^{n-1}.$$

Remark 2.25. (i) Let $f \in \Lambda^2(\mathbb{R}^n)$. Anticipating Definition 2.28 for $\operatorname{rank}_1[f]$ and Proposition 2.37 (iii) for the equivalence, we have that $\operatorname{rank}_1[f] = 2m$ if and only if

$$f^m \neq 0 \quad \text{and} \quad f^{m+1} = 0.$$

Thus, statement (ii) in the above proposition can be rephrased as follows: Any 2-form f with $\operatorname{rank}_1[f] = 2m$ can be pulled back to *the standard symplectic form of rank* $2m$, namely

$$\omega_m = \sum_{i=1}^{m} e^{2i-1} \wedge e^{2i}.$$

Observe also that

$$(\omega_m)^m = m!\, e^1 \wedge \cdots \wedge e^{2m}.$$

(ii) In the case of $(n-2)$-forms, we have in fact (see Definition 2.28 for the notations)

$$2m = \operatorname{corank}_1(f) = \operatorname{rank}_1(*f).$$

Proof. We only prove (i), (iii) and (iv). The proof of (ii) is standard and can be found, for example, in Bryant et al. [18, p. 13], Horn and Johnson [56, p. 107], or Serre [90] (for the extra statement, see Kneuss [60]).

Step 1. We first show (i). Identifying 1-forms with elements of $\mathbb{R}^n$, we let $B \in \mathrm{GL}(n)$ be a matrix whose ith row is f_i, $1 \leq i \leq k$. Note that this is possible since $f_1 \wedge \cdots \wedge f_k \neq 0$ (see Theorem 2.3). Then noticing that

$$B^*(e^i) = f_i \quad \text{for every } 1 \leq i \leq k,$$

the matrix $A = B^{-1}$ has the required property.

Step 2. We split the proof of (iii) into three steps.

Step 2.1. We show that there exists $A \in \mathrm{GL}(n)$ such that

$$A^*(f) = \pm\left[*(\omega_m)\right].$$

Indeed, since $(*f) \in \Lambda^2(\mathbb{R}^n)$, using (ii), there exists $B \in \mathrm{GL}(n)$ such that

$$B^*(*f) = \omega_m\,.$$

Therefore, using (2.10), we get

$$\det B\left[*(((B^{-1})^t)^*(f))\right] = \omega_m.$$

We thus obtain

$$\det B\left[((B^{-1})^t)^*(f)\right] = *(\omega_m).$$

Letting

$$A = |\det B|^{\frac{1}{n-2}}(B^{-1})^t,$$

we have the claim, namely $A^*(f) = \pm\left[*(\omega_m)\right]$.

Step 2.2. Let us show that if $n > 2m$ or if $\{2m = n$ and m even$\}$, there exists $A \in \mathrm{GL}(n)$ such that

$$A^*\left(*(\omega_m)\right) = -\left[*(\omega_m)\right]. \tag{2.15}$$

(i) If $2m < n$, then the diagonal matrix A defined by $A_{ii} = 1$ for $1 \le i \le n-1$ and $A_{nn} = -1$ verifies (2.15).

(ii) If $2m = n$ and m even, it is easily seen that the diagonal matrix A defined by

$$A_{ii} = \begin{cases} 1 & \text{if } 1 \le i \le n \text{ and } i \text{ even} \\ -1 & \text{if } 1 \le i \le n \text{ and } i \text{ odd} \end{cases}$$

satisfies (2.15).

Step 2.3. Assume $2m = n$ and m odd. We first prove that if $A \in \mathrm{GL}(n)$ is such that $A^*(f) = \varepsilon\,[*(\omega_m)]$, with $\varepsilon \in \{-1,1\}$, then necessarily

$$\varepsilon = \operatorname{sign}[(*f)^m].$$

Let $B = (A^{-1})^t$. Observe that since $n = 2m$, we have that $(*f)^m$ is a n-form. Identifying, as usual, n-forms with scalars and using Proposition 2.19, we find that

$$B^*((*f)^m) = \det(B)\,(*f)^m = \frac{1}{\det A}\,(*f)^m.$$

Using Theorem 2.10 and (2.10), we have the following implications:

$$\begin{aligned} A^*(f) = \varepsilon\,[*(\omega_m)] &\Leftrightarrow *(A^*(**f)) = \varepsilon\,[\omega_m] \\ &\Leftrightarrow \det(A)\,[B^*(*f)] = \varepsilon\,[\omega_m] \\ &\Rightarrow \det(A)^m\,[B^*((*f)^m)] = \varepsilon^m\,[(\omega_m)^m] \\ &\Leftrightarrow \det(A)^{m-1}\,[(*f)^m] = \varepsilon^m\,[(\omega_m)^m]. \end{aligned}$$

Hence, recalling that m is odd and that $\operatorname{sign}[(\omega_m)^m] = 1$, we get the result, namely

$$\varepsilon = \operatorname{sign}[(*f)^m].$$

Combining these three steps proves the main assertion. Applying the result of Step 2.3 with $f = (*\omega_m)$, we have the extra claim, namely there exists no $A \in \mathrm{GL}(n)$ such that

$$A^*(*\omega_m) = -\,[*(\omega_m)].$$

The proof of (iii) is therefore complete.

Step 3. We finally prove (iv). Identifying $(*f) \in \Lambda^1(\mathbb{R})$ with a vector in $\mathbb{R}^n$, we let $B \in \mathrm{GL}(n)$ be a matrix whose nth row is $*f$ and with determinant equal to 1. Since

$$B^*(e^n) = *f,$$

we have, using Theorem 2.10 and Proposition 2.19,

$$\begin{aligned} f &= (-1)^{n-1}(*(*f)) = (-1)^{n-1} * (B^*(e^n)) = (-1)^{n-1}\left(\left(B^{-1}\right)^t\right)^*(*e^n) \\ &= (-1)^{n-1}\left(\left(B^{-1}\right)^t\right)^*((-1)^{n-1}e^1\wedge\cdots\wedge e^{n-1}) \\ &= \left(\left(B^{-1}\right)^t\right)^*(e^1\wedge\cdots\wedge e^{n-1}). \end{aligned}$$

Therefore, recalling that $\left(B^{-1}\right)^t = \left(B^t\right)^{-1}$, the matrix $A = B^t$ has all the required properties. □

2.2 Annihilators, Rank and Corank

In this section we will closely follow Dacorogna and Kneuss [31], in which all of the following notions are introduced. Prior contributions can also be found in Bandyopadhyay, Dacorogna and Kneuss [9].

2.2.1 *Exterior and Interior Annihilators*

We define the different annihilating spaces and give some elementary properties. We point out that these notions of annihilators, rank and corank seem very well adapted to the pullback, since they are invariants by pullback (see Propositions 2.27(vii) and 2.33(vi)).

Definition 2.26. Let $0 \leq k \leq n$ and $f \in \Lambda^k(\mathbb{R}^n)$.

(i) The *space of exterior annihilators of f of order s* is the vector space

$$\operatorname{Anh}_\wedge(f,s) = \{h \in \Lambda^s(\mathbb{R}^n) : f \wedge h = 0\}.$$

(ii) The *space of interior annihilators of f of order s* is the vector space

$$\operatorname{Anh}_\lrcorner(f,s) = \{h \in \Lambda^s(\mathbb{R}^n) : h \lrcorner f = 0\}.$$

Proposition 2.27. *Let $0 \leq k \leq n$ and $f, g \in \Lambda^k(\mathbb{R}^n)$.*

(i) The following hold, if $1 \leq k \leq n-1$:

$$\operatorname{Anh}_\wedge(f,n-k) \neq \{0\} \quad \text{and} \quad \operatorname{Anh}_\lrcorner(f,k) \neq \{0\}.$$

(ii) The following equivalences hold:

$$\begin{aligned} f = 0 \Leftrightarrow \operatorname{Anh}_\wedge(f,0) \neq \{0\} &\Leftrightarrow \operatorname{Anh}_\wedge(f,n-k) = \Lambda^{n-k}(\mathbb{R}^n) \\ \Leftrightarrow \operatorname{Anh}_\lrcorner(f,0) \neq \{0\} &\Leftrightarrow \operatorname{Anh}_\lrcorner(f,k) = \Lambda^k(\mathbb{R}^n). \end{aligned}$$

(iii) If $0 \leq s \leq t \leq n$, then

$$\operatorname{Anh}_{\wedge}(f,s) \wedge \Lambda^{t-s}(\mathbb{R}^n), \subset \operatorname{Anh}_{\wedge}(f,t),$$

$$\operatorname{Anh}_{\lrcorner}(f,s) \wedge \Lambda^{t-s}(\mathbb{R}^n) \subset \operatorname{Anh}_{\lrcorner}(f,t).$$

(iv) If $0 \leq s \leq n$, then

$$\operatorname{Anh}_{\wedge}(f,s) = \operatorname{Anh}_{\lrcorner}(*f,s).$$

(v) The following inclusion holds:

$$\operatorname{Anh}_{\wedge}(f,s) \cup \operatorname{Anh}_{\wedge}(g,s) \subset \operatorname{Anh}_{\wedge}(f \wedge g,s).$$

(vi) If $0 \leq s \leq t \leq n-k$ and

$$\operatorname{Anh}_{\wedge}(g,t) \subset \operatorname{Anh}_{\wedge}(f,t),$$

then

$$\operatorname{Anh}_{\wedge}(g,s) \subset \operatorname{Anh}_{\wedge}(f,s).$$

(vii) Let $A \in \mathrm{GL}(n)$. Then, for every $1 \leq s \leq n$,

$$A^*(\operatorname{Anh}_{\wedge}(f,s)) = \operatorname{Anh}_{\wedge}(A^*(f),s),$$

$$((A^{-1})^t)^*(\operatorname{Anh}_{\lrcorner}(f,s)) = \operatorname{Anh}_{\lrcorner}(A^*(f),s).$$

In particular, if $A \in O(n)$, then

$$A^*(\operatorname{Anh}_{\lrcorner}(f,s)) = \operatorname{Anh}_{\lrcorner}(A^*(f),s).$$

(viii) If $\lambda \neq 0$, then

$$\operatorname{Anh}_{\wedge}(\lambda f,s) = \operatorname{Anh}_{\wedge}(f,s),$$

$$\operatorname{Anh}_{\lrcorner}(\lambda f,s) = \operatorname{Anh}_{\lrcorner}(f,s).$$

Proof. Step 1. The proofs of (i), (ii), (iv), (v) and (viii) are immediate. The first property in (iii) is also easy, whereas the second one in (iii) follows from the first statement of Proposition 2.16.

Step 2. We now prove (vi). First, we notice that if $h \in \operatorname{Anh}_{\wedge}(g,s)$, then

$$h \wedge e^{i_1} \wedge \cdots \wedge e^{i_{t-s}} \in \operatorname{Anh}_{\wedge}(g,t) \quad \text{for every } 1 \leq i_1 < \cdots < i_{t-s} \leq n.$$

Thus, by hypothesis,

$$f \wedge h \wedge e^{i_1} \wedge \cdots \wedge e^{i_{t-s}} = 0 \quad \text{for every } 1 \leq i_1 < \cdots < i_{t-s} \leq n,$$

which easily implies the claim, namely

$$f \wedge h = 0.$$

Step 3. Property (vii) is a direct consequence of (2.8) and (2.9). This concludes the proof of the proposition. □

2.2.2 Rank and Corank

The next important concept is the notion of rank and corank of a form and it is related to the dimension of the corresponding annihilating spaces.

Definition 2.28. Let $0 \le k \le n$ be integers and $f \in \Lambda^k(\mathbb{R}^n)$.

(i) The *rank of order s*, $0 \le s \le k$, of $f \in \Lambda^k(\mathbb{R}^n)$ is given by

$$\operatorname{rank}_s[f] = \binom{n}{s} - \dim(\operatorname{Anh}_{\lrcorner}(f,s)).$$

(ii) The *corank of order s*, $0 \le s \le n-k$, of $f \in \Lambda^k(\mathbb{R}^n)$ is defined by

$$\operatorname{corank}_s[f] = \binom{n}{s} - \dim(\operatorname{Anh}_{\wedge}(f,s)).$$

Remark 2.29. (i) In the literature (see Bandyopadhyay and Dacorogna [8, p. 1720], Definition 2.2 in Bandyopadhyay, Dacorogna and Kneuss [9], Definition 7.11 in Godbillon [51], Marcus [74, pp. 85–88], Martinet [71] and Sternberg [93, p. 25]) the only notion of rank, for an exterior form that is used, is the above rank of order 1. In [74] and [93], a similar notion to our interior annihilator of order 1 is given. However, the rank of order 1 is not always defined as above; but all of the definitions are equivalent, as will be seen in Propositions 2.32 and 2.37. However, before that, let us introduce the following notations.

(ii) Since the most important notion is the one of rank of order 1, we will write rank instead of rank_1 when no ambiguity occurs. This will be the case throughout Part IV, except for few instances.

Notation 2.30. *Throughout the book we identify a k-form with a vector of* $\mathbb{R}^{\binom{n}{k}}$ *and, to fix the order of the elements of the vector, we adopt the lexicographical order. Let* $0 \le k \le n$ *and* $f \in \Lambda^k(\mathbb{R}^n)$.

(i) Let $0 \le s \le k$. *To the linear map*

$$g \in \Lambda^s(\mathbb{R}^n) \to g \lrcorner f \in \Lambda^{k-s}(\mathbb{R}^n)$$

we associate a matrix $\overline{f}_{\lrcorner,s} \in \mathbb{R}^{\binom{n}{k-s} \times \binom{n}{s}}$ *such that, by abuse of notations,*

$$g \lrcorner f = \overline{f}_{\lrcorner,s}\, g \quad \text{for every } g \in \Lambda^s(\mathbb{R}^n).$$

More explicitly, using the lexicographical order for the columns (index below) and the rows (index above) of the matrix $\overline{f}_{\lrcorner,s}$, *we have (cf. Proposition 2.12(i))*

$$(\overline{f}_{\lrcorner,s})_{i_1\cdots i_s}^{j_1\cdots j_{k-s}} = (-1)^{k(k-s)} f_{i_1\cdots i_s j_1\cdots j_{k-s}}$$

for $1 \le i_1 < \cdots < i_s \le n$ *and* $1 \le j_1 < \cdots < j_{k-s} \le n$.

(ii) Similarly let $0 \le s \le n-k$. *To the linear map*

$$g \in \Lambda^s(\mathbb{R}^n) \to f \wedge g \in \Lambda^{s+k}(\mathbb{R}^n)$$

we associate a matrix $\overline{f}_{\wedge,s} \in \mathbb{R}^{\binom{n}{s+k}\times\binom{n}{s}}$ *such that, by abuse of notations,*

$$f \wedge g = \overline{f}_{\wedge,s}\, g \quad \textit{for every } g \in \Lambda^s(\mathbb{R}^n).$$

As above, the components of the matrix $\overline{f}_{\wedge,s}$ *can be written as (cf. Proposition 2.6(i))*

$$(\overline{f}_{\wedge,s})_{i_1\cdots i_s}^{j_1\cdots j_{s+k}} = \left(f \wedge e^{i_1} \wedge \cdots \wedge e^{i_s}\right)_{j_1\cdots j_{s+k}}$$

for $1 \le i_1 < \cdots < i_s \le n$ *and* $1 \le j_1 < \cdots < j_{s+k} \le n$.

Remark 2.31. As already said, in the next chapters of the book we will write, when there is no ambiguity, $\operatorname{rank}[f]$ instead of $\operatorname{rank}_1[f]$. Similarly, we will denote the matrix $\overline{f}_{\lrcorner,1}$ only by $\overline{f}$.

Proposition 2.32. *Let* $0 \le k \le n$ *and* $f \in \Lambda^k(\mathbb{R}^n)$.

(i) Let $0 \le s \le k$. *Then*

$$\operatorname{rank}_s[f] = \operatorname{rank}\left(\overline{f}_{\lrcorner,s}\right).$$

(ii) Let $0 \le s \le n-k$. *Then*

$$\operatorname{corank}_s[f] = \operatorname{rank}\left(\overline{f}_{\wedge,s}\right).$$

Proof. We only show (i), the proof of (ii) being similar. Using the definition of $\overline{f}_{\lrcorner,s}$, we see that

$$\ker(\overline{f}_{\lrcorner,s}) = \operatorname{Anh}_{\lrcorner}(f,s).$$

We thus obtain the result, since

$$\operatorname{rank}(\overline{f}_{\lrcorner,s}) = \binom{n}{s} - \dim\left[\ker(\overline{f}_{\lrcorner,s})\right].$$

This concludes the proof of the proposition. □

We now gather some elementary properties of the rank and corank.

Proposition 2.33. *Let* $0 \le k \le n$ *and* $f \in \Lambda^k(\mathbb{R}^n)$.

(i) If $f = 0$, *then for every* s,

$$\operatorname{corank}_s[f] = \operatorname{rank}_s[f] = 0.$$

(ii) If $f \neq 0$, then

$$\operatorname{corank}_{n-k}[f] = \operatorname{rank}_k[f] = 1.$$

(iii) If $0 \leq s \leq n-k$, then

$$\operatorname{corank}_s[f] = \operatorname{rank}_s[*f].$$

(iv) If $0 \leq s \leq k$, then

$$\operatorname{rank}_s[f] = \operatorname{rank}_{k-s}[f],$$

and if $f \neq 0$, then

$$\operatorname{rank}_s[f] \geq \binom{k}{s}.$$

(v) If $0 \leq s \leq n-k$, then

$$\operatorname{corank}_s[f] = \operatorname{corank}_{n-(s+k)}[f],$$

and if $f \neq 0$, then

$$\operatorname{corank}_s[f] \geq \binom{n-k}{s}.$$

(vi) Let $A \in \mathrm{GL}(n)$. If $0 \leq s \leq k$, then

$$\operatorname{rank}_s[A^*(f)] = \operatorname{rank}_s[f],$$

whereas if $0 \leq s \leq n-k$, then

$$\operatorname{corank}_s[A^*(f)] = \operatorname{corank}_s(f).$$

(vii) If $\lambda \neq 0$, then for every $0 \leq s \leq k$,

$$\operatorname{rank}_s[\lambda f] = \operatorname{rank}_s[f],$$

whereas for every $0 \leq s \leq n-k$,

$$\operatorname{corank}_s[\lambda f] = \operatorname{corank}_s(f).$$

Proof. The proofs of (i)–(iii) and (vii) are elementary.

Step 1. We now discuss assertion (iv). We here use Notation 2.30. Let us show that

$$\overline{f}_{\lrcorner,s} = (-1)^{k+s+ks}\left(\overline{f}_{\lrcorner,k-s}\right)^t,$$

which will prove the assertion using Proposition 2.32(i). Indeed, by definition, for any $1 \leq i_1 < \cdots < i_s \leq n$ and $1 \leq j_1 < \cdots < j_{k-s} \leq n$, we have

$$\begin{aligned}
\left(\overline{f}_{\lrcorner,s}\right)_{i_1\cdots i_s}^{j_1\cdots j_{k-s}} &= (-1)^{k(k-s)} f_{i_1\cdots i_s j_1\cdots j_{k-s}} = (-1)^{k(k-s)}(-1)^{s(k-s)} f_{j_1\cdots j_{k-s} i_1\cdots i_s} \\
&= (-1)^{k+s} f_{j_1\cdots j_{k-s} i_1\cdots i_s} = (-1)^{k+s+ks}\left(\overline{f}_{\lrcorner,k-s}\right)_{j_1\cdots j_{k-s}}^{i_1\cdots i_s}
\end{aligned}$$

and thus the claim is proved. We now prove that $\operatorname{rank}_s[f] \geq \binom{k}{s}$. Since $f \neq 0$, there exists $(i_1,\ldots,i_k) \in \mathcal{T}_k$ such that $f_{i_1\cdots i_k} \neq 0$. Therefore, there are at least $\binom{k}{s}$ linearly independent s-forms which are not in $\text{Anh}_{\lrcorner}(f,s)$, namely

$$e^{i_{m_1}} \wedge \cdots \wedge e^{i_{m_s}} \notin \text{Anh}_{\lrcorner}(f,s) \quad \text{for every } 1 \leq m_1 < \cdots < m_s \leq k.$$

This implies the claim.

Step 2. We then discuss (v). Recalling that $*f \in \Lambda^{n-k}(\mathbb{R}^n)$ and using (iii) and (iv), we have the assertion, since

$$\text{corank}_s[f] = \text{rank}_s[*f] = \text{rank}_{n-k-s}[*f] = \text{corank}_{n-(s+k)}[f].$$

The assertion on the lower bound for the corank follows from (iii) and (iv).

Step 3. Claim (vi) is a direct consequence of Proposition 2.27(vii). □

Before proceeding further, we give some examples.

Example 2.34. Let $f \in \Lambda^k(\mathbb{R}^n)$ with $f \neq 0$.

(i) We start with the case $k = 1$. We claim, for $0 \leq s \leq n-1$, that

$$\text{rank}_1[f] = 1 \quad \text{and} \quad \text{corank}_s[f] = \binom{n-1}{s}.$$

The first equation is just a particular case of statement (ii) of the previous proposition. To show the second equation we proceed as follows. According to Proposition 2.24(i) and Proposition 2.33(vi), we can assume that $f = e^1$. Notice that

$$\text{Anh}_{\wedge}(e^1,s) = \text{span}\{e^1 \wedge e^{i_2} \wedge \cdots \wedge e^{i_s};\, 2 \leq i_2 < \cdots < i_s \leq n\}$$

and hence

$$\dim\left(\text{Anh}_{\wedge}(e^1,s)\right) = \binom{n-1}{s-1} = \binom{n}{s} - \binom{n-1}{s}$$

as claimed.

(ii) We now turn to the case $k = 2$. The only invariant that matters is $\text{rank}_1[f]$ and, as will be seen in Proposition 2.37(iii), it is even. It determines the corank of any order (cf. Proposition 2.24(ii) and Proposition 2.33(vi)) and, according to Proposition 2.33(ii),

$$\text{rank}_2[f] = 1.$$

(iii) When $k = n$, then

$$\text{rank}_s[f] = \binom{n}{s},$$

whereas if $k = n-1$, then (cf. Proposition 2.33(iii) and (i) of the present example)

$$\text{rank}_s[f] = \binom{n-1}{s}.$$

(iv) Consider the case $k = 3$. We automatically have, according to Proposition 2.33(iv),

$$\text{rank}_2[f] = \text{rank}_1[f].$$

However, the coranks are not uniquely determined by the rank of order 1. Indeed, let $n = 7$ and

$$f = e^1 \wedge e^2 \wedge e^3 + e^2 \wedge e^4 \wedge e^5 + e^3 \wedge e^6 \wedge e^7,$$
$$g = e^1 \wedge e^2 \wedge e^3 + e^1 \wedge e^4 \wedge e^5 + e^1 \wedge e^6 \wedge e^7.$$

Then

$$\operatorname{rank}_1[f] = \operatorname{rank}_1[g] = 7$$

and

$$\operatorname{corank}_1[f] = 7 \quad \text{and} \quad \operatorname{corank}_1[g] = 6.$$

(v) We finally give an example showing that the rank of order 1 does not determine the rank of higher orders. We let $k = 4$, $n = 8$ and

$$f = e^1 \wedge e^2 \wedge e^3 \wedge e^4 + e^5 \wedge e^6 \wedge e^7 \wedge e^8,$$

$$g = e^1 \wedge \left[e^2 \wedge e^3 \wedge e^4 + e^5 \wedge e^6 \wedge e^7 + e^2 \wedge e^5 \wedge e^8 \right].$$

We have

$$\operatorname{rank}_1[f] = \operatorname{rank}_1[g] = 8,$$
$$\operatorname{rank}_2[f] = 12 \quad \text{and} \quad \operatorname{rank}_2[g] = 14.$$

We now turn to two interesting examples showing that in order that a form g be the pullback of a form f, it is not enough that they have all their ranks and coranks equal.

Example 2.35. Let m be odd, $n = 2m$ and

$$f = *(\omega_m) = * \left(\sum_{i=1}^{m} e^{2i-1} \wedge e^{2i} \right)$$
$$= \sum_{i=1}^{m} e^1 \wedge e^2 \wedge \cdots \wedge \widehat{e^{2i-1}} \wedge \widehat{e^{2i}} \wedge \cdots \wedge e^{2m-1} \wedge e^{2m}.$$

Then f and $-f$ have all their ranks and coranks equal (cf. Proposition 2.33(vii)); however, there is no $A \in \mathrm{GL}(n)$ such that (cf. Proposition 2.24(iii))

$$A^*(f) = -f.$$

This result applies, in particular, to $n = 6$ and

$$f = e^1 \wedge e^2 \wedge e^3 \wedge e^4 + e^1 \wedge e^2 \wedge e^5 \wedge e^6 + e^3 \wedge e^4 \wedge e^5 \wedge e^6.$$

Example 2.36. The same phenomenon also occurs when $n = 6$ and $k = 3$. Indeed, the forms

$$f = e^1 \wedge e^2 \wedge e^3 + e^4 \wedge e^5 \wedge e^6,$$
$$g = e^1 \wedge e^2 \wedge e^3 + e^1 \wedge e^4 \wedge e^5 + e^2 \wedge e^4 \wedge e^6 + e^3 \wedge e^5 \wedge e^6$$

have all of their ranks and coranks equal, namely

$$\operatorname{rank}_1[f] = \operatorname{rank}_1[g] = \operatorname{rank}_2[f] = \operatorname{rank}_2[g] = 6,$$

$$\operatorname{corank}_1[f] = \operatorname{corank}_1[g] = \operatorname{corank}_2[f] = \operatorname{corank}_2[g] = 6,$$

$$\operatorname{rank}_3[f] = \operatorname{rank}_3[g] = 1 \quad \text{and} \quad \operatorname{corank}_3[f] = \operatorname{corank}_3[g] = 1.$$

However (cf. Kneuss [60] for details), there is no $A \in \mathrm{GL}(6)$ so that

$$A^*(g) = f.$$

2.2.3 Properties of the Rank of Order 1

Since the most essential notion is the one of rank of order 1, we gather below some properties of this rank; for related results, see Martinet [71]. We also recall that in all of the other chapters, the rank of order 1 of a form f is just referred to as the rank of f and is denoted by $\operatorname{rank}[f]$ instead of $\operatorname{rank}_1[f]$.

Proposition 2.37. *Let $f \in \Lambda^k(\mathbb{R}^n)$ and $1 \leq k \leq n$.*

(i) If

$$\Lambda_f^1 = \left\{u \in \Lambda^1(\mathbb{R}^n) : \exists g \in \Lambda^{k-1}(\mathbb{R}^n) \ \text{ with } g \lrcorner f = u\right\},$$

then

$$\operatorname{rank}_1[f] = \dim\left(\Lambda_f^1\right),$$

$$\Lambda_f^1 = \operatorname{Range}\left(\overline{f}_{\lrcorner,k-1}\right).$$

(ii) If $f \neq 0$ and $3 \leq k \leq n$, then

$$\operatorname{rank}_1[f] \in \{k, k+2, \ldots, n\}$$

and any of the values in $\{k, k+2, \ldots, n\}$ can be achieved by the rank of order 1 of a k-form.

(iii) If $k = 2$, then the rank of order 1 of f, $f \neq 0$, is even and any even value less than or equal to n can be achieved by the rank of order 1 of a 2-form. Moreover, $\operatorname{rank}_1[f] = 2m$ if and only if

$$f^m \neq 0 \quad \text{and} \quad f^{m+1} = 0,$$

where $f^m = \underbrace{f \wedge \cdots \wedge f}_{m \text{ times}}$. Furthermore, if n is even, the following identity holds, identifying n-forms with 0-forms:

$$\left|\det \overline{f}_{\lrcorner,1}\right|^{1/2} = \frac{1}{(n/2)!}\left|f^{n/2}\right|.$$

(iv) If $g \in \Lambda^l(\mathbb{R}^n)$*, then*

$$\operatorname{rank}_1[f \wedge g] \leq \operatorname{rank}_1[f] + \operatorname{rank}_1[g] - \dim\left(\Lambda_f^1 \cap \Lambda_g^1\right).$$

Moreover,

$$\operatorname{rank}_1[f \wedge g] = \operatorname{rank}_1[f] + \operatorname{rank}_1[g] \quad \Leftrightarrow \quad \Lambda_f^1 \cap \Lambda_g^1 = \{0\}.$$

(v) Let $f \in \Lambda^2(\mathbb{R}^n)$ *and* $g \in \Lambda^1(\mathbb{R}^n)$ *be such that* $f \wedge g \neq 0$*. Then*

$$\operatorname{rank}_1[f \wedge g] = \begin{cases} \operatorname{rank}_1[f] + 1 & \text{if } g \notin \Lambda_f^1 \\ \operatorname{rank}_1[f] - 1 & \text{if } g \in \Lambda_f^1. \end{cases}$$

Remark 2.38. (i) For $f \in \Lambda^k(\mathbb{R}^n)$, the rank of order 1 of f can never be $(k+1)$. In particular, when $k = n-1$ and $f \neq 0$, then

$$\operatorname{rank}_1[f] = n - 1.$$

(ii) From Proposition 2.37(iv), we can infer that if $f \neq 0$, then

$$\operatorname{rank}_1[*f] \geq n - \operatorname{rank}_1[f].$$

When $k = 1$ or $k = n-1$, then the inequality becomes an equality. In general, however, as soon as $2 \leq k \leq n-2$, the inequality can be strict.

(iii) Let $m \geq n$ be integers; then $\operatorname{rank}_1[f]$ is independent of whether we consider $f \in \Lambda^k(\mathbb{R}^n)$ or $f \in \Lambda^k(\mathbb{R}^m)$, in view of the above proposition and Theorem 2.22 (cf. Remark 2.23(ii)). This is, however, not true for all of the other ranks and coranks.

Proof. We split the proof into five steps.

Step 1. We prove statement (i). The fact that

$$\Lambda_f^1 = \operatorname{Range}\left(\overline{f}_{\lrcorner, k-1}\right)$$

follows from the definition. We use claim (i) of Proposition 2.32 and assertion (iv) of Proposition 2.33 to get that

$$\dim \Lambda_f^1 = \operatorname{rank}(\overline{f}_{\lrcorner, k-1}) = \operatorname{rank}(\overline{f}_{\lrcorner, 1}) = \operatorname{rank}_1[f].$$

Step 2. We show (ii). Using Theorem 2.22, statement (i) of the present proposition and Remark 2.21, we may assume that $\operatorname{rank}_1[f] = n$ and thus that

$$\Lambda_f^1 = \operatorname{span}\{e^1, \ldots, e^n\}.$$

1) Since $\Lambda^k(\mathbb{R}^n) = \{0\}$ if $k > n$ and $f \neq 0$, we must have $k \leq n$ and thus

$$k \leq \operatorname{rank}_1[f] = n.$$

2) Let us show that $n \neq k+1$. Suppose, for the sake of contradiction, that $n = k+1$ and we then show the existence of $u \in \Lambda^1(\mathbb{R}^n)$, $u \neq 0$, with

$$u \lrcorner f = 0.$$

This will be the desired contradiction since

$$\operatorname{rank}_1[f] = n \quad \Leftrightarrow \quad \operatorname{Anh}_{\lrcorner}(f,1) = \{0\}.$$

Indeed, since $*f \in \Lambda^1(\mathbb{R}^n)$, we have that

$$0 = (*f) \wedge (*f) = * \left[(-1)^{n(n-2)} (*f) \lrcorner f \right]$$

and, therefore, $u = *f$ is the required 1-form.

3) Finally, we show the last part of (ii). We have to prove that for any $n \geq k$, $n \neq k+1$, there exists $f \in \Lambda^k(\mathbb{R}^n)$ with $\operatorname{rank}_1[f] = n$. This will be sufficient to show the assertion. Let $s \geq 1$ and $l \in \{0, \ldots, k-1\}$ be such that

$$n = sk + l.$$

We now define a k-form f having the required properties. We consider three cases.

Case 1. $l = 0$. We let

$$f = \sum_{t=1}^{s} e^{(t-1)k+1} \wedge \cdots \wedge e^{tk}.$$

Case 2. $l = 1$ (and thus $s \geq 2$ since $n \neq k+1$). We let

$$f = \sum_{t=1}^{s-1} e^{(t-1)k+1} \wedge \cdots \wedge e^{tk} + e^{(s-1)k} \wedge \cdots \wedge e^{sk-1} + e^{(s-1)k+2} \wedge \cdots \wedge e^{sk+1}.$$

Case 3. $2 \leq l \leq k-1$. We let

$$f = \sum_{t=1}^{s} e^{(t-1)k+1} \wedge \cdots \wedge e^{tk} + e^{(s-1)k+l+1} \wedge \cdots \wedge e^{sk+l}.$$

In the three cases, we notice that f is a sum of terms having two by two at least two distinct e^i. From this observation it follows immediately that if $u \in \Lambda^1(\mathbb{R}^n)$ verifies

$$u \lrcorner f = 0,$$

then $u = 0$. This shows that $\operatorname{rank}_1[f] = n$ and ends the proof of (ii).

Step 3. Let us show (iii) and first prove that the rank of order 1 of f is even. From Proposition 2.24(ii), we get that there exists a unique integer m, with $2 \leq 2m \leq n$, such that

$$f^m \neq 0 \quad \text{and} \quad f^{m+1} = 0$$

and there exists $A \in \mathrm{GL}(n)$ such that

$$A^*(f) = \omega_m = \sum_{i=1}^{m} e^{2i-1} \wedge e^{2i}.$$

Since the rank is invariant by pullback (cf. Proposition 2.33(vi)), we have the result since we clearly have that

$$\mathrm{rank}_1[f] = \mathrm{rank}_1[\omega_m] = 2m.$$

The fact that any even value less than or equal to n can be achieved by the rank of order 1 of a 2-form follows from the above argument. We now prove the statement concerning the determinant. Note first that if $2m < n$, the result is trivial, since both sides of the equation are 0. So let $n = 2m$. One easily sees by induction that $\det[\overline{(\omega_m)}_{\lrcorner,1}] = 1$. Note also that for any $B \in \mathbb{R}^{n\times n}$ and any $g \in \Lambda^2(\mathbb{R}^n)$, we have

$$\begin{aligned}\left(\overline{B^*(g)}_{\lrcorner,1}\right)_p^q &= (B^*(g))_{pq} = \sum_{1\le i<j\le n} g_{ij}\left(B_p^i B_q^j - B_q^i B_p^j\right)\\ &= \sum_{i,j=1}^{n} g_{ij} B_p^i B_q^j = \sum_{i,j=1}^{n} \left(\overline{g}_{\lrcorner,1}\right)_i^j B_p^i B_q^j = \left(B^t \overline{g}_{\lrcorner,1} B\right)_p^q\end{aligned}$$

for every $1 \le p,q \le n$. Next, let A be such that

$$A^*(f) = \omega_m = \sum_{i=1}^{m} e^{2i-1} \wedge e^{2i}.$$

Therefore choosing $B = A^{-1}$, we get

$$\begin{aligned}\left|\det \overline{f}_{\lrcorner,1}\right|^{1/2} &= \left|\det\left(B^t \overline{(\omega_m)}_{\lrcorner,1} B\right)\right|^{1/2} = |\det B|\\ &= \left|B^*\left(e^1 \wedge \cdots \wedge e^n\right)\right| = \frac{1}{m!}\left|B^*\left((\omega_m)^m\right)\right| = \frac{1}{(n/2)!}\left|f^{n/2}\right|.\end{aligned}$$

Step 4. For the proof of (iv), we refer to [60].

Step 5. We finally prove statement (v). First, note that if $g \notin \Lambda_f^1$, then the result is a direct application of statement (iv) since $\mathrm{rank}_1[g] = 1$. We thus assume $g \in \Lambda_f^1$. Using Proposition 2.24(ii) and Remark 2.25, there exists $A \in \mathrm{GL}(n)$ such that, writing $2m = \mathrm{rank}_1[f]$,

$$A^*(f) = \sum_{i=1}^{m} e^{2i-1} \wedge e^{2i} \quad \text{and} \quad A^*(g) = e^{2m}.$$

Therefore,

$$A^*(f \wedge g) = \sum_{i=1}^{m} e^{2i-1} \wedge e^{2i} \wedge e^{2m} = \sum_{i=1}^{m-1} e^{2i-1} \wedge e^{2i} \wedge e^{2m}.$$

Since

$$\operatorname{rank}_1\left[\sum_{i=1}^{m-1} e^{2i-1}\wedge e^{2i}\wedge e^{2m}\right]=2(m-1)+1=\operatorname{rank}_1[f]-1,$$

we obtain the result using Proposition 2.33(vi). □

2.3 Divisibility

In this section we will follow the presentation of Dacorogna and Kneuss [31] (see also Bandyopadhyay, Dacorogna and Kneuss [9]).

2.3.1 Definition and First Properties

Definition 2.39. Let $1\leq k\leq n$ and $f\in\Lambda^k(\mathbb{R}^n)$.

(i) Let $0\leq l\leq k$. We say that f is *l-divisible* if there exist $a\in\Lambda^l(\mathbb{R}^n)$ and $b\in\Lambda^{k-l}(\mathbb{R}^n)$ such that

$$f=a\wedge b.$$

We say that f is *prime* (or *indecomposable*) if it is not l-divisible for any $1\leq l\leq k-1$.

(ii) We say that f is *totally divisible* (or *totally decomposable*) if there exist $f_1,\ldots,f_k\in\Lambda^1(\mathbb{R}^n)$ such that

$$f=f_1\wedge\cdots\wedge f_k.$$

Remark 2.40. (i) In the literature the second definition is standard; it goes back to Cartan and such a form is, sometimes, also called *pure* or *decomposable*.

(ii) Let $f\in\Lambda^k(\mathbb{R}^n)$ and $A\in\mathrm{GL}(n)$. Using (2.8), we see that f is l-divisible if and only if $A^*(f)$ is l-divisible and that f is prime if and only if $A^*(f)$ is prime.

Remark 2.41. We should point out that a form is *not uniquely* decomposable into prime forms. Indeed, consider

$$f=\left[e^1\wedge e^2+e^3\wedge e^4\right]\wedge e^3=e^1\wedge e^2\wedge e^3$$

and observe that it is a product of one prime 2-form of rank 4 and one (prime) 1-form and, at the same time, a product of three (prime) 1-forms. However, only the second one is an optimal decomposition of f, in the sense that

$$f=f_1\wedge\cdots\wedge f_s$$

with $f_i\in\Lambda^{k_i}(\mathbb{R}^n)$ prime, $k_1+\cdots+k_s=k$, and

$$\operatorname{rank}_1[f]=\sum_{i=1}^{s}\operatorname{rank}_1[f_i].$$

An optimal decomposition of the above type does not always exist, as the following example shows. Let $f \in \Lambda^4(\mathbb{R}^6)$ given by

$$f = \left[e^1 \wedge e^2 + e^3 \wedge e^4\right] \wedge \left[e^1 \wedge e^2 + e^5 \wedge e^6\right]$$
$$= e^1 \wedge e^2 \wedge e^3 \wedge e^4 + e^1 \wedge e^2 \wedge e^5 \wedge e^6 + e^3 \wedge e^4 \wedge e^5 \wedge e^6.$$

Note that $\operatorname{rank}_1[f] = 6$. Let us show that f is not optimally decomposable in the above sense.

1) Observe that f is not 1-divisible. Indeed, a simple calculation shows that there exists no $a \in \Lambda^1(\mathbb{R}^6)$ with $a \neq 0$, such that

$$f \wedge a = 0.$$

Therefore, we have the assertion using Theorem 2.42.

2) By construction, f is not prime. Therefore, two cases can happen.

Case 1. $f = a \wedge b$ with $a \in \Lambda^3(\mathbb{R}^6)$ and $b \in \Lambda^1(\mathbb{R}^6)$ and this is impossible, since f is not 1-divisible.

Case 2. $f = a \wedge b$ with $a, b \in \Lambda^2(\mathbb{R}^6)$ and

$$\operatorname{rank}_1[f] = \operatorname{rank}_1[a] + \operatorname{rank}_1[b] = 6.$$

Since $\operatorname{rank}_1[a]$ and $\operatorname{rank}_1[b]$ are even numbers, then one of them is 4 and the other one is 2, say $\operatorname{rank}_1[b] = 2$. Since $b \in \Lambda^2$ with $\operatorname{rank}_1[b] = 2$, we deduce (see Proposition 2.43(ii)) that there exist $b_1, b_2 \in \Lambda^1(\mathbb{R}^6)$ such that

$$b = b_1 \wedge b_2$$

and, hence,

$$f = a \wedge b = a \wedge b_1 \wedge b_2,$$

which is also impossible since f is not 1-divisible.

We now gather some properties about divisibility and total divisibility. The first result is known as the Cartan lemma (cf., for example, Bryant et al. [18, p. 11]).

Theorem 2.42 (Cartan lemma). *Let $1 \leq k \leq n$ and $f \in \Lambda^k(\mathbb{R}^n)$ with $f \neq 0$. Let $1 \leq l \leq k$ and $g_1, \ldots, g_l \in \Lambda^1(\mathbb{R}^n)$ be such that*

$$g_1 \wedge \cdots \wedge g_l \neq 0.$$

Then there exists $u \in \Lambda^{k-l}(\mathbb{R}^n)$ verifying

$$f = g_1 \wedge \cdots \wedge g_l \wedge u$$

if and only if

$$f \wedge g_1 = \cdots = f \wedge g_l = 0.$$

Proof. For the sake of completeness we provide here a proof of the theorem, although it directly follows from Corollary 2.46 below.

Let us first show the necessary part. Let $u \in \Lambda^{k-l}(\mathbb{R}^n)$ be such that

$$f = u \wedge g_1 \wedge \cdots \wedge g_l;$$

then, clearly,

$$f \wedge g_1 = \cdots = f \wedge g_l = 0.$$

So we now turn to the sufficient part. In view of Proposition 2.19, proving the theorem for f is equivalent to proving it for $A^*(f)$ for any $A \in \mathrm{GL}(n)$. We may therefore assume, using Proposition 2.24(i), that

$$g_i = e^i, \quad 1 \leq i \leq l.$$

Since $f \wedge e^i = 0$, $1 \leq i \leq l$, implies that

$$f = \sum_{l+1 \leq j_{l+1} < \cdots < j_k \leq n} f_{1 \cdots l j_{l+1} \cdots j_k} e^1 \wedge \cdots \wedge e^l \wedge e^{j_{l+1}} \wedge \cdots \wedge e^{j_k},$$

we have the result by letting

$$u = \sum_{l+1 \leq j_{l+1} < \cdots < j_k \leq n} f_{1 \cdots l j_{l+1} \cdots j_k} e^{j_{l+1}} \wedge \cdots \wedge e^{j_k}.$$

This finishes the proof. □

We now gather some other elementary facts established in Bandyopadhyay, Dacorogna and Kneuss [9].

Proposition 2.43. *Let $1 \leq k \leq n$ and $f \in \Lambda^k(\mathbb{R}^n)$ with $f \neq 0$.*

(i) Let $a \in \Lambda^1(\mathbb{R}^n)$, $a \neq 0$, be such that

$$f \wedge a = 0.$$

Then

$$a \in \Lambda_f^1 = \left\{ u \in \Lambda^1(\mathbb{R}^n) : \exists h \in \Lambda^{k-1}(\mathbb{R}^n) \ \text{ with } h \lrcorner f = u \right\}.$$

(ii) The form f is totally divisible, meaning that there exist $f_1, \ldots, f_k \in \Lambda^1(\mathbb{R}^n)$ such that

$$f = f_1 \wedge \cdots \wedge f_k$$

if and only if

$$\mathrm{rank}_1[f] = k$$

if and only if

$$f \wedge b = 0, \quad \text{for every } b \in \Lambda_f^1$$

if and only if

$$\dim(\operatorname{Anh}_\wedge(f,1)) = k.$$

(iii) If k is odd and if $\operatorname{rank}_1[f] = k+2$, *then* f *is* 1*-divisible.*

Remark 2.44. (i) Statement (iii) in the proposition is, in general, false when k is even. Indeed, the form $f \in \Lambda^4(\mathbb{R}^n)$ given by

$$f = e^1 \wedge e^2 \wedge e^3 \wedge e^4 + e^1 \wedge e^2 \wedge e^5 \wedge e^6 + e^3 \wedge e^4 \wedge e^5 \wedge e^6$$

is not 1-divisible (although it is 2-divisible) while $\operatorname{rank}_1[f] = k+2 = 6$ (cf. Remark 2.41).

(ii) When $k = 3$, $f \neq 0$ and $\operatorname{rank}_1[f]$ is even, then f is prime. This easily follows from the fact that if f is 1-divisible, there exists $a \in \Lambda^1(\mathbb{R}^n)$ and $b \in \Lambda^2(\mathbb{R}^n)$ so that $f = a \wedge b$ and, therefore, $\operatorname{rank}_1[f]$ is odd, using Proposition 2.37(v).

(iii) If f is prime, then

$$\operatorname{Anh}_\wedge(f,1) = \{0\}.$$

Conversely if $k = 2$ or $k = 3$ and $\operatorname{Anh}_\wedge(f,1) = \{0\}$, then f is prime.

(iv) We always have, appealing to Theorem 2.42,

$$\operatorname{corank}_1[f] < n \Leftrightarrow \operatorname{Anh}_\wedge(f,1) \neq \{0\} \Leftrightarrow f \text{ is 1-divisible.}$$

Moreover, if l is odd, then, noticing that $h \wedge h = 0$ for every $h \in \Lambda^l(\mathbb{R}^n)$, we immediately deduce the following implication:

$$f \text{ is } l\text{-divisible} \Rightarrow \operatorname{corank}_l[f] < \binom{n}{l}. \tag{2.16}$$

The converse of (2.16) is not verified in general. Indeed, let

$$f = e^1 \wedge e^2 \wedge e^3 \wedge e^4 + e^5 \wedge e^6 \wedge e^7 \wedge e^8 \in \Lambda^4(\mathbb{R}^8).$$

It is easily seen that f is not 1-divisible (and thus not 3-divisible). Noticing that

$$f \wedge e^1 \wedge e^2 \wedge e^5 = 0,$$

we get

$$\operatorname{corank}_3[f] = 8 < \binom{8}{3},$$

which shows the assertion. Finally, we prove that (2.16) is, in general, false if l is even. Indeed, let

$$f = e^1 \wedge e^2 + e^3 \wedge e^4 + e^5 \wedge e^6 \in \Lambda^2(\mathbb{R}^6).$$

We immediately obtain that f is 2-divisible (by itself). In addition, a simple calculation shows that $\operatorname{corank}_2[f] = \binom{6}{2}$, which proves the assertion.

Proof. Since Remark 2.21, Proposition 2.33(vi) and Remark 2.40(ii) hold, proving the proposition for f is equivalent to proving the claims for $A^*(f)$ for any $A \in \mathrm{GL}(n)$. This fact will be constantly used throughout the proof.

Step 1. Let us show (i). We may, as already said, assume that $a = e^1$. Since $f \wedge e^1 = 0$, we find that

$$f = \sum_{2 \leq j_2 < \cdots < j_k \leq n} f_{1 j_2 \cdots j_k} e^1 \wedge e^{j_2} \wedge \cdots \wedge e^{j_k}.$$

Letting

$$u = \sum_{2 \leq j_2 < \cdots < j_k \leq n} f_{1 j_2 \cdots j_k} e^{j_2} \wedge \cdots \wedge e^{j_k},$$

we find that

$$u \lrcorner f = -\langle u; u \rangle e^1$$

and, thus, $e^1 \in \Lambda_f^1$ since $u \neq 0$. The proof of (i) is therefore complete.

Step 2. We next show (ii) and we divide the proof into three parts.

Step 2.1. We first show that f is totally divisible if and only if $\operatorname{rank}_1[f] = k$.

(i) Assume that f is totally divisible. Using Proposition 2.24(i), we can suppose that $f = e^1 \wedge \cdots \wedge e^k$. This directly implies that

$$\Lambda_f^1 = \operatorname{span}\{e^1, \ldots, e^k\}$$

and, therefore, $\operatorname{rank}_1[f] = k$, using Proposition 2.37(i).

(ii) Let f be such that $\operatorname{rank}_1[f] = k$. From Proposition 2.37(i), we have

$$\operatorname{rank}_1[f] = \dim \Lambda_f^1.$$

Appealing to Theorem 2.22, we can assume that $f \in \Lambda^k(\mathbb{R}^k)$ and, thus,

$$f = \lambda\, e^1 \wedge \cdots \wedge e^k$$

with $\lambda \neq 0$, which means that f is totally divisible.

Step 2.2. We now prove that $\operatorname{rank}_1[f] = k$ is equivalent to

$$f \wedge b = 0 \quad \text{for every } b \in \Lambda_f^1.$$

(i) Assume that $\operatorname{rank}_1[f] = k$. With the same argument as the one of Step 2.1(ii), we can assume that

$$f = \lambda\, e^1 \wedge \cdots \wedge e^k$$

with $\lambda \neq 0$. This implies that

$$\Lambda_f^1 = \operatorname{span}\{e^1, \ldots, e^k\}.$$

Thus, we have the result since

$$f \wedge b = 0 \quad \text{for every } b \in \Lambda_f^1 .$$

(ii) We now prove the converse and let f verify

$$f \wedge b = 0 \quad \text{for every } b \in \Lambda_f^1 .$$

Letting $p = \operatorname{rank}_1[f]$, we can assume, using Theorem 2.22 and Proposition 2.37(i), that

$$\Lambda_f^1 = \operatorname{span}\{e^1, \ldots, e^p\}.$$

To conclude, it is enough to show that $p = k$. Suppose, for the sake of contradiction, that $p > k$ and let us show that $f = 0$, which will be the desired contradiction. Indeed, for every $1 \leq j_1 < \cdots < j_k \leq p$ there exists then $1 \leq i \leq p$ such that

$$i \notin \{j_1, \ldots, j_k\}.$$

Combining this with $f \wedge e^i = 0$, we obtain that

$$f_{j_1 \cdots j_k} = 0$$

and, thus, $f = 0$, which is the desired contradiction.

Step 2.3. We finally establish that f is totally divisible if and only if

$$\dim\left(\operatorname{Anh}_\wedge(f, 1)\right) = k.$$

(i) Assume that f is totally divisible; then there exist $f_1, \ldots, f_k \in \Lambda^1(\mathbb{R}^n)$ such that

$$f = f_1 \wedge \cdots \wedge f_k .$$

This clearly shows the assertion, namely

$$\operatorname{Anh}_\wedge(f, 1) = \operatorname{span}\{f_1, \ldots, f_k\}.$$

(ii) Assume now that $\dim\left(\operatorname{Anh}_\wedge(f, 1)\right) = k$. Therefore, there exist

$$f_1, \ldots, f_k \in \Lambda^1(\mathbb{R}^n)$$

such that

$$\operatorname{Anh}_\wedge(f, 1) = \operatorname{span}\{f_1, \ldots, f_k\}.$$

It then follows from the Cartan lemma (cf. Theorem 2.42) that we can find $\lambda \neq 0$ such that

$$f = \lambda f_1 \wedge \cdots \wedge f_k$$

and, thus, f is totally divisible.

Step 3. We finally show (iii). Using once more Theorem 2.22, we can assume that $n = k + 2$ and, thus, by hypothesis, n is odd. Since $*f \in \Lambda^2(\mathbb{R}^n)$, we have, according to Proposition 2.37(iii), that

$$\operatorname{rank}_1[*f] \text{ is even.}$$

Combining this fact with the definition of the rank of order 1 and the fact that n is odd, we deduce that there exists $a \in \Lambda^1(\mathbb{R}^n)$, $a \neq 0$, such that

$$a \lrcorner (*f) = 0.$$

Since

$$a \lrcorner (*f) = 0 \quad \Leftrightarrow \quad a \wedge f = 0,$$

we have that f is 1-divisible (by a), using Theorem 2.42 with $l = 1$. The proof is therefore complete. □

2.3.2 Main Result

Given $f \in \Lambda^k(\mathbb{R}^n)$ and $g \in \Lambda^l(\mathbb{R}^n)$ with $0 \le l \le k$, we want to find $u \in \Lambda^{k-l}(\mathbb{R}^n)$ such that

$$f = g \wedge u. \tag{2.17}$$

We have already studied the previous equation when $g \neq 0$ is a product of 1-forms, namely

$$g = g_1 \wedge \cdots \wedge g_l \neq 0,$$

where $g_1, \ldots, g_l \in \Lambda^1(\mathbb{R}^n)$. Indeed, in Theorem 2.42 we proved that a necessary and sufficient condition for solving (2.17) is

$$f \wedge g_1 = \cdots = f \wedge g_l = 0.$$

We want here to extend this result to general l-forms. All of the following results have been established by Dacorogna and Kneuss [31]. Our main theorem is the following. We recall that we use Notation 2.30.

Theorem 2.45 (Dacorogna–Kneuss theorem). *Let $0 \le l \le k \le n$ be integers. Let $g \in \Lambda^l(\mathbb{R}^n)$ and $f \in \Lambda^k(\mathbb{R}^n)$. The following assertions are then equivalent:*

(i) There exists $u \in \Lambda^{k-l}(\mathbb{R}^n)$ verifying

$$f = g \wedge u.$$

(ii) For every $h \in \Lambda^{n-k}(\mathbb{R}^n)$, the following implication holds:

$$[h \wedge g = 0] \quad \Rightarrow \quad [h \wedge f = 0]$$

or, equivalently,

$$\operatorname{Anh}_\wedge(g,n-k) \subset \operatorname{Anh}_\wedge(f,n-k).$$

(iii) For every $0 \le s \le n-k$ *and* $h \in \Lambda^s(\mathbb{R}^n)$*, the following implication is valid:*

$$[h \wedge g = 0] \quad \Rightarrow \quad [h \wedge f = 0]$$

or, equivalently,

$$\operatorname{Anh}_\wedge(g,s) \subset \operatorname{Anh}_\wedge(f,s) \quad \textit{for every } 0 \le s \le n-k.$$

(iv) Let $r = \operatorname{rank}(\overline{g}_{\wedge,k-l})$*. Looking at* f *and the columns of* $\overline{g}_{\wedge,k-l} \in \mathbb{R}^{\binom{n}{k}\times\binom{n}{k-l}}$ *as* 1*-forms in* $\mathbb{R}^{\binom{n}{k}}$*, then*

$$(\overline{g}_{\wedge,k-l})_{i_1^{(1)}\cdots i_{k-l}^{(1)}} \wedge \cdots \wedge (\overline{g}_{\wedge,k-l})_{i_1^{(r)}\cdots i_{k-l}^{(r)}} \wedge f = 0$$

for every

$$1 \le i_1^{(1)} < \cdots < i_{k-l}^{(1)} \le n, \ldots, 1 \le i_1^{(r)} < \cdots < i_{k-l}^{(r)} \le n.$$

We will also write (cf. Corollary 2.48) the dual version of (2.17), where the exterior product is replaced by the interior product, namely we solve

$$u \lrcorner g = f.$$

This last equation has been studied in Bandyopadhyay, Dacorogna and Kneuss [9] in the case where $g \in \Lambda^2(\mathbb{R}^n)$ (see Proposition 2.50 below).

We now give some corollaries. The proof of the theorem is put off to Section 2.3.4.

Corollary 2.46. *Theorem 2.45 indeed generalizes the Cartan lemma (cf. Theorem 2.42).*

Proof. Let

$$g = g_1 \wedge \cdots \wedge g_l \neq 0$$

as in the Cartan lemma. We first claim that $h \in \Lambda^s(\mathbb{R}^n)$, $s \ge 1$, satisfies $g \wedge h = 0$ if and only if h is of the form

$$h = \sum_{j=1}^{l} g_j \wedge h_j$$

for some $h_j \in \Lambda^{s-1}(\mathbb{R}^n)$. The sufficient part being obvious, we only prove the necessary part. With no loss of generality, we can assume that $g_j = e_j$ and, thus,

$$e_1 \wedge \cdots \wedge e_l \wedge h = 0.$$

Writing $h = \sum_{i_1<\cdots<i_s} h_{i_1\cdots i_s} e^{i_1} \wedge \cdots \wedge e^{i_s}$, the previous equation immediately implies that

$$h_{i_1\cdots i_s} = 0 \quad \text{if } \{i_1,\ldots,i_s\} \cap \{1,\ldots,l\} = \emptyset,$$

which directly implies the claim, namely

$$h = \sum_{j=1}^{l} e_j \wedge h_j \quad \text{for some } h_j \in \Lambda^{s-1}(\mathbb{R}^n).$$

Thus, the Cartan lemma follows once noticed that for $f \in \Lambda^k(\mathbb{R}^n)$, we have

$$f \wedge g_j = 0 \quad \Leftrightarrow \quad \{f \wedge g_j \wedge h_j = 0 \quad \text{for every } h_j \in \Lambda^{n-k-1}\}.$$

This concludes the proof. □

Corollary 2.47. *A k-form f in* $\mathbb{R}^n$ *is characterized, up to a multiplicative constant, by* $\mathrm{Anh}_\wedge(f, n-k)$.

Proof. We give two proofs of the corollary: the first one as a consequence of the theorem and the second one in a constructive way.

Proof 1. Clearly, f and λf, with $\lambda \neq 0$, verify

$$\mathrm{Anh}_\wedge(f, n-k) = \mathrm{Anh}_\wedge(\lambda f, n-k).$$

So let us show the converse and let $f, g \in \Lambda^k(\mathbb{R}^n)$ with

$$\mathrm{Anh}_\wedge(f, n-k) = \mathrm{Anh}_\wedge(g, n-k).$$

Theorem 2.45(ii) implies then the existence of $\lambda \in \Lambda^0(\mathbb{R}^n)$ with $g = \lambda f$. Noting that $\lambda \neq 0$ (unless $f = g = 0$), we have the claim.

Proof 2. The sufficient part is as in the first proof. We divide the proof of the necessary part into two steps and assume that

$$\mathrm{Anh}_\wedge(f, n-k) = \mathrm{Anh}_\wedge(g, n-k)$$

and let us show that $g = \lambda f$.

Step 1. We show that if $f_{i_1 \cdots i_k} = 0$ for some $1 \leq i_1 < \cdots < i_k \leq n$, then

$$g_{i_1 \cdots i_k} = 0.$$

Note that $f_{i_1 \cdots i_k} = 0$ is equivalent, according to (2.4), to

$$*\left(e^{i_1} \wedge \cdots \wedge e^{i_k}\right) \in \mathrm{Anh}_\wedge(f, n-k).$$

Hence, by hypothesis,

$$*\left(e^{i_1} \wedge \cdots \wedge e^{i_k}\right) \in \mathrm{Anh}_\wedge(g, n-k),$$

which is equivalent to $g_{i_1 \cdots i_k} = 0$, using (2.4).

Step 2. If $f = 0$, then $g = 0$ according to Step 1 and the corollary is thus true for any $\lambda \in \mathbb{R}$. We therefore assume that $f \neq 0$ and, thus, $f_{j_1 \cdots j_k} \neq 0$ for a certain index $1 \leq j_1 < \cdots < j_k \leq n$. Let $1 \leq i_1 < \cdots < i_k \leq n$. We note that, using (2.4),

$$h = f_{i_1\cdots i_k}\left[*\left(e^{j_1}\wedge\cdots\wedge e^{j_k}\right)\right] - f_{j_1\cdots j_k}\left[*\left(e^{i_1}\wedge\cdots\wedge e^{i_k}\right)\right] \in \mathrm{Anh}_\wedge(f,n-k).$$

The hypothesis implies that

$$g\wedge h = 0,$$

which is equivalent, in view of (2.4), to

$$f_{i_1\cdots i_k}\, g_{j_1\cdots j_k} - f_{j_1\cdots j_k}\, g_{i_1\cdots i_k} = 0$$

and, thus,

$$g_{i_1\cdots i_k} = \frac{g_{j_1\cdots j_k}}{f_{j_1\cdots j_k}}\, f_{i_1\cdots i_k}\,.$$

Setting

$$\lambda = \frac{g_{j_1\cdots j_k}}{f_{j_1\cdots j_k}},$$

we have the assertion. □

By duality, we obtain from Theorem 2.45 the corresponding result for interior equations.

Corollary 2.48. *Let $0\leq l\leq k\leq n$ be integers. Let $g\in\Lambda^k(\mathbb{R}^n)$ and $f\in\Lambda^l(\mathbb{R}^n)$. The following statements are then equivalent.*

(i) There exists $u\in\Lambda^{k-l}(\mathbb{R}^n)$ satisfying

$$u\lrcorner g = f.$$

(ii) For every $h\in\Lambda^l(\mathbb{R}^n)$,

$$[h\lrcorner g = 0] \quad\Rightarrow\quad [h\lrcorner f = 0]$$

or, equivalently,

$$\mathrm{Anh}_\lrcorner(g,l)\subset\mathrm{Anh}_\lrcorner(f,l).$$

(iii) For every $0\leq s\leq l$ and every $h\in\Lambda^s(\mathbb{R}^n)$,

$$[h\lrcorner g = 0] \quad\Rightarrow\quad [h\lrcorner f = 0]$$

or, equivalently,

$$\mathrm{Anh}_\lrcorner(g,s)\subset\mathrm{Anh}_\lrcorner(f,s) \quad \textit{for every } 0\leq s\leq l.$$

(iv) Let $r=\operatorname{rank}(\overline{g}_{\lrcorner,k-l})$. Seeing f and the columns of $\overline{g}_{\lrcorner,k-l}$ as 1-forms in $\mathbb{R}^{\binom{n}{l}}$, then

$$(\overline{g}_{\lrcorner,k-l})_{i_1^{(1)}\cdots i_{k-l}^{(1)}}\wedge\cdots\wedge(\overline{g}_{\lrcorner,k-l})_{i_1^{(r)}\cdots i_{k-l}^{(r)}}\wedge f = 0$$

for every

$$1\leq i_1^{(1)}<\cdots<i_{k-l}^{(1)}\leq n,\ldots,1\leq i_1^{(r)}<\cdots<i_{k-l}^{(r)}\leq n.$$

Proof. The equivalences (i)–(iii) follow from Theorem 2.45 and from the following observations:

$$f = u \lrcorner g = (-1)^{nl} * (u \wedge (*g)) \Leftrightarrow *f = (-1)^{l} (u \wedge (*g)),$$

$$h \lrcorner g = 0 \Leftrightarrow h \wedge (*g) = 0,$$

$$h \lrcorner f = 0 \Leftrightarrow h \wedge (*f) = 0.$$

The equivalence between (i) and (iv) is just Lemma 2.54 applied to the matrix $\overline{g}_{\lrcorner,k-l}$, since

$$u \lrcorner g = f \Leftrightarrow \overline{g}_{\lrcorner,k-l} u = f.$$

This concludes the proof of the corollary. □

2.3.3 Some More Results

The main theorem (Theorem 2.45) immediately gives an equivalent condition for a k-form to be prime.

Corollary 2.49. *Let $f \in \Lambda^k(\mathbb{R}^n)$; then the two following statements are equivalent:*

(i) The k-form f is prime.

(ii) For any $1 \leq l \leq k-1$ and any $g \in \Lambda^l(\mathbb{R}^n)$, there exists $h \in \operatorname{Anh}_{\wedge}(g, n-k)$ such that

$$f \wedge h \neq 0.$$

When $k = 2$, we can express Corollary 2.48 in a different way. The following proposition is taken from Bandyopadhyay, Dacorogna and Kneuss [9].

Proposition 2.50. *Let $g \in \Lambda^2(\mathbb{R}^n)$ with $\operatorname{rank}_1[g] = 2m \leq n$ and $f \in \Lambda^1(\mathbb{R}^n)$. There exists $u \in \Lambda^1(\mathbb{R}^n)$ such that*

$$f = u \lrcorner g$$

if and only if

$$f \wedge g^m = 0.$$

Remark 2.51. (i) If $n = 2m$ and since $f \wedge g^m \in \Lambda^{n+1}(\mathbb{R}^n)$, we then always have

$$f \wedge g^m = 0.$$

Therefore, there always exists $u \in \Lambda^1(\mathbb{R}^n)$ such that

$$f = u \lrcorner g.$$

(ii) More generally, if k is even, $g \in \Lambda^k(\mathbb{R}^n)$ with $g^{m+1} = 0$ (with $mk \le n$) and there exists $u \in \Lambda^1(\mathbb{R}^n)$ such that $f = u \lrcorner g$, then necessarily (cf. (2.5))

$$f \wedge g^m = 0.$$

The converse is, however, not true in general. Indeed, let $k = 4$, $m = 2$ and

$$g = e^1 \wedge e^2 \wedge e^3 \wedge e^4 + e^5 \wedge e^6 \wedge e^7 \wedge e^8 \in \Lambda^4(\mathbb{R}^8).$$

Note that $\operatorname{rank}_1[g] = 4 \cdot 2 = 8$, $f \wedge g^2 = 0$ for every $f \in \Lambda^3(\mathbb{R}^8)$, but that there does not exist $u \in \Lambda^1(\mathbb{R}^8)$ such that

$$e^1 \wedge e^2 \wedge e^5 = u \lrcorner g.$$

This proves the claim.

We now prove the proposition.

Proof. Step 1. We start with a preliminary simplification. Using (2.9) and Proposition 2.24(ii), we can assume that g is of the form

$$g = \omega_m = \sum_{i=1}^{m} e^{2i-1} \wedge e^{2i}.$$

Note that

$$g^m = m!\, e^1 \wedge \cdots \wedge e^{2m}.$$

Writing $f = \sum_{i=1}^n f_i e^i$, we immediately deduce that

$$f \wedge g^m = 0 \quad \Leftrightarrow \quad [f_i = 0 \quad \text{for every } 2m+1 \le i \le n]. \tag{2.18}$$

Step 2. We now prove the sufficient part and assume that $f \wedge g^m = 0$. Therefore, using (2.18), we have that

$$u = \sum_{i=1}^{2m} u_i e^i \in \Lambda^1(\mathbb{R}^n),$$

where

$$u_i = (-1)^{i+1} f_{i+(-1)^{i+1}}$$

satisfies

$$f = u \lrcorner g.$$

Step 3. Let us finally show the necessary part. Assume that there exists u such that $f = u \lrcorner g$. The special structure of g immediately implies that $f_i = 0$ for every $2m+1 \le i \le n$ and thus the result according to (2.18). □

We conclude this subsection with the case $k = l+1$ and a special g.

Proposition 2.52. *Let $1 \le p \le n-1$ and $2 \le l \le n-1$ be integers verifying*

$$p+l+1, pl \le n.$$

Let $f \in \Lambda^{l+1}(\mathbb{R}^n)$ and $g_1, \ldots, g_{pl} \in \Lambda^1(\mathbb{R}^n)$ with

$$g_1 \wedge \cdots \wedge g_{pl} \neq 0,$$

$$g = [g_1 \wedge \cdots \wedge g_l] + [g_{l+1} \wedge \cdots \wedge g_{2l}] + \cdots + \left[g_{(p-1)l+1} \wedge \cdots \wedge g_{pl}\right].$$

The following two statements are then equivalent:

(i) There exists $u \in \Lambda^1(\mathbb{R}^n)$ verifying

$$f = g \wedge u.$$

(ii) For every $i_j \in \{(j-1)l+1, \ldots, jl\}$ and $j, s, t \in \{1, \ldots, p\}$ with $s < t$, the following two sets of identities hold

$$f \wedge g_{i_1} \wedge \cdots \wedge g_{i_p} = 0,$$

$$f \wedge g_{i_1} \wedge \cdots \wedge \widehat{g_{i_s}} \wedge \cdots \wedge \widehat{g_{i_t}} \wedge \cdots \wedge g_{i_p} \wedge G_{st} = 0,$$

where

$$G_{st} = \left[g_{(s-1)l+1} \wedge \cdots \wedge g_{sl}\right] + (-1)^{l+1} \left[g_{(t-1)l+1} \wedge \cdots \wedge g_{tl}\right].$$

Remark 2.53. The case $p = 1$ in the above proposition is the Cartan lemma (cf. Theorem 2.42) when $k = l+1$, since the last set of identities is then empty. If $p = 2$, the last set of identities reads as

$$f \wedge g_i \wedge g_j = 0, \quad 1 \le i \le l < j \le 2l,$$

$$f \wedge \Big[[g_1 \wedge \cdots \wedge g_l] + (-1)^{l+1} [g_{l+1} \wedge \cdots \wedge g_{2l}]\Big] = 0.$$

Proof. Step 1. We show that (i) $\Rightarrow$ (ii). For every $1 \le j \le p$, we set

$$I_j = \{(j-1)l+1, \ldots, jl\}.$$

Since we trivially have, for every $i_j \in I_j$ and $j, s, t \in \{1, \ldots, p\}$ with $s < t$,

$$g \wedge g_{i_1} \wedge \cdots \wedge g_{i_p} = 0,$$

$$g \wedge g_{i_1} \wedge \cdots \wedge \widehat{g_{i_s}} \wedge \cdots \wedge \widehat{g_{i_t}} \wedge \cdots \wedge g_{i_p} \wedge G_{st} = 0,$$

and since

$$f = g \wedge u,$$

we immediately get the result.

Step 2. We now prove that (ii) $\Rightarrow$ (i).

Step 2.1. Since

$$g_1 \wedge \cdots \wedge g_{pl} \neq 0,$$

we can assume (using Proposition 2.24(i)), without loss of generality, that $g_i = e^i$, $1 \leq i \leq pl$. Under this hypothesis, the existence of a u satisfying the equation

$$f = g \wedge u$$

will be implied by the following two sets of identities. The first one is

$$f_{j_1 \cdots j_{l+1}} = 0 \tag{2.19}$$

for every $1 \leq j_1 < \cdots < j_{l+1} \leq n$ such that $I_m \not\subset \{j_1, \ldots, j_{l+1}\}$ for every $m \in \{1, \ldots, p\}$. Since $l \geq 2$, the second one is

$$f_{I_s \nu} = f_{(s-1)l+1 \cdots (sl)\nu} = f_{(t-1)l+1 \cdots (tl)\nu} = f_{I_t \nu} \tag{2.20}$$

for every $1 \leq s < t \leq p$ and $\nu \in \{1, \ldots, n\} \backslash (I_s \cup I_t)$. The result will follow if we can show that (2.19) is implied by the first set of identities in statement (ii) of the present proposition (cf. Step 2.2) and (2.20) is implied by the second set of identities in statement (ii) of the proposition (cf. Step 2.3). The result then follows by setting

$$u = \sum_{\nu=1}^{n} u_\nu e^\nu,$$

where

$$u_\nu = \begin{cases} f_{1 \cdots l \nu} & \text{if } \nu \notin \{1, \ldots, l\} \\ f_{(l+1) \cdots (2l) \nu} & \text{if } \nu \in \{1, \ldots, l\}. \end{cases}$$

Step 2.2. By hypothesis, we have for every $i_s \in I_s$ and $s \in \{1, \ldots, p\}$,

$$f \wedge e^{i_1} \wedge \cdots \wedge e^{i_p} = 0.$$

We therefore deduce, for every $1 \leq j_1 < \cdots < j_{l+1} \leq n$, that

$$(f \wedge e^{i_1} \wedge \cdots \wedge e^{i_p})_{j_1 \cdots j_{l+1} i_1 \cdots i_p} = 0. \tag{2.21}$$

Let $1 \leq j_1 < \cdots < j_{l+1} \leq n$ with $I_m \not\subset \{j_1, \ldots, j_{l+1}\}$ for every $m \in \{1, \ldots, p\}$. We then choose, for $m \in \{1, \ldots, p\}$,

$$i_m \in I_m \backslash \{j_1, \ldots, j_{l+1}\}.$$

Applying (2.21) with these coefficients, we immediately have (2.19).

Step 2.3. We know that for every $(i_1, \ldots, i_p) \in I_1 \times \cdots \times I_p$ and $s, t \in \{1, \ldots, p\}$ with $s < t$, the following set of identities hold:

$$f \wedge e^{i_1} \wedge \cdots \wedge \widehat{e^{i_s}} \wedge \cdots \wedge \widehat{e^{i_t}} \wedge \cdots \wedge e^{i_p} \wedge E^{st} = 0, \tag{2.22}$$

where

$$E^{st} = e^{I_s} + (-1)^{l+1} e^{I_t}.$$

We have, due to Step 2.2,

$$f = \sum_{j=1}^{p} \sum_{\nu=1}^{n} f_{I_j \nu}\, e^{I_j} \wedge e^{\nu}. \tag{2.23}$$

Let $1 \leq s < t \leq p$ and $\nu \in \{1, \ldots, n\} \backslash (I_s \cup I_t)$. Note that if $i_r \in I_r$, $1 \leq r \leq p$, then we have

$$e^{I_j} \wedge e^{\nu} \wedge e^{i_1} \wedge \cdots \wedge \widehat{e^{i_s}} \wedge \cdots \wedge \widehat{e^{i_t}} \wedge \cdots \wedge e^{i_p} \wedge e^{I_s} = 0 \quad \text{if } j \neq t,$$
$$e^{I_j} \wedge e^{\nu} \wedge e^{i_1} \wedge \cdots \wedge \widehat{e^{i_s}} \wedge \cdots \wedge \widehat{e^{i_t}} \wedge \cdots \wedge e^{i_p} \wedge e^{I_t} = 0 \quad \text{if } j \neq s.$$

Since $l \geq 2$, we can chose i_r for $r \neq s, t$ such that $i_r \in I_r \backslash \{\nu\}$ to obtain

$$\begin{aligned} &e^{I_t} \wedge e^{\nu} \wedge e^{i_1} \wedge \cdots \wedge \widehat{e^{i_s}} \wedge \cdots \wedge \widehat{e^{i_t}} \wedge \cdots \wedge e^{i_p} \wedge e^{I_s} \\ &\quad = (-1)^{l} e^{I_s} \wedge e^{\nu} \wedge e^{i_1} \wedge \cdots \wedge \widehat{e^{i_s}} \wedge \cdots \wedge \widehat{e^{i_t}} \wedge \cdots \wedge e^{i_p} \wedge e^{I_t} \neq 0. \end{aligned}$$

Setting (2.23) into (2.22) and using the previous three equations, we get

$$f_{I_s \nu} = f_{I_t \nu}.$$

This concludes the proof of the theorem. □

2.3.4 Proof of the Main Theorem

In the proof of Corollary 2.48 we have used the following lemma. It will also be used in the proof of Theorem 2.45.

Lemma 2.54. *Let $m, n, r \geq 1$ be integers, $A \in \mathbb{R}^{n \times m}$ a matrix of rank r and $y \in \mathbb{R}^n$. Then there exists $x \in \mathbb{R}^m$ verifying*

$$Ax = y$$

if and only if

$$A_{i_1} \wedge \cdots \wedge A_{i_r} \wedge y = 0 \quad \textit{for every} \quad 1 \leq i_1 < \cdots < i_r \leq m,$$

where y is identified with a 1-form in $\mathbb{R}^n$ and A_k denotes the kth column of A and is identified to a 1-form in $\mathbb{R}^n$.

Proof. Step 1. We first prove the necessary part. Assume that there exists $x \in \mathbb{R}^m$ verifying $Ax = y$. Then, writing, $x = (x_1, \ldots, x_m)$, we have

$$y = \sum_{l=1}^{m} A_l x_l .$$

Since the rank of A is r, we get

$$A_{i_1} \wedge \cdots \wedge A_{i_r} \wedge y = \sum_{l=1}^{m} x_l \left(A_{i_1} \wedge \cdots \wedge A_{i_r} \wedge A_l \right) = 0,$$

which is our claim.

Step 2. We then turn to the proof of the sufficient part. Since the rank of A is r, we can find $1 \leq i_1 < \cdots < i_r \leq m$ such that

$$A_{i_1} \wedge \cdots \wedge A_{i_r} \neq 0.$$

Since we also have

$$A_{i_1} \wedge \cdots \wedge A_{i_r} \wedge y = 0,$$

it follows that y is a linear combination of the A_{i_l}. This means that there exist $w_l \in \mathbb{R}$, $1 \leq l \leq r$, so that

$$y = \sum_{l=1}^{r} w_l A_{i_l} .$$

Setting $x = (x_1, \ldots, x_m)$ where

$$x_s = \begin{cases} w_l & \text{if } s = i_l \\ 0 & \text{otherwise,} \end{cases}$$

it follows that

$$Ax = y.$$

This concludes the proof of the lemma. □

We now turn to the proof of Theorem 2.45.

Proof. Step 1. The implications (i) $\Rightarrow$ (iii) $\Rightarrow$ (ii) are obvious. The equivalence (i) $\Leftrightarrow$ (iv) is just a rewriting of Lemma 2.54, since

$$g \wedge u = f \quad \Leftrightarrow \quad \overline{g}_{\wedge, k-l} u = f.$$

Step 2. The only nontrivial implication is (ii) $\Rightarrow$ (iv). Let

$$1 \leq i_1^{(1)} < \cdots < i_{k-l}^{(1)} \leq n, \ldots, 1 \leq i_1^{(r)} < \cdots < i_{k-l}^{(r)} \leq n,$$

recalling that $r = \operatorname{rank}(\overline{g}_{\wedge,k-l})$ and let us prove that

$$(\overline{g}_{\wedge,k-l})_{i_1^{(1)}\cdots i_{k-l}^{(1)}} \wedge \cdots \wedge (\overline{g}_{\wedge,k-l})_{i_1^{(r)}\cdots i_{k-l}^{(r)}} \wedge f = 0, \tag{2.24}$$

where f and $(\overline{g}_{\wedge,k-l})_{i_1^{(m)}\cdots i_{k-l}^{(m)}}$, $1 \le m \le r$, are seen as 1-forms in $\mathbb{R}^{\binom{n}{k}}$. Equation (2.24) is equivalent to

$$\det \begin{vmatrix} (\overline{g}_{\wedge,k-l})_{i_1^{(1)}\cdots i_{k-l}^{(1)}}^{j_1^{(1)}\cdots j_k^{(1)}} & \cdots & (\overline{g}_{\wedge,k-l})_{i_1^{(r)}\cdots i_{k-l}^{(r)}}^{j_1^{(1)}\cdots j_k^{(1)}} & (f)^{j_1^{(1)}\cdots j_k^{(1)}} \\ \vdots & & \vdots & \vdots \\ (\overline{g}_{\wedge,k-l})_{i_1^{(1)}\cdots i_{k-l}^{(1)}}^{j_1^{(r+1)}\cdots j_k^{(r+1)}} & \cdots & (\overline{g}_{\wedge,k-l})_{i_1^{(r)}\cdots i_{k-l}^{(r)}}^{j_1^{(r+1)}\cdots j_k^{(r+1)}} & (f)^{j_1^{(r+1)}\cdots j_k^{(r+1)}} \end{vmatrix} = 0$$

for every $1 \le j_1^{(1)} < \cdots < j_k^{(1)} \le n, \ldots, 1 \le j_1^{(r+1)} < \cdots < j_k^{(r+1)} \le n$. Expanding the determinant with respect to the last column and writing, for every $1 \le m \le r+1$,

$$c_m = \det \begin{vmatrix} (\overline{g}_{\wedge,k-l})_{i_1^{(1)}\cdots i_{k-l}^{(1)}}^{j_1^{(1)}\cdots j_k^{(1)}} & \cdots & (\overline{g}_{\wedge,k-l})_{i_1^{(r)}\cdots i_{k-l}^{(r)}}^{j_1^{(1)}\cdots j_k^{(1)}} \\ \vdots & & \vdots \\ (\overline{g}_{\wedge,k-l})_{i_1^{(1)}\cdots i_{k-l}^{(1)}}^{j_1^{(m-1)}\cdots j_k^{(m-1)}} & \cdots & (\overline{g}_{\wedge,k-l})_{i_1^{(r)}\cdots i_{k-l}^{(r)}}^{j_1^{(m-1)}\cdots j_k^{(m-1)}} \\ (\overline{g}_{\wedge,k-l})_{i_1^{(1)}\cdots i_{k-l}^{(1)}}^{j_1^{(m+1)}\cdots j_k^{(m+1)}} & \cdots & (\overline{g}_{\wedge,k-l})_{i_1^{(r)}\cdots i_{k-l}^{(r)}}^{j_1^{(m+1)}\cdots j_k^{(m+1)}} \\ \vdots & & \vdots \\ (\overline{g}_{\wedge,k-l})_{i_1^{(1)}\cdots i_{k-l}^{(1)}}^{j_1^{(r+1)}\cdots j_k^{(r+1)}} & \cdots & (\overline{g}_{\wedge,k-l})_{i_1^{(r)}\cdots i_{k-l}^{(r)}}^{j_1^{(r+1)}\cdots j_k^{(r+1)}} \end{vmatrix},$$

we find that (2.24) is equivalent to

$$\sum_{m=1}^{r+1} (-1)^{m+1} c_m (f)^{j_1^{(m)}\cdots j_k^{(m)}} = 0.$$

The above equation is equivalent (seeing f as a k-form and appealing to (2.4)) to

$$f \wedge \left(\sum_{m=1}^{r+1} (-1)^{m+1} c_m \left[*(e^{j_1^{(m)}} \wedge \cdots \wedge e^{j_k^{(m)}}) \right] \right) = 0.$$

To prove our claim, it is sufficient to prove that

$$h = \sum_{m=1}^{r+1} (-1)^{m+1} c_m \left[*(e^{j_1^{(m)}} \wedge \cdots \wedge e^{j_k^{(m)}}) \right] \in \Lambda^{n-k}(\mathbb{R}^n)$$

satisfies

$$g \wedge h = 0,$$

which turns out to be equivalent to

$$g \wedge e^{t_1} \wedge \cdots \wedge e^{t_{k-l}} \wedge h = 0 \tag{2.25}$$

for every $1 \le t_1 < \cdots < t_{k-l} \le n$. Since the matrix $\overline{g}_{\wedge,k-l}$ has rank r, we get

$$(\overline{g}_{\wedge,k-l})_{i_1^{(1)} \cdots i_{k-l}^{(1)}} \wedge \cdots \wedge (\overline{g}_{\wedge,k-l})_{i_1^{(r)} \cdots i_{k-l}^{(r)}} \wedge (\overline{g}_{\wedge,k-l})_{t_1 \cdots t_{k-l}} = 0,$$

which implies that

$$\det \begin{vmatrix} (\overline{g}_{\wedge,k-l})^{j_1^{(1)} \cdots j_k^{(1)}}_{i_1^{(1)} \cdots i_{k-l}^{(1)}} & \cdots & (\overline{g}_{\wedge,k-l})^{j_1^{(1)} \cdots j_k^{(1)}}_{i_1^{(r)} \cdots i_{k-l}^{(r)}} & (\overline{g}_{\wedge,k-l})^{j_1^{(1)} \cdots j_k^{(1)}}_{t_1 \cdots t_{k-l}} \\ \vdots & & \vdots & \vdots \\ (\overline{g}_{\wedge,k-l})^{j_1^{(r+1)} \cdots j_k^{(r+1)}}_{i_1^{(1)} \cdots i_{k-l}^{(1)}} & \cdots & (\overline{g}_{\wedge,k-l})^{j_1^{(r+1)} \cdots j_k^{(r+1)}}_{i_1^{(r)} \cdots i_{k-l}^{(r)}} & (\overline{g}_{\wedge,k-l})^{j_1^{(r+1)} \cdots j_k^{(r+1)}}_{t_1 \cdots t_{k-l}} \end{vmatrix} = 0.$$

Expanding the above determinant with respect to the last column, we obtain

$$\sum_{m=1}^{r+1} (-1)^{m+1} c_m (\overline{g}_{\wedge,k-l})^{j_1^{(m)} \cdots j_k^{(m)}}_{t_1 \cdots t_{k-l}} = 0.$$

Let us show that this last equation is equivalent to (2.25), namely

$$g \wedge e^{t_1} \wedge \cdots \wedge e^{t_{k-l}} \wedge h = 0.$$

Noting that from Notation 2.30 and (2.4), we have, for every $1 \le m \le r+1$,

$$\begin{aligned} (\overline{g}_{\wedge,k-l})^{j_1^{(m)} \cdots j_k^{(m)}}_{t_1 \cdots t_{k-l}} &= \left(g \wedge e^{t_1} \wedge \cdots \wedge e^{t_{k-l}}\right)_{j_1^{(m)} \cdots j_k^{(m)}} \\ &= *(g \wedge e^{t_1} \wedge \cdots \wedge e^{t_{k-l}} \wedge (*(e^{j_1^{(m)}} \wedge \cdots \wedge e^{j_k^{(m)}})). \end{aligned}$$

We therefore obtain that

$$*(g \wedge e^{t_1} \wedge \cdots \wedge e^{t_{k-l}} \wedge h) = \sum_{m=1}^{r+1} (-1)^{m+1} c_m (\overline{g}_{\wedge,k-l})^{j_1^{(m)} \cdots j_k^{(m)}}_{t_1 \cdots t_{k-l}} = 0.$$

This is exactly our claim (2.25). □

Chapter 3
Differential Forms

3.1 Notations

In this section we recall the definitions and basic properties of differential forms on $\mathbb{R}^n$. Our presentation is very brief and for a detailed introduction on differential forms, we refer, for instance, to Abraham, Marsden and Ratiu [1], do Carmo [37], Lee [68], or Spivak [91]. From now on, we will denote the dual vectors e^i, $1 \leq i \leq n$, in $\Lambda^1(\mathbb{R}^n)$ by dx^i and, hence, a basis of $\Lambda^k(\mathbb{R}^n)$ is given by the $dx^{i_1} \wedge \cdots \wedge dx^{i_k}$. Throughout this section Ω will stand for an open subset of $\mathbb{R}^n$.

Notation 3.1. *Let $0 \leq k \leq n$. A differential k-form $f : \Omega \to \Lambda^k$ will be written as*

$$f = \sum_{1 \leq i_1 < \cdots < i_k \leq n} f_{i_1 \cdots i_k} dx^{i_1} \wedge \cdots \wedge dx^{i_k},$$

where $f_{i_1 \cdots i_k} : \Omega \to \mathbb{R}$, for every $1 \leq i_1 < \cdots < i_k \leq n$. When $f_{i_1 \cdots i_k} \in L^p(\Omega)$, for every $1 \leq i_1 < \cdots < i_k \leq n$, we will write $f \in L^p(\Omega;\Lambda^k)$ and similarly for $W^{1,p}(\Omega;\Lambda^k)$, $C^{r,\alpha}(\Omega;\Lambda^k)$, or $C^{r,\alpha}(\overline{\Omega};\Lambda^k)$. The norm is defined componentwise; for instance,

$$\|f\|_{L^2}^2 = \sum_{1 \leq i_1 < \cdots < i_k \leq n} \|f_{i_1 \cdots i_k}\|_{L^2}^2.$$

The differential forms obey pointwise the laws of the exterior algebra. For instance, the exterior product is defined pointwise as

$$(f \wedge g)(x) = f(x) \wedge g(x).$$

The scalar product, the Hodge duality, the interior product, rank and corank of differential forms are also defined pointwise in an analogous way.

We can now introduce the two important differential operators on differential forms.

Definition 3.2 (Exterior and interior derivative). Let $\Omega \subset \mathbb{R}^n$ be open and $f \in C^1(\Omega;\Lambda^k)$,

$$f = \sum_{1 \le i_1 < \cdots < i_k \le n} f_{i_1 \cdots i_k} dx^{i_1} \wedge \cdots \wedge dx^{i_k}.$$

(i) The *exterior derivative* of f, denoted df, belongs to $C^0\left(\Omega;\Lambda^{k+1}\right)$ and is defined by

$$df = \sum_{1 \le i_1 < \cdots < i_k \le n} \sum_{m=1}^{n} \frac{\partial f_{i_1 \cdots i_k}}{\partial x_m} dx^m \wedge dx^{i_1} \wedge \cdots \wedge dx^{i_k}.$$

If $k = n$, then $df = 0$.

(ii) The *interior derivative* or *codifferential* of f denoted δf belongs to $C^0\left(\Omega;\Lambda^{k-1}\right)$ and is defined by

$$\delta f = (-1)^{n(k-1)} * (d(*f)).$$

We will use the next formulas on several occasions.

Proposition 3.3 (Formulas for d and δ). *Let* $f \in C^1\left(\Omega;\Lambda^k\right)$,

$$f = \sum_{1 \le i_1 < \cdots < i_k \le n} f_{i_1 \cdots i_k} dx^{i_1} \wedge \cdots \wedge dx^{i_k}.$$

The following formulas hold true:

(i) If $k < n$,

$$df = \sum_{1 \le i_1 < \cdots < i_{k+1} \le n} \left(\sum_{\gamma=1}^{k+1} (-1)^{\gamma-1} \frac{\partial f_{i_1 \cdots i_{\gamma-1} i_{\gamma+1} \cdots i_{k+1}}}{\partial x_{i_\gamma}} \right) dx^{i_1} \wedge \cdots \wedge dx^{i_{k+1}}.$$

(ii) If $k > 0$,

$$\delta f = \sum_{1 \le i_1 < \cdots < i_{k-1} \le n} \left(\sum_{\gamma=1}^{k} (-1)^{\gamma-1} \sum_{i_{\gamma-1} < j < i_\gamma} \frac{\partial f_{i_1 \cdots i_{\gamma-1} j i_\gamma \cdots i_{k-1}}}{\partial x_j} \right) dx^{i_1} \wedge \cdots \wedge dx^{i_{k-1}},$$

where if $\gamma = 1$, *it is understood that* $f_{i_1 \cdots i_{\gamma-1} j i_\gamma \cdots i_{k-1}} = f_{j i_1 \cdots i_{k-1}}$, *and, similarly, when* $\gamma = k$, *it is understood that* $f_{i_1 \cdots i_{\gamma-1} j i_\gamma \cdots i_{k-1}} = f_{i_1 \cdots i_{k-1} j}$. *If* $k = 1$, *by abuse of notations, the formula can be written as*

$$\delta f = \operatorname{div} f = \sum_{j=1}^{n} \frac{\partial f_j}{\partial x_j}.$$

We now define the notions of closed, coclosed, exact and coexact forms as well as of harmonic fields.

Definition 3.4. Let $\Omega \subset \mathbb{R}^n$ be open and f be a k-form.

(i) If $f \in C^1\left(\Omega;\Lambda^k\right)$ satisfies $df = 0$ in Ω (respectively $\delta f = 0$ in Ω), then f is said to be *closed* (respectively *coclosed*) in Ω.

(ii) If there exists $g \in C^1\left(\Omega;\Lambda^{k-1}\right)$ such that $dg = f$ in Ω, then f is said to be *exact* in Ω. Similarly, if there exists $g \in C^1\left(\Omega;\Lambda^{k+1}\right)$ such that $\delta g = f$ in Ω, then f is said to be *coexact* in Ω.

(iii) A differential form $f \in C^1(\Omega;\Lambda^k)$ that satisfies both $df = 0$ and $\delta f = 0$ in Ω is called a *harmonic field.*

We next gather some well-known properties of the operators d and δ.

Theorem 3.5. *Let f be a k-form and g be a l-form, then*

$$d(f \wedge g) = df \wedge g + (-1)^k f \wedge dg,$$

$$\delta(f \lrcorner g) = (-1)^{k+l} df \lrcorner g - f \lrcorner \delta g.$$

Moreover, every exact form is closed and every coexact form is coclosed; that is,

$$ddf = 0 \quad \text{and} \quad \delta\delta f = 0.$$

Definition 3.6 (Laplacian). Let $f \in C^2(\Omega;\Lambda^k)$. The Laplacian $\Delta f \in C^0(\Omega;\Lambda^k)$ is defined by the Laplacian acting componentwise; that is,

$$\Delta\Big(\sum_{1 \le i_1 < \cdots < i_k \le n} f_{i_1 \cdots i_k} dx^{i_1} \wedge \cdots \wedge dx^{i_k}\Big) = \sum_{1 \le i_1 < \cdots < i_k \le n} \Delta f_{i_1 \cdots i_k} dx^{i_1} \wedge \cdots \wedge dx^{i_k},$$

where

$$\Delta f_{i_1 \cdots i_k} = \sum_{l=1}^{n} \frac{\partial^2 f_{i_1 \cdots i_k}}{\partial x_l^2}.$$

Theorem 3.7. *Let $f \in C^2(\Omega;\Lambda^k)$. Then*

$$d\delta f + \delta d f = \Delta f.$$

Definition 3.8 (Pullback of a differential form). Let $U \subset \mathbb{R}^m$ and $V \subset \mathbb{R}^n$ be open and $\varphi = (\varphi^1, \ldots, \varphi^n) \in C^1(U;V)$. Let

$$f = \sum_{1 \le i_1 < \cdots < i_k \le n} f_{i_1 \cdots i_k} dx^{i_1} \wedge \cdots \wedge dx^{i_k} \in C^0(V;\Lambda^k(\mathbb{R}^n)). \tag{3.1}$$

Then the *pullback* of f by φ, denoted $\varphi^*(f)$, belongs to $C^0(U;\Lambda^k(\mathbb{R}^m))$ and is defined by

$$\varphi^*(f) = \sum_{1 \le i_1 < \cdots < i_k \le n} (f_{i_1 \cdots i_k} \circ \varphi) d\varphi^{i_1} \wedge \cdots \wedge d\varphi^{i_k} = \sum_I (f_I \circ \varphi) d\varphi^I, \tag{3.2}$$

where

$$d\varphi^s = \sum_{l=1}^{m} \frac{\partial \varphi^s}{\partial x_l} dx^l.$$

Remark 3.9. (i) We see that this is a generalization of Definition 2.17. Indeed, if $\varphi(x) = Ax$, where $A \in \mathbb{R}^{n \times m}$ is a matrix, and f is constant, then

$$\varphi^*(f) = A^*(f),$$

where the right-hand side has to be understood in the sense of Definition 2.17.

(ii) We can also define, in an equivalent way,

$$\varphi^*(f)(X_1,\ldots,X_k) = f\circ\varphi(\nabla\varphi\cdot X_1,\ldots,\nabla\varphi\cdot X_k)$$

for every $X_j \in \mathbb{R}^m$ for $1 \leq j \leq k$. In the above notation, $\nabla\varphi \cdot X_j \in \mathbb{R}^n$ should be understood as the matrix $\nabla\varphi$ multiplied by the vector X_j.

The following theorem is easily proved (cf. Definition 2.17 and Proposition 2.19).

Theorem 3.10 (Properties of pullback). *Let $U \subset \mathbb{R}^m$, $V \subset \mathbb{R}^n$ and $W \subset \mathbb{R}^s$ be open, $\varphi \in C^1(U;V)$ and $\psi \in C^1(W;U)$. Let $f,g \in C^0(V;\Lambda^k)$ and $h \in C^0(V;\Lambda^l)$; then*

$$\varphi^*(f+g) = \varphi^*(f) + \varphi^*(g),$$

$$\varphi^*(f\wedge h) = \varphi^*(f)\wedge\varphi^*(h),$$

$$(\varphi\circ\psi)^*(f) = \psi^*(\varphi^*(f)).$$

Moreover, if $f \in C^1(V;\Lambda^k)$ and $\varphi \in C^2(U;V)$, then

$$\varphi^*(df) = d(\varphi^*(f)).$$

If $m = n$ and $\varphi \in \operatorname{Diff}^1(U;V)$, then

$$\varphi^*(f \lrcorner h) = \varphi^\sharp(f) \lrcorner \varphi^*(h),$$

where

$$\varphi^\sharp(f)(x) = \left(\left((\nabla\varphi(x))^{-1}\right)^t\right)^*[f(\varphi(x))]$$

for every fixed $x \in U$.

The next proposition is straightforward.

Proposition 3.11. *Let $1 \leq i \leq n$, $\varphi \in \operatorname{Diff}^1(U;V)$ and $a \in C^0(V;\mathbb{R}^n)$ be such that*

$$\frac{\partial\varphi}{\partial x_i} = a\circ\varphi \quad \text{in } U.$$

Then

$$\varphi^\sharp(a) = dx^i \quad \text{in } U,$$

where a has been identified with a 1-form.

Proof. We have to show that for every fixed $x \in U$,

$$\left(\left((\nabla\varphi(x))^{-1}\right)^t\right)^*[a(\varphi(x))] = dx^i$$

or, equivalently,

$$a(\varphi(x)) = \left((\nabla\varphi(x))^t\right)^*(dx^i).$$

Since the last equation is equivalent to

$$a(\varphi(x)) = \frac{\partial\varphi}{\partial x_i}(x),$$

we have the result. □

3.2 Tangential and Normal Components

In this section $\Omega \subset \mathbb{R}^n$ will be a bounded open C^1 set (see Definition 16.4) and $\nu = \nu(x)$ will be the exterior unit normal on $\partial\Omega$ at x. Let f be an element of $C^0\left(\overline{\Omega};\Lambda^k\right)$. We will introduce the notion of tangential and normal components of f on $\partial\Omega$ and establish the relationship between the other definitions occurring in the literature. The definition used, for instance, by Dacorogna [27], Kress [63], or Taylor [96] is the one we will adopt in this book. For this definition, we consider ν as a 1-form

$$\nu = \nu_1 dx^1 + \cdots + \nu_n dx^n \in C^0\left(\partial\Omega;\Lambda^1\right).$$

In this section we will skip back and forth between identifying ν as a 1-form and $\nu = (\nu_1,\ldots,\nu_n) \in \mathbb{R}^n$ as a vector. In that sense, we will frequently use Remark 2.15, which identifies $a \lrcorner f$ by $i_a(f)$.

Definition 3.12 (Tangential and normal component). Let f be a k-form. The *tangential component* of f on $\partial\Omega$ is the $(k+1)$-form

$$\nu \wedge f.$$

The *normal component* of f on $\partial\Omega$ is the $(k-1)$-form

$$\nu \lrcorner f.$$

Another definition (see, for instance, Schwarz [89]) for the tangential and normal components is the following.

Definition 3.13. Let f be a k-form.

(i) Let $X \in \mathbb{R}^n$. Then, for every $x \in \partial\Omega$, the vector X can be decomposed as

$$X = X^\perp + X^\parallel,$$

where $X^\parallel$ is the component of X tangential to $\partial\Omega$ at x; that is,

$$X^\parallel = X - \nu\langle\nu;X\rangle. \tag{3.3}$$

(ii) Let $X_1,\dots,X_k \in \mathbb{R}^n$. We denote by tf the k-form on $\partial\Omega$ defined by

$$tf(X_1,\dots,X_k) = f(X_1^{\|},\dots,X_k^{\|}).$$

(iii) We let nf be the k-form defined by

$$nf = f - tf.$$

It follows from (3.3) that tf and nf are in fact differential forms, as claimed in this definition.

For the third definition, which is the most classical and is used for instance by Duff and Spencer [38], Iwaniec and Martin [57], or Morrey [77], we first need to introduce the notion of an admissible boundary coordinate system for an open set Ω.

Notation 3.14. *Throughout the present section, for $y \in \mathbb{R}^n$ we write*

$$y = (y', y_n) \in \mathbb{R}^{n-1} \times \mathbb{R}, \quad \textit{where} \quad y' = (y_1,\dots,y_{n-1}).$$

Definition 3.15 (Admissible boundary coordinate system). Let U, V, $\Omega \subset \mathbb{R}^n$ be open sets. We say that

$$\varphi : U \to V$$

is an *admissible boundary coordinate system for* Ω if $\varphi \in \operatorname{Diff}^1(U;V)$,

$$\partial\Omega \cap V = \{\varphi(y',0) : (y',0) \in U\},$$

and for every $1 \le i \le n$ and every $(y',0) \in U$,

$$\left\langle \frac{\partial \varphi}{\partial y_i}(y',0); \frac{\partial \varphi}{\partial y_n}(y',0) \right\rangle = \delta_{in}\,.$$

Remark 3.16. It follows from the definition that

$$\frac{\partial \varphi}{\partial y_n}(y',0) = \pm \nu(\varphi(y',0)) \quad \text{for every } (y',0) \in U,$$

since the $\frac{\partial \varphi}{\partial y_i}(y',0)$, $1 \le i \le n-1$, form a basis of the tangent space of $\partial\Omega$ and $\frac{\partial \varphi}{\partial y_n}(y',0)$ is a unit vector.

We now prove, following Morrey [77], that every $a \in \partial\Omega$ is in the range of an admissible boundary coordinate system.

Proposition 3.17. *Let $r \ge 1$ be an integer, $0 \le \alpha \le 1$, $\Omega \subset \mathbb{R}^n$ be an open $C^{r,\alpha}$ set and let $a \in \partial\Omega$. Then there exist an open set $U \subset \mathbb{R}^n$, a neighborhood $V \subset \mathbb{R}^n$ of a and an admissible boundary coordinate system $\varphi \in \operatorname{Diff}^{r,\alpha}(U;V)$.*

Proof. We will denote

$$H = \{x \in \mathbb{R}^n : x_n = 0\} \subset \mathbb{R}^n$$

and $B \subset \mathbb{R}^n$ is the open unit ball centered at 0. By definition of a $C^{r,\alpha}$ set (see Definition 16.5) there exists a neighborhood V of a and $\psi \in \mathrm{Diff}^{r,\alpha}(B;V)$ such that $\psi(0) = a$ and

$$\psi(B \cap H) = V \cap \partial\Omega.$$

For $1 \leq i, j \leq n$, define $g_{ij} \in C^{r-1,\alpha}(B)$ by

$$g_{ij}(x) = \left\langle \frac{\partial \psi}{\partial x_i}(x); \frac{\partial \psi}{\partial x_j}(x) \right\rangle.$$

For $x \in B$, let $g(x)$ be the associated $n \times n$ matrix. Since $\det \nabla \psi(x) \neq 0$, the matrix $g(x)$ is symmetric and positive definite for every $x \in B$, thus so is $g^{-1}(x)$ and in particular $(g^{-1})_{nn}(x) > 0$ for all $x \in B$. Define $d \in C^{r-1,\alpha}(B \cap H; \mathbb{R}^n)$ by

$$d(x') = \left(0, \ldots, 0, 1/\sqrt{(g^{-1})_{nn}(x', 0)}\right)$$

and let $f \in C^{r-1,\alpha}(B \cap H; \mathbb{R}^n)$ be given by

$$f(x') = g^{-1}(x', 0)\, d(x').$$

Note that, by construction,

$$f_n(x') = \sqrt{(g^{-1})_{nn}(x', 0)}.$$

We next extend f to all of $\mathbb{R}^{n-1}$ (cf. Theorem 16.11) and we define $\phi = (\phi^1, \ldots, \phi^n)$ through

$$\phi^i(x) = x_i + x_n \int_{\mathbb{R}^{n-1}} \eta(y') f_i(x' - x_n y')\, dy', \quad 1 \leq i \leq n-1,$$

$$\phi^n(x) = x_n \int_{\mathbb{R}^{n-1}} \eta(y') f_n(x' - x_n y')\, dy',$$

where $\eta \in C_0^\infty(\mathbb{R}^{n-1})$ verifies $\int_{\mathbb{R}^{n-1}} \eta = 1$. As in Lemma 8.10, we have that $\phi \in C^{r,\alpha}(\mathbb{R}^n)$ and on H the following equations hold:

$$\frac{\partial \phi^i}{\partial x_l} = \begin{cases} \delta_{il} & \text{if } 1 \leq l \leq n-1 \\ f_i & \text{if } l = n, \end{cases} \tag{3.4}$$

whenever $1 \leq i \leq n-1$,

$$\frac{\partial \phi^n}{\partial x_l} = \delta_{nl} f_n \quad \text{and} \quad \phi = \mathrm{id}. \tag{3.5}$$

In particular, $\det \nabla \phi(0) \neq 0$. Therefore, there exists a neighborhood U of 0 small enough so that $\phi(U) \subset B$ and $\phi \in \mathrm{Diff}^{r,\alpha}(U; \phi(U))$. We now claim that $\varphi : U \to \varphi(U)$, given by

$$\varphi(x) = \psi(\phi(x)),$$

has all the desired properties of an admissible boundary coordinate system. It only remains to show the last property, namely that

$$\left\langle \frac{\partial \varphi}{\partial x_l}; \frac{\partial \varphi}{\partial x_n} \right\rangle = \delta_{ln} \quad \text{on } U \cap H.$$

Suppose first that $1 \leq l \leq n-1$. Using (3.4), (3.5) and the definition of f, we get, on $U \cap H$,

$$\begin{aligned} \left\langle \frac{\partial \varphi}{\partial x_l}; \frac{\partial \varphi}{\partial x_n} \right\rangle &= \sum_{i,j=1}^{n} g_{ij} \frac{\partial \phi^i}{\partial x_l} \frac{\partial \phi^j}{\partial x_n} = \sum_{i=1}^{n-1} \sum_{j=1}^{n} g_{ij} \frac{\partial \phi^i}{\partial x_l} \frac{\partial \phi^j}{\partial x_n} \\ &= \sum_{j=1}^{n} g_{lj} f_j = d_l = 0. \end{aligned}$$

Similarly,

$$\left\langle \frac{\partial \varphi}{\partial x_n}; \frac{\partial \varphi}{\partial x_n} \right\rangle = \sum_{i,j=1}^{n} g_{ij} f_i f_j = \langle f; gf \rangle = \langle f; d \rangle = 1,$$

which concludes the proof of the lemma. □

We are now in a position to define the third notion of tangential and normal components.

Definition 3.18. Let $\varphi \in \operatorname{Diff}^1(U;V)$ be an admissible boundary coordinate system and write, for a k-form f,

$$\varphi^*(f) = \sum_{1 \leq i_1 < \cdots < i_k \leq n} g_{i_1 \cdots i_k} dy^{i_1} \wedge \cdots \wedge dy^{i_k}.$$

We then say that f has *vanishing tangential component* at $x = \varphi(y', 0)$ if

$$g_{i_1 \cdots i_k}(y', 0) = 0 \quad \text{for every } 1 \leq i_1 < \cdots < i_{k-1} < i_k < n$$

and write, in that case, $t'f(x) = 0$. We say that f has *vanishing normal component* at $x = \varphi(y', 0)$ if

$$g_{i_1 \cdots i_k}(y', 0) = 0 \quad \text{for every } 1 \leq i_1 < \cdots < i_{k-1} < i_k = n$$

and write, in that case, $n'f(x) = 0$.

Remark 3.19. (i) The previous definition is independent of the choice of the admissible boundary coordinate system as a direct consequence of Corollary 3.21.

(ii) Decompose $\varphi^*(f)$ as $\varphi^*(f) = G_{<n} + G_n$, where

$$G_{<n} = \sum_{1 \leq i_1 < \cdots < i_k < n} g_{i_1 \cdots i_k} dy^{i_1} \wedge \cdots \wedge dy^{i_k},$$

$$G_n = \sum_{1 \le i_1 < \cdots < i_k = n} g_{i_1 \cdots i_{k-1} n}\, dy^{i_1} \wedge \cdots \wedge dy^{i_{k-1}} \wedge dy^n.$$

Note that

$$G_{<n} = dy^n \lrcorner (dy^n \wedge \varphi^*(f)) \quad \text{and} \quad G_n = dy^n \wedge (dy^n \lrcorner \varphi^*(f)).$$

Moreover, for $x = \varphi(y', 0)$,

$$t' f(x) = 0 \Leftrightarrow G_{<n}(y', 0) = (dy^n \lrcorner (dy^n \wedge \varphi^*(f)))(y', 0) = 0$$

and, similarly,

$$n' f(x) = 0 \Leftrightarrow G_n(y', 0) = (dy^n \wedge (dy^n \lrcorner \varphi^*(f)))(y', 0) = 0.$$

Proposition 3.20. *The following assertions are verified:*

(i) The following formulas for tf and nf hold:

$$tf = \nu \lrcorner (\nu \wedge f) \quad \text{and} \quad nf = \nu \wedge (\nu \lrcorner f).$$

(ii) Let φ be an admissible boundary coordinate system. Then

$$\varphi^*(\nu \lrcorner (\nu \wedge f)) = dy^n \lrcorner (dy^n \wedge \varphi^*(f)),$$

$$\varphi^*(\nu \wedge (\nu \lrcorner f)) = dy^n \wedge (dy^n \lrcorner \varphi^*(f))$$

Proof. *Step 1.* We prove (i). We first claim that

$$\nu \lrcorner tf = 0 \quad \text{and} \quad \nu \wedge nf = 0. \tag{3.6}$$

Step 1.1. We establish the first part of (3.6). Let $X_1, \ldots, X_{k-1} \in \mathbb{R}^n$. Using Remark 2.15 and the fact that $\nu^{\parallel} = 0$, we get

$$\nu \lrcorner tf(X_1, \ldots, X_{k-1}) = tf(\nu, X_1, \ldots, X_{k-1}) = f(0, X_1^{\parallel}, \ldots, X_{k-1}^{\parallel}) = 0,$$

which proves the assertion.

Step 1.2. We prove the second part of (3.6). Recalling that $f = tf + nf$, we therefore have to prove that

$$\nu \wedge f = \nu \wedge tf.$$

Let $\{X_1, \ldots, X_n\}$ be a basis of $\mathbb{R}^n$ such that

$$X_1 = \nu \quad \text{and} \quad \langle \nu; X_j \rangle = 0 \quad \text{for every } 2 \le j \le n. \tag{3.7}$$

In order to have the claim, it is enough to show that for every $1 \le i_1 < \cdots < i_{k+1} \le n$,

$$(\nu \wedge f)(X_{i_1}, \ldots, X_{i_{k+1}}) = (\nu \wedge tf)(X_{i_1}, \ldots, X_{i_{k+1}}).$$

We split the discussion into two cases.

Case 1: $i_1 \geq 2$. Using Definition 2.2 and (3.7), we get

$$(\nu \wedge f)(X_{i_1},\ldots,X_{i_{k+1}}) = \sum_{j=1}^{k+1}(-1)^{j+1}\langle \nu;X_{i_j}\rangle f(X_{i_1},\ldots,\widehat{X_{i_j}},\ldots,X_{i_{k+1}}) = 0$$

and, similarly,

$$(\nu \wedge tf)(X_{i_1},\ldots,X_{i_{k+1}}) = \sum_{j=1}^{k+1}(-1)^{j+1}\langle \nu;X_{i_j}\rangle tf(X_{i_1},\ldots,\widehat{X_{i_j}},\ldots,X_{i_{k+1}}) = 0.$$

Case 2: $i_1 = 1$. Using again Definition 2.2 and (3.7), which, in particular, implies that

$$X_{i_j}^{\|} = X_{i_j} \quad \text{for every } 2 \leq j \leq n,$$

we get

$$\begin{aligned}
(\nu \wedge f)(X_{i_1},\ldots,X_{i_{k+1}}) &= \sum_{j=1}^{k+1}(-1)^{j+1}\langle \nu;X_{i_j}\rangle f(X_{i_1},\ldots,\widehat{X_{i_j}},\ldots,X_{i_{k+1}}) \\
&= f(X_{i_2},\ldots,X_{i_{k+1}}) = f(X_{i_2}^{\|},\ldots,X_{i_{k+1}}^{\|}) = tf(X_{i_2},\ldots,X_{i_{k+1}}) \\
&= \sum_{j=1}^{k+1}(-1)^{j+1}\langle \nu;X_{i_j}\rangle tf(X_{i_1},\ldots,\widehat{X_{i_j}},\ldots,X_{i_{k+1}}) \\
&= (\nu \wedge tf)(X_{i_1},\ldots,X_{i_{k+1}}).
\end{aligned}$$

Step 1.3 (conclusion). Using Proposition 2.16, Steps 1.1 and 1.2, we obtain

$$tf = \nu \lrcorner (\nu \wedge tf) + \nu \wedge (\nu \lrcorner tf) = \nu \lrcorner (\nu \wedge f).$$

Since (by Proposition 2.16 and by the above equation)

$$f = \nu \lrcorner (\nu \wedge f) + \nu \wedge (\nu \lrcorner f) = tf + \nu \wedge (\nu \lrcorner f)$$

and recalling that $f = tf + nf$, we get from the previous equation that

$$nf = \nu \wedge (\nu \lrcorner f),$$

which ends the proof of (i).

Step 2. We now establish (ii). Applying Theorem 3.10, the result will immediately follow once it is shown that

$$\varphi^*(\nu) = \varepsilon\, dy^n \quad \text{and} \quad \varphi^\sharp(\nu) = \frac{1}{\varepsilon}\, dy^n$$

for some nonvanishing ε (in fact, $\varepsilon = \pm 1$). Using Remark 3.16, we know that

$$\frac{\partial \varphi}{\partial y_n} = \varepsilon\, \nu(\varphi). \tag{3.8}$$

Using the orthogonality property of φ in Definition 3.15 and (3.8), we get

$$\begin{aligned}\varphi^*(\nu) &= \sum_{i=1}^n \nu_i(\varphi)\, d\varphi^i = \sum_{i=1}^n \nu_i(\varphi) \sum_{j=1}^n \frac{\partial \varphi^i}{\partial y_j} dy^j \\ &= \sum_{j=1}^n \left(\sum_{i=1}^n \nu_i(\varphi) \frac{\partial \varphi^i}{\partial y_j} \right) dy^j = \sum_{i=1}^n \nu_i(\varphi) \frac{\partial \varphi^i}{\partial y_n} dy^n = \varepsilon\, dy^n.\end{aligned}$$

Combining Proposition 3.11 and (3.8), we get

$$\varphi^\sharp(\varepsilon\,\nu) = dy^n, \quad \text{which is equivalent to} \quad \varphi^\sharp(\nu) = \frac{1}{\varepsilon}\, dy^n.$$

The proof is therefore finished. □

As an immediate consequence of Proposition 3.20 we get the equivalence of the three definitions. As already said, this will prove that Definition 3.18 is independent of the choice of the admissible boundary coordinate system.

Corollary 3.21. *The following equivalence relations hold true:*

$$\begin{aligned}\nu \wedge f = 0 \Leftrightarrow tf = 0 \Leftrightarrow t'f = 0,\\ \nu \lrcorner f = 0 \Leftrightarrow nf = 0 \Leftrightarrow n'f = 0.\end{aligned}$$

Remark 3.22. Note that the equation

$$\nu \wedge f = 0 \quad \text{on } \partial\Omega$$

can be equivalently written as

$$i^*(f) = 0,$$

where $i : \partial\Omega \to \mathbb{R}^n$ is the inclusion map.

Proof. Using Proposition 2.16, we immediately deduce that

$$\nu \wedge f = 0 \Leftrightarrow \nu \lrcorner (\nu \wedge f) = 0 \quad \text{and} \quad \nu \lrcorner f = 0 \Leftrightarrow \nu \wedge (\nu \lrcorner f) = 0.$$

Therefore, using Proposition 3.20(i), we get

$$\nu \wedge f = 0 \Leftrightarrow tf = 0 \quad \text{and} \quad \nu \lrcorner f = 0 \Leftrightarrow nf = 0.$$

Recall (cf. Remark 3.19(ii)) that

$$t'f = 0 \Leftrightarrow dy^n \lrcorner (dy^n \wedge \varphi^*(f)) = 0,$$

$$n'f = 0 \Leftrightarrow dy^n \wedge (dy^n \lrcorner \varphi^*(f)) = 0.$$

Hence, using Proposition 3.20, we immediately deduce that

$$tf = 0 \Leftrightarrow t'f = 0 \quad \text{and} \quad nf = 0 \Leftrightarrow n'f = 0,$$

which ends the proof. □

We will frequently use the next theorem.

Theorem 3.23. *Let Ω be a bounded open C^2 set, $0 \le k \le n$ and $f \in C^1\left(\overline{\Omega};\Lambda^k\right)$.*
(i) If $\nu \wedge f = 0$ on $\partial\Omega$, then $\nu \wedge df = 0$ on $\partial\Omega$.
(ii) If $\nu \lrcorner f = 0$ on $\partial\Omega$, then $\nu \lrcorner \delta f = 0$ on $\partial\Omega$.

Proof. We need to prove only (i), since (ii) is obtained from (i) by duality. The result (i) follows from Corollary 3.21, Definition 3.18 and the fourth statement of Theorem 3.10 applied to the admissible boundary coordinate system. More precisely, let φ be an admissible boundary coordinate system and write

$$\varphi^*(f)(y) = \sum_{1\le i_1<\cdots<i_k\le n} g_{i_1\cdots i_k}(y)\, dy^{i_1}\wedge\cdots\wedge dy^{i_k},$$
$$\varphi^*(df)(y) = \sum_{1\le i_1<\cdots<i_{k+1}\le n} h_{i_1\cdots i_{k+1}}(y)\, dy^{i_1}\wedge\cdots\wedge dy^{i_{k+1}}.$$

Due to Corollary 3.21 and Definition 3.18, we have to show that

$$h_{i_1\cdots i_{k+1}}(y',0) = 0 \quad \text{if } i_{k+1} < n. \tag{3.9}$$

From the hypothesis and Corollary 3.21, we have $g_{i_1\cdots i_k}(y',0) = 0$ if $i_k < n$ and, hence,

$$\frac{\partial g_{i_1\cdots i_k}}{\partial y_j}(y',0) = 0 \quad \text{if } j < n \text{ and } i_k < n.$$

So we have

$$\begin{aligned}
\varphi^*(df)(y',0) &= d\left(\varphi^*(f)\right)(y',0)\\
&= \sum_{1\le i_1<\cdots<i_k\le n}\sum_{j=1}^{n} \frac{\partial g_{i_1\cdots i_k}}{\partial y_j}(y',0)\, dy^j\wedge dy^{i_1}\wedge\cdots\wedge dy^{i_k}\\
&= \sum_{1\le i_1<\cdots<i_k<n} \frac{\partial g_{i_1\cdots i_k}}{\partial y_n}(y',0)\, dy^n\wedge dy^{i_1}\wedge\cdots\wedge dy^{i_k}\\
&\quad + \sum_{1\le i_1<\cdots<i_k=n}\sum_{j=1}^{n} \frac{\partial g_{i_1\cdots i_{k-1}n}}{\partial y_j}(y',0)\, dy^j\wedge dy^{i_1}\wedge\cdots\wedge dy^{i_{k-1}}\wedge dy^n.
\end{aligned}$$

This indeed shows (3.9). □

We next define the tangential and normal components of forms belonging to Sobolev spaces. Let $1 \le p \le \infty$. If $f \in W^{1,p}\left(\Omega;\Lambda^k\right)$, then its boundary value $f \in L^p\left(\partial\Omega;\Lambda^k\right)$ is well defined in the sense of traces. Since $\nu \in C^0\left(\partial\Omega;\Lambda^1\right)$, the following functions are well defined:

$$\nu\wedge f \in L^p\left(\partial\Omega;\Lambda^{k+1}\right) \quad \text{and} \quad \nu\lrcorner f \in L^p\left(\partial\Omega;\Lambda^{k-1}\right),$$

and we can now define the following spaces.

Definition 3.24. Let $r \geq 0$ be an integer and $0 \leq \alpha \leq 1 \leq p \leq \infty$. Spaces with *vanishing tangential component* are defined by

$$C_T^{r,\alpha}(\overline{\Omega};\Lambda^k) = \{f \in C^{r,\alpha}(\overline{\Omega};\Lambda^k) : \nu \wedge f = 0 \text{ on } \partial\Omega\},$$
$$W_T^{r+1,p}(\Omega;\Lambda^k) = \{f \in W^{r+1,p}(\Omega;\Lambda^k) : \nu \wedge f = 0 \text{ on } \partial\Omega\}.$$

Similarly, spaces with a *vanishing normal component* are defined by

$$C_N^{r,\alpha}(\overline{\Omega};\Lambda^k) = \{f \in C^{r,\alpha}(\overline{\Omega};\Lambda^k) : \nu \lrcorner f = 0 \text{ on } \partial\Omega\},$$
$$W_N^{r+1,p}(\Omega;\Lambda^k) = \{f \in W^{r+1,p}(\Omega;\Lambda^k) : \nu \lrcorner f = 0 \text{ on } \partial\Omega\}.$$

We will need the following density argument; for a proof, see Iwaniec, Scott and Stroffolini [58] or, in more detail, Csató [23].

Theorem 3.25. *Let $r \geq 1$ be an integer, $1 \leq p < \infty$ and $\Omega \subset \mathbb{R}^n$ be a bounded open C^{r+1} set. Then the following two statements hold true:*

(i) $C_T^r(\overline{\Omega};\Lambda^k)$ is dense in $W_T^{1,p}(\Omega;\Lambda^k)$.

(ii) $C_N^r(\overline{\Omega};\Lambda^k)$ is dense in $W_N^{1,p}(\Omega;\Lambda^k)$.

Using Theorem 3.28 of the following section, one can give an equivalent definition of the spaces $W_T^{1,p}(\Omega;\Lambda^k)$ and $W_N^{1,p}(\Omega;\Lambda^k)$ (and similarly for the Hölder spaces). The set $W_T^{1,p}(\Omega;\Lambda^k)$ is the set of $f \in W^{1,p}(\Omega;\Lambda^k)$ satisfying

$$\int_\Omega \langle df;\varphi\rangle + \int_\Omega \langle f;\delta\varphi\rangle = 0, \quad \forall\, \varphi \in C^\infty(\overline{\Omega};\Lambda^{k+1}).$$

The set $W_N^{1,p}(\Omega;\Lambda^k)$ is the set of $f \in W^{1,p}(\Omega;\Lambda^k)$ satisfying

$$\int_\Omega \langle \delta f;\varphi\rangle + \int_\Omega \langle f;d\varphi\rangle = 0, \quad \forall\, \varphi \in C^\infty(\overline{\Omega};\Lambda^{k-1}).$$

3.3 Gauss–Green Theorem and Integration-by-Parts Formula

We will assume that $\Omega \subset \mathbb{R}^n$ is a bounded open sufficiently regular set so that integration by parts can be performed, but most of the time we will even require that Ω is at least C^2. We begin with the Gauss–Green theorem.

Theorem 3.26 (Gauss–Green theorem). *Let $0 \leq k \leq n$ and $f \in C^1(\overline{\Omega};\Lambda^k)$. Then*

$$\int_\Omega \frac{\partial f_{i_1\cdots i_k}}{\partial x_j} = \int_{\partial\Omega} \nu_j f_{i_1\cdots i_k}, \quad \forall\, 1 \leq i_1 < \cdots < i_k \leq n,\ 1 \leq j \leq n,$$

and, thus, component by component,

$$\int_{\Omega} df = \int_{\partial\Omega} \nu \wedge f \quad \text{and} \quad \int_{\Omega} \delta f = \int_{\partial\Omega} \nu \lrcorner f.$$

Remark 3.27. (i) If $k = n-1$ in the identity involving d, or if $k=1$ in the identity involving δ, this is exactly the divergence theorem and hence a special case of the Stokes formula. If $k \neq n-1$ in the identity involving d, or if $k \neq 1$ in the identity involving δ, the statement of the theorem has no connection with the Stokes theorem and is to be seen as the classical Gauss–Green theorem consisting of $\binom{n}{k+1}$ equations in the first identity, respectively $\binom{n}{k-1}$ equations in the second one.

(ii) Due to density, the theorem is also true for Sobolev spaces of forms.

Proof. The first assertion is just the classical Gauss–Green theorem. Let us prove the statement with d, the one with δ being proved analogously. It follows immediately from Proposition 2.6, Proposition 3.3 and the first assertion of the theorem, since

$$\begin{aligned}\int_{\Omega} (df)_{i_1\cdots i_{k+1}} &= \int_{\Omega} \sum_{\gamma=1}^{k+1} (-1)^{\gamma-1} \frac{\partial f_{i_1\cdots i_{\gamma-1} i_{\gamma+1}\cdots i_{k+1}}}{\partial x_{i_\gamma}} \\ &= \int_{\partial\Omega} \sum_{\gamma=1}^{k+1} (-1)^{\gamma-1} \nu_{i_\gamma} f_{i_1\cdots i_{\gamma-1} i_{\gamma+1}\cdots i_{k+1}} = \int_{\partial\Omega} (\nu \wedge f)_{i_1\cdots i_{k+1}} .\end{aligned}$$

This concludes the proof of the lemma. □

Due to density, the next theorem is also valid in Sobolev spaces.

Theorem 3.28 (Integration-by-parts formula). *Let* $1 \leq k \leq n$. *Let*

$$f \in C^1\left(\overline{\Omega};\Lambda^{k-1}\right) \quad \text{and} \quad g \in C^1\left(\overline{\Omega};\Lambda^{k}\right).$$

Then

$$\int_{\Omega} \langle df; g\rangle + \int_{\Omega} \langle f; \delta g\rangle = \int_{\partial\Omega} \langle \nu \wedge f; g\rangle = \int_{\partial\Omega} \langle f; \nu \lrcorner g\rangle.$$

Proof. The second equality is trivial, as a consequence of Proposition 2.16. Appealing to Theorems 3.26, 3.5 and 2.10, we get (in the next equations, we will overlook the difference between 0 and n-forms)

$$\begin{aligned}\int_{\partial\Omega} \langle \nu \wedge f; g\rangle &= \int_{\partial\Omega} \nu \wedge f \wedge (*g) = \int_{\Omega} d(f \wedge (*g)) \\ &= \int_{\Omega} df \wedge (*g) + \int_{\Omega} (-1)^{k-1} f \wedge (d(*g)) \\ &= \int_{\Omega} \langle df; g\rangle + \int_{\Omega} (-1)^{k-1} f \wedge (d(*g)).\end{aligned}$$

It is thus left to show that

$$(-1)^{k-1} f \wedge (d(*g)) = \langle f; \delta g\rangle.$$

Since $d(*g)$ is a $(n-k+1)$-form, Theorem 2.10 yields

$$**(d(*g)) = (-1)^{(n-k+1)(k-1)} d(*g).$$

By definition, we know that

$$\delta g = (-1)^{n(k-1)} * d(*g).$$

We therefore find

$$\begin{aligned} f \wedge (d(*g)) &= (-1)^{(n-k+1)(k-1)} f \wedge (**(d(*g))) \\ &= (-1)^{k-1} f \wedge (*\delta g) = (-1)^{k-1} \langle f; \delta g \rangle. \end{aligned}$$

The proof is therefore complete. □

The following result is obtained from the previous theorem by density.

Corollary 3.29. *Let $1 \leq k \leq n$ and $1 < p, q < \infty$ with $\frac{1}{p} + \frac{1}{q} = 1$. Let $\Omega \subset \mathbb{R}^n$ be a bounded open C^2 set. Let $f \in W^{1,p}(\Omega; \Lambda^{k-1})$, $g \in W^{1,q}(\Omega; \Lambda^{k+1})$ and $\nu \wedge f = 0$ or $\nu \lrcorner g = 0$ on $\partial\Omega$. Then*

$$\int_\Omega \langle df; \delta g \rangle = 0.$$

We finally deduce the following corollary, which will turn out to be useful in the next chapter.

Corollary 3.30. *Let $1 \leq k \leq n-1$ and $f, g \in C^1(\overline{\Omega}; \Lambda^k)$. Then*

$$\begin{aligned} &\int_\Omega \langle df; dg \rangle + \int_\Omega \langle \delta f; \delta g \rangle \\ &\quad = \int_\Omega \langle \nabla f; \nabla g \rangle - \sum_I \int_{\partial\Omega} g_I \langle \nabla f_I; \nu \rangle + \int_{\partial\Omega} \langle df; \nu \wedge g \rangle + \int_{\partial\Omega} \langle \delta f; \nu \lrcorner g \rangle. \end{aligned}$$

In the corollary and in the next chapters, we have adopted the following notation.

Notation 3.31. *Let $1 \leq k \leq n-1$ and $f, g \in C^1(\overline{\Omega}; \Lambda^k)$; we then define*

$$\langle \nabla f; \nabla g \rangle = \sum_I \langle \nabla f_I; \nabla g_I \rangle = \sum_I \sum_{i=1}^n \frac{\partial f_I}{\partial x_i} \frac{\partial g_I}{\partial x_i}.$$

Proof. By density, it is enough to prove the corollary for $f, g \in C^2(\overline{\Omega}; \Lambda^k)$. Note first that

$$\begin{aligned} -\int_\Omega \langle \Delta f; g \rangle &= -\sum_I \int_\Omega \Delta f_I g_I = \sum_I \int_\Omega \langle \nabla f_I; \nabla g_I \rangle - \sum_I \int_{\partial\Omega} g_I \langle \nabla f_I; \nu \rangle \\ &= \int_\Omega \langle \nabla f; \nabla g \rangle - \sum_I \int_{\partial\Omega} g_I \langle \nabla f_I; \nu \rangle. \end{aligned}$$

The claim then follows from Theorems 3.28 and 3.7, since

$$\begin{aligned}
&\int_\Omega \langle df; dg\rangle + \int_\Omega \langle \delta f; \delta g\rangle \\
&\quad = -\int_\Omega \langle d\delta f + \delta d f; g\rangle + \int_{\partial\Omega} \langle df; \nu \wedge g\rangle + \int_{\partial\Omega} \langle \delta f; \nu \lrcorner g\rangle \\
&\quad = -\int_\Omega \langle \Delta f; g\rangle + \int_{\partial\Omega} \langle df; \nu \wedge g\rangle + \int_{\partial\Omega} \langle \delta f; \nu \lrcorner g\rangle.
\end{aligned}$$

The corollary is thus proved. □

Chapter 4
Dimension Reduction

We turn our attention to a very useful result, which is well known in the case of 2-forms. However, it can be extended in a straightforward way to the case of k-forms; it seems, however, that this extension has never been noticed. We will provide two proofs of the theorem. The first one is based on the Frobenius theorem and the second one is much more elementary and self-contained. Both versions are the same when $k = n-1$ and the first one is better from the point of view of regularity when $2 \leq k \leq n-2$.

4.1 Frobenius Theorem

We begin by recalling a few notions and results related to the theory of differential forms. For details, see, for example, Abraham, Marsden and Ratiu [1], Lee [68] and Taylor [96].

Notation 4.1 (Lie derivative and involutive family). *Let $U \subset \mathbb{R}^n$ be an open set, $a,b \in C^1(U;\mathbb{R}^n)$ and $\omega \in C^1(U;\Lambda^k)$.*

(i) $\mathcal{L}_a\omega$ stands for the Lie derivative *of ω with respect to a. It is given by*

$$\mathcal{L}_a\omega = \frac{d}{dt}\Big|_{t=0} \varphi_t^*(\omega),$$

where φ_t is the flow associated to the vector field a; that is,

$$\begin{cases} \dfrac{d}{dt}\varphi_t = a \circ \varphi_t, \\ \varphi_0 = \mathrm{id}. \end{cases}$$

(ii) Let $[a,b] = ([a,b]_1, \dots, [a,b]_n)$, *where*

$$[a,b]_i = \sum_{j=1}^{n} \left(a_j \frac{\partial b_i}{\partial x_j} - b_j \frac{\partial a_i}{\partial x_j} \right).$$

$[a,b]$ *is sometimes referred to as the* Lie bracket *of a and b.*

(iii) The Cartan formula, *which is a direct consequence of Theorem 12.5 (see Remark 12.6), states that*

$$\mathscr{L}_a \omega = a \lrcorner d\omega + d(a \lrcorner \omega). \tag{4.1}$$

Moreover, the following equality holds:

$$[a,b] \lrcorner \omega = \mathscr{L}_a (b \lrcorner \omega) - b \lrcorner (\mathscr{L}_a \omega). \tag{4.2}$$

(iv) For $a_1, \dots, a_m \in C^1(U; \mathbb{R}^n)$, *we say that* $\{a_1, \dots, a_m\}$ *is an* involutive family *in* U *if for every* $1 \le i, j \le m$, *there exist* $c_{ij}^p \in C^0(U)$, $1 \le p \le m$, *satisfying*

$$[a_i, a_j](x) = \sum_{p=1}^{m} c_{ij}^p(x) a_p(x) \text{ for every } x \in U.$$

We now recall the Frobenius theorem.

Theorem 4.2 (Frobenius theorem). *Let* $r \ge 1$ *and* $1 \le m \le n$ *be integers. Let* $x_0 \in \mathbb{R}^n$. *Let* $a_1, \dots, a_m \in C^r$ *be an involutive family in a neighborhood of* x_0, *with*

$$\{a_1(x_0), \dots, a_m(x_0)\} \quad \text{linearly independent.}$$

Then there exist a neighborhood $U \subset \mathbb{R}^n$ *of* x_0 *and* $\varphi \in \operatorname{Diff}^r(U; \varphi(U))$ *such that* $\varphi(x_0) = x_0$ *and, for every* $1 \le i \le m$,

$$\frac{\partial \varphi}{\partial x_i} \in \operatorname{span}\{(a_1 \circ \varphi), \dots, (a_m \circ \varphi)\} \text{ in } U.$$

Remark 4.3. (i) The result is still valid in Hölder spaces.

(ii) The Frobenius theorem has a sharper form, if we assume, in addition to the linear independence, the following stronger hypothesis:

$$[a_i, a_j] = 0 \quad \text{in a neighborhood of } x_0 \text{ and for every } 1 \le i, j \le m.$$

Indeed, in that case, there exist a neighborhood $U \subset \mathbb{R}^n$ of x_0 and $\varphi \in \operatorname{Diff}^r(U; \varphi(U))$ such that $\varphi(x_0) = x_0$ and for every $1 \le i \le m$,

$$\frac{\partial \varphi}{\partial x_i} = a_i \circ \varphi \text{ in } U.$$

In particular, if $a(x_0) \neq 0$ and since we always have $[a,a] = 0$, then there exist a neighborhood $U \subset \mathbb{R}^n$ of x_0 and $\varphi \in \operatorname{Diff}^r(U; \varphi(U))$ such that

$$\frac{\partial \varphi}{\partial x_n} = a \circ \varphi \text{ in } U \quad \text{and} \quad \varphi(x_0) = x_0.$$

In other words, a nonvanishing vector field can always be straightened out locally. In fact, this last observation will be achieved in Step 2 of the second proof of Theorem 4.5 below.

4.2 Reduction Theorem

Notation 4.4. *We recall, from Chapter 2, some notations that we will use throughout this section. As usual, when necessary, we identify in a natural way 1-forms with vector fields in $\mathbb{R}^n$.*

(i) Let $1 \leq k \leq n$. Given $f \in \Lambda^k(\mathbb{R}^n)$, the matrix $\overline{f} \in \mathbb{R}^{\binom{n}{k-1} \times n}$ (written in Notation 2.30 as $\overline{f}_{\lrcorner,1}$) is defined, by abuse of notations, as

$$\overline{f}\, u = u \lrcorner f \text{ for every } u \in \Lambda^1(\mathbb{R}^n) \approx \mathbb{R}^n.$$

(ii) The rank *of $f \in \Lambda^k(\mathbb{R}^n)$ (denoted by* rank_1 *in Proposition 2.32(i)) is defined by*

$$\operatorname{rank}[f] = \operatorname{rank}(\overline{f}).$$

In particular, note that $\operatorname{rank}[f] = l$ *is equivalent to* $\dim\ker(\overline{f}) = n - l$ *and that* $u \in \ker(\overline{f})$ *is equivalent to* $u \lrcorner f = 0$.

We now state the result on reduction of dimension and we follow Bandyopadhyay, Dacorogna and Kneuss [9].

Theorem 4.5 (Reduction of dimension). *Let $r \geq 1$ and $1 \leq k \leq l \leq n-1$ be integers and $x_0 \in \mathbb{R}^n$. Let g be a C^r k-form verifying*

$$dg = 0 \quad \text{and} \quad \operatorname{rank}[g] = l \quad \text{in a neighborhood of } x_0.$$

Then there exist a neighborhood U of x_0 and $\varphi \in \operatorname{Diff}^r(U; \varphi(U))$ with $\varphi(x_0) = x_0$ and such that for every $x = (x_1, \ldots, x_n) \in U$,

$$\begin{aligned} \varphi^*(g)(x_1, \ldots, x_n) &= f(x_1, \ldots, x_l) \\ &= \sum_{1 \leq i_1 < \cdots < i_k \leq l} f_{i_1 \cdots i_k}(x_1, \ldots, x_l)\, dx^{i_1} \wedge \cdots \wedge dx^{i_k}. \end{aligned}$$

Thus $f = \varphi^*(g)$ can be seen, by abuse of notations, as a k-form on $\mathbb{R}^l$ with maximal rank (i.e., $\operatorname{rank}[f] = l$*).*

Remark 4.6. (i) The result is still valid in Hölder spaces.

(ii) Note that $\varphi^*(g)$ is only in C^{r-1} although $g \in C^r$.

Before starting with the two proofs of the theorem, we need the following simple lemma.

Lemma 4.7. *Let $V \subset \mathbb{R}^n$ be an open set, $g \in C^0\left(V;\Lambda^k\right)$ and $a \in C^0\left(V;\mathbb{R}^n\right)$ be such that*

$$a \lrcorner g = 0 \quad \textit{in } V.$$

Let $U \subset \mathbb{R}^n$ be an open set, $\varphi \in \operatorname{Diff}^1\left(U;\varphi\left(U\right)\right)$ be such that $\varphi\left(U\right) \subset V$, $1 \le j \le n$ and

$$\frac{\partial \varphi}{\partial x_j} = a \circ \varphi \quad \textit{in } U.$$

Then, in U and for every $1 \le i_1 < \cdots < i_k \le n$,

$$\left(\varphi^*\left(g\right)\right)_{i_1 \cdots i_k} = 0 \quad \textit{if } j \in \left\{i_1, \ldots, i_k\right\}.$$

Proof. We assume, without loss of generality, that $j = n$. Use Theorem 3.10 and Proposition 3.11 and get

$$0 = \varphi^*(a \lrcorner g) = dx^n \lrcorner \varphi^*(g) \quad \text{in } U,$$

which directly implies the claim. □

We now turn our attention to the first proof of our theorem.

Proof (First proof of Theorem 4.5). We divide the proof into four steps.

Step 1. Since $\operatorname{rank}[g] = l \le n-1$, it is easy to find a neighborhood V of x_0 and $a_i \in C^r\left(V;\mathbb{R}^n\right)$ for every $l+1 \le i \le n$ such that for every $x \in V$,

$$\left\{a_{l+1}(x), \ldots, a_n(x)\right\} \text{ are linearly independent} \tag{4.3}$$

and

$$\operatorname{span}\left\{a_{l+1}(x), \ldots, a_n(x)\right\} = \ker \overline{g}(x). \tag{4.4}$$

Then we have, in particular, for every $l+1 \le i \le n$,

$$a_i \lrcorner g = 0 \quad \text{in } V.$$

Step 2. We now show that the family $\left\{a_{l+1}, \ldots, a_n\right\}$ is involutive in V; that is, for every $l+1 \le i, j \le n$, there exist $c_{ij}^p \in C^0(V)$, $l+1 \le p \le n$, satisfying

$$[a_i, a_j](x) = \sum_{p=l+1}^{n} c_{ij}^p(x) a_p(x) \quad \text{for every } x \in V.$$

Indeed, since $dg = 0$ and (4.1), (4.2) and (4.4) hold, it follows that

$$\begin{aligned}[a_i, a_j] \lrcorner g &= \mathscr{L}_{a_i}(a_j \lrcorner g) - a_j \lrcorner (\mathscr{L}_{a_i} g) = -a_j \lrcorner (\mathscr{L}_{a_i} g) \\ &= -a_j \lrcorner (a_i \lrcorner dg + d(a_i \lrcorner g)) = 0 \quad \text{in } V.\end{aligned}$$

Hence, we have $[a_i, a_j](x) \in \ker \overline{g}(x)$, for every $x \in V$, from where, using (4.3) and (4.4), the existence of unique coefficients $c_{ij}^p(x)$, for every $x \in V$, follows. It is easy to check that $c_{ij}^p \in C^r(V)$.

Step 3. Therefore, using Theorem 4.2, there exist a neighborhood U of x_0 and $\varphi \in \text{Diff}^r(U; \varphi(U))$ with $\varphi(U) \subset V$ such that $\varphi(x_0) = x_0$ and

$$\frac{\partial \varphi}{\partial x_i} \in \ker \overline{g} \circ \varphi \quad \text{in } U \text{ and for every } l+1 \leq i \leq n. \tag{4.5}$$

Let us show that φ has all of the desired properties. We have to show that

$$\varphi^*(g)_{i_1 \cdots i_k} = 0 \quad \text{in } U \tag{4.6}$$

for every $1 \leq i_1 < \cdots < i_k \leq n$ with $i_k \geq l+1$ and that

$$\varphi^*(g)(x_1, \ldots, x_n) = \varphi^*(g)(x_1, \ldots, x_l) \quad \text{for every } x \in U.$$

Indeed, (4.6) comes directly from Lemma 4.7 and (4.5). Finally, since $dg = 0$, we have $d\varphi^*(g) = 0$. Hence, on writing (using (4.6))

$$\varphi^*(g) = \sum_{1 \leqslant i_1 < \cdots < i_k \leqslant l} r_{i_1 \cdots i_k} dx^{i_1} \wedge \cdots \wedge dx^{i_k},$$

we obtain, for every $s \geq l+1$,

$$\frac{\partial r_{i_1 \cdots i_k}}{\partial x_s} = (d\varphi^*(g))_{i_1 \cdots i_k s} = 0,$$

which shows the second claim and ends the proof. □

We finish this chapter with the second proof of the theorem. As already stated, it is much more elementary, but it gives a less sharp regularity; indeed we will only be able to establish that

$$\varphi \in \text{Diff}^{r+l-n+1}(U; \varphi(U)).$$

The proof therefore requires $r \geq n-l$. Note that when $k = n-1$ (and, thus, $l = n-1$), both proofs are the same.

Proof (Second proof of Theorem 4.5). Without loss of generality, we assume that $x_0 = 0$. In the following, U stands for a generic neighborhood of 0.

Step 1. Since $\operatorname{rank}[g] = l \leq n-1$, there exists $a \in C^r(U;\mathbb{R}^n)$ satisfying

$$a(x) \neq 0 \quad \text{and} \quad a(x) \lrcorner g(x) = 0 \quad \text{for every } x \in U. \tag{4.7}$$

Step 2. We claim that we can find a neighborhood U of 0 and $\varphi_n \in \operatorname{Diff}^r(U;\varphi_n(U))$ such that $\varphi_n(0) = 0$ and

$$\frac{\partial \varphi_n}{\partial x_n} = a \circ \varphi_n \quad \text{in } U. \tag{4.8}$$

Indeed, using well-known results on ordinary differential equations (see, e.g., [22]), there exist a neighborhood U of 0, $\varepsilon > 0$ small enough and $\psi \in C^r(U \times (-\varepsilon,\varepsilon);\mathbb{R}^n)$ verifying, for every $(x,t) \in U \times (-\varepsilon,\varepsilon)$,

$$\frac{\partial}{\partial t}\psi(x,t) = a(\psi(x,t)) \quad \text{and} \quad \psi(x,0) = x.$$

Since $a(0) \neq 0$, we can choose $b_1,\ldots,b_{n-1} \in \mathbb{R}^n$ so that

$$\{b_1,\ldots,b_{n-1},a(0)\} \text{ are linearly independent.}$$

Let $B \in \mathbb{R}^{n\times(n-1)}$ be the matrix whose ith column is b_i. Finally, define

$$\varphi_n(x) = \varphi_n(x_1,\ldots,x_n) = \psi(B(x_1,\ldots,x_{n-1}),x_n)$$

and observe that

$$\varphi_n(0) = 0, \quad \det \nabla \varphi_n(0) \neq 0 \quad \text{and} \quad \frac{\partial \varphi_n}{\partial x_n} = a \circ \varphi_n \quad \text{in } U,$$

which shows the claim.

Step 3. From (4.7), (4.8) and Lemma 4.7, it follows that for every $1 \leq i_1 < \cdots < i_{k-1} \leq n-1$,

$$\varphi_n^*(g)_{i_1\cdots i_{k-1}n} = 0 \quad \text{in } U.$$

Finally, since $dg = 0$, we have $d\varphi_n^*(g) = 0$. Hence, writing

$$\varphi_n^*(g) = \sum_{1 \leqslant i_1 < \cdots < i_k \leq n-1} r_{i_1\cdots i_k}\, dx^{i_1} \wedge \cdots \wedge dx^{i_k},$$

we obtain

$$\frac{\partial r_{i_1\cdots i_k}}{\partial x_n} = (d\varphi_n^*(g))_{i_1\cdots i_k n} = 0,$$

which implies $\varphi_n^*(g)(x) = \varphi_n^*(g)(x_1,\ldots,x_{n-1})$.

Step 4 (conclusion). If $l = n-1$, the proof is finished in view of Steps 2 and 3. Henceforth, we will assume that $1 \leq l < n-1$. Since (using Proposition 17.1) $\operatorname{rank}(\varphi_n^*(g)) = l$ and since $\varphi_n^*(g) \in C^{r-1}$, repeating the aforementioned argument we find $\varphi_{n-1} \in \operatorname{Diff}^{r-1}(U;\varphi_{n-1}(U))$ satisfying $\varphi_{n-1}(0) = 0$,

$$\varphi_{n-1}^*(\varphi_n^*(g))_{i_1\cdots i_k} = 0 \quad \text{in } U$$

for every $1 \leq i_1 < \cdots < i_k \leq n$ with $i_k \geq n-1$ and

$$\varphi_{n-1}^*(\varphi_n^*(g))(x) = \varphi_{n-1}^*(\varphi_n^*(g))(x_1, \ldots, x_{n-2})) \quad \text{for every } x \in U.$$

After repeating the same argument $n-l$ times, we set

$$\varphi = \varphi_n \circ \cdots \circ \varphi_{l+1}$$

and we get $\varphi \in \mathrm{Diff}^{r+l-n+1}(U; \varphi(U))$, $\varphi(0) = 0$,

$$\varphi^*(g)_{i_1 \cdots i_k} = 0 \quad \text{in } U,$$

for every $1 \leq i_1 < \cdots < i_k \leq n$ with $i_k \geq l+1$ and

$$\varphi^*(g)(x_1, \ldots, x_n) = \varphi^*(g)(x_1, \ldots, x_l), \quad \text{for every } x \in U.$$

This finishes the proof. □

Part II
Hodge–Morrey Decomposition and Poincaré Lemma

Chapter 5

An Identity Involving Exterior Derivatives and Gaffney Inequality

5.1 Introduction

The aim of this chapter is twofold. In Section 5.2 we prove a very general identity (cf. Theorem 5.7) involving the operators d, δ and ∇. In Section 5.3 we show how the above identity leads to a very simple proof of the classical Gaffney inequality (cf. Theorem 5.16). This inequality will be one of the key points for Hodge–Morrey decomposition (cf. Chapter 6). Let us now describe in more detail the results.

Let $\Omega \subset \mathbb{R}^n$ be a bounded open smooth set and ν be the exterior unit normal to the boundary. We will first prove, following Csató and Dacorogna [24] (cf. Theorem 5.7 for a more general statement), that for k-forms ω (cf. Notation 3.31 for $|\nabla\omega|^2$),

$$\begin{aligned}
&\int_\Omega \left(|d\omega|^2 + |\delta\omega|^2 - |\nabla\omega|^2\right) \\
&= -\int_{\partial\Omega} \left(\langle \nu \wedge d(\nu \lrcorner \omega); \nu \wedge \omega\rangle + \langle \nu \lrcorner \delta(\nu \wedge \omega); \nu \lrcorner \omega\rangle\right) \\
&\quad + \int_{\partial\Omega} \left(\langle L^\nu(\nu \wedge \omega); \nu \wedge \omega\rangle + \langle K^\nu(\nu \lrcorner \omega); \nu \lrcorner \omega\rangle\right) \qquad (5.1)
\end{aligned}$$

where L^ν and K^ν are given in Definition 5.1; they can also be seen as matrices acting on $(k+1)$-forms and $(k-1)$-forms, respectively (where we identify a k-form with a $\binom{n}{k}$ vector). They depend only on the geometry of Ω and on the degree k of the form. They can easily be calculated explicitly for general k-forms and when Ω is a ball of radius R (cf. Corollary 5.9 for a more general statement), it turns out that

$$L^\nu(\nu \wedge \omega) = \frac{k}{R}\, \nu \wedge \omega \quad \text{and} \quad K^\nu(\nu \lrcorner \omega) = \frac{n-k}{R}\, \nu \lrcorner \omega.$$

We therefore have

$$\langle L^\nu(\nu \wedge \omega); \nu \wedge \omega\rangle = \frac{k}{R}|\nu \wedge \omega|^2 \quad \text{and} \quad \langle K^\nu(\nu \lrcorner \omega); \nu \lrcorner \omega\rangle = \frac{n-k}{R}|\nu \lrcorner \omega|^2.$$

We will also give general formulas (cf. Proposition 5.11 and Corollary 5.12) in the case of 1-forms and for general domains Ω; in this case, K^{ν} is a scalar and it is a multiple of κ, the mean curvature of the hypersurface $\partial\Omega$, namely

$$K^{\nu} = (n-1)\,\kappa.$$

The advantages of this formula, besides its generality and elegancy, are the following.

1) The right-hand side of the identity is expressed solely in terms of the tangential and normal components of ω. Therefore, if either $\nu \wedge \omega = 0$ or $\nu \lrcorner \omega = 0$, then the right-hand side of (5.1) does not depend on derivatives of ω. It hence leads to an elementary proof of the classical *Gaffney inequality* (cf. Theorem 5.16 below). This inequality states that there exists a constant $C = C(\Omega) > 0$ such that for every k-form ω with either $\nu \wedge \omega = 0$ or $\nu \lrcorner \omega = 0$,

$$C \int_{\Omega} |\nabla\omega|^2 \leq \int_{\Omega} |d\omega|^2 + \int_{\Omega} |\delta\omega|^2 + \int_{\Omega} |\omega|^2. \tag{5.2}$$

The classical proof of (5.2) by Morrey [76, 77] (see also, e.g., Iwaniec, Scott and Stroffolini [58]), generalizing results of Gaffney [44, 45], is more technical. It requires the use of local rectification of the boundary, partition of unity and some estimates concerning $d\omega$, $\delta\omega$ and $\nabla\omega$.

2) The formula is valid with no restriction on the behavior of ω on $\partial\Omega$. This observation will allow us to obtain (cf. Theorem 5.19) Gaffney-type inequalities for more general boundary conditions than the classical ones, which are $\nu \wedge \omega = 0$ or $\nu \lrcorner \omega = 0$. If one assumes $\nu \wedge \omega = 0$ (and similarly if $\nu \lrcorner \omega = 0$), then an identity in the same spirit as (5.1) can be found in Amrouche, Bernardi, Dauge and Girault [6] and Duvaut and Lions [39] (cf. proof of Theorem 6.1 in Chapter 7) for the special case of a 1-form in $\mathbb{R}^3$ and in Schwarz [89] (cf. Theorem 2.1.5). However, in this last book, the actual K^{ν} is only very implicitly defined.

The proof of the formula is as follows. We start, as in classical proofs of the Gaffney inequality, by expressing the left-hand side of (5.1) by a boundary integral through some quite simple integrations by parts, together with the formula $d\delta + \delta d = \Delta$ (cf. Corollary 3.30), obtaining that

$$\int_{\Omega} \left(|d\omega|^2 + |\delta\omega|^2 - |\nabla\omega|^2\right) = \int_{\partial\Omega} \left(\langle d\omega; \nu \wedge \omega\rangle + \langle \delta\omega; \nu \lrcorner \omega\rangle - \textstyle\sum_I \langle \nabla\omega_I; \nu\rangle \omega_I\right).$$

We then transform the right-hand side through algebraic manipulations only and no more integration by parts, so as to get our formula.

The L^p versions for $p \neq 2$ of Gaffney-type inequalities have been treated by Iwaniec, Scott and Stroffolini [58] and by Bolik [13], who also deals with Hölder spaces $C^{r,\alpha}$. These results will be cited at the end of this chapter (cf. Theorem 5.21).

5.2 An Identity Involving Exterior Derivatives

5.2.1 Preliminary Formulas

Recall the notation for a differential form ω,

$$\omega = \sum_{1 \le i_1 < \cdots < i_k \le n} \omega_{i_1 \cdots i_k} dx^{i_1} \wedge \cdots \wedge dx^{i_k} = \sum_I \omega_I dx^I.$$

Definition 5.1. Let $U \subset \mathbb{R}^n$ be open and $v \in C^1\left(U;\Lambda^1\right)$. We define for every $0 \le k \le n$, the two maps

$$L^v, K^v : C^0\left(U;\Lambda^k\right) \to C^0\left(U;\Lambda^k\right)$$

by

$$L^v(\omega) = \sum_I \omega_I \, d\left(v \lrcorner dx^I\right) \quad \text{if } k \ge 1,$$

$$K^v(\omega) = \sum_I \omega_I \, \delta\left(v \wedge dx^I\right) \quad \text{if } k \le n-1,$$

whereas $L^v(\omega) = 0$ if $k = 0$ and $K^v(\omega) = 0$ if $k = n$.

Remark 5.2. Note that $L^v(\omega)$ and $K^v(\omega)$ are linear in ω and v.

The next two lemmas present some elementary properties of the maps L^v and K^v and their connection with the Lie derivative. We therefore recall the following facts. Depending on the context, we will identify 1-forms with vector fields. Let $v \in C^1(U;\mathbb{R}^n)$ and $\omega \in C^1(U;\Lambda^k)$. The Lie derivative $\mathscr{L}_v\omega$ is defined by

$$\mathscr{L}_v\omega = \left.\frac{d}{dt}\right|_{t=0} \varphi_t^*(\omega),$$

where $\varphi = \varphi(t,x) = \varphi_t(x)$ is the flow associated to the vector field v; that is,

$$\begin{cases} \dfrac{d}{dt}\varphi_t = v \circ \varphi_t, \\ \varphi_0 = \mathrm{id} \end{cases} \tag{5.3}$$

for t small enough. The Cartan formula (see Notation 4.1 and Remark 12.6) states that

$$\mathscr{L}_v\omega = v \lrcorner d\omega + d(v \lrcorner \omega). \tag{5.4}$$

Its dual version is

$$(-1)^{k(n-k)} * \mathscr{L}_v(*\omega) = v \wedge \delta\omega + \delta(v \wedge \omega). \tag{5.5}$$

Lemma 5.3. *Let $U \subset \mathbb{R}^n$ be open, $0 \le k \le n$, $\omega \in C^0\left(U;\Lambda^k\right)$ and $v \in C^1\left(U;\Lambda^1\right)$.*

(i) The following duality relations hold true:

$$K^v(\omega) = (-1)^{k(n-k)}\left(*L^v(*\omega)\right) \quad \textit{and} \quad L^v(\omega) = (-1)^{k(n-k)}\left(*K^v(*\omega)\right),$$

$$\langle \nu \lrcorner K^{\nu}(\omega); \nu \lrcorner \omega \rangle = \langle \nu \wedge L^{\nu}(*\omega); \nu \wedge (*\omega) \rangle.$$

(ii) If φ is the flow associated to ν, then

$$L^{\nu}(\omega) = \sum_I \omega_I \frac{d}{dt}\bigg|_{t=0} d\varphi_t^I$$

if $k \geq 1$ and if $k \leq n-1$, then

$$K^{\nu}(\omega) = (-1)^{k(n-k)} * \left(\sum_I \omega_I(x) \frac{d}{dt}\bigg|_{t=0} d\varphi_t^{I^c} (-1)^r \right),$$

where I^c is the complement of I in the set $\{1,\ldots,n\}$ and (cf. Theorem 2.10)

$$dx^I \wedge dx^{I^c} = (-1)^r dx^1 \wedge \cdots \wedge dx^n.$$

Proof. (i) Due to linearity, one can assume that $\omega = dx^I$ for some $I \in \mathscr{T}_k$. The statement follows directly from the definitions of the interior derivative and the interior product and Theorem 2.10.

(ii) Standard results on ordinary differential equations (cf. (7.13) in the proof of Theorem 7.2 in Chapter 1 of Coddington and Levinson [22]) give that $\nabla\varphi_t$ is differentiable in t and satisfies

$$\frac{d}{dt}[\nabla\varphi_t] = \nabla\nu(\varphi_t)\,\nabla\varphi_t\,.$$

This is indeed what is immediately obtained by formal differentiation of (5.3). In particular, for every $1 \leq i \leq n$, we have

$$\frac{d}{dt}\bigg|_{t=0} [d\varphi_t^i] = d\nu_i\,.$$

As in (i), we can assume that $\omega = dx^I$ and, thus,

$$\begin{aligned}
\frac{d}{dt}\bigg|_{t=0} d\varphi_t^I &= \frac{d}{dt}\bigg|_{t=0} d\varphi_t^{i_1} \wedge \ldots \wedge d\varphi_t^{i_k} \\
&= \sum_{\gamma=1}^{k} dx^{i_1} \wedge \ldots \wedge dx^{i_{\gamma-1}} \wedge d\nu_{i_\gamma} \wedge dx^{i_{\gamma+1}} \wedge \ldots \wedge dx^{i_k} \\
&= \sum_{\gamma=1}^{k} (-1)^{\gamma-1} d\left(\nu_{i_\gamma} dx^{i_1} \wedge \cdots \wedge \widehat{dx^{i_\gamma}} \wedge \cdots \wedge dx^{i_k} \right) \\
&= d\left(\nu \lrcorner dx^I\right).
\end{aligned}$$

This proves the equality concerning L^{ν}. The corresponding equality for K^{ν} follows from (i). □

Lemma 5.4. *Let $U \subset \mathbb{R}^n$ be open, $0 \leq k \leq n$, $\omega \in C^1(U;\Lambda^k)$ and $v \in C^1(U;\Lambda^1)$; then*

$$\mathscr{L}_v \omega = \sum_I \langle \nabla \omega_I ; v \rangle dx^I + L^v(\omega),$$

$$(-1)^{k(n-k)} * \mathscr{L}_v(*\omega) = \sum_I \langle \nabla \omega_I ; v \rangle dx^I + K^v(\omega),$$

$$\begin{aligned}\sum_I \langle \nabla \omega_I ; v \rangle dx^I &= v \lrcorner d\omega + d(v \lrcorner \omega) - L^v(\omega) \\ &= v \wedge \delta\omega + \delta(v \wedge \omega) - K^v(\omega).\end{aligned}$$

Proof. Using the definition of the Lie derivative and (ii) of Lemma 5.3, we get, if $k \geq 1$,

$$\begin{aligned}\mathscr{L}_v \omega &= \frac{d}{dt}\Big|_{t=0} \sum_I \omega_I(\varphi_t)\, d\varphi_t^I \\ &= \sum_I \langle \nabla \omega_I ; v \rangle dx^I + \sum_I \omega_I \frac{d}{dt}\Big|_{t=0} d\varphi_t^I \\ &= \sum_I \langle \nabla \omega_I ; v \rangle dx^I + L^v(\omega).\end{aligned}$$

If $k = 0$, the statement is immediate, since then $L^v(\omega) = 0$. The proof of the corresponding equality for K^v is completely analogous. The third statement of the lemma follows from the first two identities of the lemma and the Cartan formula (5.4), respectively (5.5). □

We also have the following useful property of the operators L^v and K^v.

Lemma 5.5. *Let $0 \leq k \leq n$, U be an open subset of $\mathbb{R}^n$, $\omega \in C^1(U;\Lambda^k)$ and $v \in C^1(U;\Lambda^1)$; then, the following equations hold true in U:*

$$\begin{aligned} L^v(v \wedge \omega) &= \frac{1}{2} d(|v|^2) \wedge \omega + v \wedge L^v(\omega), \\ K^v(v \lrcorner \omega) &= \frac{1}{2} d(|v|^2) \lrcorner \omega + v \lrcorner K^v(\omega).\end{aligned}$$

In particular, if $|v|$ is constant in U, then

$$L^v(v \wedge \omega) = v \wedge L^v(\omega) \quad \textit{and} \quad K^v(v \lrcorner \omega) = v \lrcorner K^v(\omega).$$

Proof. We prove the first equality. The second result concerning K^v follows from the first one using Lemma 5.3(i). The two extra assertions are trivial. First, noticing that L^v is linear, it is enough to prove the claim for $\omega = dx^I$ for any $I = (i_1, \ldots, i_k) \in \mathscr{T}_k$. By definition of L^v, we get

$$L^v(v \wedge dx^I) = L^v\left(\sum_{j=1}^n v_j dx^j \wedge dx^I\right) = \sum_{j=1}^n v_j d\left(v \lrcorner (dx^j \wedge dx^I)\right).$$

Since, by Proposition 2.16,

$$\nu \lrcorner (dx^j \wedge dx^I) = \nu_j dx^I - dx^j \wedge (\nu \lrcorner dx^I),$$

we deduce that

$$\begin{aligned} L^\nu(\nu \wedge dx^I) &= \sum_{j=1}^n \nu_j d\nu_j \wedge dx^I + \sum_{j=1}^n \nu_j dx^j \wedge d(\nu \lrcorner dx^I) \\ &= \frac{1}{2} d(|\nu|^2) \wedge dx^I + \nu \wedge d(\nu \lrcorner dx^I) \\ &= \frac{1}{2} d(|\nu|^2) \wedge dx^I + \nu \wedge L^\nu(dx^I). \end{aligned}$$

The proof is therefore finished. □

Our last lemma will turn out to be the key point in our main result.

Lemma 5.6. *Let $0 \le k \le n$, U be an open subset of $\mathbb{R}^n$, $\nu \in C^1(U;\Lambda^1)$, α, $\beta \in C^1(U;\Lambda^k)$ and $x \in U$ be such that $|\nu(x)| = 1$. Then the following equation holds true, for every such x:*

$$\begin{aligned} &\langle d\alpha; \nu \wedge \beta \rangle + \langle \delta\alpha; \nu \lrcorner \beta \rangle - \sum_I \langle \nabla \alpha_I; \nu \rangle \beta_I \\ &\quad = -\langle \nu \wedge d(\nu \lrcorner \alpha); \nu \wedge \beta \rangle - \langle \nu \lrcorner \delta(\nu \wedge \alpha); \nu \lrcorner \beta \rangle \\ &\qquad + \langle \nu \wedge L^\nu(\alpha); \nu \wedge \beta \rangle + \langle \nu \lrcorner K^\nu(\alpha); \nu \lrcorner \beta \rangle. \end{aligned} \tag{5.6}$$

Proof. Essential in the proof are the results of Proposition 2.16 and the fact that $|\nu(x)| = 1$. We split $\sum \langle \nabla \alpha_I; \nu \rangle \beta_I$ in the following way:

$$\begin{aligned} \sum_I \langle \nabla \alpha_I; \nu \rangle \beta_I &= \left\langle \sum_I \langle \nabla \alpha_I; \nu \rangle dx^I; \beta \right\rangle \\ &= \left\langle \sum_I \langle \nabla \alpha_I; \nu \rangle dx^I; \nu \wedge (\nu \lrcorner \beta) \right\rangle + \left\langle \sum_I \langle \nabla \alpha_I; \nu \rangle dx^I; \nu \lrcorner (\nu \wedge \beta) \right\rangle \end{aligned}$$

and, similarly,

$$\begin{aligned} \langle d\alpha; \nu \wedge \beta \rangle &= \langle \nu \wedge (\nu \lrcorner d\alpha); \nu \wedge \beta \rangle + \langle \nu \lrcorner (\nu \wedge d\alpha); \nu \wedge \beta \rangle \\ &= \langle \nu \wedge (\nu \lrcorner d\alpha); \nu \wedge \beta \rangle. \end{aligned}$$

Using this and Proposition 2.16, we obtain

$$\begin{aligned} &\langle d\alpha; \nu \wedge \beta \rangle - \left\langle \sum_I \langle \nabla \alpha_I; \nu \rangle dx^I; \nu \lrcorner (\nu \wedge \beta) \right\rangle \\ &\quad = \langle \nu \wedge (\nu \lrcorner d\alpha); \nu \wedge \beta \rangle - \left\langle \nu \wedge \left(\sum_I \langle \nabla \alpha_I; \nu \rangle dx^I \right); \nu \wedge \beta \right\rangle \end{aligned}$$

and, thus, according to Lemma 5.4 applied to α,

$$\begin{aligned}\langle d\alpha;\nu\wedge\beta\rangle-\Big\langle\sum_I\langle\nabla\alpha_I;\nu\rangle\,dx^I;\nu\lrcorner(\nu\wedge\beta)\Big\rangle\\=-\langle\nu\wedge d(\nu\lrcorner\alpha);\nu\wedge\beta\rangle+\langle\nu\wedge L^\nu(\alpha);\nu\wedge\beta\rangle.\end{aligned}\tag{5.7}$$

We now carry out the analogue computations for $\langle\delta\alpha;\nu\lrcorner\beta\rangle$. Note first that

$$\langle\delta\alpha;\nu\lrcorner\beta\rangle=\langle\nu\lrcorner(\nu\wedge\delta\alpha);\nu\lrcorner\beta\rangle.$$

Using this fact and again Proposition 2.16, we find

$$\begin{aligned}\langle\delta\alpha;\nu\lrcorner\beta\rangle-\Big\langle\sum_I\langle\nabla\alpha_I;\nu\rangle\,dx^I;\nu\wedge(\nu\lrcorner\beta)\Big\rangle\\=\langle\nu\lrcorner(\nu\wedge\delta\alpha);\nu\lrcorner\beta\rangle-\Big\langle\nu\lrcorner\Big(\sum_I\langle\nabla\alpha_I;\nu\rangle\,dx^I\Big);\nu\lrcorner\beta\Big\rangle\end{aligned}$$

and, hence, according to Lemma 5.4 applied to α,

$$\begin{aligned}\langle\delta\alpha;\nu\lrcorner\beta\rangle-\Big\langle\sum_I\langle\nabla\alpha_I;\nu\rangle\,dx^I;\nu\wedge(\nu\lrcorner\beta)\Big\rangle\\=-\langle\nu\lrcorner\delta(\nu\wedge\alpha);\nu\lrcorner\beta\rangle+\langle\nu\lrcorner K^\nu(\alpha);\nu\lrcorner\beta\rangle.\end{aligned}\tag{5.8}$$

We now combine (5.7) and (5.8) to conclude the proof. □

5.2.2 The Main Theorem

The following theorem has been established by Csató and Dacorogna [24].

Theorem 5.7 (A general identity). *Let $0\le k\le n$ and let $\Omega\subset\mathbb{R}^n$ be a bounded open C^2 set with exterior unit normal ν. Then every $\alpha,\,\beta\in C^1\left(\overline{\Omega};\Lambda^k\right)$ satisfy the equation*

$$\begin{aligned}\int_\Omega(\langle d\alpha;d\beta\rangle+\langle\delta\alpha;\delta\beta\rangle-\langle\nabla\alpha;\nabla\beta\rangle)\\=-\int_{\partial\Omega}(\langle\nu\wedge d(\nu\lrcorner\alpha);\nu\wedge\beta\rangle+\langle\nu\lrcorner\delta(\nu\wedge\alpha);\nu\lrcorner\beta\rangle)\\+\int_{\partial\Omega}(\langle L^\nu(\nu\wedge\alpha);\nu\wedge\beta\rangle+\langle K^\nu(\nu\lrcorner\alpha);\nu\lrcorner\beta\rangle).\end{aligned}$$

Remark 5.8. (i) In the above theorem and in the sequel, we have always assumed that the exterior unit normal ν has been extended to $\mathbb{R}^n$ in a C^1 way with $|\nu|=1$ in a neighborhood of $\partial\Omega$. This is, of course, always possible. The formulas here and below will be seen to be independent of the extension.

(ii) An alternative version to formulate the theorem would be (see Lemma 5.5)

$$\begin{aligned}\int_\Omega (\langle d\alpha; d\beta\rangle + \langle \delta\alpha; \delta\beta\rangle - \langle \nabla\alpha; \nabla\beta\rangle) \\ = -\int_{\partial\Omega} (\langle \nu \wedge d(\nu \lrcorner \alpha); \nu \wedge \beta\rangle + \langle \nu \lrcorner \delta(\nu \wedge \alpha); \nu \lrcorner \beta\rangle) \\ + \int_{\partial\Omega} (\langle \nu \wedge L^\nu(\alpha); \nu \wedge \beta\rangle + \langle \nu \lrcorner K^\nu(\alpha); \nu \lrcorner \beta\rangle).\end{aligned} \tag{5.9}$$

In that case, we do not need to extend ν, since all four terms in the boundary integral depend only on the values of ν on the boundary. This follows from Theorem 3.23.

(iii) If $\alpha = \beta$, the first boundary integral could be expressed more compactly, since by taking an arbitrary extension of ν onto the whole Ω, we obtain, appealing to Theorem 3.28,

$$\begin{aligned}\int_\Omega \langle d(\nu \lrcorner \alpha); \delta(\nu \wedge \alpha)\rangle &= \int_{\partial\Omega} \langle \nu \lrcorner \delta(\nu \wedge \alpha); \nu \lrcorner \alpha\rangle \\ &= \int_{\partial\Omega} \langle \nu \wedge d(\nu \lrcorner \alpha); \nu \wedge \alpha\rangle.\end{aligned}$$

(iv) In the special cases $k = 0$ or $k = n$, the proof is much more immediate than the one we will provide below, since then all terms in the boundary integral vanish.

As an example, we first present the following corollary.

Corollary 5.9. *Let $\Omega = B_R(a)$ be the ball of radius R centered at a with exterior unit normal ν. Then*

$$L^\nu(\nu \wedge \alpha) = \frac{k}{R}\nu \wedge \alpha \quad \text{and} \quad K^\nu(\nu \lrcorner \alpha) = \frac{n-k}{R}\nu \lrcorner \alpha$$

and, thus, every $\alpha, \beta \in C^1(\overline{\Omega}; \Lambda^k)$ satisfy the equation

$$\begin{aligned}\int_\Omega (\langle d\alpha; d\beta\rangle + \langle \delta\alpha; \delta\beta\rangle - \langle \nabla\alpha; \nabla\beta\rangle) \\ = -\int_{\partial\Omega} (\langle \nu \wedge d(\nu \lrcorner \alpha); \nu \wedge \beta\rangle + \langle \nu \lrcorner \delta(\nu \wedge \alpha); \nu \lrcorner \beta\rangle) \\ + \int_{\partial\Omega} \left(\frac{k}{R}\langle \nu \wedge \alpha; \nu \wedge \beta\rangle + \frac{n-k}{R}\langle \nu \lrcorner \alpha; \nu \lrcorner \beta\rangle\right).\end{aligned}$$

We first prove the corollary.

Proof. Without loss of generality we can assume $a = 0$. We use Lemma 5.5 and Proposition 5.10 below to obtain

$$L^\nu(\nu \wedge \alpha) = \nu \wedge L^\nu(\alpha) = \mu \wedge L^\mu(\alpha),$$

where $\mu(x) = x/R$. We use Lemma 5.3(ii) to calculate $L^\mu(\alpha)$. Let φ_t be the flow associated to μ, namely

$$\varphi_t = \varphi_t(x) = e^{\frac{1}{R}t}x.$$

We therefore obtain

$$d\varphi_t^I = e^{\frac{k}{R}t} dx^I \quad \Rightarrow \quad \frac{d}{dt}\bigg|_{t=0} d\varphi_t^I = \frac{k}{R} dx^I$$

and, thus,

$$L^\mu(\alpha) = \frac{k}{R}\alpha \quad \text{and} \quad \mu \wedge L^\mu(\alpha) = \frac{k}{R}\mu \wedge \alpha.$$

It now follows from Lemma 5.3(i) that

$$K^\mu(\alpha) = \frac{n-k}{R}\alpha \quad \text{and} \quad \mu \lrcorner K^\mu(\alpha) = \frac{n-k}{R}\mu \lrcorner \alpha.$$

The corollary is therefore proved. □

We now continue with the proof of Theorem 5.7.

Proof. We assume $\alpha, \beta \in C^2(\overline{\Omega}; \Lambda^k)$, since the result for $\alpha, \beta \in C^1(\overline{\Omega}; \Lambda^k)$ follows by a density argument. We apply Corollary 3.30 to obtain

$$\begin{aligned}\int_\Omega \langle d\alpha; d\beta \rangle + \int_\Omega \langle \delta\alpha; \delta\beta \rangle = \sum_I \int_\Omega \langle \nabla\alpha_I; \nabla\beta_I \rangle - \sum_I \int_{\partial\Omega} \beta_I \langle \nabla\alpha_I; \nu \rangle \\ + \int_{\partial\Omega} \langle d\alpha; \nu \wedge \beta \rangle + \int_{\partial\Omega} \langle \delta\alpha; \nu \lrcorner \beta \rangle.\end{aligned}$$

We next apply Lemma 5.6, which proves the alternative version (5.9). The theorem now follows from Lemma 5.5. □

Consider the tangent vectors $E_{ij} = (E_{ij}^1, \ldots, E_{ij}^n)$ at $x \in \partial\Omega$ defined in the following way:

$$E_{ij}(x) = \begin{pmatrix} 0 \\ \vdots \\ \nu_j(x) \\ \vdots \\ -\nu_i(x) \\ \vdots \\ 0 \end{pmatrix} \begin{matrix} \\ \\ \leftarrow i\text{th coordinate position} \\ \\ \leftarrow j\text{th coordinate position,} \\ \\ \\ \end{matrix}$$

the dots standing for zeros. We define $E_{ii} = 0$. For $f \in C^1(\partial\Omega)$, we denote its derivative in direction of E_{ij} by $\partial_{ij}[f]$; that is,

$$\partial_{ij}[f](x) = \begin{cases} \left[\dfrac{d}{dt} f(c_{ij}(t))\right]_{t=0} & \text{if } i \neq j \\ 0 & \text{if } i = j, \end{cases}$$

where $c_{ij}(t)$ is any curve lying in $\partial\Omega$, which satisfies $c_{ij}(0) = x$ and $\frac{d}{dt}c_{ij}(0) = E_{ij}$. It turns out that if f has been extended to $\mathbb{R}^n$, then

$$\partial_{ij}[f] = \frac{\partial f}{\partial x_i}\nu_j - \frac{\partial f}{\partial x_j}\nu_i = (df \wedge \nu)_{ij}\,.$$

Let us denote by $\mathbb{I}_x(\cdot,\cdot)$ the second fundamental form of the hypersurface $\partial\Omega$ at x. We recall that given two tangent vectors Y and Z, the second fundamental form is defined by

$$\mathbb{I}_x(Y,Z) = -\left\langle \left[\frac{d}{dt}\nu(c_Y(t))\right]_{t=0};Z\right\rangle,$$

where c_Y is any curve lying in $\partial\Omega$, which satisfies

$$c_Y(0) = x \quad \text{and} \quad \frac{d}{dt}c_Y(0) = Y.$$

Recall also that the second fundamental form is a symmetric bilinear form. A straightforward consequence of our definition is the following identity:

$$\mathbb{I}_x(E_{ij},E_{kl}) = \partial_{ij}[\nu_l]\nu_k - \partial_{ij}[\nu_k]\nu_l = \partial_{kl}[\nu_j]\nu_i - \partial_{kl}[\nu_i]\nu_j\,. \tag{5.10}$$

With this notation, we can prove the following.

Proposition 5.10. *Let Ω be a bounded open C^2 set of $\mathbb{R}^n$ with exterior unit normal ν. Then for every k-form α, the expressions $\nu \wedge L^\nu(\alpha)$ and $\nu \lrcorner K^\nu(\alpha)$ depend only on the values of ν on $\partial\Omega$. In particular, the following formula holds:*

$$\nu \wedge L^\nu(\alpha) = \sum_I \sum_{\gamma=1}^{k} \sum_{1\le r<s\le n} (-1)^\gamma \alpha_I \partial_{rs}[\nu_{i_\gamma}] dx^r \wedge dx^s \wedge dx^{I\setminus\{i_\gamma\}}.$$

Proof. The fact that $\nu \wedge L^\nu$ and $\nu \lrcorner K^\nu$ do not depend on the extension of ν onto a neighborhood of $\partial\Omega$ follows from the definition of L^ν and K^ν and Theorem 3.23. Thus, we only have to show the formula. Observe that

$$\begin{aligned}\nu \wedge L^\nu(\alpha) &= \sum_I \alpha_I\, \nu \wedge d\left(\nu \lrcorner dx^I\right)\\ &= \sum_I \alpha_I\, \nu \wedge d\left(\sum_{\gamma=1}^{k}(-1)^{\gamma-1}\nu_{i_\gamma} dx^{i_1}\wedge\cdots\wedge\widehat{dx^{i_\gamma}}\wedge\cdots\wedge dx^{i_k}\right)\end{aligned}$$

and, thus,

$$\begin{aligned}&\nu \wedge L^\nu(\alpha)\\ &= \sum_I \alpha_I\, \nu \wedge \left(\sum_{\gamma=1}^{k}\sum_{s=1}^{n}(-1)^{\gamma-1}\frac{\partial \nu_{i_\gamma}}{\partial x_s} dx^s \wedge dx^{i_1}\wedge\cdots\wedge\widehat{dx^{i_\gamma}}\wedge\cdots\wedge dx^{i_k}\right)\\ &= \sum_I \alpha_I \sum_{\gamma=1}^{k}\sum_{r,s=1}^{n}(-1)^{\gamma-1}\frac{\partial \nu_{i_\gamma}}{\partial x_s}\nu_r\, dx^r\wedge dx^s \wedge dx^{i_1}\wedge\ldots\wedge\widehat{dx^{i_\gamma}}\wedge\cdots\wedge dx^{i_k}.\end{aligned}$$

We now split the sum over the r and s as

$$\sum_{r,s=1}^{n} = \sum_{r<s} + \sum_{r>s}.$$

In the second sum of these two, we interchange the roles of r and s. Recalling that $dx^r \wedge dx^s = -dx^s \wedge dx^r$, the desired formula follows. □

Proposition 5.11. *Let $\Omega \subset \mathbb{R}^n$ be a bounded open C^2 set and α be a 0-form. Then*

$$K^\nu(\alpha) = [(n-1)\,\kappa]\,\alpha,$$

where κ is the mean curvature of the hypersurface $\partial\Omega$.

Proof. Recall that the exterior unit normal ν has been extended on a neighborhood of $\partial\Omega$ so that $|\nu| = 1$. Let α be a 0-form. Due to the definition of K^ν, we obtain for a zero form α,

$$K^\nu(\alpha) = \alpha\,\delta\nu = \alpha \operatorname{div}\nu.$$

Since the divergence of ν is equal (see, for instance, Krantz and Parks [61]) to $(n-1)\kappa$, if $|\nu| = 1$ near $\partial\Omega$, the proposition follows. □

Using Proposition 5.11 and doing some manipulations on $L^\nu(\omega)$, we can rewrite Theorem 5.7 in the case of 1-forms as follows.

Corollary 5.12 (General identity for 1-forms). *Let $\Omega \subset \mathbb{R}^n$ be a bounded open C^2 set with exterior unit normal ν. Every $\alpha, \beta \in C^1(\overline{\Omega};\Lambda^1)$ satisfy*

$$\begin{aligned}
&\int_\Omega (\langle d\alpha; d\beta\rangle + \langle\delta\alpha;\delta\beta\rangle - \langle\nabla\alpha;\nabla\beta\rangle) \\
&\quad = -\int_{\partial\Omega} (\langle\nu\wedge d(\nu\lrcorner\alpha);\nu\wedge\beta\rangle + \langle\nu\lrcorner\delta(\nu\wedge\alpha);\nu\lrcorner\beta\rangle) \\
&\qquad + \int_{\partial\Omega} (\langle B^\nu(\nu\wedge\alpha);\nu\wedge\beta\rangle + (n-1)\,\kappa\,\langle\nu\lrcorner\alpha;\nu\lrcorner\beta\rangle),
\end{aligned}$$

where κ is the mean curvature and B^ν acts on 2-forms and is defined by

$$B^\nu(\omega) = -\sum_{i<j}\left(\sum_{r,s=1}^{n} \partial_{ij}[\nu_r]\,\nu_s\,\omega_{sr}\right) dx^i \wedge dx^j.$$

In particular, if $\nu\wedge\alpha = \nu\wedge\beta = 0$, then

$$\int_\Omega (\langle d\alpha; d\beta\rangle + \langle\delta\alpha;\delta\beta\rangle - \langle\nabla\alpha;\nabla\beta\rangle) = \int_{\partial\Omega} (n-1)\,\kappa\,\langle\alpha;\beta\rangle.$$

Proof. Let ν be extended to a neighborhood of U of $\partial\Omega$ such that $|\nu| = 1$ in U. Note that (cf. (2.7) in Proposition 2.16)

$$\nu\wedge\alpha = \nu\wedge(\nu\lrcorner(\nu\wedge\alpha)).$$

Due to Lemma 5.5 we obtain

$$\nu \wedge L^{\nu}(\alpha) = L^{\nu}(\nu \wedge \alpha) = L^{\nu}(\nu \wedge (\nu \lrcorner (\nu \wedge \alpha))) = \nu \wedge L^{\nu}(\nu \lrcorner (\nu \wedge \alpha)). \quad (5.11)$$

Let $\omega = \nu \wedge \alpha$. Applying Proposition 2.12 yields

$$(\nu \lrcorner \omega)_i = \sum_{t=1}^{n} \omega_{ti}\, \nu_t \, .$$

Setting this into the formula given by Proposition 5.10 gives

$$\nu \wedge L^{\nu}(\alpha) = \nu \wedge L^{\nu}(\nu \lrcorner \omega) = -\sum_{i=1}^{n} \sum_{r<s} \sum_{t=1}^{n} \omega_{ti} \nu_t \partial_{rs}[\nu_i] dx^r \wedge dx^s = B^{\nu}(\omega).$$

We see now that the corollary follows from Theorem 5.7 together with Proposition 5.11. □

We present another possibility to express Theorem 5.7 for 1-forms.

Proposition 5.13. *Let $\alpha = \sum \alpha_i dx^i$ and $\beta = \sum \beta_i dx^i$ be 1-forms. Then for every $x \in \partial\Omega$, the identity*

$$\langle \nu \wedge L^{\nu}(\alpha); \nu \wedge \beta \rangle + \langle \nu \lrcorner K^{\nu}(\alpha); \nu \lrcorner \beta \rangle = -\sum_{i,j=1}^{n} \alpha_i \beta_j \sum_{r=1}^{n} \mathbb{I}_x(E_{ir}, E_{jr})$$

is valid, where $\mathbb{I}_x$ is the second fundamental form of the hypersurface $\partial\Omega$ at x.

Remark 5.14. It can be shown that also for $k > 1$ and any k-form ω,

$$\langle \nu \wedge L^{\nu}(\omega); \nu \wedge \omega \rangle + \langle \nu \lrcorner K^{\nu}(\omega); \nu \lrcorner \omega \rangle$$

can be expressed in terms of the second fundamental form and the tangent vectors E_{ij}; see Csató [23]. However, the formulas in the case $k > 1$ do not turn out to be as simple and elegant as for 1-forms.

Proof. Proposition 5.10 gives

$$\nu \wedge L^{\nu}(\alpha) = \sum_{j=1}^{n} \sum_{i<r} \alpha_j \partial_{ri}[\nu_j] dx^i \wedge dx^r$$

and, hence,

$$\begin{aligned}\langle \nu \wedge L^{\nu}(\alpha); \nu \wedge \beta \rangle &= \sum_{j=1}^{n} \sum_{i<r} \alpha_j \partial_{ri}[\nu_j](\nu_i \beta_r - \nu_r \beta_i) \\ &= \sum_{j=1}^{n} \sum_{i<r} \alpha_j \beta_r \partial_{ri}[\nu_j] \nu_i - \sum_{j=1}^{n} \sum_{i<r} \beta_i \alpha_j \partial_{ri}[\nu_j] \nu_r \, .\end{aligned}$$

Interchanging the roles of i and r in the second sum and noticing that $\partial_{ir} = -\partial_{ri}$ gives

$$\langle \nu \wedge L^{\nu}(\alpha); \nu \wedge \beta \rangle = -\sum_{i,j,r} \alpha_j \beta_i \partial_{ri}[\nu_j] \nu_r \,. \tag{5.12}$$

We know from Lemma 5.5 and the definition of K^{ν} that

$$\langle \nu \lrcorner K^{\nu}(\alpha); \nu \lrcorner \beta \rangle = \sum_{i,j,r} \alpha_j \beta_i \nu_i \nu_j \frac{\partial \nu_r}{\partial x_r}$$

for any ν, whose extension satisfies $|\nu| = 1$. This implies in particular that

$$\sum_r \nu_r \frac{\partial \nu_r}{\partial x_i} = 0 \quad \text{for every } 1 \leq i \leq n.$$

We therefore get

$$\begin{aligned} \langle \nu \lrcorner K^{\nu}(\alpha); \nu \lrcorner \beta \rangle &= \sum_{i,j,r} \alpha_j \beta_i \nu_i \nu_j \frac{\partial \nu_r}{\partial x_r} - \sum_{i,j,r} \alpha_j \beta_i \nu_r \nu_j \frac{\partial \nu_r}{\partial x_i} \\ &= \sum_{i,j,r} \alpha_j \beta_i \partial_{ri}[\nu_r] \nu_j \,. \end{aligned} \tag{5.13}$$

Adding (5.12) to (5.13) gives the desired result, using the identity (5.10). □

5.3 Gaffney Inequality

5.3.1 An Elementary Proof

The Gaffney inequality is essentially based on the fact that the first boundary integral in Theorem 5.7 drops, whenever $\alpha = \beta = \omega$ and one of the conditions $\nu \wedge \omega = 0$ or $\nu \lrcorner \omega = 0$ is satisfied.

The Gaffney inequality will be essential for the proof of the Hodge–Morrey decomposition theorem. Due to Theorem 5.7, the proof of the Gaffney inequality will be very simple (cf. also Duvaut and Lions [39]) (see Theorem 6.1 in Chapter 7) or Schwarz [89]. The inequality was first proved by Gaffney in [44] and [45] for compact manifolds without boundary. The generalization of that proof to the case of manifolds with boundary proved by Morrey [77] in Chapter 7 (cf. also Iwaniec, Scott and Stroffolini [58]) is different from the one presented here. A third version of the proof can be found in Taylor [96, Chapter 5.9], which involves more geometric arguments.

To obtain the Gaffney inequality from Theorem 5.7 we need the following elementary result.

Proposition 5.15. *Let $\Omega \subset \mathbb{R}^n$ be a bounded open C^2 set. Then there exists $C = C(\Omega) > 0$ such that for any $0 < \varepsilon < 1$,*

$$\int_{\partial\Omega} u^2 \leq \varepsilon \int_\Omega |\nabla u|^2 + \frac{C}{\varepsilon} \int_\Omega |u|^2$$

for every $u \in W^{1,2}(\Omega)$.

Proof. Due to the density of $C^1(\overline{\Omega})$ in $W^{1,2}(\Omega)$ and the continuous imbedding $W^{1,2}(\Omega) \hookrightarrow L^2(\partial\Omega)$, it suffices to show the inequality for every $u \in C^1(\overline{\Omega})$. As Ω is a C^2 set, we can extend the exterior unit normal ν to a $C^1(\overline{\Omega};\mathbb{R}^n)$ map. Hence, $|\nu|$ and the divergence $|\operatorname{div}\nu|$ are bounded on Ω by some $C > 0$. Using the divergence theorem, we get

$$\begin{aligned}
\int_{\partial\Omega} u^2 &= \int_{\partial\Omega} u^2 \sum_{i=1}^n \nu_i^2 = \int_{\partial\Omega} \langle u^2\nu;\nu\rangle \\
&= \int_\Omega \operatorname{div}(u^2\nu) = \int_\Omega u^2 \operatorname{div}\nu + \int_\Omega \langle \nu;\nabla u^2\rangle \\
&\leq \int_\Omega |\operatorname{div}\nu| u^2 + \int_\Omega |\nu|\,|\nabla u^2| \leq C\int_\Omega u^2 + C\int_\Omega 2|u|\,|\nabla u|.
\end{aligned}$$

Since

$$2C|u|\,|\nabla u| \leq \varepsilon|\nabla u|^2 + \frac{C^2}{\varepsilon}u^2,$$

we have the desired result. □

Theorem 5.16 (Gaffney inequality). *Let $\Omega \subset \mathbb{R}^n$ be a bounded open C^2 set. Then there exists a constant $C = C(\Omega) > 0$ such that*

$$\|\omega\|^2_{W^{1,2}} \leq C\left(\|d\omega\|^2_{L^2} + \|\delta\omega\|^2_{L^2} + \|\omega\|^2_{L^2}\right)$$

for every $\omega \in W_T^{1,2}(\Omega;\Lambda^k) \cup W_N^{1,2}(\Omega;\Lambda^k)$.

Proof. By density (cf. Theorem 3.25), it is enough to prove the result for $\omega \in C_T^1(\overline{\Omega};\Lambda^k) \cup C_N^1(\overline{\Omega};\Lambda^k)$. Appealing to Theorem 5.7 and the properties of L^ν and K^ν, there exist continuous functions $f_{IJ} \in C^0(\partial\Omega)$, depending only on the geometry of $\partial\Omega$ and on k, such that

$$\int_\Omega \left(|d\omega|^2 + |\delta\omega|^2\right) = \int_\Omega |\nabla\omega|^2 + \int_{\partial\Omega} \sum_{I,J} f_{IJ}\,\omega_I\,\omega_J$$

for every $\omega \in C_T^1(\overline{\Omega};\Lambda^k) \cup C_N^1(\overline{\Omega};\Lambda^k)$. In particular, since $\partial\Omega$ is compact, there exists a constant $C = C(\Omega) > 0$ such that

$$\int_\Omega \left(|d\omega|^2 + |\delta\omega|^2\right) \geq \int_\Omega |\nabla\omega|^2 - C\int_{\partial\Omega} |\omega|^2. \tag{5.14}$$

Combining this with Proposition 5.15, we have the Gaffney inequality. □

5.3.2 A Generalization of the Boundary Condition

We just saw that the proof of the Gaffney inequality is essentially based on the fact that the first boundary integral in Theorem 5.7 drops whenever $\alpha = \beta = \omega$ and the tangential or normal component of ω vanishes. In that case, no derivatives of ω occur in the boundary integral and one obtains the estimate (5.14). We now discuss the possibility of extending Theorem 5.16 to more general conditions than those of vanishing tangential or normal components. We give in Theorem 5.19 two ways of generalizing the Gaffney inequality. However, before proceeding further we need the following algebraic lemma.

Lemma 5.17. *(i) Let $2k \leq n$, k odd,*

$$\omega \in C^1\left(\mathbb{R}^n;\Lambda^k\right), \quad \nu \in C^1\left(\mathbb{R}^n;\Lambda^1\right) \quad \text{and} \quad \lambda \in C^1\left(\mathbb{R}^n;\Lambda^{n-2k}\right)$$

such that

$$*[\nu \wedge \omega] = \lambda \wedge (\nu \lrcorner \omega).$$

Then

$$\langle \nu \wedge d(\nu \lrcorner \omega); \nu \wedge \omega\rangle + \langle \nu \lrcorner \delta(\nu \wedge \omega); \nu \lrcorner \omega\rangle = -\langle \nu \wedge d\lambda \wedge (\nu \lrcorner \omega); *(\nu \lrcorner \omega)\rangle.$$

(ii) Let $2k \geq n$, $(n-k)$ odd,

$$\omega \in C^1\left(\mathbb{R}^n;\Lambda^k\right), \quad \nu \in C^1\left(\mathbb{R}^n;\Lambda^1\right) \quad \text{and} \quad \lambda \in C^1\left(\mathbb{R}^n;\Lambda^{2k-n}\right)$$

such that

$$\nu \lrcorner \omega = \lambda \wedge *(\nu \wedge \omega).$$

Then

$$\langle \nu \wedge d(\nu \lrcorner \omega); \nu \wedge \omega\rangle + \langle \nu \lrcorner \delta(\nu \wedge \omega); \nu \lrcorner \omega\rangle = \langle \nu \wedge d\lambda \wedge (*(\nu \wedge \omega)); \nu \wedge \omega\rangle.$$

Proof. Step 1. We first show that if ω is a k-form, ν is a 1-form and $dV = dx^1 \wedge \cdots \wedge dx^n$, we have

$$\begin{aligned} &\langle \nu \wedge d(\nu \lrcorner \omega); \nu \wedge \omega\rangle dV + \langle \nu \lrcorner \delta(\nu \wedge \omega); \nu \lrcorner \omega\rangle dV \\ &\quad = \nu \wedge \left[d(\nu \lrcorner \omega) \wedge (*(\nu \wedge \omega)) + (-1)^k (\nu \lrcorner \omega) \wedge d(*(\nu \wedge \omega))\right]. \end{aligned} \tag{5.15}$$

Indeed, from the definitions of the interior product and the interior derivative and from Theorem 2.10, we have

$$\begin{aligned} \nu \lrcorner \delta(\nu \wedge \omega) &= (-1)^{n(k-1)} * [\nu \wedge (*(\delta(\nu \wedge \omega)))] \\ &= (-1)^{nk+n(k-1)} * [\nu \wedge (** (d * (\nu \wedge \omega)))] \\ &= (-1)^{nk+n(k-1)+k(n-k)} * [\nu \wedge (d * (\nu \wedge \omega))] \end{aligned}$$

and, thus, using again Theorem 2.10,

$$\begin{aligned}
&\langle \nu\wedge d(\nu\lrcorner\omega);\nu\wedge\omega\rangle dV+\langle \nu\lrcorner\delta(\nu\wedge\omega);\nu\lrcorner\omega\rangle dV\\
&\quad=\nu\wedge d(\nu\lrcorner\omega)\wedge(*(\nu\wedge\omega))+(\nu\lrcorner\omega)\wedge[**(\nu\wedge d(*(\nu\wedge\omega)))](-1)^{n+k+nk}\\
&\quad=\nu\wedge d(\nu\lrcorner\omega)\wedge(*(\nu\wedge\omega))-(\nu\lrcorner\omega)\wedge\nu\wedge d(*(\nu\wedge\omega))\\
&\quad=\nu\wedge\left[d(\nu\lrcorner\omega)\wedge(*(\nu\wedge\omega))+(-1)^{k}(\nu\lrcorner\omega)\wedge d(*(\nu\wedge\omega))\right].
\end{aligned}$$

We have therefore obtained (5.15).

Step 2. We first prove (i). We set the equality $*[\nu\wedge\omega]=\lambda\wedge(\nu\lrcorner\omega)$ into the right-hand side of (5.15), which yields, since k is odd,

$$\begin{aligned}
&\nu\wedge\left[d(\nu\lrcorner\omega)\wedge\lambda\wedge(\nu\lrcorner\omega)-(\nu\lrcorner\omega)\wedge d(\lambda\wedge(\nu\lrcorner\omega))\right]\\
&\quad=-\nu\wedge(\nu\lrcorner\omega)\wedge d\lambda\wedge(\nu\lrcorner\omega)+\nu\wedge A,
\end{aligned}$$

where

$$A=d(\nu\lrcorner\omega)\wedge\lambda\wedge(\nu\lrcorner\omega)-(-1)^{n-2k}(\nu\lrcorner\omega)\wedge\lambda\wedge d(\nu\lrcorner\omega).$$

Using again that k is odd, we have that $A=0$. It therefore follows from (5.15) and the above two identities that

$$\begin{aligned}
&\langle \nu\wedge d(\nu\lrcorner\omega);\nu\wedge\omega\rangle dV+\langle \nu\lrcorner\delta(\nu\wedge\omega);\nu\lrcorner\omega\rangle dV\\
&\quad=-\nu\wedge(\nu\lrcorner\omega)\wedge d\lambda\wedge(\nu\lrcorner\omega)=-\nu\wedge d\lambda\wedge(\nu\lrcorner\omega)\wedge(**(\nu\lrcorner\omega))\\
&\quad=-\langle \nu\wedge d\lambda\wedge(\nu\lrcorner\omega);*(\nu\lrcorner\omega)\rangle dV.
\end{aligned}$$

Step 3. The proof of (ii) is analogous to that of (i) by setting the equality

$$\nu\lrcorner\omega=\lambda\wedge*(\nu\wedge\omega)$$

into the right-hand side of (5.15). □

Remark 5.18. The hypothesis of the lemma can be relaxed. It follows from Step 2 of the proof that the lemma remains valid at all points x where the following two equations hold true:

$$\begin{aligned}
*[\nu\wedge\omega]&=\lambda\wedge(\nu\lrcorner\omega),\\
\nu\wedge d(*[\nu\wedge\omega])&=\nu\wedge d(\lambda\wedge(\nu\lrcorner\omega)).
\end{aligned}$$

If the first identity is true not just at one point x but also in an open set, then it trivially implies the second one. This implication remains valid if ν is the exterior unit normal on some sufficiently regular hypersurface and the first identity holds true on that surface, due to Theorem 3.23. We will use the lemma exactly in this setting.

Theorem 5.19. *Let $0 \leq k \leq n$ be integers and $\Omega \subset \mathbb{R}^n$ be a bounded open C^2 set with exterior unit normal ν.*

(i) Let $2k \leq n$ with k odd. Let $\lambda \in C^1\left(\partial\Omega;\Lambda^{n-2k}\right)$. Then there exists a constant $C = C(\lambda,\Omega)$ such that

$$\|\nabla\omega\|_{W^{1,2}}^2 \leq C\left(\|d\omega\|_{L^2}^2 + \|\delta\omega\|_{L^2}^2 + \|\omega\|_{L^2}^2\right)$$

for every $\omega \in C^1(\overline{\Omega};\Lambda^k)$ satisfying

$$*(\nu\wedge\omega) = \lambda\wedge(\nu\lrcorner\omega) \quad on\ \partial\Omega.$$

(ii) Let $2k \geq n$ with $(n-k)$ odd. Let $\lambda \in C^1\left(\partial\Omega;\Lambda^{2k-n}\right)$. Then there exists a constant $C = C(\lambda,\Omega)$ such that

$$\|\nabla\omega\|_{W^{1,2}}^2 \leq C\left(\|d\omega\|_{L^2}^2 + \|\delta\omega\|_{L^2}^2 + \|\omega\|_{L^2}^2\right)$$

for every $\omega \in C^1(\overline{\Omega};\Lambda^k)$ verifying

$$\nu\lrcorner\omega = \lambda\wedge*(\nu\wedge\omega) \quad on\ \partial\Omega.$$

Proof. We prove (i). The proof of (ii) is completely analogous. Due to Theorem 5.7, Lemma 5.17(i) and the remark thereafter,

$$\begin{aligned}&\int_\Omega |d\omega|^2 + \int_\Omega |\delta\omega|^2 - \int_\Omega |\nabla\omega|^2\\ &\quad = \int_{\partial\Omega} \left(\langle\nu\wedge d\lambda\wedge(\nu\lrcorner\omega);*(\nu\lrcorner\omega)\rangle + \langle L^\nu(\nu\wedge\omega);\nu\wedge\omega\rangle + \langle K^\nu(\nu\lrcorner\omega);\nu\lrcorner\omega\rangle\right).\end{aligned}$$

The regularity assumption $\lambda \in C^1\left(\partial\Omega;\Lambda^{n-2k}\right)$ implies that $\nu\wedge d\lambda$, which is well defined by Theorem 3.23, is a continuous function on $\partial\Omega$. One can now proceed exactly as in the proof of the Gaffney inequality. □

We give the following example to part (i) of Theorem 5.19.

Example 5.20. Let $k = 1$ and $n \geq 3$. Hence, $n-2k = n-2$. It will be more convenient to use $*\lambda$ than λ, so we suppose that $*\lambda \in C^1\left(\partial\Omega;\Lambda^{n-2}\right)$. In that case, the condition $*(\nu\wedge\omega) = (*\lambda)\wedge(\nu\lrcorner\omega)$ can be written as

$$\nu\wedge\omega = (\nu\lrcorner\omega)\lambda \quad \text{on } \partial\Omega,$$

which consists of the $\binom{n}{2}$ equations

$$\nu_i\,\omega_j - \nu_j\,\omega_i = \lambda_{ij}\sum_{l=1}^n \nu_l\,\omega_l \quad \text{for } 1 \leq i < j \leq n.$$

To make the example even simpler, assume that

$$H = \partial\Omega\cap\{x\in\mathbb{R}^n : x_n = 0\}$$

contains a relatively open set. Furthermore, suppose that, for every $x \in H$,

$$\lambda_{ij}(x) = 0 \quad \text{if } j \neq n \text{ and } 1 \leq i < j.$$

Since $\nu = (0, \ldots, 0, 1)$ at every $x \in H$, we have

$$\nu \wedge \omega = 0 \quad \Leftrightarrow \quad \omega_1 = \cdots = \omega_{n-1} = 0,$$

$$\nu \lrcorner \omega = 0 \quad \Leftrightarrow \quad \omega_n = 0,$$

whereas

$$\nu \wedge \omega = (\nu \lrcorner \omega)\lambda \quad \Leftrightarrow \quad \omega_i + \lambda_{in}\omega_n = 0 \quad \text{for} \quad 1 \leq i \leq n-1.$$

5.3.3 Gaffney-Type Inequalities in L^p and Hölder Spaces

The following result follows from Theorem 2 in Bolik [13].

Theorem 5.21. *Let $n > 2$, $r \geq 1$ and $1 \leq k \leq n-1$ be integers and $0 < \alpha < 1 < p < \infty$. Let $\Omega \subset \mathbb{R}^n$ be a bounded open smooth set with exterior unit normal ν. Then there exist $C_1 = C_1(r, p, \Omega)$ and $C_2 = C_2(r, \alpha, \Omega)$ such that for every $\omega \in W^{r,p}(\Omega; \Lambda^k)$,*

$$\|\omega\|_{W^{r,p}(\Omega)} \leq C_1 \left(\|d\omega\|_{W^{r-1,p}(\Omega)} + \|\delta\omega\|_{W^{r-1,p}(\Omega)} \right) + C_1 \left(\|\nu \wedge \omega\|_{W^{r-\frac{1}{p},p}(\partial\Omega)} + \|\nu \lrcorner \omega\|_{L^1(\partial\Omega)} \right),$$

$$\|\omega\|_{W^{r,p}(\Omega)} \leq C_1 \left(\|d\omega\|_{W^{r-1,p}(\Omega)} + \|\delta\omega\|_{W^{r-1,p}(\Omega)} \right) + C_1 \left(\|\nu \lrcorner \omega\|_{W^{r-\frac{1}{p},p}(\partial\Omega)} + \|\nu \wedge \omega\|_{L^1(\partial\Omega)} \right),$$

whereas for every $\omega \in C^{r,\alpha}(\overline{\Omega}; \Lambda^k)$,

$$\|\omega\|_{C^{r,\alpha}(\overline{\Omega})} \leq C_2 \left(\|d\omega\|_{C^{r-1,\alpha}(\overline{\Omega})} + \|\delta\omega\|_{C^{r-1,\alpha}(\overline{\Omega})} \right) + C_2 \left(\|\nu \wedge \omega\|_{C^{r,\alpha}(\partial\Omega)} + \|\nu \lrcorner \omega\|_{C^0(\partial\Omega)} \right),$$

$$\|\omega\|_{C^{r,\alpha}(\overline{\Omega})} \leq C_2 \left(\|d\omega\|_{C^{r-1,\alpha}(\overline{\Omega})} + \|\delta\omega\|_{C^{r-1,\alpha}(\overline{\Omega})} \right) + C_2 \left(\|\nu \lrcorner \omega\|_{C^{r,\alpha}(\partial\Omega)} + \|\nu \wedge \omega\|_{C^0(\partial\Omega)} \right).$$

Remark 5.22. (i) We see that this is a generalization of the Gaffney inequality (cf. Theorem 5.16). Indeed, if $\omega \in W_T^{1,2}(\Omega; \Lambda^k) \cup W_N^{1,2}(\Omega; \Lambda^k)$, then the first two inequalities reduce to

$$\|\omega\|_{W^{1,2}(\Omega)} \leq C\left(\|d\omega\|_{L^2(\Omega)} + \|\delta\omega\|_{L^2(\Omega)} + \|\omega\|_{L^1(\partial\Omega)}\right).$$

(ii) The actual theorem, as stated in Bolik [13], is more precise. Before explaining the refinement, we need to introduce the spaces

$$\mathscr{H}_T\left(\Omega^c;\Lambda^k\right) \quad \text{and} \quad \mathscr{H}_N\left(\Omega^c;\Lambda^k\right), \tag{5.16}$$

where Ω^c is the complement of Ω in $\mathbb{R}^n$. In Definition 6.1, we will introduce the sets

$$\mathscr{H}_T\left(\Omega;\Lambda^k\right) \quad \text{and} \quad \mathscr{H}_N\left(\Omega;\Lambda^k\right).$$

In Theorem 6.5, it will be proved that

$$\dim \mathscr{H}_T\left(\Omega;\Lambda^k\right) < \infty \quad \text{and} \quad \dim \mathscr{H}_N\left(\Omega;\Lambda^k\right) < \infty.$$

The sets in (5.16) are defined analogously, requiring that $\omega(x) \to 0$ uniformly as $|x| \to \infty$. In Kress [63] (cf. Satz 7.5), it is proved that

$$B_{n-k} = \dim \mathscr{H}_T\left(\Omega;\Lambda^k\right) = \dim \mathscr{H}_N\left(\Omega^c;\Lambda^{k-1}\right), \quad 1 \leq k \leq n,$$
$$B_k = \dim \mathscr{H}_N\left(\Omega;\Lambda^k\right) = \dim \mathscr{H}_T\left(\Omega^c;\Lambda^{k+1}\right), \quad 0 \leq k \leq n-1,$$

where B_k are the Betti numbers (for more details, see Duff and Spencer [38] or Kress [63]). Let z^i, $i = 1,\ldots,B_{n-k}$, be a basis of $\mathscr{H}_N\left(\Omega^c;\Lambda^{k-1}\right)$ and y^i, $i = 1,\ldots,B_k$, be a basis of $\mathscr{H}_T\left(\Omega^c;\Lambda^{k+1}\right)$. The sharper version of Theorem 5.21 is now obtained by replacing in the first inequality the term $\|\nu \lrcorner \omega\|_{L^1(\partial\Omega)}$ (or the term $\|\nu \lrcorner \omega\|_{C^0(\partial\Omega)}$ in the third inequality) by

$$\sum_{i=1}^{B_{n-k}} \left|\int_{\partial\Omega} \langle \omega; \nu \wedge z^i \rangle\right| = \sum_{i=1}^{B_{n-k}} \left|\int_{\partial\Omega} \langle \nu \lrcorner \omega; z^i \rangle\right|$$

and replacing in the second inequality $\|\nu \wedge \omega\|_{L^1(\partial\Omega)}$ (or the term $\|\nu \wedge \omega\|_{C^0(\partial\Omega)}$ in the fourth inequality) by

$$\sum_{i=1}^{B_k} \left|\int_{\partial\Omega} \langle \omega; \nu \lrcorner y^i \rangle\right| = \sum_{i=1}^{B_k} \left|\int_{\partial\Omega} \langle \nu \wedge \omega; y^i \rangle\right|.$$

To obtain Theorem 5.21 from [13], we have estimated these terms by taking into account that the z^i and y^i are smooth up to the boundary, according to a result similar to Theorem 6.3.

(iii) Note that if Ω is contractible, $1 \leq k \leq n-1$ and $\omega \in W_T^{r,p}\left(\Omega;\Lambda^k\right) \cup W_N^{r,p}\left(\Omega;\Lambda^k\right)$, then

$$\|\omega\|_{W^{r,p}(\Omega)} \leq C_1\left(\|d\omega\|_{W^{r-1,p}(\Omega)} + \|\delta\omega\|_{W^{r-1,p}(\Omega)}\right),$$

and, similarly, if $\omega \in C_T^{r,\alpha}(\overline{\Omega};\Lambda^k) \cup C_N^{r,\alpha}(\overline{\Omega};\Lambda^k)$, then

$$\|\omega\|_{C^{r,\alpha}(\overline{\Omega})} \leq C_2 \left(\|d\omega\|_{C^{r-1,\alpha}(\overline{\Omega})} + \|\delta\omega\|_{C^{r-1,\alpha}(\overline{\Omega})} \right).$$

This follows from the previous remark and Theorem 6.5, since if Ω is contractible and $1 \leq k \leq n-1$, then $\mathscr{H}_T(\Omega;\Lambda^k) = \mathscr{H}_N(\Omega;\Lambda^k) = \{0\}$.

Chapter 6
The Hodge–Morrey Decomposition

6.1 Properties of Harmonic Fields

We recall the definition of harmonic fields and of contractible sets. Let $0 \leq k \leq n$ be an integer.

Definition 6.1. (i) The set of *harmonic fields* is defined by

$$\mathscr{H}\left(\Omega;\Lambda^k\right) = \left\{\omega \in W^{1,2}\left(\Omega;\Lambda^k\right) : d\omega = 0 \text{ and } \delta\omega = 0\right\},$$

and we will write

$$\mathscr{H}_T(\Omega;\Lambda^k) = \mathscr{H}(\Omega;\Lambda^k) \cap W_T^{1,2}(\Omega;\Lambda^k),$$
$$\mathscr{H}_N(\Omega;\Lambda^k) = \mathscr{H}(\Omega;\Lambda^k) \cap W_N^{1,2}(\Omega;\Lambda^k);$$

that is,

$$\mathscr{H}_T\left(\Omega;\Lambda^k\right) = \left\{\omega \in \mathscr{H}\left(\Omega;\Lambda^k\right) : \nu \wedge \omega = 0 \text{ on } \partial\Omega\right\},$$
$$\mathscr{H}_N\left(\Omega;\Lambda^k\right) = \left\{\omega \in \mathscr{H}\left(\Omega;\Lambda^k\right) : \nu \lrcorner \omega = 0 \text{ on } \partial\Omega\right\}.$$

(ii) The set $\Omega \subset \mathbb{R}^n$ is said to be *contractible* if there exist $x_0 \in \Omega$ and $F \in C^\infty\left([0,1] \times \Omega;\Omega\right)$ such that for every $x \in \Omega$,

$$F(0,x) = x_0 \quad \text{and} \quad F(1,x) = x.$$

Remark 6.2. (i) Note that a contractible set is necessarily simply connected.

(ii) The set $\mathscr{H}\left(\Omega;\Lambda^k\right)$ can be equivalently defined as

$$\mathscr{H}\left(\Omega;\Lambda^k\right) = \left\{\omega \in L^1_{\text{loc}}\left(\Omega;\Lambda^k\right) : d\omega = 0 \text{ and } \delta\omega = 0\right\},$$

where we understand the equations $d\omega = 0$ and $\delta\omega = 0$ in the sense of distributions, namely

$$\int_\Omega \langle \omega; \delta\varphi \rangle = 0, \quad \text{for every } \varphi \in C_0^\infty\left(\Omega; \Lambda^{k+1}\right),$$

$$\int_\Omega \langle \omega; d\varphi \rangle = 0, \quad \text{for every } \varphi \in C_0^\infty\left(\Omega; \Lambda^{k-1}\right).$$

The proof of Theorem 6.3 shows that the two definitions are equivalent.

We now list some properties of these fields.

Theorem 6.3. *Let $\Omega \subset \mathbb{R}^n$ be an open set. Then*

$$\mathcal{H}\left(\Omega; \Lambda^k\right) \subset C^\infty\left(\Omega; \Lambda^k\right).$$

Moreover, if Ω is bounded and smooth, then

$$\mathcal{H}_T\left(\Omega; \Lambda^k\right) \cup \mathcal{H}_N\left(\Omega; \Lambda^k\right) \subset C^\infty\left(\overline{\Omega}; \Lambda^k\right).$$

Furthermore, if $r \geq 1$ is an integer, then there exists $C = C(r, \Omega)$ such that for every $\omega \in \mathcal{H}_T\left(\Omega; \Lambda^k\right) \cup \mathcal{H}_N\left(\Omega; \Lambda^k\right)$,

$$\|\omega\|_{W^{r,2}} \leq C \|\omega\|_{L^2}. \tag{6.1}$$

Remark 6.4. If $r \geq 0$ is an integer and $0 \leq \alpha \leq 1$, then there exists $C = C(r, \Omega)$ such that for every $\omega \in \mathcal{H}_T\left(\Omega; \Lambda^k\right) \cup \mathcal{H}_N\left(\Omega; \Lambda^k\right)$,

$$\|\omega\|_{C^{r,\alpha}} \leq C \|\omega\|_{C^0}.$$

Indeed, we have, by the Morrey imbedding theorem, that for s sufficiently large,

$$\|\omega\|_{C^{r,\alpha}} \leq C_1 \|\omega\|_{W^{s,2}}.$$

Since trivially

$$\|\omega\|_{L^2} \leq C_2 \|\omega\|_{C^0},$$

we have the result by combining the theorem with the above two estimates.

Proof. Step 1. The inclusion

$$\mathcal{H}\left(\Omega; \Lambda^k\right) \subset C^\infty\left(\Omega; \Lambda^k\right)$$

follows from the Weyl lemma (cf. e.g., [29]). Indeed, let $\phi \in C_0^\infty\left(\Omega; \Lambda^k\right)$;

$$\int_\Omega \langle \omega; \Delta\phi \rangle = \int_\Omega \langle \omega; \delta d\phi + d\delta\phi \rangle = \int_\Omega \langle d\omega; d\phi \rangle + \int_\Omega \langle \delta\omega; \delta\phi \rangle = 0.$$

Choose $\phi = \varphi\, dx^I$ and $\varphi \in C_0^\infty(\Omega)$ and, thus, $\omega_I \in C^\infty(\Omega)$.

Step 2. The extra statements are direct consequences of Theorem 6.11. □

In the sequel we will sometimes omit the brackets $\left(\Omega; \Lambda^k\right)$ in the expressions $W^{1,2}\left(\Omega; \Lambda^k\right)$, $\mathcal{H}_T\left(\Omega; \Lambda^k\right)$ ... whenever the degree k of the form is evident.

Theorem 6.5. *Let $\Omega \subset \mathbb{R}^n$ be a bounded open C^2 set.*

(i) The spaces $\mathscr{H}_T(\Omega;\Lambda^k)$ and $\mathscr{H}_N(\Omega;\Lambda^k)$ are finite dimensional.

(ii) The sets $\mathscr{H}_T(\Omega;\Lambda^k)$ and $\mathscr{H}_N(\Omega;\Lambda^k)$ are closed in $L^2(\Omega;\Lambda^k)$.

(iii) Furthermore, if Ω is contractible, then

$$\mathscr{H}_T(\Omega;\Lambda^k) = \{0\} \quad \text{if } 0 \le k \le n-1$$
$$\mathscr{H}_N(\Omega;\Lambda^k) = \{0\} \quad \text{if } 1 \le k \le n.$$

(iv) If $k = 0$ or $k = n$ and $h \in \mathscr{H}(\Omega;\Lambda^k)$, then h is constant on each connected component. In particular,

$$\mathscr{H}_T(\Omega;\Lambda^0) = \{0\} \quad \text{and} \quad \mathscr{H}_N(\Omega;\Lambda^n) = \{0\}.$$

(v) Let $(\mathscr{H}_T)^\perp$ be the orthogonal complement of $\mathscr{H}_T$ with respect to the L^2-inner product, then

$$L^2 = \mathscr{H}_T \oplus (\mathscr{H}_T)^\perp.$$

More precisely, for any $f \in L^2$, there exist unique $h \in \mathscr{H}_T$ and $g \in (\mathscr{H}_T)^\perp$ such that

$$f = h + g, \tag{6.2}$$

$$\|h\|_{L^2}, \ \|g\|_{L^2} \le \|f\|_{L^2}. \tag{6.3}$$

A similar result holds for $\mathscr{H}_N$.

Remark 6.6. Statement (iii) can be improved, since it is a special case of the de Rham theorem (cf. [68], for instance). For example, if $k = 1$, then

$$\mathscr{H}_T(\Omega;\Lambda^1) = \mathscr{H}_N(\Omega;\Lambda^1) = \{0\}$$

if Ω is only simply connected.

Proof. (i) We only prove the statement for $\mathscr{H}_T$; the proof for $\mathscr{H}_N$ is similar. Let

$$E = \{\omega \in \mathscr{H}_T(\Omega;\Lambda^k) : \|\omega\|_{W^{1,2}} \le 1\}.$$

If we can prove that E is compact, then the result will follow from the Riesz theorem (cf. [17]). Let $\{\omega_l\}_{l\in\mathbb{N}}$ be a sequence in E. Then there is a subsequence also denoted by ω_l which converges weakly in $W^{1,2}$ to some $\omega \in W^{1,2}(\Omega;\Lambda^k)$. The compact imbedding $W^{1,2}(\Omega) \hookrightarrow L^2(\Omega)$ implies that the convergence is strong in L^2. The Gaffney inequality (cf. Theorem 5.16) implies that the convergence is strong in $W^{1,2}$. Hence, $\omega \in E$.

(ii) We only do the proof for $\mathscr{H}_T(\Omega;\Lambda^k)$; the other case is completely analogous. Let $\{\omega_l\}_{l\in\mathbb{N}} \subset \mathscr{H}_T$ be such that $\omega_l \to \omega$ in $L^2(\Omega;\Lambda^k)$. Then $\{\omega_l\}_{l\in\mathbb{N}}$ is a Cauchy sequence in $L^2(\Omega;\Lambda^k)$. From the Gaffney inequality it follows that it is also a Cauchy sequence in $W^{1,2}(\Omega;\Lambda^k)$. So it converges also in $W^{1,2}(\Omega;\Lambda^k)$ to the same limit ω. The trace theorem for Sobolev functions yields $\omega \in \mathscr{H}_T(\Omega;\Lambda^k)$.

(iii) Let $h \in \mathcal{H}_N\left(\Omega;\Lambda^k\right)$. Since $dh=0$, $k \geq 1$ and Ω is contractible, it follows from the Poincaré lemma (Theorem 8.1) that there exists a $(k-1)$-form g such that $h = dg$. So we have

$$\int_\Omega \langle h;h\rangle = \int_\Omega \langle dg;h\rangle = -\int_\Omega \langle g;\delta h\rangle + \int_{\partial\Omega} \langle g;\nu \lrcorner h\rangle = 0.$$

This proves that $\mathcal{H}_N = \{0\}$. The claim concerning $\mathcal{H}_T$ follows by duality.

(iv) This is obvious, since a function with vanishing gradient on a connected set is constant.

(v) Since L^2 is a Hilbert space and $\mathcal{H}_T$ is closed, we have the claim. □

6.2 Existence of Minimizers and Euler–Lagrange Equation

Let $0 \leq k \leq n$. We now establish the first step in the Hodge–Morrey decomposition. Recall that

$$W_T^{1,2}(\Omega;\Lambda^k) = \{f \in W^{1,2}(\Omega;\Lambda^k) : \nu \wedge f = 0 \text{ on } \partial\Omega\}.$$

Theorem 6.7 (Existence of minimizer). *Let $\Omega \subset \mathbb{R}^n$ be a bounded open C^2 set. Let $g \in L^2(\Omega;\Lambda^k)$ and*

$$D_g(\omega) = \int_\Omega \left(\frac{1}{2}|d\omega|^2 + \frac{1}{2}|\delta\omega|^2 + \langle g;\omega\rangle\right),$$

$$X = W_T^{1,2}(\Omega;\Lambda^k) \cap (\mathcal{H}_T(\Omega;\Lambda^k))^\perp.$$

Then there exists a unique $\overline{\omega} \in X$ such that

$$D_g(\overline{\omega}) \leq D_g(\omega) \quad \text{for every } \omega \in X$$

and satisfying the weak form of the Euler–Lagrange equation, namely

$$\int_\Omega \langle d\overline{\omega};du\rangle + \int_\Omega \langle \delta\overline{\omega};\delta u\rangle = -\int_\Omega \langle g;u\rangle \quad \text{for every } u \in X. \tag{6.4}$$

Moreover, if $g \in \left(\mathcal{H}_T(\Omega;\Lambda^k)\right)^\perp$, then $\overline{\omega}$ verifies

$$\int_\Omega \langle d\overline{\omega};du\rangle + \int_\Omega \langle \delta\overline{\omega};\delta u\rangle = -\int_\Omega \langle g;u\rangle \quad \text{for every } u \in W_T^{1,2}(\Omega;\Lambda^k). \tag{6.5}$$

The same result holds true by replacing everywhere the subscript T by N.

Remark 6.8. It can be easily shown that the correspondence $g \to \overline{\omega}$ is linear, since $g \to D_g$ is linear.

Proof. The statement with N instead of T is obtained in a completely analogous way. The existence of a minimizer follows from the direct methods of the calculus of variations (cf. e.g., Dacorogna [28]) and is established in Steps 1–3. The second statement, namely (6.4), is just the weak form of the Euler–Lagrange equation. The uniqueness is discussed in Step 4. The last identity (6.5) will be obtained in Step 5.

Step 1. We first claim that there exists a $C_1 > 0$ such that

$$\|\omega\|^2_{W^{1,2}} \leq C_1 \int_\Omega \left(|d\omega|^2 + |\delta\omega|^2\right) \quad \text{for every } \omega \in X. \tag{6.6}$$

Step 1.1. If this were not the case, then there would exist for every $m \in \mathbb{N}$, a $\omega_m \in X$ such that

$$\|\omega_m\|^2_{W^{1,2}} \geq m \int_\Omega \left(|d\omega_m|^2 + |\delta\omega_m|^2\right). \tag{6.7}$$

According to the Gaffney inequality (cf. Theorem 5.16), there exists a $C_2 > 0$ such that

$$\|\omega_m\|^2_{W^{1,2}} \leq C_2 \int_\Omega \left(|d\omega_m|^2 + |\delta\omega_m|^2 + |\omega_m|^2\right).$$

Taking $\|\omega_m\|_{L^2} = 1$ for every m, the two inequalities yield, for m large,

$$\|\omega_m\|^2_{W^{1,2}} \leq \frac{C_2}{m}\|\omega_m\|^2_{W^{1,2}} + C_2 \quad \Rightarrow \quad \|\omega_m\|^2_{W^{1,2}} \leq C_3.$$

So the $\|\omega_m\|_{W^{1,2}}$ are uniformly bounded in the reflexive space $W^{1,2}(\Omega;\Lambda^k)$. We can therefore extract a subsequence, that we do not relabel and find $u \in W^{1,2}(\Omega;\Lambda^k)$ such that

$$\begin{aligned} \omega_m &\rightharpoonup u \quad \text{in } W^{1,2} \quad \text{and} \quad \omega_m \to u \quad \text{in } L^2, \\ \delta\omega_m &\rightharpoonup \delta u \quad \text{in } L^2 \quad \text{and} \quad d\omega_m \rightharpoonup du \quad \text{in } L^2. \end{aligned}$$

The strong convergence yields $\|u\|_{L^2} = 1$. As $\omega_m \in (\mathcal{H}_T)^\perp$, we find that $u \in (\mathcal{H}_T)^\perp$.

Step 1.2. We now assert that

$$du = 0 \text{ in } \Omega, \quad \delta u = 0 \text{ in } \Omega \quad \text{and} \quad \nu \wedge u = 0 \text{ on } \partial\Omega.$$

From (6.7) and the bound on the $W^{1,2}$ norm for the ω_m, we obtain, for m sufficiently large,

$$\|\delta\omega_m\|^2_{L^2} + \|d\omega\|^2_{L^2} \leq \frac{C_3}{m}$$

and, consequently,

$$\delta\omega_m \to 0 = \delta u, \quad d\omega_m \to 0 = du \quad \text{in } L^2.$$

Due to the compact imbedding of $W^{1,2}(\Omega)$ into $L^2(\partial\Omega)$, we find that $\nu \wedge u = 0$ on $\partial\Omega$, since $\nu \wedge \omega_m = 0$ on $\partial\Omega$. Steps 1.1 and 1.2 yield $u \in \mathscr{H}_T \cap (\mathscr{H}_T)^\perp = \{0\}$, which is in contradiction with $\|u\|_{L^2} = 1$.

Step 2. We next prove that there exists a constant C_4 such that

$$\|\omega\|_{W^{1,2}} \le 4C_1\, D_g(\omega) + C_4 \quad \text{for every } \omega \in X. \tag{6.8}$$

From (6.6) we obtain

$$\begin{aligned}\|\omega\|_{W^{1,2}}^2 &\le 2C_1 \int_\Omega \left(|d\omega|^2 + |\delta\omega|^2\right) - \|\omega\|_{W^{1,2}}^2 \\ &= 4C_1\, D_g(\omega) - 4C_1 \int_\Omega \langle g;\omega\rangle - \|\nabla\omega\|_{L^2}^2 - \|\omega\|_{L^2}^2 \\ &\le 4C_1\, D_g(\omega) + 4C_1 \|g\|_{L^2}\|\omega\|_{L^2} - \|\omega\|_{L^2}^2 .\end{aligned}$$

The claim of Step 2 follows, since the sum of the last two terms on the right-hand side is bounded from above by a constant C_4.

Step 3. The fact that D_g is weakly lower semicontinuous in $W^{1,2}$ is immediate. So to conclude at the existence of a minimizer $\overline{\omega} \in X$, it is enough to show that any minimizing sequence $\{\omega_m\} \subset X$ has a subsequence that converges weakly in $W^{1,2}$ to a limit $\overline{\omega} \in X$. From (6.8) we obtain that $\|\omega_m\|_{W^{1,2}}$ is bounded and, therefore, up to the extraction of a subsequence that we do not relabel, there exists $\overline{\omega} \in W^{1,2}$ such that

$$\omega_m \rightharpoonup \overline{\omega} \quad \text{in } W^{1,2} \;\Rightarrow\; \omega_m \to \overline{\omega} \quad \text{in } L^2.$$

Since all of the ω_m are in X and the imbedding $W^{1,2}(\Omega)$ into $L^2(\partial\Omega)$ is compact, we obtain that $\overline{\omega} \in W_T^{1,2}$. Similarly, since the ω_m are all in $(\mathscr{H}_T)^\perp$ and $(\mathscr{H}_T)^\perp$ is closed in L^2, we obtain that $\overline{\omega} \in (\mathscr{H}_T)^\perp$. Thus, $\overline{\omega} \in W_T^{1,2} \cap (\mathscr{H}_T)^\perp = X$. The existence part of the proof is then complete. Moreover, we also have immediately (6.4).

Step 4. The uniqueness is easily obtained. Indeed, let $u, v \in X$ be two minimizers. Due to the strict convexity of the map $\omega \to \int(|d\omega|^2 + |\delta\omega|^2)$, we find that

$$\delta u = \delta v \quad \text{and} \quad du = dv.$$

Applying (6.6) to $u - v$, we obtain that

$$\|u - v\|_{W^{1,2}}^2 \le C_1 \left(\int_\Omega |du - dv|^2 + \int_\Omega |\delta u - \delta v|^2 \right) = 0$$

and thus the claim.

Step 5. It remains to establish (6.5). Let u in $W_T^{1,2}(\Omega;\Lambda^k) \subset L^2(\Omega;\Lambda^k)$. We then write, according to (6.2), $u = w + v$ with $w \in \mathscr{H}_T$ and $v \in (\mathscr{H}_T)^\perp$. We therefore deduce that

$$\delta w = 0, \quad dw = 0 \quad \text{and} \quad \int_\Omega \langle g, w\rangle = 0.$$

In addition, $v \in X$ because $v \in (\mathscr{H}_T)^\perp$ and $v = u - w \in W_T^{1,2}$. We hence find that

$$\int_\Omega \langle d\overline{\omega}; du\rangle + \int_\Omega \langle \delta\overline{\omega}; \delta u\rangle - \int_\Omega \langle g, u\rangle = \int_\Omega \langle d\overline{\omega}; dv\rangle + \int_\Omega \langle \delta\overline{\omega}; \delta v\rangle - \int_\Omega \langle g, v\rangle = 0,$$

which is precisely what had to be shown. □

6.3 The Hodge–Morrey Decomposition

We now turn to the main result of the present chapter.

Theorem 6.9 (Hodge–Morrey decomposition). *Let $\Omega \subset \mathbb{R}^n$ be a bounded open C^3 set with exterior unit normal ν. Let $0 \le k \le n$ and $f \in L^2(\Omega;\Lambda^k)$.*

(i) There exist $\alpha \in W_T^{1,2}(\Omega;\Lambda^{k-1})$, $\beta \in W_T^{1,2}(\Omega;\Lambda^{k+1})$, $h \in \mathscr{H}_T(\Omega;\Lambda^k)$ and $\omega \in W_T^{2,2}(\Omega;\Lambda^k)$ such that, in Ω,

$$f = d\alpha + \delta\beta + h, \quad \alpha = \delta\omega \quad \text{and} \quad \beta = d\omega.$$

(ii) There exist $\alpha \in W_N^{1,2}(\Omega;\Lambda^{k-1})$, $\beta \in W_N^{1,2}(\Omega;\Lambda^{k+1})$, $h \in \mathscr{H}_N(\Omega;\Lambda^k)$ and $\omega \in W_N^{2,2}(\Omega;\Lambda^k)$ such that, in Ω,

$$f = d\alpha + \delta\beta + h, \quad \alpha = \delta\omega \quad \text{and} \quad \beta = d\omega.$$

(iii) There exist $\alpha \in W_T^{1,2}(\Omega;\Lambda^{k-1})$, $\beta \in W_N^{1,2}(\Omega;\Lambda^{k+1})$, $h \in \mathscr{H}(\Omega;\Lambda^k)$, $\omega^1 \in W_T^{2,2}(\Omega;\Lambda^k)$ and $\omega^2 \in W_N^{2,2}(\Omega;\Lambda^k)$ such that, in Ω,

$$f = d\alpha + \delta\beta + h, \quad \alpha = \delta\omega^1 \quad \text{and} \quad \beta = d\omega^2.$$

(iv) In addition, in each of the three cases, $d\alpha$, $\delta\beta$ and h are mutually orthogonal with respect to the L^2-inner product. Moreover, in each of the three cases, there exists a positive constant $C = C(\Omega)$ such that

$$\|\omega\|_{W^{2,2}} + \|h\|_{L^2} \le C\|f\|_{L^2}.$$

Remark 6.10. (i) We recall that if $r \ge 1$ is an integer,

$$W_T^{r,2}(\Omega;\Lambda^k) = \{f \in W^{r,2}(\Omega;\Lambda^k) : \nu \wedge f = 0 \text{ on } \partial\Omega\},$$
$$W_N^{r,2}(\Omega;\Lambda^k) = \{f \in W^{r,2}(\Omega;\Lambda^k) : \nu \lrcorner f = 0 \text{ on } \partial\Omega\}.$$

(ii) If $k = 0$, then statement (i) of the theorem is the simplest of the three decompositions and it reads as

$$f = \delta\beta = \delta d\omega = \Delta\omega \text{ in } \Omega \quad \text{with} \quad \omega = 0 \text{ on } \partial\Omega.$$

If $k=n$, then statement (ii) of the theorem is the simplest of the three decompositions and it reads as

$$f=d\alpha=d\delta\omega=\Delta\omega \text{ in } \Omega \quad \text{with} \quad \omega=0 \text{ on } \partial\Omega.$$

(iii) If Ω is contractible, then in statement (i), $h=0$ if $k\leq n-1$ and in statement (ii), $h=0$ if $k\geq 1$, as seen in Theorem 6.5.

(iv) There exists a much simpler decomposition, namely

$$f=d\alpha+\delta\beta \text{ in } \Omega, \quad \nu\lrcorner\alpha=0 \text{ on } \partial\Omega \quad \text{and} \quad \nu\wedge\beta=0 \text{ on } \partial\Omega.$$

This can be immediately obtained by solving componentwise the Poisson equation with Dirichlet boundary condition; that is,

$$\begin{cases} \Delta\omega=d\delta\omega+\delta d\omega=f & \text{in } \Omega, \\ \omega=0 & \text{on } \partial\Omega \end{cases}$$

Setting $\alpha=\delta\omega$ and $\beta=d\omega$, we get the claim according to Theorem 3.23. However, this decomposition turns out to be much less useful (in particular, it is not an orthogonal decomposition) than the Hodge–Morrey one, since $\nu\lrcorner\alpha$, respectively $\nu\wedge\beta$, do not give any information about the boundary behavior of $d\alpha$, respectively $\delta\beta$.

We will deduce Theorem 6.9 from Theorem 6.7. Indeed, after proving with usual arguments (see Theorem 6.11) that the minimizer ω found in Theorem 6.7 is in fact $W^{2,2}$, the Hodge–Morrey decomposition will be seen to be, in fact, a rewriting of the strong Euler–Lagrange equation. We now proceed with the proof of Theorem 6.9.

Proof. We divide the proof into three steps, each one corresponding to one of the statements.

Step 1. We start by proving (i).

Step 1.1. Let $f=h+g$, $h\in\mathscr{H}_T$ and $g\in(\mathscr{H}_T)^{\perp}$ as in (6.2). We use Theorem 6.7 for g and obtain $\omega\in W_T^{1,2}\cap(\mathscr{H}_T)^{\perp}$ such that

$$\int_\Omega (\langle d\omega;du\rangle+\langle\delta\omega;\delta u\rangle)=-\int_\Omega\langle g;u\rangle, \quad \forall u\in W_T^{1,2}.$$

In view of the regularity Theorem 6.11, we conclude that $\omega\in W^{2,2}$ and we can integrate by parts to obtain, for every $u\in W_T^{1,2}$,

$$-\int_\Omega (\langle\delta d\omega;u\rangle+\langle d\delta\omega;u\rangle)+\int_{\partial\Omega}(\langle d\omega;\nu\wedge u\rangle+\langle\nu\wedge\delta\omega;u\rangle)=-\int_\Omega\langle g;u\rangle. \tag{6.9}$$

Taking first $u\in C_0^\infty(\Omega;\Lambda^k)$, we obtain

$$\delta d\omega+d\delta\omega=g \quad \text{in } \Omega. \tag{6.10}$$

We set $\alpha = \delta\omega$ and $\beta = d\omega$. We immediately obtain from Theorem 3.23 that $\nu \wedge \beta = 0$. It is left to show that $\nu \wedge \alpha = \nu \wedge \delta\omega = 0$. From (6.9) and (6.10) we conclude that the integrals both over Ω and $\partial\Omega$ vanish separately. So we have

$$0 = \int_{\partial\Omega} \langle d\omega, \nu \wedge u\rangle + \int_{\partial\Omega} \langle \nu \wedge \delta\omega, u\rangle = \int_{\partial\Omega} \langle \nu \wedge \delta\omega, u\rangle, \quad \forall u \in W_T^{1,2}. \tag{6.11}$$

In fact, this is also true for all $u \in W^{1,2}(\Omega;\Lambda^k)$, as we will show right now. So let $u \in W^{1,2}(\Omega;\Lambda^k)$ be arbitrary. We use in (6.11) the test function $\nu \wedge (\nu \lrcorner u) \in W_T^{1,2}(\Omega;\Lambda^k)$, where we have extended ν to a $C^1(\overline{\Omega};\Lambda^1)$ function. We obtain, using Proposition 2.16,

$$\begin{aligned} 0 &= \int_{\partial\Omega} \langle \nu \wedge \delta\omega; \nu \wedge (\nu \lrcorner u)\rangle = \int_{\partial\Omega} \langle \nu \wedge \delta\omega; u\rangle - \int_{\partial\Omega} \langle \nu \wedge \delta\omega; \nu \lrcorner (\nu \wedge u)\rangle \\ &= \int_{\partial\Omega} \langle \nu \wedge \delta\omega; u\rangle. \end{aligned}$$

Hence, since $u \in W^{1,2}(\Omega;\Lambda^k)$ is arbitrary, we deduce that $\nu \wedge \delta\omega = 0$ on $\partial\Omega$.

Step 1.2. We next prove the orthogonality of the decomposition (i). We have to show that h, $d\alpha$, and $\delta\beta$ are mutually orthogonal with respect to the L^2-inner product. As $d\alpha + \delta\beta = \Delta\omega = g \in (\mathscr{H}_T)^\perp$, we already know that

$$\int_\Omega \langle h; d\alpha + \delta\beta\rangle = 0.$$

Using the boundary condition on α and the fact that h is a harmonic field, we obtain

$$\int_\Omega \langle h; d\alpha\rangle = -\int_\Omega \langle \delta h; \alpha\rangle + \int_{\partial\Omega} \langle h; \nu \wedge \alpha\rangle = 0 \quad \Rightarrow \quad \int_\Omega \langle h; \delta\beta\rangle = 0.$$

The orthogonality of $d\alpha$ and $\delta\beta$ follows immediately from Corollary 3.29.

Step 1.3. The estimate immediately follows from (6.3) and Theorem 6.11.

Step 2. The proof of (ii) is completely analogous to that of (i) and we skip the details.

Step 3. We prove (iii).

Step 3.1. We use decomposition (i) to get $\alpha^1 \in W_T^{1,2}(\Omega;\Lambda^{k-1})$, $\beta^1 \in W_T^{1,2}(\Omega;\Lambda^{k+1})$, $h^1 \in \mathscr{H}_T(\Omega;\Lambda^k)$ and $\omega^1 \in W_T^{2,2}(\Omega;\Lambda^k)$ such that, in Ω,

$$f = d\alpha^1 + \delta\beta^1 + h^1, \quad \alpha^1 = \delta\omega^1 \quad \text{and} \quad \beta^1 = d\omega^1. \tag{6.12}$$

Similarly, appealing to decomposition (ii), we get $\alpha^2 \in W_N^{1,2}(\Omega;\Lambda^{k-1})$, $\beta^2 \in W_N^{1,2}(\Omega;\Lambda^{k+1})$, $h^2 \in \mathscr{H}_N(\Omega;\Lambda^k)$ and $\omega^2 \in W_N^{2,2}(\Omega;\Lambda^k)$ such that, in Ω,

$$f = d\alpha^2 + \delta\beta^2 + h^2, \quad \alpha^2 = \delta\omega^2 \quad \text{and} \quad \beta^2 = d\omega^2. \tag{6.13}$$

We set

$$h = \delta\beta^1 - \delta\beta^2 + h^1 = d\alpha^2 - d\alpha^1 + h^2$$

and observe that

$$f = d\alpha^1 + \delta\beta^2 + h.$$

It therefore remains to prove that $h \in \mathscr{H}(\Omega;\Lambda^k)$. Let us first prove that for every $\psi \in C_0^\infty(\Omega;\Lambda^{k-1})$ and $\chi \in C_0^\infty(\Omega;\Lambda^{k+1})$,

$$\int_\Omega \langle h; d\psi\rangle = \int_\Omega \langle h; \delta\chi\rangle = 0. \tag{6.14}$$

We prove only the first identity, the other one being established analogously. We have, from Theorem 3.28,

$$\begin{aligned}\int_\Omega \langle h; d\psi\rangle &= \int_\Omega \langle \delta\beta^1 - \delta\beta^2 + h^1; d\psi\rangle \\ &= \int_\Omega \langle \beta^1 - \beta^2; dd\psi\rangle + \int_\Omega \langle \delta h^1; \psi\rangle = 0.\end{aligned}$$

We now choose, for any $\phi \in C_0^\infty(\Omega;\Lambda^k)$, $\psi = \delta\phi$ and $\chi = d\phi$ in (6.14). We therefore obtain

$$\int_\Omega \langle h; \Delta\phi\rangle = 0 \quad \text{for every } \phi \in C_0^\infty(\Omega;\Lambda^k).$$

The Weyl lemma (cf. [29]) implies that $h \in C^\infty(\Omega;\Lambda^k)$ and, therefore, by (6.14), $h \in \mathscr{H}(\Omega;\Lambda^k)$.

Step 3.2. The orthogonality of the decomposition is obtained in the same way as in Step 1.2.

Step 3.3. The estimate follows from the fact that $\alpha^1, \alpha^2, \beta^1, \beta^2, h^1$ and h^2 satisfy the corresponding inequality. □

6.4 Higher Regularity

The following theorems have been established by Morrey [76] (cf. also Theorems 7.7.4 and 7.7.8 in [77]), see also Agmon, Douglis and Nirenberg [4] (for the regularity), Bolik [13], Iwaniec, Scott and Stroffolini [58] and Schwarz [89].

Theorem 6.11 ($W^{r,2}$-regularity). *Let $r \geq 0$, $0 \leq k \leq n$ be integers and $\Omega \subset \mathbb{R}^n$ be a bounded open C^{r+3} set. Let $f \in W^{r,2}(\Omega;\Lambda^k)$ and $\omega \in W_T^{1,2}(\Omega;\Lambda^k)$ be such that*

$$\int_\Omega (\langle d\omega; du\rangle + \langle \delta\omega; \delta u\rangle) = \int_\Omega \langle f; u\rangle \quad \forall u \in W_T^{1,2}(\Omega;\Lambda^k). \tag{6.15}$$

Then there exists a constant $C = C(r,\Omega) > 0$ such that the inequality

$$\|\omega\|_{W^{r+2,2}} \leq C(\|\omega\|_{L^2} + \|f\|_{W^{r,2}})$$

holds and if, in addition, $\omega \in W_T^{1,2}(\Omega;\Lambda^k) \cap (\mathcal{H}_T(\Omega;\Lambda^k))^{\perp}$, *then*

$$\|\omega\|_{W^{r+2,2}} \le C\|f\|_{W^{r,2}}. \tag{6.16}$$

The same theorem holds true by replacing the subscript T by N.

Proof. Note that the interior regularity for a solution ω of (6.15) is exactly that of the Laplacian, since by choosing $u = \varphi\, dx^{i_1} \wedge \cdots \wedge dx^{i_k}$ in (6.15), we obtain by partial integration, for every $\varphi \in C_0^\infty(\Omega)$,

$$\begin{aligned}\int_\Omega \langle \nabla \omega_{i_1\cdots i_k}; \nabla \varphi\rangle &= -\int_\Omega \omega_{i_1\cdots i_k}\, \Delta\varphi = -\int_\Omega \langle \omega; \Delta u\rangle \\ &= \int_\Omega \langle d\omega; du\rangle + \int_\Omega \langle \delta\omega; \delta u\rangle = \int_\Omega f_{i_1\cdots i_k}\, \varphi.\end{aligned}$$

We will not prove the boundary regularity. We, however, illustrate the idea in the following simplified setting, which is essentially obtained after having locally rectified the boundary by an admissible boundary coordinate system. Suppose Ω is of the form

$$\Omega = \{x = (x_1,\ldots,x_n) \in \mathbb{R}^n : x_n < 0\} \cap U$$

for some open set $U \subset \mathbb{R}^n$, $\operatorname{supp}\omega \subset U$ and that

$$\int_\Omega (\langle d\omega; du\rangle + \langle \delta\omega; \delta u\rangle) = \int_\Omega \langle f; u\rangle, \quad \forall u \in W_T^{1,2}(\Omega;\Lambda^k),$$

with $\operatorname{supp} u \subset U$. Then, invoking Theorem 5.7, we find

$$\int_\Omega \langle \nabla\omega; \nabla u\rangle = \int_\Omega \langle f; u\rangle,$$

since K^ν and L^ν vanish on the hyperplane $\{x_n = 0\}$. Thus, exactly the same methods can be applied as for the Laplacian. We refer, for a detailed proof, to Csató [23]. □

The second theorem gives now the appropriate regularity for the Hodge–Morrey decomposition theorem, cf. Theorem 6.9.

Theorem 6.12. *Let $r \ge 0$ be an integer, $0 < q < 1 < p < \infty$ and $\Omega \subset \mathbb{R}^n$ be a bounded open smooth set. Let $f \in W^{r,p}(\Omega;\Lambda^k)$, respectively $f \in C^{r,q}(\overline{\Omega};\Lambda^k)$.*

(i) There exist

$$\alpha \in W_T^{r+1,p}(\Omega;\Lambda^{k-1}), \quad \beta \in W_T^{r+1,p}(\Omega;\Lambda^{k+1}),$$

$$h \in \mathcal{H}_T(\Omega;\Lambda^k) \quad \text{and} \quad \omega \in W_T^{r+2,p}(\Omega;\Lambda^k),$$

respectively

$$\alpha \in C_T^{r+1,q}(\overline{\Omega};\Lambda^{k-1}), \quad \beta \in C_T^{r+1,q}(\overline{\Omega};\Lambda^{k+1}),$$

$$h \in \mathcal{H}_T(\Omega;\Lambda^k) \quad \text{and} \quad \omega \in C_T^{r+2,q}(\overline{\Omega};\Lambda^k),$$

such that, in Ω,

$$f = d\alpha + \delta\beta + h, \quad \alpha = \delta\omega \quad \textit{and} \quad \beta = d\omega.$$

Moreover, there exist constants $C_1 = C_1(r,p,\Omega)$ *and* $C_2 = C_2(r,q,\Omega)$ *such that*

$$\|\omega\|_{W^{r+2,p}} + \|h\|_{W^{r,p}} \le C_1 \|f\|_{W^{r,p}},$$

$$\|\omega\|_{C^{r+2,q}} + \|h\|_{C^{r,q}} \le C_2 \|f\|_{C^{r,q}}.$$

(ii) There exist

$$\alpha \in W_N^{r+1,p}(\Omega;\Lambda^{k-1}), \quad \beta \in W_N^{r+1,p}(\Omega;\Lambda^{k+1}),$$

$$h \in \mathscr{H}_N(\Omega;\Lambda^k) \quad \textit{and} \quad \omega \in W_N^{r+2,p}(\Omega;\Lambda^k),$$

respectively

$$\alpha \in C_N^{r+1,q}(\overline{\Omega};\Lambda^{k-1}), \quad \beta \in C_N^{r+1,q}(\overline{\Omega};\Lambda^{k+1}),$$

$$h \in \mathscr{H}_N(\Omega;\Lambda^k) \quad \textit{and} \quad \omega \in C_N^{r+2,q}(\overline{\Omega};\Lambda^k),$$

such that, in Ω,

$$f = d\alpha + \delta\beta + h, \quad \alpha = \delta\omega \quad \textit{and} \quad \beta = d\omega.$$

Moreover, there exist constants $C_1 = C_1(r,p,\Omega)$ *and* $C_2 = C_2(r,q,\Omega)$ *such that*

$$\|\omega\|_{W^{r+2,p}} + \|h\|_{W^{r,p}} \le C_1 \|f\|_{W^{r,p}},$$

$$\|\omega\|_{C^{r+2,q}} + \|h\|_{C^{r,q}} \le C_2 \|f\|_{C^{r,q}}.$$

(iii) There exist

$$\alpha \in W_T^{r+1,p}(\Omega;\Lambda^{k-1}), \quad \beta \in W_N^{r+1,p}(\Omega;\Lambda^{k+1}),$$

$$h \in \mathscr{H}(\Omega;\Lambda^k), \quad \omega^1 \in W_T^{r+2,p}(\Omega;\Lambda^k) \quad \textit{and} \quad \omega^2 \in W_N^{r+2,p}(\Omega;\Lambda^k),$$

respectively

$$\alpha \in C_T^{r+1,q}(\overline{\Omega};\Lambda^{k-1}), \quad \beta \in C_N^{r+1,q}(\overline{\Omega};\Lambda^{k+1}),$$

$$h \in \mathscr{H}(\Omega;\Lambda^k), \quad \omega^1 \in C_T^{r+2,q}(\overline{\Omega};\Lambda^k) \quad \textit{and} \quad \omega^2 \in C_N^{r+2,q}(\overline{\Omega};\Lambda^k),$$

such that, in Ω,

$$f = d\alpha + \delta\beta + h, \quad \alpha = \delta\omega^1 \quad \textit{and} \quad \beta = d\omega^2.$$

Moreover, there exist constants $C_1 = C_1(r,p,\Omega)$ *and* $C_2 = C_2(r,q,\Omega)$ *such that*

$$\|\omega^1\|_{W^{r+2,p}} + \|\omega^2\|_{W^{r+2,p}} + \|h\|_{W^{r,p}} \le C_1 \|f\|_{W^{r,p}},$$

$$\|\omega^1\|_{C^{r+2,q}} + \|\omega^2\|_{C^{r+2,q}} + \|h\|_{C^{r,q}} \leq C_2\|f\|_{C^{r,q}}\,.$$

Remark 6.13. (i) When $1 < p < 2$, decomposition (i) reads as a direct sum:

$$L^p = dW_T^{1,p}(\Omega;\Lambda^{k-1}) \oplus \delta W_T^{1,p}(\Omega;\Lambda^{k+1}) \oplus \mathscr{H}_T(\Omega;\Lambda^k),$$

and similarly for the other two decompositions. If $p \geq 2$, then the decomposition is even orthogonal with respect to the L^2-inner product (cf. Theorem 6.9(iv)).

(ii) The above results remain valid if Ω is C^{r+3} for the Sobolev case and $C^{r+3,q}$ for the Hölder case.

(iii) The correspondence $f \to (\alpha,\beta,h,\omega)$ is linear. Furthermore, the construction is universal in the sense that all of the assertions remain valid if (r,p,q) is replaced by (r',p',q') as far as $f \in W^{r',p'}(\Omega;\Lambda^k)$, respectively $f \in C^{r',q'}(\overline{\Omega};\Lambda^k)$, with the same (α,β,h,ω) and with constants $C_1' = C_1'(r',p',\Omega)$ and $C_2' = C_2'(r',q',\Omega)$.

Chapter 7
First-Order Elliptic Systems of Cauchy–Riemann Type

7.1 System with Prescribed Tangential Component

We first deal with the following boundary value problem:

$$\begin{cases} d\omega = f \quad \text{and} \quad \delta\omega = g & \text{in } \Omega, \\ \nu \wedge \omega = \nu \wedge \omega_0 & \text{on } \partial\Omega, \end{cases}$$

where f, g and ω_0 are given and ν is the exterior unit normal. If there is a solution and ω, ω_0, f and g are sufficiently regular, then Theorem 3.5 and Theorem 3.23 imply that

$$df = 0 \text{ in } \Omega, \quad \delta g = 0 \text{ in } \Omega, \quad \nu \wedge d\omega_0 = \nu \wedge f \quad \text{on } \partial\Omega.$$

We will in fact show that these conditions are also sufficient to guarantee the existence of ω if Ω is contractible (cf. Remark 7.3(v)).

Our method is similar to that applied by Schwarz [89, Theorem 3.1.1] in the framework of Sobolev spaces; see also Borchers and Sohr [14] and Von Wahl [103, 104] for the case of 1-forms. The problem has also been treated by Georgescu [47, Theorem 4.2.2] and Kress [63, Satz 8.1] in the setting of Hölder spaces. We start our analysis with an extension theorem.

Lemma 7.1. *Let $r \geq 1$ be an integer, $0 \leq q \leq 1 < p < \infty$ and $\Omega \subset \mathbb{R}^n$ be a bounded open smooth set with exterior unit normal ν. Let $\omega_0 : \partial\Omega \to \Lambda^k$.*

(i) Suppose

$$\nu \wedge \omega_0 \in W^{r-\frac{1}{p},p}\left(\partial\Omega; \Lambda^{k+1}\right), \quad \textit{respectively} \quad \nu \wedge \omega_0 \in C^{r,q}\left(\partial\Omega; \Lambda^{k+1}\right).$$

Then there exists

$$\omega \in W^{r,p}\left(\Omega; \Lambda^k\right), \quad \textit{respectively} \quad \omega \in C^{r,q}\left(\overline{\Omega}; \Lambda^k\right),$$

such that

$$\nu \wedge \omega = \nu \wedge \omega_0 \quad \text{on } \partial\Omega.$$

Moreover, there exists a constant $C_1 = C_1(r,p,\Omega)$, *respectively* $C_2 = C_2(r,q,\Omega)$, *such that*

$$\|\omega\|_{W^{r,p}(\Omega)} \le C_1 \|\nu \wedge \omega_0\|_{W^{r-\frac{1}{p},p}(\partial\Omega)}, \qquad \|\omega\|_{C^{r,q}(\overline{\Omega})} \le C_2 \|\nu \wedge \omega_0\|_{C^{r,q}(\partial\Omega)}.$$

(ii) Suppose

$$\nu \lrcorner \omega_0 \in W^{r-\frac{1}{p},p}(\partial\Omega;\Lambda^{k-1}), \quad \text{respectively} \quad \nu \lrcorner \omega_0 \in C^{r,q}(\partial\Omega;\Lambda^{k-1}).$$

Then there exists

$$\omega \in W^{r,p}(\Omega;\Lambda^k), \quad \text{respectively} \quad \omega \in C^{r,q}(\overline{\Omega};\Lambda^k),$$

such that

$$\nu \lrcorner \omega = \nu \lrcorner \omega_0 \quad \text{on } \partial\Omega.$$

Moreover, there exists a constant $C_1 = C_1(r,p,\Omega)$, *respectively* $C_2 = C_2(r,q,\Omega)$, *such that*

$$\|\omega\|_{W^{r,p}(\Omega)} \le C_1 \|\nu \lrcorner \omega_0\|_{W^{r-\frac{1}{p},p}(\partial\Omega)}, \qquad \|\omega\|_{C^{r,q}(\overline{\Omega})} \le C_2 \|\nu \lrcorner \omega_0\|_{C^{r,q}(\partial\Omega)}.$$

Proof. We only discuss statement (i) concerning the exterior product, the other one being handled similarly. The extension theorem for functions is well known (cf. Adams [2] and Gilbarg and Trudinger [49]) and we get, if $f \in W^{r-\frac{1}{p},p}(\partial\Omega)$, respectively $C^{r,q}(\partial\Omega)$, that one can extend f by $\tilde{f}$ so that

$$\|\tilde{f}\|_{W^{r,p}(\Omega)} \le C_1 \|f\|_{W^{r-\frac{1}{p},p}(\partial\Omega)}, \quad \text{respectively} \quad \|\tilde{f}\|_{C^{r,q}(\overline{\Omega})} \le C_2 \|f\|_{C^{r,q}(\partial\Omega)},$$

for some constants C_1 and C_2 independent of f. We now let $\alpha = \nu \wedge \omega_0$ and extend it so as to have, without relabeling, $\alpha \in W^{r,p}(\Omega;\Lambda^k)$, respectively $C^{r,q}(\overline{\Omega};\Lambda^k)$. Extending ν in such a way that the extension, still denoted ν, belongs to $C^\infty(\overline{\Omega};\mathbb{R}^n)$ and setting

$$\omega = \nu \lrcorner \alpha,$$

we have the claim. Indeed, in view of Proposition 2.16, we find, on $\partial\Omega$,

$$\nu \wedge \omega = \nu \wedge (\nu \lrcorner \alpha) = \alpha - \nu \lrcorner (\nu \wedge \alpha) = \alpha = \nu \wedge \omega_0,$$

which is the assertion. □

We now state the main theorem of the present section.

Theorem 7.2. *Let* $r \ge 0$ *and* $0 \le k \le n$ *be integers,* $0 < q < 1$, $2 \le p < \infty$ *and* $\Omega \subset \mathbb{R}^n$ *be a bounded open smooth set with exterior unit normal* ν. *Let* $f : \overline{\Omega} \to \Lambda^{k+1}$, $g : \overline{\Omega} \to \Lambda^{k-1}$ *and* $\omega_0 : \partial\Omega \to \Lambda^k$. *Then the following statements are equivalent:*

(i) Let

$$f \in W^{r,p}\left(\Omega;\Lambda^{k+1}\right),\ g \in W^{r,p}\left(\Omega;\Lambda^{k-1}\right) \text{ and } \nu \wedge \omega_0 \in W^{r+1-\frac{1}{p},p}\left(\partial\Omega;\Lambda^{k+1}\right),$$

respectively

$$f \in C^{r,q}\left(\overline{\Omega};\Lambda^{k+1}\right),\ g \in C^{r,q}\left(\overline{\Omega};\Lambda^{k-1}\right) \text{ and } \nu \wedge \omega_0 \in C^{r+1,q}\left(\partial\Omega;\Lambda^{k+1}\right),$$

satisfying the conditions

$$df = 0 \text{ in } \Omega, \quad \delta g = 0 \text{ in } \Omega, \quad \nu \wedge d\omega_0 = \nu \wedge f \text{ on } \partial\Omega \tag{A1}$$

and, for every $\chi \in \mathcal{H}_T\left(\Omega;\Lambda^{k+1}\right)$ *and* $\psi \in \mathcal{H}_T\left(\Omega;\Lambda^{k-1}\right)$,

$$\int_\Omega \langle f;\chi\rangle - \int_{\partial\Omega} \langle \nu \wedge \omega_0;\chi\rangle = 0 \quad \text{and} \quad \int_\Omega \langle g;\psi\rangle = 0. \tag{A2}$$

(ii) There exists $\omega \in W^{r+1,p}\left(\Omega;\Lambda^k\right)$, *respectively* $\omega \in C^{r+1,q}\left(\overline{\Omega};\Lambda^k\right)$, *such that*

$$\begin{cases} d\omega = f \quad \text{and} \quad \delta\omega = g & \text{in } \Omega, \\ \nu \wedge \omega = \nu \wedge \omega_0 & \text{on } \partial\Omega. \end{cases}$$

In addition, there exists a constant $C_1 = C_1(r,p,\Omega)$ *such that*

$$\|\omega\|_{W^{r+1,p}(\Omega)} \le C_1\left(\|f\|_{W^{r,p}(\Omega)} + \|g\|_{W^{r,p}(\Omega)} + \|\nu \wedge \omega_0\|_{W^{r+1-\frac{1}{p},p}(\partial\Omega)}\right),$$

respectively $C_2 = C_2(r,q,\Omega)$ *such that*

$$\|\omega\|_{C^{r+1,q}(\overline{\Omega})} \le C_2\left(\|f\|_{C^{r,q}(\overline{\Omega})} + \|g\|_{C^{r,q}(\overline{\Omega})} + \|\nu \wedge \omega_0\|_{C^{r+1,q}(\partial\Omega)}\right).$$

Remark 7.3. (i) When $k = 0$, all statements of the theorem have to be understood as if g were not present. For example, statement (ii) reads then as

$$\begin{cases} d\omega = f & \text{in } \Omega, \\ \omega = \omega_0 & \text{on } \partial\Omega. \end{cases}$$

Moreover, the result is also valid when $q = 0,1$ (see Theorem 8.16 and the remark following it). Furthermore, when $k = n$, then all statements of the theorem have to be understood as if f and ω_0 were not present. In this case, statement (ii) becomes

$$\delta\omega = g \quad \text{in } \Omega.$$

(ii) If $r \ge 1$, conditions (A1) are well defined. This is obvious for the first two equations. The third one makes sense in $W^{r-\frac{1}{p},p}\left(\partial\Omega;\Lambda^{k+2}\right)$, respectively $C^{r,q}\left(\partial\Omega;\Lambda^{k+2}\right)$, due to Theorem 3.23 and Lemma 7.1.

(iii) If $r=0$, then the conditions $df=0$ and $\delta g=0$ are understood in the sense of distributions. The third statement in (A1) is well defined in the Hölder case and is understood, in the Sobolev case, in the weak sense, namely

$$\int_\Omega \langle f;\delta\varphi\rangle - \int_{\partial\Omega} \langle \nu\wedge\omega_0;\delta\varphi\rangle = 0 \tag{B1}$$

for every $\varphi\in C^\infty\left(\overline{\Omega};\Lambda^{k+2}\right)$.

(iv) If $r\geq 1$, then (B1) is equivalent to the first and third conditions in (A1). This can be shown by several partial integrations (cf. Step 2.1 in the proof of Theorem 7.2).

(v) If Ω is contractible and $k\leq n-2$, then condition (A2) drops due to Theorem 6.5.

(vi) The above results remain valid if the set Ω is C^{r+3} for the Sobolev case and $C^{r+3,q}$ for the Hölder case.

(vii) We will prove the result for $C^{r,q}$ and the same proof is valid for $W^{r,p}$ when $p\geq 2$.

(viii) The construction is linear and universal in the sense of Remark 6.13.

Proof. We deal with the statement in Hölder spaces.

Step 1. We start by showing that (ii) implies (i).

Suppose first that $r\geq 1$. Theorems 3.5 and 3.23 immediately imply the conditions in (A1). The first condition in (A2) follows by partial integration; indeed, for any $\chi\in\mathcal{H}_T\left(\Omega;\Lambda^{k+1}\right)$,

$$\begin{aligned}\int_\Omega\langle f;\chi\rangle-\int_{\partial\Omega}\langle\nu\wedge\omega_0;\chi\rangle&=\int_\Omega\langle d\omega;\chi\rangle-\int_{\partial\Omega}\langle\nu\wedge\omega_0;\chi\rangle\\&=\int_{\partial\Omega}\langle\nu\wedge(\omega-\omega_0);\chi\rangle-\int_\Omega\langle\omega;\delta\chi\rangle=0.\end{aligned}$$

The second condition in (A2) follows in a similar way.

If $r=0$, the first two conditions in (A1) are understood in the sense of distributions (cf. Proposition 7.6) and follow by partial integration, since

$$\int_\Omega\langle f;\delta\varphi\rangle=\int_\Omega\langle d\omega;\delta\varphi\rangle=-\int_\Omega\langle f;\delta\delta\varphi\rangle=0$$

for every $\varphi\in C_0^\infty\left(\Omega;\Lambda^{k+2}\right)$. The second condition in (A1) follows in the same way. Condition (A2) and the third condition in (A1) follow exactly as in the case $r\geq 1$.

Step 2. We next turn to the implication (i) $\Rightarrow$ (ii). We first extend, according to Lemma 7.1, ω_0 by $\tilde{\omega}_0$ so that $\nu\wedge\omega_0=\nu\wedge\tilde{\omega}_0$.

Step 2.1. We now show that (A1) implies the following two equations:

$$\int_\Omega\langle f;\delta\varphi\rangle-\int_\Omega\langle d\tilde{\omega}_0;\delta\varphi\rangle=0,\quad\forall\varphi\in C^\infty\left(\overline{\Omega};\Lambda^{k+2}\right), \tag{7.1}$$

$$\int_\Omega \langle g; d\psi \rangle = 0, \quad \forall \psi \in C_T^\infty(\overline{\Omega}; \Lambda^{k-2}). \tag{7.2}$$

Let us first assume that $r \geq 1$. Equation (7.1) follows by several partial integrations and the first and third identity in (A1) in the following way:

$$\begin{aligned}
\int_\Omega \langle f; \delta\varphi \rangle &= -\int_\Omega \langle df; \varphi \rangle + \int_{\partial\Omega} \langle \nu \wedge f; \varphi \rangle = \int_{\partial\Omega} \langle \nu \wedge d\omega_0; \varphi \rangle \\
&= \int_\Omega \langle dd\tilde{\omega}_0; \varphi \rangle + \int_\Omega \langle d\tilde{\omega}_0; \delta\varphi \rangle = \int_\Omega \langle d\tilde{\omega}_0; \delta\varphi \rangle.
\end{aligned}$$

If $r = 0$, we can apply Proposition 7.6 twice, since f and $d\tilde{\omega}_0$ are closed in the sense of distributions. This gives

$$\int_\Omega \langle f; \delta\varphi \rangle = \int_{\partial\Omega} \langle \nu \wedge f; \varphi \rangle = \int_{\partial\Omega} \langle \nu \wedge d\omega_0; \varphi \rangle = \int_\Omega \langle d\tilde{\omega}_0; \delta\varphi \rangle.$$

Equation (7.2) follows immediately from the second condition in (A1) by a single partial integration if $r \geq 1$, respectively from Proposition 7.8 if $r = 0$.

Step 2.2. We apply the Hodge–Morrey decomposition (cf. Theorem 6.12(i)) to decompose $f - d\tilde{\omega}_0$ and obtain (if $k = n$, we do not need this construction)

$$\begin{gathered}
f - d\tilde{\omega}_0 = d\alpha_f + \delta\beta_f + \chi_f \quad \text{in } \Omega, \\
\delta\alpha_f = 0, \; d\beta_f = 0 \quad \text{in } \Omega, \\
\nu \wedge \alpha_f = 0, \; \nu \wedge \beta_f = 0 \quad \text{on } \partial\Omega,
\end{gathered}$$

where $\chi_f \in \mathscr{H}_T\left(\Omega; \Lambda^{k+1}\right)$. Moreover, there exists a positive constant $C = C(r, q, \Omega)$ such that

$$\|\alpha_f\|_{C^{r+1,q}(\overline{\Omega})} \leq C\left(\|f\|_{C^{r,q}(\overline{\Omega})} + \|\tilde{\omega}_0\|_{C^{r+1,q}(\overline{\Omega})}\right).$$

We claim that $\delta\beta_f$ and χ_f vanish. Using the orthogonality of the decomposition and partial integration, we obtain

$$\int_\Omega |\delta\beta_f|^2 = \int_\Omega \langle \delta\beta_f; f - d\tilde{\omega}_0 \rangle = 0.$$

In the last equality we have used (7.1) and a density argument. The claim $\chi_f = 0$ follows in the same way using partial integration and the first condition in (A2), namely

$$\int_\Omega |\chi_f|^2 = \int_\Omega \langle \chi_f; f - d\tilde{\omega}_0 \rangle = \int_\Omega \langle \chi_f; f \rangle - \int_{\partial\Omega} \langle \chi_f; \nu \wedge \omega_0 \rangle = 0.$$

Hence, we have found $\alpha_f \in C^{r+1,q}\left(\overline{\Omega}; \Lambda^k\right)$ satisfying (if $k = n$, we take $\alpha_f = 0$)

$$\begin{cases}
d\alpha_f = f - d\tilde{\omega}_0 \quad \text{and} \quad \delta\alpha_f = 0 & \text{in } \Omega, \\
\nu \wedge \alpha_f = 0 & \text{on } \partial\Omega.
\end{cases} \tag{7.3}$$

We now apply the same decomposition to $g-\delta\tilde{\omega}_0$ (if $k=0$, we do not need this construction) and get

$$\begin{aligned} g-\delta\tilde{\omega}_0 &= d\alpha_g+\delta\beta_g+\psi_g \quad \text{in } \Omega, \\ \delta\alpha_g &= 0,\ d\beta_g=0 \quad \text{in } \Omega, \\ \nu\wedge\alpha_g &= 0,\ \nu\wedge\beta_g=0 \quad \text{on } \partial\Omega, \end{aligned}$$

where $\psi_g\in\mathscr{H}_T\left(\Omega;\Lambda^{k-1}\right)$. Moreover, there exists a positive constant $C=C(r,q,\Omega)$ such that

$$\|\beta_g\|_{C^{r+1,q}(\overline{\Omega})}\leq C\left(\|g\|_{C^{r,q}(\overline{\Omega})}+\|\tilde{\omega}_0\|_{C^{r+1,q}(\overline{\Omega})}\right).$$

Using (7.2), the second condition in (A2) and the same argument as before, we have that $d\alpha_g$ and ψ_g vanish (cf. Theorem 3.25 and Corollary 3.29). Hence, we have found $\beta_g\in C^{r+1,q}(\Omega;\Lambda^k)$ satisfying (if $k=0$, we take $\beta_g=0$)

$$\begin{cases} d\beta_g=0 \quad \text{and} \quad \delta\beta_g=g-\delta\tilde{\omega}_0 & \text{in } \Omega, \\ \nu\wedge\beta_g=0 & \text{on } \partial\Omega. \end{cases} \tag{7.4}$$

We now set

$$\omega=\alpha_f+\beta_g+\tilde{\omega}_0,$$

which satisfies, due to (7.3) and (7.4),

$$\begin{cases} d\omega=d\alpha_f+d\tilde{\omega}_0=f \quad \text{and} \quad \delta\omega=\delta\beta_g+\delta\tilde{\omega}_0=g & \text{in } \Omega, \\ \nu\wedge\omega=\nu\wedge\tilde{\omega}_0=\nu\wedge\omega_0 & \text{on } \partial\Omega. \end{cases}$$

This concludes the proof. □

7.2 System with Prescribed Normal Component

Using statement (ii) instead of (i) in Theorem 6.12, we obtain the following theorem in a completely analogous way as in Theorem 7.2.

Theorem 7.4. *Let $r\geq 0$ and $0\leq k\leq n$ be integers, $0<q<1$, $2\leq p<\infty$ and $\Omega\subset\mathbb{R}^n$ be a bounded open smooth set with exterior unit normal ν. Let $f:\overline{\Omega}\to\Lambda^{k+1}$, $g:\overline{\Omega}\to\Lambda^{k-1}$ and $\omega_0:\partial\Omega\to\Lambda^k$. Then the following statements are equivalent:*

(i) Let

$$f\in W^{r,p}\left(\Omega;\Lambda^{k+1}\right),\ g\in W^{r,p}\left(\Omega;\Lambda^{k-1}\right) \text{ and } \nu\lrcorner\omega_0\in W^{r+1-\frac{1}{p},p}\left(\partial\Omega;\Lambda^{k-1}\right),$$

respectively

$$f\in C^{r,q}\left(\overline{\Omega};\Lambda^{k+1}\right),\ g\in C^{r,q}\left(\overline{\Omega};\Lambda^{k-1}\right) \text{ and } \nu\lrcorner\omega_0\in C^{r+1,q}\left(\partial\Omega;\Lambda^{k-1}\right),$$

satisfying the conditions

$$df = 0 \text{ in } \Omega, \quad \delta g = 0 \text{ in } \Omega, \quad \nu \lrcorner \delta \omega_0 = \nu \lrcorner g \text{ on } \partial\Omega \tag{C1}$$

and, for every $\chi \in \mathscr{H}_N\left(\Omega;\Lambda^{k-1}\right)$ *and* $\psi \in \mathscr{H}_N\left(\Omega;\Lambda^{k+1}\right)$,

$$\int_\Omega \langle g;\chi\rangle - \int_{\partial\Omega} \langle \nu \lrcorner \omega_0;\chi\rangle = 0 \quad \text{and} \quad \int_\Omega \langle f;\psi\rangle = 0. \tag{C2}$$

(ii) There exists $\omega \in W^{r+1,p}\left(\Omega;\Lambda^k\right)$*, respectively* $\omega \in C^{r+1,q}\left(\overline{\Omega};\Lambda^k\right)$*, such that*

$$\begin{cases} d\omega = f \quad \text{and} \quad \delta\omega = g & \text{in } \Omega, \\ \qquad \nu \lrcorner \omega = \nu \lrcorner \omega_0 & \text{on } \partial\Omega. \end{cases}$$

In addition, there exists a constant $C_1 = C_1(r,p,\Omega)$ *such that*

$$\|\omega\|_{W^{r+1,p}(\Omega)} \leq C_1 \left(\|f\|_{W^{r,p}(\Omega)} + \|g\|_{W^{r,p}(\Omega)} + \|\nu \lrcorner \omega_0\|_{W^{r+1-\frac{1}{p},p}(\partial\Omega)}\right),$$

respectively $C_2 = C_2(r,q,\Omega)$ *such that*

$$\|\omega\|_{C^{r+1,q}(\overline{\Omega})} \leq C_2 \left(\|f\|_{C^{r,q}(\overline{\Omega})} + \|g\|_{C^{r,q}(\overline{\Omega})} + \|\nu \lrcorner \omega_0\|_{C^{r+1,q}(\partial\Omega)}\right).$$

Remark 7.5. (i) When $k = 0$, all statements of the theorem have to be understood as if g and ω_0 were not present. For example, statement (ii) reads then as

$$d\omega = f \quad \text{in } \Omega.$$

Similarly, when $k = n$, then all statements of the theorem have to be understood as if f was not present. In this case, statement (ii) becomes

$$\begin{cases} \delta\omega = g & \text{in } \Omega, \\ \omega = \omega_0 & \text{on } \partial\Omega. \end{cases}$$

Moreover, the result is also valid when $q = 0,1$ (see Theorem 8.18 and the remark following it).

(ii) If $r \geq 1$, conditions (C1) are well defined. This is obvious for the first two equations. The third one makes sense in $W^{r-\frac{1}{p},p}\left(\partial\Omega;\Lambda^{k-2}\right)$, respectively $C^{r,q}\left(\partial\Omega;\Lambda^{k-2}\right)$, due to Theorem 3.23 and Lemma 7.1.

(iii) If $r = 0$, then the conditions $df = 0$ and $\delta g = 0$ are understood in the sense of distributions. The third condition in (C1) is well defined in the Hölder case and is understood, in the Sobolev case, in the weak sense, namely

$$\int_\Omega \langle g;d\varphi\rangle - \int_{\partial\Omega} \langle \nu \lrcorner \omega_0;d\varphi\rangle = 0 \tag{D1}$$

for every $\varphi \in C^\infty\left(\overline{\Omega};\Lambda^{k-2}\right)$.

(iv) If $r \geq 1$, it can be easily shown, as in Remark 7.3, that (D1) is equivalent to the second and third conditions in (C1).

(v) If Ω is contractible and $k \geq 2$, then condition (C2) drops due to Theorem 6.5.

(vi) The above results remain valid if the set Ω is C^{r+3} for the Sobolev case and $C^{r+3,q}$ for the Hölder case.

(vii) The construction is linear and universal in the sense of Remark 6.13.

7.3 Weak Formulation for Closed Forms

We now establish two propositions that allow one to express the conditions $df = 0$ and $\delta g = 0$ in the sense of distributions in equivalent ways. We have used them to prove Theorems 7.2 and 7.4 when $r = 0$. The proof below, however, uses implicitly (through Theorem 8.18) the theorems when $r = \infty$.

Proposition 7.6. *Let $0 \leq k \leq n-1$ be integers and $\Omega \subset \mathbb{R}^n$ be a bounded open smooth set with exterior unit normal ν.*

Part 1. Let $f \in L^1\left(\Omega;\Lambda^k\right)$. Then the following statements are equivalent:

(i) f is closed in the sense of distributions, namely

$$\int_{\Omega} \langle f;\delta\varphi\rangle = 0 \quad \textit{for every } \varphi \in C_0^\infty\left(\Omega;\Lambda^{k+1}\right).$$

(ii) The following holds:

$$\int_{\Omega} \langle f;\delta\varphi\rangle = 0 \quad \textit{for every } \varphi \in C_N^\infty\left(\overline{\Omega};\Lambda^{k+1}\right).$$

Part 2. Let $f \in C^{0,q}\left(\overline{\Omega};\Lambda^k\right)$ with $0 < q < 1$. Then the two statements of part 1 are equivalent to the two conditions below.

(iii) The following holds:

$$\int_{\Omega} \langle f;\delta\varphi\rangle - \int_{\partial\Omega} \langle \nu \wedge f;\varphi\rangle = 0 \quad \textit{for every } \varphi \in C_T^\infty\left(\overline{\Omega};\Lambda^{k+1}\right).$$

(iv) The following identity is valid:

$$\int_{\Omega} \langle f;\delta\varphi\rangle - \int_{\partial\Omega} \langle \nu \wedge f;\varphi\rangle = 0 \quad \textit{for every } \varphi \in C^\infty\left(\overline{\Omega};\Lambda^{k+1}\right).$$

Remark 7.7. Since (i) is equivalent to (ii), the following statement is also equivalent with f being closed in the sense of distributions. For every open smooth set $O \subset \Omega$,

$$\int_{O} \langle f;\delta\varphi\rangle = 0 \quad \text{for every } \varphi \in C_N^\infty\left(\overline{O};\Lambda^{k+1}\right).$$

Proof. Part 1. The implication (ii) $\Rightarrow$ (i) is trivial, so let us prove (i) $\Rightarrow$ (ii).

Step 1. Let $\varphi \in C_N^\infty\left(\overline{\Omega};\Lambda^{k+1}\right)$ and find, using Theorem 8.18,

$$\widetilde{\varphi} \in C^\infty\left(\overline{\Omega};\Lambda^{k+1}\right) \quad \text{and} \quad \|\widetilde{\varphi}\|_{C^{1,1/2}} \leq C,$$

where $C = C\left(\|\varphi\|_{C^{1,1/2}}, \Omega\right) > 0$ such that

$$\begin{cases} \delta\widetilde{\varphi} = \delta\varphi & \text{in } \Omega, \\ \widetilde{\varphi} = 0 & \text{on } \partial\Omega. \end{cases}$$

This is possible using Theorem 3.23 and since $\varphi \in C_N^\infty\left(\overline{\Omega};\Lambda^{k+1}\right)$. Next, let $\varepsilon > 0$ be small enough. We can then find an open set Ω_ε such that

$$\Omega_\varepsilon \subset \Omega \quad \text{and} \quad \operatorname{dist}(\Omega_\varepsilon;\partial\Omega) \leq \varepsilon$$

and $\rho_\varepsilon \in C_0^\infty(\Omega)$ such that

$$\rho_\varepsilon \equiv 1 \text{ in } \Omega_\varepsilon \quad \text{and} \quad \|\operatorname{grad}\rho_\varepsilon\|_{C^0} \leq 2/\varepsilon.$$

We then let $\varphi_\varepsilon = \rho_\varepsilon\widetilde{\varphi} \in C_0^\infty\left(\Omega;\Lambda^{k+1}\right)$ and observe that since $\widetilde{\varphi} = 0$ on $\partial\Omega$, there exists a constant $C = C\left(\|\varphi\|_{C^{1,1/2}}, \Omega\right) > 0$ independent of ε such that

$$\|\varphi_\varepsilon\|_{C^1} \leq C \quad \text{and} \quad \varphi_\varepsilon \to \widetilde{\varphi} \text{ in } W^{1,1} \text{ as } \varepsilon \to 0.$$

We can therefore assume that, up to a subsequence, we also have

$$\delta\varphi_\varepsilon \to \delta\widetilde{\varphi} = \delta\varphi \quad \text{a.e. as } \varepsilon \to 0.$$

It therefore follows from the dominated convergence theorem that

$$\int_\Omega \langle f;\delta\varphi - \delta\varphi_\varepsilon\rangle \to 0 \quad \text{as } \varepsilon \to 0.$$

Step 2. Let $\varphi \in C_N^\infty\left(\overline{\Omega};\Lambda^{k+1}\right)$ and let $\varphi_\varepsilon \in C_0^\infty\left(\Omega;\Lambda^{k+1}\right)$ be as in Step 1. The hypothesis (i) and Step 1 lead, as $\varepsilon \to 0$, to

$$\int_\Omega \langle f;\delta\varphi\rangle = \int_\Omega \langle f;\delta\varphi_\varepsilon\rangle + \int_\Omega \langle f;\delta\varphi - \delta\varphi_\varepsilon\rangle = \int_\Omega \langle f;\delta\varphi - \delta\varphi_\varepsilon\rangle \to 0,$$

which implies (ii), namely

$$\int_\Omega \langle f;\delta\varphi\rangle = 0.$$

Part 2. The implications (iv) $\Rightarrow$ (iii) $\Rightarrow$ (i) are obvious. So let us show (i) $\Rightarrow$ (iv). In view of part 1, (i) implies

$$\int_\Omega \langle f;\delta\varphi\rangle = 0 \quad \text{for every } \varphi \in C_N^\infty\left(\overline{\Omega};\Lambda^{k+1}\right). \tag{7.5}$$

Applying the Hodge–Morrey decomposition theorem 6.12(ii) to f, we get $\alpha \in C_N^{1,q}\left(\overline{\Omega};\Lambda^{k-1}\right)$, $\beta \in C_N^{1,q}\left(\overline{\Omega};\Lambda^{k+1}\right)$ and $h \in \mathscr{H}_N\left(\Omega;\Lambda^k\right)$ such that

$$f = d\alpha + \delta\beta + h \quad \text{in } \Omega.$$

Due to (7.5) and by a density argument (cf. Theorem 3.25), we obtain

$$\int_\Omega \langle f;\delta\beta\rangle = 0.$$

The L^2-orthogonality of the Hodge–Morrey decomposition and the previous equation give

$$f = d\alpha + h \quad \text{in } \Omega.$$

Let now $\{\alpha_\varepsilon\} \subset C^\infty\left(\overline{\Omega};\Lambda^{k-1}\right)$ be such that

$$\alpha_\varepsilon \to \alpha \quad \text{in } C^1\left(\overline{\Omega};\Lambda^{k-1}\right) \quad \text{as } \varepsilon \to 0.$$

Define $f_\varepsilon = d\alpha_\varepsilon + h$. By construction, it has the following properties:

$$df_\varepsilon = 0 \text{ in } \Omega \quad \text{and} \quad f_\varepsilon \to f \text{ uniformly as } \varepsilon \to 0.$$

Since f_ε is in C^1 and closed, we obtain by partial integration

$$\int_\Omega \langle f_\varepsilon;\delta\varphi\rangle - \int_{\partial\Omega} \langle \nu \wedge f_\varepsilon;\varphi\rangle = 0 \quad \text{for every } \varphi \in C^\infty\left(\overline{\Omega};\Lambda^{k+1}\right).$$

Taking the limit as $\varepsilon \to 0$, we have (iv). □

We also have the dual version.

Proposition 7.8. *Let $1 \le k \le n$ be integers and $\Omega \subset \mathbb{R}^n$ be a bounded open smooth set with exterior unit normal ν.*

Part 1. Let $f \in L^1\left(\Omega;\Lambda^k\right)$. Then the following statements are equivalent:

(i) f is coclosed in the sense of distributions, namely

$$\int_\Omega \langle f;d\varphi\rangle = 0 \quad \textit{for every } \varphi \in C_0^\infty\left(\Omega;\Lambda^{k-1}\right).$$

(ii) The following holds:

$$\int_\Omega \langle f;d\varphi\rangle = 0 \quad \textit{for every } \varphi \in C_T^\infty\left(\overline{\Omega};\Lambda^{k-1}\right).$$

Part 2. Let $f \in C^{0,q}\left(\overline{\Omega};\Lambda^k\right)$ with $0 < q < 1$. Then the two statements of part 1 are equivalent to the two conditions below.

(iii) The following holds:

$$\int_\Omega \langle f;d\varphi\rangle - \int_{\partial\Omega} \langle \nu \lrcorner f;\varphi\rangle = 0 \quad \textit{for every } \varphi \in C_N^\infty\left(\overline{\Omega};\Lambda^{k-1}\right).$$

(iv) The following identity is valid:

$$\int_{\Omega}\langle f;d\varphi\rangle-\int_{\partial\Omega}\langle\nu\lrcorner f;\varphi\rangle=0\quad\text{for every }\varphi\in C^{\infty}\left(\overline{\Omega};\Lambda^{k-1}\right).$$

7.4 Equivalence Between Hodge Decomposition and Cauchy–Riemann-Type Systems

We now show that Theorem 7.2, respectively Theorem 7.4, is in fact equivalent to the Hodge–Morrey decomposition theorem (cf. part (i), respectively part (ii), of Theorem 6.12). We already saw that part (i) of Theorem 6.12 implies Theorem 7.2 (and part (ii) implies Theorem 7.4); we now show the converse. We establish this fact only in Hölder spaces, but the same result holds in Sobolev spaces.

Proposition 7.9. *Let $\Omega\subset\mathbb{R}^n$ be a bounded open smooth set. Let $0\le k\le n$ and $r\ge 1$ be integers, $0<q<1$ and $f\in C^{r,q}\left(\overline{\Omega};\Lambda^k\right)$. Then Theorem 7.2 implies the Hodge–Morrey decomposition Theorem 6.12(i); more precisely, it implies the existence of $\alpha\in C_T^{r+1,q}\left(\overline{\Omega};\Lambda^{k-1}\right)$, $\beta\in C_T^{r+1,q}\left(\overline{\Omega};\Lambda^{k+1}\right)$ and $h\in\mathscr{H}_T\left(\Omega;\Lambda^k\right)$ such that, in Ω,*

$$f=d\alpha+\delta\beta+h,\qquad\delta\alpha=0\quad\text{and}\quad d\beta=0,$$

with α, β and h mutually orthogonal with respect to the L^2-inner product. It also implies the existence of a constant $C=C(r,q,\Omega)$ such that

$$\|\alpha\|_{C^{r+1,q}}+\|\beta\|_{C^{r+1,q}}+\|h\|_{C^{r,q}}\le C\|f\|_{C^{r,q}}.$$

Remark 7.10. The proposition does not, however, establish the existence of a ω such that $\alpha=\delta\omega$ and $\beta=d\omega$.

Proof. We use Theorem 7.2 three times. We first find a solution $\phi\in C^{r,q}\left(\overline{\Omega};\Lambda^k\right)$ of

$$\begin{cases}d\phi=0\quad\text{and}\quad\delta\phi=\delta f & \text{in }\Omega,\\ \nu\wedge\phi=0 & \text{on }\partial\Omega.\end{cases}\tag{7.6}$$

The solvability conditions (A1) (or equivalently (B1)) and (A2) are easily verified. We next write

$$\phi=\phi^{\perp}+\chi,\quad\phi^{\perp}\in(\mathscr{H}_T)^{\perp},\quad\chi\in\mathscr{H}_T.$$

Since $\mathscr{H}_T$ and $(\mathscr{H}_T)^{\perp}$ are closed in $C^{r,q}\left(\overline{\Omega};\Lambda^k\right)$, we can apply the closed complement theorem (cf. for instance, Alt [5, Theorem 7.15]) and find that the projections onto $\mathscr{H}_T$ and $(\mathscr{H}_T)^{\perp}$ are continuous, namely

$$\|\phi^{\perp}\|_{C^{r,q}}+\|\chi\|_{C^{r,q}}\le C\|\phi\|_{C^{r,q}},$$

for some constant C independent of ϕ. We now find a solution $\alpha\in C^{r+1,q}\left(\overline{\Omega};\Lambda^{k-1}\right)$ such that

$$\begin{cases} d\alpha = \phi^{\perp} \quad \text{and} \quad \delta\alpha = 0 & \text{in } \Omega, \\ \nu \wedge \alpha = 0 & \text{on } \partial\Omega. \end{cases}$$

The solvability conditions (A1) and (A2) are satisfied due to (7.6) and the fact that $\phi^{\perp} \in (\mathscr{H}_T)^{\perp}$. We now apply again the L^2-orthogonal decomposition to $f - d\alpha$ and obtain

$$f - d\alpha = g + h, \quad g \in (\mathscr{H}_T)^{\perp} \quad \text{and} \quad h \in \mathscr{H}_T.$$

As above,

$$\|g\|_{C^{r,q}} + \|h\|_{C^{r,q}} \le C\|f - d\alpha\|_{C^{r,q}}.$$

At last, we find a solution $\beta \in C^{r+1,q}\left(\overline{\Omega}; \Lambda^{k+1}\right)$ of

$$\begin{cases} d\beta = 0 \quad \text{and} \quad \delta\beta = g & \text{in } \Omega, \\ \nu \wedge \beta = 0 & \text{on } \partial\Omega. \end{cases}$$

This is possible since

$$\delta g = \delta(f - d\alpha) = \delta\phi - \delta d\alpha = \delta\phi^{\perp} - \delta d\alpha = 0$$

and $g \in (\mathscr{H}_T)^{\perp}$. By construction, α, β and h have all of the required properties. □

Chapter 8
Poincaré Lemma

8.1 The Classical Poincaré Lemma

Our first result is the classical Poincaré lemma. Its proof is elementary and does not use the Hodge–Morrey decomposition. Its drawback (compare with Theorem 8.3) is that it does not provide the expected gain in regularity and is restricted to contractible sets.

Theorem 8.1 (Poincaré lemma). *Let $r \geq 1$ and $0 \leq k \leq n-1$ be integers and $\Omega \subset \mathbb{R}^n$ be an open contractible set. Let $g \in C^r\left(\Omega;\Lambda^{k+1}\right)$ with $dg = 0$ in Ω. Then there exists $G \in C^r\left(\Omega;\Lambda^k\right)$ such that*

$$dG = g \quad \text{in } \Omega.$$

Remark 8.2. When $k = 0$, the theorem gives immediately that $G \in C^{r+1}(\Omega)$.

Proof. Since Ω is contractible, we have that there exist $x_0 \in \Omega$ and

$$F \in C^\infty([0,1] \times \Omega; \Omega)$$

such that for every $x \in \Omega$,

$$F(0,x) = x_0 \quad \text{and} \quad F(1,x) = x.$$

We then apply Theorem 17.3 to F to get that there exists $G \in C^r\left(\Omega;\Lambda^k\right)$ such that

$$dG = F_1^*(g) - F_0^*(g) = g \quad \text{in } \Omega.$$

This achieves the proof of the theorem. □

8.2 Global Poincaré Lemma with Optimal Regularity

We now have a global version of Poincaré lemma with optimal regularity, as well as its dual version.

Theorem 8.3. *Let $r \geq 0$ and $0 \leq k \leq n-1$ be integers, $0 < \alpha < 1$, $2 \leq p < \infty$ and $\Omega \subset \mathbb{R}^n$ be a bounded open smooth set. The following statements are equivalent:*

(i) Let $f \in W^{r,p}\left(\Omega;\Lambda^{k+1}\right)$, respectively $C^{r,\alpha}\left(\overline{\Omega};\Lambda^{k+1}\right)$, be such that

$$df = 0 \text{ in } \Omega \quad \text{and} \quad \int_\Omega \langle f;\psi\rangle = 0 \text{ for every } \psi \in \mathscr{H}_N\left(\Omega;\Lambda^{k+1}\right).$$

(ii) There exists $\omega \in W^{r+1,p}\left(\Omega;\Lambda^{k}\right)$, respectively $C^{r+1,\alpha}\left(\overline{\Omega};\Lambda^{k}\right)$, such that

$$d\omega = f \quad \text{in } \Omega.$$

Moreover, there exists a constant $C_1 = C_1(r,p,\Omega)$ such that

$$\|\omega\|_{W^{r+1,p}} \leq C_1 \|f\|_{W^{r,p}},$$

respectively there exists a constant $C_2 = C_2(r,\alpha,\Omega)$ such that

$$\|\omega\|_{C^{r+1,\alpha}} \leq C_2 \|f\|_{C^{r,\alpha}}.$$

Theorem 8.4. *Let $r \geq 0$ and $1 \leq k \leq n$ be integers, $0 < \alpha < 1$, $2 \leq p < \infty$ and $\Omega \subset \mathbb{R}^n$ be a bounded open smooth set. The following statements are equivalent:*

(i) Let $g \in W^{r,p}\left(\Omega;\Lambda^{k-1}\right)$, respectively $C^{r,\alpha}\left(\overline{\Omega};\Lambda^{k-1}\right)$, be such that

$$\delta g = 0 \text{ in } \Omega \quad \text{and} \quad \int_\Omega \langle g;\psi\rangle = 0 \text{ for every } \psi \in \mathscr{H}_T\left(\Omega;\Lambda^{k-1}\right).$$

(ii) There exists $\omega \in W^{r+1,p}\left(\Omega;\Lambda^{k}\right)$, respectively $C^{r+1,\alpha}\left(\overline{\Omega};\Lambda^{k}\right)$, such that

$$\delta\omega = g \quad \text{in } \Omega.$$

Moreover, there exists a constant $C_1 = C_1(r,p,\Omega)$ such that

$$\|\omega\|_{W^{r+1,p}} \leq C_1 \|g\|_{W^{r,p}},$$

respectively there exists a constant $C_2 = C_2(r,\alpha,\Omega)$ such that

$$\|\omega\|_{C^{r+1,\alpha}} \leq C_2 \|g\|_{C^{r,\alpha}}.$$

Remark 8.5. (i) When $k = n-1$ in Theorem 8.3 or $k = 1$ in Theorem 8.4, there is no restriction on the solvability of $d\omega = f$ or $\delta\omega = g$ (cf. Theorem 6.5).

(ii) If $r = 0$, then the conditions $df = 0$ or $\delta g = 0$ have to be understood in the sense of distributions.

(iii) The above results remain valid if Ω is C^{r+3} for the Sobolev case and $C^{r+3,\alpha}$ for the Hölder case.

(iv) If Ω is contractible, then (cf. Theorem 6.5)

$$\mathscr{H}_T(\Omega;\Lambda^k) = \{0\} \quad \text{if } 0 \le k \le n-1,$$
$$\mathscr{H}_N(\Omega;\Lambda^k) = \{0\} \quad \text{if } 1 \le k \le n.$$

(v) The construction is linear and universal in the sense of Remark 6.13.

We only prove the first theorem and only in the Sobolev case; all of the other statements are obtained by trivial adaptation of the proof below.

Proof. (ii) $\Rightarrow$ (i). Suppose first that there exists $\omega \in W^{r+1,p}(\Omega;\Lambda^k)$ such that $f = d\omega$. Clearly, $df = 0$ and the other assertion follows by partial integration, since, for every $\psi \in \mathscr{H}_N$,

$$\int_\Omega \langle f;\psi\rangle = \int_\Omega \langle d\omega;\psi\rangle = -\int_\Omega \langle \omega;\delta\psi\rangle + \int_{\partial\Omega} \langle \omega;\nu \lrcorner \psi\rangle = 0.$$

(i) $\Rightarrow$ (ii). Suppose now that

$$df = 0 \text{ in } \Omega \quad \text{and} \quad \int_\Omega \langle f;\psi\rangle = 0 \text{ for every } \psi \in \mathscr{H}_N(\Omega;\Lambda^{k+1}).$$

We then appeal to Theorem 7.4 to solve the problem

$$\begin{cases} d\omega = f \quad \text{and} \quad \delta\omega = 0 & \text{in } \Omega, \\ \nu \lrcorner \omega = 0 & \text{on } \partial\Omega. \end{cases}$$

This concludes the proof. ☐

When $k = 0$ in Theorem 8.3 or $k = n$ in Theorem 8.4, then the result can be refined so as to include the limit cases $\alpha = 0, 1$.

Corollary 8.6. *Let $r \ge 0$ be an integer, $0 \le \alpha \le 1$ and $\Omega \subset \mathbb{R}^n$ be a bounded open smooth set. Let $f \in C^{r,\alpha}(\overline{\Omega};\Lambda^1)$ be such that*

$$df = 0 \text{ in } \Omega \quad \text{and} \quad \int_\Omega \langle f;\psi\rangle = 0 \text{ for every } \psi \in \mathscr{H}_N(\Omega;\Lambda^1).$$

Then there exist $\omega \in C^{r+1,\alpha}(\overline{\Omega})$ and a constant $C = C(r,\Omega)$ such that

$$d\omega = f \text{ in } \Omega \quad \text{and} \quad \|\omega\|_{C^{r+1,\alpha}} \le C\,\|f\|_{C^{r,\alpha}}.$$

Corollary 8.7. *Let $r \ge 0$ be an integer, $0 \le \alpha \le 1$ and $\Omega \subset \mathbb{R}^n$ be a bounded open smooth set. Let $g \in C^{r,\alpha}(\overline{\Omega};\Lambda^{n-1})$ be such that*

$$\delta g = 0 \text{ in } \Omega \quad \text{and} \quad \int_\Omega \langle g;\psi\rangle = 0 \text{ for every } \psi \in \mathscr{H}_T(\Omega;\Lambda^{n-1}).$$

Then there exist $\omega \in C^{r+1,\alpha}\left(\overline{\Omega};\Lambda^n\right)$ *and a constant* $C = C(r,\Omega)$ *such that*

$$\delta\omega = g \text{ in } \Omega \quad \text{and} \quad \|\omega\|_{C^{r+1,\alpha}} \leq C\,\|g\|_{C^{r,\alpha}}.$$

We only prove the first corollary, the second one being obtained by duality.

Proof. Choose p such that $n < p < \infty$. Since $C^{r,\alpha}\left(\overline{\Omega}\right) \subset W^{r,p}(\Omega)$, we can apply Theorem 8.3 and find $\omega \in W^{r+1,p}(\Omega)$ such that

$$d\omega = \operatorname{grad}\omega = f \quad \text{in } \Omega.$$

Note first that, since $p > n$, we get, from the Morrey imbedding theorem,

$$\omega \in C^r\left(\overline{\Omega}\right).$$

Using again that $p > n$, we obtain, since $\nabla^r\omega \in W^{1,p}$, that $\nabla^r\omega$ is differentiable in Ω and its gradient equals its weak gradient (cf. e.g., Theorem 5 in Section 5.8.3 in [41]). Note that in our case, we have everywhere (and not just almost everywhere) differentiability, since $f \in C^{r,\alpha}$. Hence, we have obtained that $\omega \in C^{r+1,\alpha}\left(\overline{\Omega}\right)$ and the proof is complete. □

It is possible to give a more classical and direct proof of the theorem without appealing to Hodge–Morrey decomposition. We discuss only the case where Ω is star-shaped with respect to 0. It can be easily seen that if f is a closed 1-form, then

$$\omega(x) = \int_0^1 \langle f(tx);x\rangle\, dt$$

has all the desired properties. This is elementary if $r \geq 1$ and can easily be established even when $r = 0$ (for more details, we refer to Csató [23]).

8.3 Some Preliminary Lemmas

We start with a slight improvement of a lemma proved in Dacorogna and Moser [33].

Lemma 8.8. *Let* $r \geq 0$ *be an integer,* $0 \leq \alpha \leq 1$ *and* $\Omega \subset \mathbb{R}^n$ *be a bounded open* $C^{r+1,\alpha}$ *set with exterior unit normal* ν. *Let* $c \in C^{r,\alpha}(\partial\Omega)$. *Then there exists*

$$b \in C^{r+1,\alpha}\left(\overline{\Omega}\right)$$

satisfying, all over $\partial\Omega$,

$$\operatorname{grad} b = c\,\nu \quad \text{and} \quad b = 0.$$

Furthermore, there exists a constant $C = C(r,\Omega) > 0$ *such that*

$$\|b\|_{C^{r+1,\alpha}\left(\overline{\Omega}\right)} \leq C\,\|c\|_{C^{r,\alpha}(\partial\Omega)}.$$

Remark 8.9. The above result is valid (cf. Proof 1 below) in Sobolev spaces and reads as follows. Let $r \geq 1$ be an integer, $1 < p < \infty$ and $\Omega \subset \mathbb{R}^n$ be a bounded open C^{r+1} set with exterior unit normal ν. Let $c \in W^{r-\frac{1}{p},p}(\partial\Omega)$; then there exists

$$b \in W^{r+1,p}(\Omega)$$

satisfying, all over $\partial\Omega$,

$$\operatorname{grad} b = c\,\nu \quad \text{and} \quad b = 0.$$

Moreover, there exists a constant $C = C(r, p, \Omega) > 0$ such that

$$\|b\|_{W^{r+1,p}(\Omega)} \leq C\,\|c\|_{W^{r-1/p,p}(\partial\Omega)}\,.$$

We start by proving the lemma in the particular case of the half-space.

Lemma 8.10. *Let $n \geq 2$, $r \geq 0$ be integers, $0 \leq \alpha \leq 1$ and $f \in C^{r,\alpha}(\mathbb{R}^{n-1})$. Let $\delta > 0$ and $\varphi \in C_0^\infty(\mathbb{R}^{n-1})$ be such that*

$$\operatorname{supp}\varphi \subset B'_\delta \quad \textit{and} \quad \int_{\mathbb{R}^{n-1}} \varphi = 1,$$

where $B'_\delta \subset \mathbb{R}^{n-1}$ denotes the open ball centered at 0 and of radius δ. Then $F : \mathbb{R}^n \to \mathbb{R}$ defined by

$$F(x) = F(x', x_n) = x_n \int_{\mathbb{R}^{n-1}} \varphi(y')\, f(x' - x_n y')\, dy'$$

belongs to $C^{r+1,\alpha}(\mathbb{R}^n) \cap C^\infty(\mathbb{R}^n \setminus \{x_n = 0\})$ and satisfies, for every $x' \in \mathbb{R}^{n-1}$,

$$F(x', 0) = 0 \quad \textit{and} \quad \operatorname{grad} F(x', 0) = (0, \ldots, 0, f(x')).$$

Moreover, for every $R > 0$, there exists $C = C(r, R, \varphi) > 0$ such that

$$\|F\|_{C^{r+1,\alpha}(\overline{B}_R)} \leq C \|f\|_{C^{r,\alpha}(\overline{B'}_{R(1+\delta)})}\,.$$

Proof. We first compute the derivatives of F when $x_n \neq 0$. Since

$$\begin{aligned} F(x', x_n) &= x_n \int_{\mathbb{R}^{n-1}} \varphi(y')\, f(x' - x_n y')\, dy' \\ &= x_n \int_{\mathbb{R}^{n-1}} \frac{1}{x_n^{n-1}}\, \varphi\left(\frac{x' - y'}{x_n}\right) f(y')\, dy', \end{aligned}$$

we find, for $1 \leq i \leq n-1$,

$$\begin{aligned} \frac{\partial F}{\partial x_i}(x', x_n) &= \int_{\mathbb{R}^{n-1}} \frac{1}{x_n^{n-1}} \frac{\partial \varphi}{\partial x_i}\left(\frac{x' - y'}{x_n}\right) f(y')\, dy' \\ &= \int_{\mathbb{R}^{n-1}} \frac{\partial \varphi}{\partial x_i}(y')\, f(x' - x_n y')\, dy', \end{aligned}$$

whereas, for $i = n$,

$$\begin{aligned}\frac{\partial F}{\partial x_n}(x', x_n) &= (2-n)\int_{\mathbb{R}^{n-1}} \frac{1}{x_n^{n-1}}\varphi\left(\frac{x'-y'}{x_n}\right) f(y')\,dy' \\ &\quad - \int_{\mathbb{R}^{n-1}} \frac{1}{x_n^{n-1}} \sum_{j=1}^{n-1} \frac{x_j - y_j}{x_n}\frac{\partial \varphi}{\partial x_j}\left(\frac{x'-y'}{x_n}\right) f(y')\,dy' \\ &= (2-n)\int_{\mathbb{R}^{n-1}} \varphi(y') f(x' - x_n y')\,dy' \\ &\quad - \int_{\mathbb{R}^{n-1}} \langle y'; \operatorname{grad}\varphi\,(y')\rangle\, f(x' - x_n y')\,dy'.\end{aligned}$$

These formulas extend continuously to $x_n = 0$ and we get, since $\operatorname{supp}\varphi \subset B'_\delta$ and $\int_{\mathbb{R}^{n-1}} \varphi = 1$, that

$$\frac{\partial F}{\partial x_i}(x', 0) = f(x')\int_{\mathbb{R}^{n-1}} \frac{\partial \varphi}{\partial x_i}(y')\,dy' = 0, \quad 1 \le i \le n-1,$$

$$\begin{aligned}\frac{\partial F}{\partial x_n}(x', 0) &= (2-n) f(x')\int_{\mathbb{R}^{n-1}} \varphi(y')\,dy' \\ &\quad - f(x')\int_{\mathbb{R}^{n-1}} \langle y'; \operatorname{grad}\varphi\,(y')\rangle\,dy' = f(x').\end{aligned}$$

From the above formulas we immediately infer that

$$\|F\|_{C^{r+1,\alpha}(\overline{B}_R)} \le C\|f\|_{C^{r,\alpha}(\overline{B'}_{R(1+\delta)})}.$$

This concludes the proof of the lemma. □

We next prove Lemma 8.8.

Proof. If one is not interested in the sharp regularity result, a solution of the problem is given by

$$b(x) = -c(x)\,\zeta\,(d(x, \partial\Omega)),$$

where c has been extended to $\overline{\Omega}$ and $d(x, \partial\Omega)$ stands for the distance from x to the boundary (recalling that the distance function is as regular as the set Ω near the boundary; see, e.g., Gilbarg and Trudinger [49]) and ζ is a smooth function so that $\zeta(0) = 0$, $\zeta'(0) = 1$ and $\zeta \equiv 0$ outside a small neighborhood of 0.

We will provide two proofs of the lemma. The first one uses elliptic regularity and hence only works whenever $0 < \alpha < 1$ (and also works in L^p for $1 < p < \infty$); in this case, the constant obtained depends also on α. The second one, which works also when $\alpha = 0, 1$, uses admissible boundary coordinate systems and the previous lemma.

Proof 1. The desired solution b is obtained by solving

$$\begin{cases} \Delta^2 b = 0 & \text{in } \Omega, \\ b = 0 \text{ and } \frac{\partial b}{\partial \nu} = c & \text{on } \partial\Omega. \end{cases}$$

The solution

$$b \in C^\infty(\Omega) \cap C^{r+1,\alpha}\left(\overline{\Omega}\right)$$

satisfies the estimate (see Agmon, Douglis and Nirenberg [3], Theorem 12.10 for the existence and Theorem 7.3 and the remarks following for the estimate)

$$\|b\|_{C^{r+1,\alpha}(\overline{\Omega})} \leq C \|c\|_{C^{r,\alpha}(\partial\Omega)}.$$

Clearly, b solves on $\partial\Omega$,

$$\operatorname{grad} b = c\,\nu \quad \text{and} \quad b = 0.$$

Proof 2. Let $m > 0$ be an integer and, for $1 \leq i \leq m$, let U_i, V_i be open sets in $\mathbb{R}^n$ and $\varphi_i \in \operatorname{Diff}^{r+1,\alpha}(U_i; V_i)$ be admissible boundary coordinate systems, as given in Proposition 3.17, such that

$$\partial\Omega \subset \bigcup_{i=1}^{m} V_i.$$

Without loss of generality, we can assume that U_i are open balls centered at 0, $\varphi_i \in \operatorname{Diff}^{r+1,\alpha}\left(\overline{U}_i; \overline{V}_i\right)$ and (see Remark 3.16)

$$\frac{\partial \varphi_i}{\partial y_n} = \nu(\varphi_i).$$

Moreover, let $\{\theta_i\}_{i=1}^m$ be a smooth partition of unity of $\partial\Omega$ subordinate to $\{V_i\}_{i=1}^m$. In the sequel, C_1, C_2 and C_3 will denote generic constants depending on r, Ω, U_i and θ_i. We also let

$$H = \{(y', 0) \in \mathbb{R}^n; y' \in \mathbb{R}^{n-1}\} \subset \mathbb{R}^n.$$

We define for $(y', 0) \in \overline{U}_i$,

$$g_i(y') = c(\varphi_i(y', 0)).$$

Note that $g_i \in C^{r,\alpha}\left(\overline{U}_i \cap H\right)$ in view of Definition 16.7. Using Theorem 16.11 and Definition 16.7 again, we can extend g_i such that $g_i \in C_0^{r,\alpha}\left(\mathbb{R}^{n-1}\right)$, satisfying the estimate

$$\|g_i\|_{C^{r,\alpha}(\mathbb{R}^{n-1})} \leq C_1 \|c\|_{C^{r,\alpha}(\partial\Omega)}.$$

According to Lemma 8.10, there exists $f_i \in C^{r+1,\alpha}(\mathbb{R}^n)$ such that on H,

$$f_i = 0 \quad \text{and} \quad \operatorname{grad} f_i = g_i\, e_n,$$

where $e_n = (0, \ldots, 0, 1)$ and we have

$$\|f_i\|_{C^{r+1,\alpha}(\overline{U}_i)} \leq C_2 \|g_i\|_{C^{r,\alpha}(\mathbb{R}^{n-1})} \leq C_3 \|c\|_{C^{r,\alpha}(\partial\Omega)}. \tag{8.1}$$

We claim that b given by

$$b(x) = \sum_{i=1}^{m} \theta_i(x) f_i(\varphi_i^{-1}(x))$$

has all of the desired properties. If $x \in \partial\Omega$, then $\varphi_i^{-1}(x) \in H$ and, thus, we immediately obtain that $f_i(\varphi_i^{-1}(x)) = 0$. Therefore, we get all over $\partial\Omega$,

$$\begin{aligned} \operatorname{grad} b(x) &= \sum_{i=1}^{m} \theta_i(x) \nabla (f_i \circ \varphi_i^{-1})(x) = \sum_{i=1}^{m} \theta_i(x) \nabla f_i(\varphi_i^{-1}(x)) \nabla \varphi_i^{-1}(x) \\ &= \sum_{i=1}^{m} \theta_i(x) g_i(\varphi_i^{-1}(x)) e_n \nabla \varphi_i^{-1}(x) = \sum_{i=1}^{m} \theta_i(x) c(x) e_n \nabla \varphi_i^{-1}(x). \end{aligned}$$

It remains to show that $e_n \nabla \varphi_i^{-1}(x) = \nu(x)$, which is equivalent to, setting $y = \varphi_i^{-1}(x)$,

$$e_n = \nu(\varphi_i(y)) \nabla \varphi_i(y) \quad \text{for every } y \in H \cap U_i.$$

This follows from the fact that φ_i is an admissible boundary coordinate system, namely for $1 \leq l \leq n$,

$$[\nu(\varphi_i(y)) \nabla \varphi_i(y)]_l = \left\langle \nu(\varphi_i(y)); \frac{\partial \varphi_i}{\partial y_l}(y) \right\rangle = \left\langle \frac{\partial \varphi_i}{\partial y_n}(y); \frac{\partial \varphi_i}{\partial y_l}(y) \right\rangle = \delta_{nl}.$$

The estimate of the Hölder norm of b follows from (8.1), Theorem 16.28 and Theorem 16.31. □

We now need a generalization of the above lemma to differential forms, as achieved in Dacorogna [27].

Lemma 8.11. *Let $r \geq 0$ and $1 \leq k \leq n-1$ be integers, $0 \leq \alpha \leq 1$ and $\Omega \subset \mathbb{R}^n$ be a bounded open $C^{r+1,\alpha}$ set with exterior unit normal ν.*

(i) If $c \in C^{r,\alpha}(\partial\Omega; \Lambda^k)$ is such that

$$\nu \wedge c = 0 \quad \text{on } \partial\Omega,$$

then there exists $b \in C^{r+1,\alpha}(\overline{\Omega}; \Lambda^{k-1})$ satisfying all over $\partial\Omega$,

$$db = c, \quad \delta b = 0 \quad \text{and} \quad b = 0.$$

Moreover, there exists a constant $C = C(r, \Omega) > 0$ such that

$$\|b\|_{C^{r+1,\alpha}(\overline{\Omega})} \leq C \|c\|_{C^{r,\alpha}(\partial\Omega)}.$$

(ii) If $c \in C^{r,\alpha}(\partial\Omega;\Lambda^k)$ *is such that*

$$\nu \lrcorner c = 0 \quad \text{on } \partial\Omega,$$

then there exists $b \in C^{r+1,\alpha}(\overline{\Omega};\Lambda^{k+1})$ *satisfying all over* $\partial\Omega$,

$$\delta b = c, \quad db = 0 \quad \text{and} \quad b = 0.$$

Furthermore, there exists a constant $C = C(r,\Omega) > 0$ *such that*

$$\|b\|_{C^{r+1,\alpha}(\overline{\Omega})} \leq C\,\|c\|_{C^{r,\alpha}(\partial\Omega)}\,.$$

Remark 8.12. (i) If $k = 0$ in statement (ii) (and analogously if $k = n$ in statement (i)) and $0 < \alpha < 1$, then it is easy to find b such that (and without any restriction on c)

$$\delta b = c \quad \text{and} \quad db = 0 \quad \text{in } \overline{\Omega},$$

where c has been extended to $\overline{\Omega}$ with the appropriate regularity. Indeed, choose $b = \operatorname{grad} B$, where B solves

$$\begin{cases} \Delta B = c & \text{in } \Omega, \\ B = 0 & \text{on } \partial\Omega. \end{cases}$$

(ii) The above result remains valid, with the same proof, in the Sobolev setting. More precisely, statement (i) (and similarly for statement (ii)) reads as follows. Let $r \geq 1$ be an integer, $1 < p < \infty$ and $\Omega \subset \mathbb{R}^n$ be a bounded open C^{r+1} set with exterior unit normal ν. Let $c \in W^{r-1/p,p}(\partial\Omega;\Lambda^k)$; then there exists

$$b \in W^{r+1,p}(\Omega;\Lambda^{k-1})$$

satisfying all over $\partial\Omega$,

$$db = c, \quad \delta b = 0 \quad \text{and} \quad b = 0.$$

Moreover, there exists a constant $C = C(r,p,\Omega) > 0$ such that

$$\|b\|_{W^{r+1,p}(\Omega)} \leq C\,\|c\|_{W^{r-1/p,p}(\partial\Omega)}\,.$$

Proof. Step 1. We start with case (i). First, solve with Lemma 8.8 the problem, on $\partial\Omega$,

$$\operatorname{grad} b_{i_1\cdots i_{k-1}} = (\nu \lrcorner c)_{i_1\cdots i_{k-1}}\,\nu \quad \text{and} \quad b_{i_1\cdots i_{k-1}} = 0$$

for every multi-index $1 \leq i_1 < \cdots < i_{k-1} \leq n$ and set

$$b = \sum_{1 \leq i_1 < \ldots < i_{k-1} \leq n} b_{i_1\cdots i_{k-1}}\,dx^{i_1} \wedge \cdots \wedge dx^{i_{k-1}}.$$

The formulas of Propositions 2.6 and 3.3 immediately imply that, on $\partial\Omega$,

$$db = \nu \wedge (\nu \lrcorner c) \quad \text{and} \quad \delta b = \nu \lrcorner (\nu \lrcorner c) = 0.$$

We combine the first equation with the hypothesis $\nu \wedge c = 0$ and use (2.7) to get

$$db = \nu \wedge (\nu \lrcorner c) = \nu \wedge (\nu \lrcorner c) + \nu \lrcorner (\nu \wedge c) = c \quad \text{on } \partial\Omega.$$

We have therefore proved the assertion.

Step 2. For (ii), we first solve, on $\partial\Omega$,

$$\operatorname{grad} b_{i_1 \ldots i_{k+1}} = (\nu \wedge c)_{i_1 \ldots i_{k+1}} \nu \quad \text{and} \quad b_{i_1 \cdots i_{k+1}} = 0$$

and then proceed exactly as in Step 1. This concludes the proof of the lemma. □

If Ω is contractible, then Lemma 8.11 can be generalized and gives a global version of the Poincaré lemma on the manifold $\partial\Omega$.

Corollary 8.13. *Let $\Omega \subset \mathbb{R}^n$ be a bounded open contractible smooth set with exterior unit normal ν. Let $r \geq 1$, $1 \leq k \leq n-1$ be two integers and $0 < \alpha < 1$. Then the following are equivalent:*

(i) Let $c \in C^{r,\alpha}\left(\partial\Omega; \Lambda^k\right)$ satisfy

$$\begin{aligned} &\nu \wedge dc = 0 \text{ on } \partial\Omega && \text{if } k \leq n-2, \\ &\int_{\partial\Omega} \nu \wedge c = 0 && \text{if } k = n-1. \end{aligned}$$

(ii) There exists $b \in C^{r+1,\alpha}\left(\overline{\Omega}; \Lambda^{k-1}\right)$ satisfying all over $\partial\Omega$,

$$db = c \quad \text{and} \quad \delta b = 0.$$

Remark 8.14. (i) $\nu \wedge dc \in C^{r-1,\alpha}\left(\partial\Omega; \Lambda^{k+2}\right)$ is well defined in view of Theorem 3.23.

(ii) If $k = n$, then the problem is solvable without any condition on c and the topology of Ω (cf. Remark 8.12(i)).

(iii) The corollary is indeed a generalization of Lemma 8.11 since $\nu \wedge c = 0$ implies $\nu \wedge dc = 0$, appealing again to Theorem 3.23.

(iv) We cannot require the solution b to satisfy $b = 0$ on $\partial\Omega$ as in Lemma 8.11. This would imply $0 = \nu \wedge db = \nu \wedge c$ on $\partial\Omega$, but $\nu \wedge dc = 0$ does not imply, in general, $\nu \wedge c = 0$.

(v) An analogous result holds true in Sobolev spaces.

Proof. The implication (ii) ⇒ (i) follows immediately from Theorems 3.23 and 3.26. So let us show the reverse implication. We first apply Theorem 7.2 to find a solution $u \in C^{r,\alpha}\left(\overline{\Omega}; \Lambda^k\right)$ of the problem

$$\begin{cases} du = 0 \quad \text{and} \quad \delta u = 0 & \text{in } \Omega, \\ \nu \wedge u = \nu \wedge c & \text{on } \partial\Omega. \end{cases}$$

Conditions (A1) and (A2) of Theorem 7.2 are satisfied, since $\mathcal{H}_T\left(\Omega;\Lambda^{k+1}\right)$ is equal to $\{0\}$ if $k \leq n-2$ and is equal to the set of constant volume forms if $k = n-1$ (cf. Theorem 6.5). We next use Theorem 7.4 to find a solution $\omega \in C^{r+1,\alpha}\left(\overline{\Omega};\Lambda^{k-1}\right)$ of

$$\begin{cases} d\omega = u \quad \text{and} \quad \delta\omega = 0 & \text{in } \Omega, \\ \nu \lrcorner \omega = 0 & \text{on } \partial\Omega. \end{cases}$$

Conditions (C1) and the first equation in (C2) are obviously satisfied. The second equation in (C2) follows from the fact that $\mathcal{H}_N\left(\Omega;\Lambda^k\right) = \{0\}$, cf. Theorem 6.5. We now use Lemma 8.11 to find $v \in C^{r+1,\alpha}\left(\overline{\Omega};\Lambda^{k-1}\right)$ satisfying all over $\partial\Omega$

$$dv = c - u \quad \text{and} \quad \delta v = 0.$$

At last we set $b = \omega + v$. It can be easily seen that b has the required properties. □

We also have the dual version of the corollary.

Corollary 8.15. *Let $\Omega \subset \mathbb{R}^n$ be a bounded open contractible smooth set with exterior unit normal ν. Let $r \geq 1$, $1 \leq k \leq n-1$ be two integers and $0 < \alpha < 1$. Then the following are equivalent:*

(i) Let $c \in C^{r,\alpha}\left(\partial\Omega;\Lambda^k\right)$ satisfy

$$\begin{aligned} \nu \lrcorner \delta c = 0 \text{ on } \partial\Omega \quad & \text{if } k \geq 2, \\ \textstyle\int_{\partial\Omega} \nu \lrcorner c = 0 \quad & \text{if } k = 1. \end{aligned}$$

(ii) There exists $b \in C^{r+1,\alpha}\left(\overline{\Omega};\Lambda^{k+1}\right)$ satisfying all over $\partial\Omega$,

$$\delta b = c \quad \text{and} \quad db = 0.$$

8.4 Poincaré Lemma with Dirichlet Boundary Data

We now consider the boundary value problems

$$\begin{cases} d\omega = f & \text{in } \Omega, \\ \omega = \omega_0 & \text{on } \partial\Omega \end{cases} \quad \text{and} \quad \begin{cases} \delta\omega = g & \text{in } \Omega, \\ \omega = \omega_0 & \text{on } \partial\Omega. \end{cases}$$

In contrast to the problems of Section 7.1 (respectively Section 7.2), $\delta\omega$ (respectively $d\omega$) is not prescribed; however both the tangential and normal components of ω are given on the boundary. It turns out that the problems can be solved under exactly the same hypotheses on f, g and ω_0 as in Theorem 7.2 (respectively Theorem 7.4). We follow exactly the construction in Dacorogna [27] for Hölder spaces; a very similar method is used in Schwarz [89] for Sobolev spaces.

Theorem 8.16. *Let $r \geq 0$ and $0 \leq k \leq n-1$ be integers, $0 < \alpha < 1$, $2 \leq p < \infty$ and $\Omega \subset \mathbb{R}^n$ be a bounded open smooth set with exterior unit normal ν. Let $f : \overline{\Omega} \to \Lambda^{k+1}$ and $\omega_0 : \partial\Omega \to \Lambda^k$. Then the following statements are equivalent:*

(i) Let $f \in C^{r,\alpha}\left(\overline{\Omega};\Lambda^{k+1}\right)$ and $\omega_0 \in C^{r+1,\alpha}\left(\partial\Omega;\Lambda^k\right)$, respectively $f \in W^{r,p}(\Omega;\Lambda^{k+1})$ and $\omega_0 \in W^{r+1-\frac{1}{p},p}\left(\partial\Omega;\Lambda^k\right)$, satisfy

$$df = 0 \text{ in } \Omega, \quad \nu \wedge d\omega_0 = \nu \wedge f \text{ on } \partial\Omega \tag{A1}$$

and, for every $\chi \in \mathscr{H}_T\left(\Omega;\Lambda^{k+1}\right)$,

$$\int_\Omega \langle f;\chi\rangle - \int_{\partial\Omega} \langle \nu \wedge \omega_0;\chi\rangle = 0. \tag{A2}$$

(ii) There exists $\omega \in C^{r+1,\alpha}\left(\overline{\Omega};\Lambda^k\right)$, respectively $\omega \in W^{r+1,p}\left(\Omega;\Lambda^k\right)$, such that

$$\begin{cases} d\omega = f & \text{in } \Omega, \\ \omega = \omega_0 & \text{on } \partial\Omega \end{cases}$$

and there exists a constant $C_1 = C_1(r,\alpha,\Omega)$ such that

$$\|\omega\|_{C^{r+1,\alpha}(\overline{\Omega})} \leq C_1 \left(\|f\|_{C^{r,\alpha}(\overline{\Omega})} + \|\omega_0\|_{C^{r+1,\alpha}(\partial\Omega)}\right),$$

respectively there exists a constant $C_2 = C_2(r,p,\Omega)$ such that

$$\|\omega\|_{W^{r+1,p}(\Omega)} \leq C_2 \left(\|f\|_{W^{r,p}(\Omega)} + \|\omega_0\|_{W^{r+1-1/p,p}(\partial\Omega)}\right).$$

Remark 8.17. (i) In the case $k = n-1$, conditions (A1) are trivially satisfied and (A2) reads as

$$\int_\Omega f = \int_{\partial\Omega} \nu \wedge \omega_0$$

if Ω is connected (cf. Theorem 6.5).

(ii) When $k = 0$, then the result is still valid for $\alpha = 0, 1$ with an argument completely analogous to thc one of Corollary 8.6.

(iii) If $r \geq 1$, conditions (A1) are well defined. This is obvious for the first equation. The second one makes sense in $W^{r-\frac{1}{p},p}\left(\partial\Omega;\Lambda^{k+2}\right)$, respectively $C^{r,\alpha}(\partial\Omega;\Lambda^{k+2})$, due to Theorem 3.23.

(iv) If $r = 0$, then the condition $df = 0$ is understood in the sense of distributions. The second condition in (A1) is well defined in the Hölder setting and is to be understood, in the Sobolev setting, in the weak sense, namely

$$\int_\Omega \langle f;\delta\varphi\rangle - \int_{\partial\Omega} \langle \nu \wedge \omega_0;\delta\varphi\rangle = 0, \quad \forall \varphi \in C^\infty\left(\overline{\Omega};\Lambda^{k+2}\right), \tag{B1}$$

which is equivalent with (A1) whenever $r \geq 1$ (cf. Section 7.1).

(v) The above results remain valid if the set Ω is C^{r+3} for the Sobolev case and $C^{r+3,\alpha}$ for the Hölder case.

(vi) If Ω is contractible, then (cf. Theorem 6.5)

$$\mathscr{H}_T\left(\Omega;\Lambda^{k+1}\right)=\{0\} \quad \text{if } 0\le k\le n-2,$$

whereas if $k=n-1$,

$$\mathscr{H}_T\left(\Omega;\Lambda^{k+1}\right)=\Lambda^n(\mathbb{R}^n)\cong\mathbb{R}.$$

(vii) The construction is linear and universal in the sense of Remark 6.13.

Proof. We deal only with the case of Hölder spaces, the case of Sobolev spaces being handled similarly. The implication (ii) $\Rightarrow$ (i) is straightforward using partial integration (cf. Theorem 3.28). To show the other implication, we first use Theorem 7.2 to find a solution $u\in C^{r+1,\alpha}\left(\overline{\Omega};\Lambda^k\right)$ of the problem

$$\begin{cases} du=f \quad \text{and} \quad \delta u=0 & \text{in } \Omega,\\ \nu\wedge u=\nu\wedge\omega_0 & \text{on } \partial\Omega.\end{cases}$$

If $k=0$, then $\omega=u$ has already all the claimed properties. So we next assume that $k\ge 1$. Since $\nu\wedge(\omega_0-u)=0$, we can apply Lemma 8.11(i) to find $\beta\in C^{r+2,\alpha}\big(\overline{\Omega};\Lambda^{k-1}\big)$ such that

$$d\beta=\omega_0-u \quad \text{on } \partial\Omega.$$

We finally set

$$\omega=u+d\beta$$

to obtain the result. □

Due to Theorem 7.4 and Lemma 8.11(ii), we can prove the dual version in the same way.

Theorem 8.18. *Let $r\ge 0$ and $1\le k\le n$ be integers, $0<\alpha<1$, $2\le p<\infty$ and $\Omega\subset\mathbb{R}^n$ be a bounded open smooth set with exterior unit normal ν. Let $g:\overline{\Omega}\to\Lambda^{k-1}$ and $\omega_0:\partial\Omega\to\Lambda^k$. Then the following claims are equivalent:*

(i) Let $g\in C^{r,\alpha}\left(\overline{\Omega};\Lambda^{k-1}\right)$ and $\omega_0\in C^{r+1,\alpha}\left(\partial\Omega;\Lambda^k\right)$, respectively $g\in W^{r,p}(\Omega;\Lambda^{k-1})$ and $\omega_0\in W^{r+1-\frac{1}{p},p}\left(\partial\Omega;\Lambda^k\right)$, satisfy

$$\delta g=0 \ \text{in } \Omega, \quad \nu\lrcorner\delta\omega_0=\nu\lrcorner g \ \text{on } \partial\Omega \tag{C1}$$

and, for every $\chi\in\mathscr{H}_N\left(\Omega;\Lambda^{k-1}\right)$,

$$\int_\Omega\langle g;\chi\rangle-\int_{\partial\Omega}\langle\nu\lrcorner\omega_0;\chi\rangle=0. \tag{C2}$$

(ii) There exists $\omega \in C^{r+1,\alpha}\left(\overline{\Omega};\Lambda^k\right)$*, respectively* $\omega \in W^{r+1,p}\left(\Omega;\Lambda^k\right)$*, such that*

$$\begin{cases} \delta\omega = g & \text{in } \Omega, \\ \omega = \omega_0 & \text{on } \partial\Omega \end{cases}$$

and there exists a constant $C_1 = C_1(r,\alpha,\Omega)$ *such that*

$$\|\omega\|_{C^{r+1,\alpha}(\overline{\Omega})} \le C_1\left(\|g\|_{C^{r,\alpha}(\overline{\Omega})} + \|\omega_0\|_{C^{r+1,\alpha}(\partial\Omega)}\right),$$

respectively there exists a constant $C_2 = C_2(r,p,\Omega)$ *such that*

$$\|\omega\|_{W^{r+1,p}(\Omega)} \le C_2\left(\|g\|_{W^{r,p}(\Omega)} + \|\omega_0\|_{W^{r+1-1/p,p}(\partial\Omega)}\right).$$

Remark 8.19. (i) In the case $k=1$, conditions (C1) are trivially satisfied and (C2) reads as

$$\int_\Omega g = \int_{\partial\Omega} \nu \lrcorner \omega_0$$

if Ω is connected (cf. Theorem 6.5).

(ii) When $k=n$, then the result is still valid for $\alpha = 0,1$ with an argument completely analogous to the one of Corollary 8.7.

(iii) If $r \ge 1$, conditions (C1) are well defined. This is obvious for the first equation. The second one makes sense in $W^{r-\frac{1}{p},p}\left(\partial\Omega;\Lambda^{k-2}\right)$, respectively $C^{r,\alpha}(\partial\Omega;\Lambda^{k-2})$, due to Theorem 3.23.

(iv) If $r=0$, then the condition $\delta g = 0$ is understood in the sense of distributions. The second condition in (C1) is well defined in the Hölder setting and is to be understood, in the Sobolev setting, in the weak sense, namely

$$\int_\Omega \langle g; d\varphi\rangle - \int_{\partial\Omega} \langle \nu \lrcorner \omega_0; d\varphi\rangle = 0, \quad \forall \varphi \in C^\infty\left(\overline{\Omega};\Lambda^{k-2}\right), \tag{D1}$$

which is equivalent to (C1) whenever $r \ge 1$ (cf. Section 7.2).

(v) The above results remains valid if Ω is C^{r+3} for the Sobolev case and $C^{r+3,\alpha}$ for the Hölder case.

(vi) If Ω is contractible, then (cf. Theorem 6.5)

$$\mathcal{H}_N\left(\Omega;\Lambda^{k-1}\right) = \{0\} \quad \text{if } 2 \le k \le n,$$

whereas if $k=1$,

$$\mathcal{H}_N\left(\Omega;\Lambda^{k-1}\right) = \Lambda^0(\mathbb{R}^n) = \mathbb{R}.$$

(vii) The construction is linear and universal in the sense of Remark 6.13.

8.5 Poincaré Lemma with Constraints

8.5.1 A First Result

Our first proposition is the Poincaré lemma under a constraint on the scalar product.

Proposition 8.20. *Let $x_0 \in \mathbb{R}^n$ and f be a C^∞ closed 2-form. Let a be a C^∞ 1-form such that $a(x_0) \neq 0$ and let b be a C^∞ function. Then there exist a neighborhood U of x_0 and $\omega \in C^\infty(U;\Lambda^1)$ such that*

$$d\omega = f \quad \text{and} \quad \langle \omega; a \rangle = b \text{ in } U.$$

Moreover, if $b(x_0) = 0$, then ω can be chosen so that, in addition to the previous equation, $\omega(x_0) = 0$.

Remark 8.21. If f, a and b depend in a smooth way on a parameter t, we find exactly in the same way a ω depending smoothly on t and with the same properties, provided there exists $1 \leq i \leq n$ so that

$$a_i(t, x_0) \neq 0 \quad \text{for every } t.$$

Proof. Without loss of generality, we can assume $x_0 = 0$ and that $a_n(0) \neq 0$. Using Theorem 8.1, there exist a neighborhood V of 0 and $u \in C^\infty\left(V;\Lambda^1\right)$ such that

$$du = f \quad \text{in } V.$$

Replacing u by $u - u(0)$, we can assume without loss of generality that $u(0) = 0$. By the methods of characteristics, recalling that $a_n(0) \neq 0$, there exist a neighborhood $U \subset V$ of 0 and $v \in C^\infty(U)$ such that

$$\begin{cases} \langle dv; a \rangle = b - \langle u; a \rangle & \text{in } U, \\ v(x_1, \ldots, x_{n-1}, 0) = 0 & \text{for every } (x_1, \ldots, x_{n-1}, 0) \in U. \end{cases}$$

Letting $\omega = u + dv$, we have the main result. Finally, let us show that the same ω fulfills the extra assertion. Indeed, since in that case, $b(0) = 0$, $u(0) = 0$ and $a_n(0) \neq 0$, we immediately deduce that $dv(0) = 0$ and, hence, $\omega(0) = 0$, which concludes the proof. □

8.5.2 A Second Result

We now give a theorem that will be used in the second proof of Theorem 14.3 (cf. Bandyopadhyay, Dacorogna and Kneuss [9]).

Theorem 8.22. *Let $2 \leq 2m \leq n$ be integers and $x_0 \in \mathbb{R}^n$. Let f and g be two C^∞ closed 2-forms such that*

$$g^m(x_0) \neq 0$$

and, in a neighborhood of x_0,

$$f \wedge g^m = 0 \quad \text{and} \quad g^{m+1} = 0.$$

Then there exist a neighborhood U *of* x_0 *and* $\omega \in C^\infty(U;\Lambda^1)$ *such that* $\omega(x_0) = 0$,

$$d\omega = f \quad \text{and} \quad \omega \wedge g^m = 0, \quad \text{in } U.$$

Remark 8.23. (i) We can easily replace C^∞ by C^r, but a refined version of our construction finds ω only in C^{r-1} (see [60]).

(ii) Note that the hypotheses imply that in a neighborhood of x_0,

$$\operatorname{rank}[g] = 2m.$$

(iii) If f and g depend in a smooth way on a parameter t, we find exactly in the same way a ω depending smoothly on t and with the same properties.

Proof. The following proof becomes much simpler if we can invoke Theorem 14.3. However, since, later, we want to use the present theorem to give a second proof of Theorem 14.3, we have to find an independent proof. The proof will rely on several technical results that are gathered in Section 8.5.3.

Without loss of generality, we can assume that $x_0 = 0$, and in what follows, U will be a generic neighborhood of 0.

Step 1. Appealing to the classical Poincaré lemma (see Theorem 8.1), we can find a neighborhood U of 0 and $u \in C^\infty(U;\Lambda^1)$ such that $du = f$ in U. Replacing u by $u - u(0)$, we can assume that $u(0) = 0$. We then set

$$\omega = u - dv.$$

Our result will follow if we can find $v \in C^\infty(U)$ verifying

$$\begin{cases} dv \wedge g^m = u \wedge g^m & \text{in } U, \\ dv(0) = 0. \end{cases} \tag{8.2}$$

Step 2 (simplification of g). It follows from Proposition 8.31 that there exist a neighborhood U of 0 and $\varphi \in \operatorname{Diff}^\infty(U;\varphi(U))$ such that $\varphi(0) = 0$ and

$$\varphi^*(g) = \sum_{i=1}^m dx^{2i-1} \wedge dx^{2i} + \sum_{\substack{1 \le i < j \le n \\ 2m < j}} r_{ij} dx^i \wedge dx^j$$

for some $r_{ij} \in C^\infty(U)$. Problem (8.2) is then equivalent to finding $w \in C^\infty(U)$ such that

$$\begin{cases} dw \wedge (\varphi^*(g))^m = \varphi^*(u) \wedge (\varphi^*(g))^m & \text{in } U, \\ dw(0) = 0. \end{cases}$$

Indeed, it is enough to set $v = (\varphi^{-1})^*(w)$ to have a solution of (8.2). So from now on, we will assume, upon substitution of $\varphi^*(g)$ and $\varphi^*(u)$ by g and u, that

$$g = \sum_{i=1}^{m} dx^{2i-1} \wedge dx^{2i} + \sum_{\substack{1 \le i < j \le n \\ 2m < j}} r_{ij} dx^i \wedge dx^j \tag{8.3}$$

and we therefore have to find $v \in C^\infty(U)$ satisfying (8.2) only for g as in (8.3). Note that for such a g, we have

$$(g^m)_{1 \cdots (2m)} = m! \neq 0.$$

Step 3. We then solve (8.2) by induction on n, m being fixed. In the case $n = 2m$, nothing is to be proved; just choose $v = 0$. So we assume that the result has been proven for $n = 2m + j$, $j \ge 0$, and let us prove it for $n = 2m + j + 1$. We therefore assume that we can find a neighborhood $\widetilde{U} \subset \mathbb{R}^{n-1}$ of $0 \in \mathbb{R}^{n-1}$ and $h \in C^\infty(\widetilde{U})$ with

$$\begin{cases} dh \wedge \widetilde{g}^m = \widetilde{u} \wedge \widetilde{g}^m, \\ dh(0) = 0 \end{cases}$$

whenever $\widetilde{g} \in C^\infty(\widetilde{U}; \Lambda^2(\mathbb{R}^{n-1}))$ and $\widetilde{u} \in C^\infty(\widetilde{U}; \Lambda^1(\mathbb{R}^{n-1}))$ verify

$$d\widetilde{u} \wedge \widetilde{g}^m = 0, \quad d\widetilde{g} = 0, \quad \widetilde{g}^{m+1} = 0,$$

$$\widetilde{g} = \sum_{i=1}^{m} dx^{2i-1} \wedge dx^{2i} + \sum_{\substack{1 \le i < j \le n-1 \\ 2m < j}} \widetilde{r}_{ij} dx^i \wedge dx^j$$

for some $\widetilde{r}_{ij} \in C^\infty(\widetilde{U})$ and let us prove that it holds for n. To establish this result, we proceed in four substeps.

Step 3.1. We first solve, by the method of characteristics, the Cauchy problem for the first-order partial differential equation

$$\begin{cases} (dv \wedge g^m)_{1 \cdots (2m)n} = (u \wedge g^m)_{1 \cdots (2m)n}, \\ v(x_1, \ldots, x_{n-1}, 0) = h(x_1, \ldots, x_{n-1}), \end{cases} \tag{8.4}$$

where $h \in C^r(\widetilde{U})$ is a solution, which exists by hypothesis of induction, of

$$\begin{cases} dh \wedge i_n^*(g)^m = i_n^*(u) \wedge i_n^*(g)^m, \\ dh(0) = 0 \end{cases} \tag{8.5}$$

and $i_n : \mathbb{R}^{n-1} \to \mathbb{R}^n$ is defined by

$$i_n(x_1, \ldots, x_{n-1}) = (x_1, \ldots, x_{n-1}, 0).$$

Indeed we can apply the method of characteristics since the first equation of (8.4) is equivalent to, recalling that $(g^m)_{1 \cdots (2m)} = m!$,

$$\frac{\partial v}{\partial x_n} \cdot m! + \sum_{i=1}^{2m} (-1)^{i+1} \frac{\partial v}{\partial x_i} \cdot (g^m)_{1 \cdots \widehat{i} \cdots (2m)n} = (u \wedge g^m)_{1 \cdots (2m)n} . \tag{8.6}$$

Observe also that

$$i_n^*(v) = h \Leftrightarrow v(x_1, \ldots, x_{n-1}, 0) = h(x_1, \ldots, x_{n-1}).$$

Finally, note that we can apply the hypothesis of induction since

$$dg = 0, \quad du \wedge g^m = 0, \quad g^{m+1} = 0,$$

$$g = \sum_{i=1}^{m} dx^{2i-1} \wedge dx^{2i} + \sum_{\substack{1 \le i < j \le n \\ 2m < j}} r_{ij} dx^i \wedge dx^j$$

for some $r_{ij} \in C^\infty$ imply

$$d(i_n^*(g)) = 0, \quad d(i_n^*(u)) \wedge i_n^*(g)^m = 0, \quad i_n^*(g)^{m+1} = 0,$$

$$i_n^*(g) = \sum_{i=1}^{m} dx^{2i-1} \wedge dx^{2i} + \sum_{\substack{1 \le i < j \le n-1 \\ 2m < j}} \widetilde{r}_{ij} dx^i \wedge dx^j$$

for some $\widetilde{r}_{ij} \in C^\infty$.

Step 3.2. It now remains to prove that the solution v of (8.4) is indeed a solution of (8.2). First, we claim that $dv(0) = 0$. Indeed, combining the last equation of (8.4) and (8.5), we deduce that

$$\frac{\partial v}{\partial x_i}(0) = 0, \quad 1 \le i \le n-1.$$

Next, inserting the previous equation in (8.6) and using the fact that $u(0) = 0$, we immediately deduce that

$$\frac{\partial v}{\partial x_n}(0) = 0,$$

which gives the claim. Hence, it only remains to show that

$$dv \wedge g^m = u \wedge g^m. \tag{8.7}$$

Lemma 8.28 implies that to show (8.7), it is enough to establish

$$(dv \wedge g^m)_{1 \cdots (2m)k} = (u \wedge g^m)_{1 \cdots (2m)k}, \quad 2m+1 \le k \le n. \tag{8.8}$$

Step 3.3. We now prove (8.8). Define, for every $2m+1 \le k \le n$,

$$L_k(v) = (dv \wedge g^m)_{1 \cdots (2m)k} \quad \text{and} \quad w_k = L_k(v) - (u \wedge g^m)_{1 \cdots (2m)k} .$$

Since we already have from (8.4) that $w_n = 0$, our claim (8.8) reduces to proving that

$$w_k = 0 \quad \text{for every } 2m+1 \leq k \leq n-1. \tag{8.9}$$

Since

$$0 = f \wedge g^m = du \wedge g^m$$

and (8.3) holds, we can apply Lemma 8.29 and Lemma 8.30 to obtain

$$\begin{aligned} L_n(w_k) &= L_n L_k(v) - L_n\left((u \wedge g^m)_{1\cdots(2m)k}\right) \\ &= L_k L_n(v) - L_k\left((u \wedge g^m)_{1\cdots(2m)n}\right) = L_k(w_n) = 0. \end{aligned}$$

Assume (cf. Step 3.4), that we can prove that

$$w_k(x_1, \ldots, x_{n-1}, 0) = 0; \tag{8.10}$$

we will then have, by uniqueness of the solutions of the Cauchy problem, that the only solution of

$$\begin{cases} L_n(w_k) = 0, \\ \quad w_k = 0 \quad \text{on } x_n = 0 \end{cases}$$

is $w_k = 0$. This is exactly our claim (8.9).

Step 3.4. Finally, we show (8.10), which is equivalent to proving that $i_n^*(w_k) = 0$. We have that, recalling that $i_n^*(v) = h$,

$$\begin{aligned} i_n^*(w_k) &= i_n^*\left((dv \wedge g^m - u \wedge g^m)_{1\cdots(2m)k}\right) = \left(i_n^*(dv \wedge g^m - u \wedge g^m)\right)_{1\cdots(2m)k} \\ &= \left(d(i_n^*(v)) \wedge i_n^*(g^m)) - i_n^*(u) \wedge i_n^*(g^m)\right)_{1\cdots(2m)k} \end{aligned}$$

and thus, appealing to (8.5),

$$i_n^*(w_k) = \left(dh \wedge i_n^*(g^m) - i_n^*(u) \wedge i_n^*(g^m)\right)_{1\cdots(2m)k} = 0.$$

This conclues the proof of the theorem. □

With substantially the same proof we can get a global result (see Kneuss [60]).

Theorem 8.24. *Let* $2 \leq 2m < n$ *and* $g, f \in C^\infty(\mathbb{R}^n; \Lambda^2)$ *be closed. Assume that* g *is of the form*

$$g = \sum_{i=1}^{m} dx^{2i-1} \wedge dx^{2i} + \sum_{\substack{1 \leq i < j \leq n \\ 2m < j}} g_{ij}\, dx^i \wedge dx^j$$

for some $g_{ij} \in C^\infty(\mathbb{R}^n)$ *and, for every* $x \in \mathbb{R}^n$,

$$f(x) \wedge g^m(x) = 0, \quad g^{m+1}(x) = 0 \quad \textit{and} \quad |g(x)| \leq a|x| + b,$$

where $a,b>0$ are constants. Then there exists $w\in C^{\infty}\left(\mathbb{R}^n;\Lambda^1\right)$ so that the following equations are satisfied in $\mathbb{R}^n$:

$$dw=f \quad \text{and} \quad w\wedge g^m=0.$$

8.5.3 Some Technical Lemmas

In this subsection we gather all algebraic lemmas that we have used in the proof of Theorem 8.22.

Lemma 8.25. *Let $2\le 2m<n$ be integers and $g\in\Lambda^2(\mathbb{R}^n)$ with* $\operatorname{rank}[g]=2m$ *and of the form*

$$g=\sum_{i=1}^{m}dx^{2i-1}\wedge dx^{2i}+\sum_{\substack{1\le i<j\le n\\ 2m<j}}g_{ij}\,dx^i\wedge dx^j.$$

Then, for every $1\le i,j\le 2m<k\le n$, the following hold:

$$(g^m)_{1\cdots\widehat{i}\cdots(2m)k}=\begin{cases} m!\,g_{(i+1)k} & \text{if } i \text{ is odd}\\ m!\,g_{(i-1)k} & \text{if } i \text{ is even,}\end{cases} \tag{8.11}$$

$$\begin{gathered}(g^m)_{1\cdots\widehat{i}\cdots(2m)n}(g^m)_{1\cdots\widehat{j}\cdots(2m)k}-(g^m)_{1\cdots\widehat{i}\cdots(2m)k}(g^m)_{1\cdots\widehat{j}\cdots(2m)n}\\ =\begin{cases}0 & \text{if } i=j\\ m!\,(g^m)_{1\cdots\widehat{i}\cdots\widehat{j}\cdots(2m)kn} & \text{if } i<j\\ -m!\,(g^m)_{1\cdots\widehat{j}\cdots\widehat{i}\cdots(2m)kn} & \text{if } i>j.\end{cases}\end{gathered}$$

Remark 8.26. When $m=1$, the two conclusions of the lemma are immediate and the last one reads as (since $g^2=0$)

$$g_{2n}g_{1k}-g_{2k}g_{1n}=g_{kn}=g_{12}g_{kn}\,.$$

For the proof of the lemma we will need the following result, whose proof is elementary (see Kneuss [60] for details).

Lemma 8.27. *Let $2\le 2m\le n$, $f\in\Lambda^2(\mathbb{R}^n)$, $1\le i_1<\cdots<i_{2m}\le n$ and $1\le l\le 2m$. Then*

$$\begin{aligned}(f^m)_{i_1\cdots i_{2m}}&=m\sum_{j=1}^{l-1}(-1)^{j+l+1}f_{i_ji_l}(f^{m-1})_{i_1\cdots\widehat{i_j}\cdots\widehat{i_l}\cdots i_{2m}}\\ &\quad+m\sum_{j=l+1}^{2m}(-1)^{j+l+1}f_{i_li_j}(f^{m-1})_{i_1\cdots\widehat{i_l}\cdots\widehat{i_j}\cdots i_{2m}}\,.\end{aligned}$$

We now prove Lemma 8.25.

Proof. We split the proof into two steps.

Step 1. We first show (8.11). We assume that i is odd and thus $i = 2s-1$ for some $1 \leq s \leq m$ (the case that i is even can be handled exactly in the same way). From Lemma 8.27 (choosing $l = 2s-1$ and hence $i_l = 2s$ in the computation below), it follows that

$$\begin{aligned}(g^m)_{1\cdots\widehat{2s-1}\cdots(2m)k} &= m\sum_{j=1}^{2s-2}(-1)^{j+2s} g_{j(2s)}\,(g^{m-1})_{1\cdots\widehat{j}\cdots\widehat{2s-1}\widehat{2s}\cdots(2m)k} \\ &\quad + m\sum_{j=2s+1}^{2m}(-1)^{j+2s+1} g_{(2s)j}\,(g^{m-1})_{1\cdots\widehat{2s-1}\widehat{2s}\cdots\widehat{j}\cdots(2m)k} \\ &\quad + m\, g_{(2s)k}\,(g^{m-1})_{1\cdots\widehat{2s-1}\widehat{2s}\cdots\cdots(2m)}\,.\end{aligned}$$

Since by the special structure of g, $g_{j(2s)} = 0$ for $1 \leq j \leq 2s-2$ and $g_{(2s)j} = 0$ for $2s+1 \leq j \leq 2m$ and

$$(g^{m-1})_{1\cdots\widehat{2s-1}\widehat{2s}\cdots\cdots(2m)} = (m-1)!\,,$$

the previous equation reduces to

$$(g^m)_{1\cdots\widehat{2s-1}\cdots(2m)k} = m!\, g_{(2s)k}\,,$$

which is exactly the claim.

Step 2. We now prove the second statement. When $i = j$, the proof is trivial. We prove the result for $i < j$, which, in turn, immediately implies the case $i > j$. Moreover, we assume that i is odd and thus $i = 2s-1$ for some $1 \leq s \leq m$ (the case that i is even being handled exactly in the same way). Using (8.11), it is enough to show that

$$(g^m)_{1\cdots\widehat{2s-1}\cdots\widehat{j}\cdots(2m)kn} = \begin{cases} m!\, g_{(2s)n}\, g_{(j+1)k} - m!\, g_{(2s)k}\, g_{(j+1)n} & \text{if } j \text{ is odd} \\ m!\, g_{(2s)n}\, g_{(j-1)k} - m!\, g_{(2s)k}\, g_{(j-1)n} & \text{if } j \text{ is even.}\end{cases} \tag{8.12}$$

We consider two cases to establish (8.12).

Case 1: $2s+1 \leq j \leq 2m$. From Lemma 8.27 (choosing $l = 2s-1$ and hence $i_l = 2s$ in the computation below), it follows that

$$\begin{aligned}(g^m)_{1\cdots\widehat{2s-1}\cdots\widehat{j}\cdots(2m)kn} &= m\sum_{t=1}^{2s-2}(-1)^{t+2s} g_{t(2s)}\,(g^{m-1})_{1\cdots\widehat{t}\cdots\widehat{2s-1}\widehat{2s}\cdots\widehat{j}\cdots(2m)kn} \\ &\quad + m\sum_{t=2s+1}^{j-1}(-1)^{t+2s+1} g_{(2s)t}\,(g^{m-1})_{1\cdots\widehat{2s-1}\widehat{2s}\cdots\widehat{t}\cdots\widehat{j}\cdots(2m)kn} \\ &\quad + m\sum_{t=j+1}^{2m}(-1)^{t+2s} g_{(2s)t}\,(g^{m-1})_{1\cdots\widehat{2s-1}\widehat{2s}\cdots\widehat{j}\cdots\widehat{t}\cdots(2m)kn} \\ &\quad - m\, g_{(2s)k}\,(g^{m-1})_{1\cdots\widehat{2s-1}\widehat{2s}\cdots\widehat{j}\cdots(2m)n} \\ &\quad + m\, g_{(2s)n}\,(g^{m-1})_{1\cdots\widehat{2s-1}\widehat{2s}\cdots\widehat{j}\cdots(2m)k}\,.\end{aligned}$$

Since $g_{t(2s)} = 0$ for $1 \le t \le 2s-2$ and $g_{(2s)t} = 0$ for $2s+1 \le t \le 2m$, the previous equation reduces to

$$(g^m)_{1\cdots\widehat{2s-1}\cdots\widehat{j}\cdots(2m)kn} = -m\, g_{(2s)k}\,(g^{m-1})_{1\cdots\widehat{2s-1}\widehat{2s}\cdots\widehat{j}\cdots(2m)n} + m\, g_{(2s)n}\,(g^{m-1})_{1\cdots\widehat{2s-1}\widehat{2s}\cdots\widehat{j}\cdots(2m)k}.$$

Moreover, proceeding exactly as in Step 1, we can show that

$$(g^{m-1})_{1\cdots\widehat{2s-1}\widehat{2s}\cdots\widehat{j}\cdots(2m)k} = \begin{cases} (m-1)!\, g_{(j+1)k} & \text{if } j \text{ is odd} \\ (m-1)!\, g_{(j-1)k} & \text{if } j \text{ is even} \end{cases}$$

and the same for k replaced by n. Hence, combining the last two equations, we get (8.12).

Case 2: $j = 2s$. First note that $(g^{m+1})_{1\cdots(2m)kn} = 0$ since $\operatorname{rank}(g) = 2m$ (see Proposition 2.37(iii)). Hence, using Lemma 8.27 once more (with $m+1$, $l = 2s$ and hence $i_l = 2s$ in the computation below), we obtain

$$\begin{aligned} 0 = (g^{m+1})_{1\cdots(2m)kn} &= (m+1)\sum_{t=1}^{2s-2}(-1)^{t+2s+1} g_{t(2s)}\,(g^m)_{1\cdots\widehat{t}\cdots\widehat{2s}\cdots(2m)kn} \\ &\quad + (m+1) g_{(2s-1)(2s)}\,(g^m)_{1\cdots\widehat{2s-1}\widehat{2s}\cdots(2m)kn} \\ &\quad + (m+1)\sum_{t=2s+1}^{2m}(-1)^{t+2s+1} g_{(2s)t}\,(g^m)_{1\cdots\widehat{2s}\cdots\widehat{t}\cdots(2m)kn} \\ &\quad + (m+1) g_{(2s)k}\,(g^m)_{1\cdots\widehat{2s}\cdots(2m)n} \\ &\quad - (m+1) g_{(2s)n}\,(g^m)_{1\cdots\widehat{2s}\cdots(2m)k}. \end{aligned}$$

Since $g_{(2s-1)(2s)} = 1$, $g_{t(2s)} = 0$ for $1 \le t \le 2s-2$ and $g_{(2s)t} = 0$ for $2s+1 \le t \le 2m$, the previous equation rewrites as

$$(g^m)_{1\cdots\widehat{2s-1}\widehat{2s}\cdots(2m)kn} = g_{(2s)n}\,(g^m)_{1\cdots\widehat{2s}\cdots(2m)k} - g_{(2s)k}\,(g^m)_{1\cdots\widehat{2s}\cdots(2m)n}.$$

Using (8.11), we immediately deduce that

$$(g^m)_{1\cdots\widehat{2s-1}\widehat{2s}\cdots(2m)kn} = m!\, g_{(2s)n}\, g_{(2s-1)k} - m!\, g_{(2s)k}\, g_{(2s-1)n},$$

which is exactly (8.12). This finishes the proof. □

The next lemma has been used in Step 3.2 of the proof of Theorem 8.22.

Lemma 8.28. *Let $2 \le 2m \le n$ be integers, $\omega \in \Lambda^1(\mathbb{R}^n)$ and $g \in \Lambda^2(\mathbb{R}^n)$ with* $\operatorname{rank}[g] = 2m$ *and of the form*

$$g = \sum_{i=1}^{m} dx^{2i-1} \wedge dx^{2i} + \sum_{\substack{1 \le i < j \le n \\ 2m < j}} g_{ij} dx^i \wedge dx^j .$$

Suppose that

$$(\omega \wedge g^m)_{1 \cdots (2m) k} = 0, \quad \textit{for every } 2m+1 \le k \le n.$$

Then

$$\omega \wedge g^m = 0.$$

Proof. According to Proposition 2.50, it is enough to show that there exists $u \in \Lambda^1(\mathbb{R}^n)$ such that

$$\omega = u \lrcorner g.$$

We claim that this is satisfied by

$$u = \sum_{s=1}^{m} \left(\omega_{2s} dx^{2s-1} - \omega_{2s-1} dx^{2s} \right).$$

We will show that $(u \lrcorner g)_k = \omega_k$ for $1 \le k \le n$ by considering two cases.

Case 1: $k \le 2m$. Hence, we have

$$\left(dx^l \lrcorner \sum_{\substack{1 \le i < j \le n \\ 2m < j}} g_{ij} dx^i \wedge dx^j \right)_k = 0 \quad \text{if } l \le 2m.$$

We therefore obtain

$$\begin{aligned} (u \lrcorner g)_k &= \left(\sum_{l=1}^{2m} u_l dx^l \lrcorner \sum_{i=1}^{m} dx^{2i-1} \wedge dx^{2i} \right)_k \\ &= \left\{ \begin{array}{ll} u_{2s-1} & \text{if } k = 2s \\ -u_{2s} & \text{if } k = 2s-1 \end{array} \right\} = \omega_k . \end{aligned}$$

Case 2: $k \ge 2m+1$. Using the hypothesis $(\omega \wedge g^m)_{1 \cdots (2m) k} = 0$ and (8.11), we obtain

$$\begin{aligned} 0 &= (\omega \wedge g^m)_{1 \cdots (2m) k} \\ &= \sum_{\gamma=1}^{2m} (-1)^{\gamma-1} \omega_\gamma (g^m)_{1 \cdots \widehat{\gamma} \cdots (2m) k} + \omega_k (g^m)_{1 \cdots (2m)} \end{aligned}$$

$$= \sum_{s=1}^{m} \left(\omega_{2s-1} \left(g^m\right)_{1 \cdots \widehat{(2s-1)} \cdots (2m)k} - \omega_{2s} \left(g^m\right)_{1 \cdots \widehat{(2s)} \cdots (2m)k} \right) + m!\, \omega_k$$

$$= m! \left(\sum_{s=1}^{m} \left(\omega_{2s-1} g_{(2s)k} - \omega_{2s} g_{(2s-1)k} \right) + \omega_k \right).$$

Using that $k \geq 2m+1$, the definition of u and the previous equation yield

$$(u \lrcorner g)_k = \left(\left(\sum_{l=1}^{m} u_l dx^l \right) \lrcorner \left(\sum_{s=1}^{m} dx^{2s-1} \wedge dx^{2s} + \sum_{\substack{1 \leq i < j \leq n \\ 2m < j}} g_{ij} dx^i \wedge dx^j \right) \right)_k$$

$$= \left(\sum_{l=1}^{m} u_l dx^l \lrcorner \sum_{\substack{1 \leq i < j \leq n \\ 2m < j}} g_{ij} dx^i \wedge dx^j \right)_k = \sum_{l=1}^{m} u_l g_{lk}$$

$$= \sum_{s=1}^{m} \left(u_{2s-1} g_{(2s-1)k} + u_{2s} g_{(2s)k} \right) = \sum_{s=1}^{m} \left(\omega_{2s} g_{(2s-1)k} - \omega_{2s-1} g_{(2s)k} \right)$$

$$= \omega_k\,,$$

which concludes the proof of the lemma. □

The following two lemmas have been used in Step 3.3 of Theorem 8.22.

Lemma 8.29. *Let $2 \leq 2m < n$ be integers and $\Omega \subset \mathbb{R}^n$ be an open set. Let $g \in C^\infty\left(\Omega; \Lambda^2\right)$ be closed with* $\operatorname{rank}[g] = 2m$ *in Ω and of the form*

$$g(x) = \sum_{i=1}^{m} dx^{2i-1} \wedge dx^{2i} + \sum_{\substack{1 \leq i < j \leq n \\ 2m < j}} g_{ij}(x)\, dx^i \wedge dx^j, \quad x \in \Omega,$$

where $g_{ij} \in C^\infty(\Omega)$. Then, for every $2m+1 \leq k \leq n$,

$$L_n L_k = L_k L_n, \tag{8.13}$$

where $L_k : C^\infty(\Omega) \to C^\infty(\Omega)$, $2m+1 \leq k \leq n$, is defined by

$$L_k(z) = \left(dz \wedge g^m\right)_{1 \cdots (2m)k}.$$

Proof. We begin by noting that the structure of g implies that

$$\left(g^m\right)_{1 \cdots (2m)} = m! \quad \text{in } \Omega. \tag{8.14}$$

For $z \in C^\infty(\Omega)$, we have

$$L_k(z) = \left(dz \wedge g^m\right)_{1 \cdots (2m)k} = z_{x_k}\, m! + \sum_{i=1}^{2m} (-1)^{i+1} z_{x_i} \left(g^m\right)_{1 \cdots \widehat{i} \cdots (2m)k},$$

where we have denoted partial differentiation of z by x_k by z_{x_k}. We therefore find

$$
\begin{aligned}
&L_n L_k(z) \\
&\quad = L_n\left(z_{x_k} m! + \sum_{i=1}^{2m}(-1)^{i+1} z_{x_i} (g^m)_{1\cdots\widehat{i}\cdots(2m)k}\right) \\
&\quad = \left(z_{x_k} m! + \sum_{i=1}^{2m}(-1)^{i+1} z_{x_i} (g^m)_{1\cdots\widehat{i}\cdots(2m)k}\right)_{x_n} m! \\
&\qquad + \sum_{j=1}^{2m}(-1)^{j+1}\left(z_{x_k} m! + \sum_{i=1}^{2m}(-1)^{i+1} z_{x_i} (g^m)_{1\cdots\widehat{i}\cdots(2m)k}\right)_{x_j} (g^m)_{1\cdots\widehat{j}\cdots(2m)n}.
\end{aligned}
$$

Setting

$$
\begin{aligned}
A_1 &= m!^2 z_{x_k x_n}, \quad A_2 = m! \sum_{i=1}^{2m}(-1)^{i+1} z_{x_i x_n} (g^m)_{1\cdots\widehat{i}\cdots(2m)k}, \\
A_3 &= m! \sum_{i=1}^{2m}(-1)^{i+1} z_{x_i}\left((g^m)_{1\cdots\widehat{i}\cdots(2m)k}\right)_{x_n}, \\
A_4 &= m! \sum_{j=1}^{2m}(-1)^{j+1} z_{x_k x_j} (g^m)_{1\cdots\widehat{j}\cdots(2m)n}, \\
A_5 &= \sum_{i,j=1}^{2m}(-1)^{i+1}(-1)^{j+1} z_{x_i x_j} (g^m)_{1\cdots\widehat{i}\cdots(2m)k} (g^m)_{1\cdots\widehat{j}\cdots(2m)n}, \\
A_6 &= \sum_{i,j=1}^{2m}(-1)^{i+1}(-1)^{j+1} z_{x_i}\left((g^m)_{1\cdots\widehat{i}\cdots(2m)k}\right)_{x_j} (g^m)_{1\cdots\widehat{j}\cdots(2m)n},
\end{aligned}
$$

we find that

$$L_n L_k(z) = A_1 + A_2 + A_3 + A_4 + A_5 + A_6.$$

Note that A_1, $A_2 + A_4$ and A_5 are symmetric in k and n. Therefore, for proving that $L_k L_n(z) = L_n L_k(z)$, it is enough to show that $A_3 + A_6$ is symmetric in k and n, which is equivalent to

$$
\begin{aligned}
&\sum_{i=1}^{2m}(-1)^{i+1} z_{x_i}\left[\begin{array}{c} m!\left((g^m)_{1\cdots\widehat{i}\cdots(2m)k}\right)_{x_n} \\ +\sum_{j=1}^{2m}(-1)^{j+1}\left((g^m)_{1\cdots\widehat{i}\cdots(2m)k}\right)_{x_j}(g^m)_{1\cdots\widehat{j}\cdots(2m)n}\end{array}\right] \\
&\quad = \sum_{i=1}^{2m}(-1)^{i+1} z_{x_i}\left[\begin{array}{c} m!\left((g^m)_{1\cdots\widehat{i}\cdots(2m)n}\right)_{x_k} \\ +\sum_{j=1}^{2m}(-1)^{j+1}\left((g^m)_{1\cdots\widehat{i}\cdots(2m)n}\right)_{x_j}(g^m)_{1\cdots\widehat{j}\cdots(2m)k}\end{array}\right].
\end{aligned} \tag{8.15}
$$

To prove this, note first that, for every $2m+1 \le k \le n$,

$$\sum_{j=1}^{2m}(-1)^{j+1}\left((g^m)_{1\cdots\widehat{j}\cdots(2m)k}\right)_{x_j} = 0 \tag{8.16}$$

since $(dg^m)_{1\cdots(2m)k}=0$, g being closed, and $\left((g^m)_{1\cdots(2m)}\right)_{x_k}=0$ according to (8.14). Hence, it follows from (8.16) that (8.15) is equivalent to

$$\sum_{i=1}^{2m}(-1)^{i+1}z_{x_i}C_i=0,$$

where

$$\begin{aligned}C_i=&\,m!\left((g^m)_{1\cdots\widehat{i}\cdots(2m)k}\right)_{x_n}-m!\left((g^m)_{1\cdots\widehat{i}\cdots(2m)n}\right)_{x_k}\\&+\sum_{j=1}^{2m}(-1)^{j+1}\left((g^m)_{1\cdots\widehat{i}\cdots(2m)k}(g^m)_{1\cdots\widehat{j}\cdots(2m)n}\right)_{x_j}\\&-\sum_{j=1}^{2m}(-1)^{j+1}\left((g^m)_{1\cdots\widehat{i}\cdots(2m)n}(g^m)_{1\cdots\widehat{j}\cdots(2m)k}\right)_{x_j}.\end{aligned}$$

To finish the proof, it is enough to prove that $C_i=0$ for every $1\le i\le 2m$. Indeed, using Lemma 8.25, we deduce that

$$\begin{aligned}C_i=&\,m!\left((g^m)_{1\cdots\widehat{i}\cdots(2m)k}\right)_{x_n}-m!\left((g^m)_{1\cdots\widehat{i}\cdots(2m)n}\right)_{x_k}\\&+m!\sum_{j=1}^{i-1}(-1)^{j+1}\left((g^m)_{1\cdots\widehat{j}\cdots\widehat{i}\cdots(2m)kn}\right)_{x_j}\\&+m!\sum_{j=i+1}^{2m}(-1)^{j}\left((g^m)_{1\cdots\widehat{i}\cdots\widehat{j}\cdots(2m)kn}\right)_{x_j}\\=&\,m!\,(dg^m)_{1\cdots\widehat{i}\cdots(2m)kn}=0\end{aligned}$$

since g is closed. This finishes the proof of the lemma. □

Lemma 8.30. *Let $2\le 2m<n$ be integers and $\Omega\subset\mathbb{R}^n$ be an open set. Let $g\in C^\infty\left(\Omega;\Lambda^2\right)$ be closed with* $\operatorname{rank}[g]=2m$ *in Ω and of the form*

$$g(x)=\sum_{i=1}^{m}dx^{2i-1}\wedge dx^{2i}+\sum_{\substack{1\le i<j\le n\\2m<j}}g_{ij}(x)\,dx^i\wedge dx^j,\quad x\in\Omega,$$

where $g_{ij}\in C^\infty(\Omega)$. Let $u\in C^\infty\left(\Omega;\Lambda^1\right)$ be such that

$$du\wedge g^m=0\quad in\ \Omega.$$

Then for every integer $2m+1\le k\le n$, the following holds:

$$L_n((u\wedge g^m)_{1\cdots(2m)k})=L_k((u\wedge g^m)_{1\cdots(2m)n}),$$

where $L_k : C^\infty(\Omega) \to C^\infty(\Omega)$, $2m+1 \le k \le n$, *is given by*

$$L_k(z) = (dz \wedge g^m)_{1\cdots(2m)k}.$$

Proof. We divide the proof of the lemma into three steps.

Step 1. We have, since $(g^m)_{1\cdots(2m)} = m!$ in Ω,

$$\begin{aligned} L_k((u\wedge g^m)_{1\cdots(2m)n}) &= (d[(u\wedge g^m)_{1\cdots(2m)n}]\wedge g^m)_{1\cdots(2m)k} \\ &= m!\left((u\wedge g^m)_{1\cdots(2m)n}\right)_{x_k} \\ &\quad + \sum_{i=1}^{2m}(-1)^{i+1}\left((u\wedge g^m)_{1\cdots(2m)n}\right)_{x_i}(g^m)_{1\cdots\widehat{i}\cdots(2m)k}. \end{aligned}$$

Since $(dg^m)_{1\cdots(2m)k} = 0$, g being closed, and $((g^m)_{1\cdots(2m)})_{x_k} = 0$, it follows that, for every $2m+1 \le k \le n$,

$$\sum_{i=1}^{2m}(-1)^{i+1}\left((g^m)_{1\cdots\widehat{i}\cdots(2m)k}\right)_{x_i} = 0$$

and therefore

$$\begin{aligned} L_k((u\wedge g^m)_{1\cdots(2m)n}) &= m!\left((u\wedge g^m)_{1\cdots(2m)n}\right)_{x_k} \\ &\quad + \sum_{i=1}^{2m}(-1)^{i+1}\left((u\wedge g^m)_{1\cdots(2m)n}(g^m)_{1\cdots\widehat{i}\cdots(2m)k}\right)_{x_i}. \end{aligned}$$

Similarly, we have

$$\begin{aligned} L_n((u\wedge g^m)_{1\cdots(2m)k}) &= m!\left((u\wedge g^m)_{1\cdots(2m)k}\right)_{x_n} \\ &\quad + \sum_{i=1}^{2m}(-1)^{i+1}\left((u\wedge g^m)_{1\cdots(2m)k}(g^m)_{1\cdots\widehat{i}\cdots(2m)n}\right)_{x_i}. \end{aligned}$$

We then set, for $1 \le i \le 2m$,

$$A_i = (u\wedge g^m)_{1\cdots(2m)n}(g^m)_{1\cdots\widehat{i}\cdots(2m)k} - (u\wedge g^m)_{1\cdots(2m)k}(g^m)_{1\cdots\widehat{i}\cdots(2m)n}.$$

In order to prove the lemma, we therefore have to show the following:

$$m!\left((u\wedge g^m)_{1\cdots(2m)n}\right)_{x_k} - m!\left((u\wedge g^m)_{1\cdots(2m)k}\right)_{x_n} + \sum_{i=1}^{2m}(-1)^{i+1}(A_i)_{x_i} = 0. \quad (8.17)$$

Step 2. In this step, we prove that, for every $1 \le i \le 2m$,

$$A_i = m!\,(u\wedge g^m)_{1\cdots\widehat{i}\cdots(2m)kn}. \quad (8.18)$$

To show this, we note that

$$A_i = \left(\sum_{j=1}^{2m} (-1)^{j+1} u^j (g^m)_{1\cdots\widehat{j}\cdots(2m)n} + m!\,u^n \right) (g^m)_{1\cdots\widehat{i}\cdots(2m)k}$$
$$- \left(\sum_{j=1}^{2m} (-1)^{j+1} u^j (g^m)_{1\cdots\widehat{j}\cdots(2m)k} + m!\,u^k \right) (g^m)_{1\cdots\widehat{i}\cdots(2m)n}.$$

We therefore get

$$\begin{aligned} A_i = {} & m!\,u^n (g^m)_{1\cdots\widehat{i}\cdots(2m)k} - m!\,u^k (g^m)_{1\cdots\widehat{i}\cdots(2m)n} \\ & + \sum_{j=1}^{2m} (-1)^{j+1} u^j \Big[(g^m)_{1\cdots\widehat{j}\cdots(2m)n} (g^m)_{1\cdots\widehat{i}\cdots(2m)k} \\ & \qquad - (g^m)_{1\cdots\widehat{j}\cdots(2m)k} (g^m)_{1\cdots\widehat{i}\cdots(2m)n} \Big]. \end{aligned}$$

Invoking Lemma 8.25 at this point, it follows that

$$\begin{aligned} A_i &= m!\,u^n (g^m)_{1\cdots\widehat{i}\cdots(2m)k} - m!\,u^k (g^m)_{1\cdots\widehat{i}\cdots(2m)n} \\ &\quad + \sum_{j=1}^{i-1} (-1)^{j+1} m!\,u^j (g^m)_{1\cdots\widehat{j}\cdots\widehat{i}\cdots(2m)kn} + \sum_{j=i+1}^{2m} (-1)^{j} m!\,u^j (g^m)_{1\cdots\widehat{i}\cdots\widehat{j}\cdots(2m)kn} \\ &= m!\,(u \wedge g^m)_{1\cdots\widehat{i}\cdots(2m)kn}. \end{aligned}$$

Step 3. We finally use (8.18) in the left-hand side of (8.17) to deduce that

$$\begin{aligned} & m!\, \left((u \wedge g^m)_{1\cdots(2m)n} \right)_{x_k} - m!\, \left((u \wedge g^m)_{1\cdots(2m)k} \right)_{x_n} \\ & + m! \sum_{i=1}^{2m} (-1)^{i+1} \left((u \wedge g^m)_{1\cdots\widehat{i}\cdots(2m)kn} \right)_{x_i} \\ & = m!\, \left(d(u \wedge g^m) \right)_{1\cdots(2m)kn} = 0, \end{aligned}$$

the last equality coming from the fact that $dg = 0$ and $du \wedge g^m = 0$. The proof is finished. □

The final proposition has been used in Step 3.1 of Theorem 8.22.

Proposition 8.31. *Let $2 \leq 2m < n$ be integers and $x_0 \in \mathbb{R}^n$. Let g be a C^∞ closed 2-form and such that, in a neighborhood of x_0,*

$$\operatorname{rank}[g] = 2m. \tag{8.19}$$

Then there exist a neighborhood U of x_0, $r_{ij} \in C^\infty(U)$ and $\varphi \in \operatorname{Diff}^\infty(U; \varphi(U))$ such that $\varphi(x_0) = x_0$ and, in U,

$$\varphi^*(g) = \sum_{i=1}^{m} dx^{2i-1} \wedge dx^{2i} + \sum_{\substack{1 \leq i < j \leq n \\ 2m < j}} r_{ij}\, dx^i \wedge dx^j.$$

Proof. Step 1. Without loss of generality, we can assume that $x_0 = 0$. In addition, using Proposition 2.24(ii), we can also assume that

$$g(0) = \omega_m = \sum_{i=1}^{m} dx^{2i-1} \wedge dx^{2i}.$$

We next introduce some notations. We let

$$x = (y,z) = (x_1, \ldots, x_{2m}, x_{2m+1}, \ldots, x_n) \in \mathbb{R}^{2m} \times \mathbb{R}^{n-2m}$$

and we define, for every $z \in \mathbb{R}^{n-2m}$, the map $i_z : \mathbb{R}^{2m} \to \mathbb{R}^n$ through

$$i_z(y) = (y,z) = x.$$

Step 2. We define for every $z \in \mathbb{R}^{n-2m}$ with $|z|$ small and every $t \in [0,1]$ the 2-form

$$g_{z,t} : W \to \Lambda^2\left(\mathbb{R}^{2m}\right) \quad \text{by} \quad g_{z,t}(y) = i_z^*\left[tg + (1-t)\omega_m\right](y),$$

where W is a small neighborhood of 0 in $\mathbb{R}^{2m}$. Note that

$$g_{z,0}(y) = i_z^*[\omega_m] = \omega_m \quad \text{and} \quad g_{z,1}(y) = \sum_{1 \le i < j \le 2m} g_{ij}(x)\, dx^i \wedge dx^j.$$

Our assumption in Step 1 leads to $g_{z,t}(0) = \omega_m$ and therefore, in a sufficiently small cube Q centered at $0 \in \mathbb{R}^{2m} \times \mathbb{R}^{n-2m}$, we can ensure that

$$\operatorname{rank}\left[g_{z,t}(y)\right] = 2m \quad \text{for every } (y,z) \in Q \text{ and } t \in [0,1]. \tag{8.20}$$

Furthermore, $g_{z,t}$ has the property that

$$d_y g_{z,t} = 0 \quad \text{in } Q, \quad \text{for every } t \in [0,1], \tag{8.21}$$

where d_y is understood as the exterior differential operator involving only the variable $y = (x_1, \ldots, x_{2m})$; namely $d_y g_{z,t} = 0$ is equivalent to

$$\frac{\partial (g_{z,t})_{ij}}{\partial x_k} - \frac{\partial (g_{z,t})_{ik}}{\partial x_j} + \frac{\partial (g_{z,t})_{jk}}{\partial x_i} = 0 \quad \text{for every } 1 \le i,j,k \le 2m.$$

Step 3. Using (8.20), (8.21) and the Poincaré lemma (see Theorem 8.1), we find a C^∞ vector field $u_{z,t} : \mathbb{R}^{2m} \to \mathbb{R}^{2m}$ such that

$$d_y\left(u_{z,t} \lrcorner\, g_{z,t}\right) = -\frac{d}{dt} g_{z,t} = \omega_m - i_z^*[g] \quad \text{in } Q, \quad \text{for every } t \in [0,1].$$

We now consider the initial value problem, for every $x = (y,z) \in Q$,

$$\frac{d}{dt}\varphi_{z,t} = u_{z,t} \circ \varphi_{z,t} \quad \text{and} \quad \varphi_{z,0}(y) = y.$$

Using Theorem 12.8, we deduce that, up to restricting the set Q,

$$\varphi_{z,1}^{*}\left(g_{z,1}\right)=g_{z,0}=\omega_m,$$

which means that

$$\sum_{1\le i<j\le 2m}\left(g_{z,1}\right)_{ij}\left(\varphi_{z,1}\left(y\right)\right)d_y\varphi_{z,1}^{i}\wedge d_y\varphi_{z,1}^{j}=\omega_m, \tag{8.22}$$

where for $u:\mathbb{R}^n\to\mathbb{R}$, we have set

$$d_y u=\sum_{i=1}^{2m}\frac{\partial u}{\partial x_i}dx^i \quad\text{and}\quad d_z u=\sum_{i=2m+1}^{n}\frac{\partial u}{\partial x_i}dx^i.$$

We finally let, for $x=(y,z)\in Q$,

$$\varphi\left(x\right)=\left(\varphi_{z,1}\left(y\right),z\right)$$

and we claim that this is the diffeomorphism we are looking for. Indeed, first observe that

$$\sum_{1\le i<j\le 2m}\left(g_{z,1}\right)_{ij}\left(\varphi_{z,1}\left(y\right)\right)d_y\varphi_{z,1}^{i}\wedge d_y\varphi_{z,1}^{j}=\sum_{1\le i<j\le 2m}g_{ij}\left(\varphi\left(x\right)\right)d_y\varphi^{i}\wedge d_y\varphi^{j}. \tag{8.23}$$

We, moreover, have

$$\begin{aligned}
\varphi^{*}\left(g\right)&=\sum_{1\le i<j\le n}g_{ij}\left(\varphi\left(y\right)\right)d\varphi^{i}\wedge d\varphi^{j}\\
&=\sum_{1\le i<j\le 2m}g_{ij}\left(\varphi\left(y\right)\right)d\varphi^{i}\wedge d\varphi^{j}+\sum_{\substack{1\le i<j\le n\\2m<j}}g_{ij}\left(\varphi\left(y\right)\right)d\varphi^{i}\wedge d\varphi^{j}\\
&=\sum_{1\le i<j\le 2m}g_{ij}\left(\varphi\left(y\right)\right)\left(d_y\varphi^{i}+d_z\varphi^{i}\right)\wedge\left(d_y\varphi^{j}+d_z\varphi^{j}\right)\\
&\quad+\sum_{\substack{1\le i<j\le n\\2m<j}}g_{ij}\left(\varphi\left(y\right)\right)d\varphi^{i}\wedge dx^{j}
\end{aligned}$$

and thus

$$\begin{aligned}
\varphi^{*}\left(g\right)=&\sum_{1\le i<j\le 2m}g_{ij}\left(\varphi\left(y\right)\right)d_y\varphi^{i}\wedge d_y\varphi^{j}\\
&+\sum_{1\le i<j\le 2m}g_{ij}\left(\varphi\left(y\right)\right)d_y\varphi^{i}\wedge d_z\varphi^{j}+\sum_{1\le i<j\le 2m}g_{ij}\left(\varphi\left(y\right)\right)d_z\varphi^{i}\wedge d_y\varphi^{j}\\
&+\sum_{1\le i<j\le 2m}g_{ij}\left(\varphi\left(y\right)\right)d_z\varphi^{i}\wedge d_z\varphi^{j}+\sum_{\substack{1\le i<j\le n\\2m<j}}g_{ij}\left(\varphi\left(y\right)\right)d\varphi^{i}\wedge dx^{j}.
\end{aligned} \tag{8.24}$$

Appealing to (8.22)–(8.24), we get

$$\varphi^*(g) = \omega_m + \sum_{\substack{1 \le i < j \le n \\ 2m < j}} r_{ij}\, dx^i \wedge dx^j$$

for appropriate r_{ij}. This finishes the proof. □

Chapter 9
The Equation $\operatorname{div} u = f$

We now study the equation

$$\operatorname{div} u = f,$$

which is constantly used in Chapter 10. Of course, most of the results can be found in Chapter 8. However, the proofs are much more elementary in this case and, in most cases, do not require the sophisticated machinery of Hodge–Morrey decomposition. They use only standard properties of the Laplacian. Therefore, for the convenience of the reader, we have gathered and proved the results in the present chapter.

9.1 The Main Theorem

We first introduce the following notations.

Notation 9.1. *(i) For a C^1 vector field $u : \mathbb{R}^n \to \mathbb{R}^n$ we let*

$$\operatorname{div} u = \sum_{i=1}^{n} \frac{\partial u^i}{\partial x_i}.$$

In terms of differential forms, u is seen as a 1-form and the divergence operator is seen as the δ operator on 1-forms.

(ii) For a C^1 vector field $v : \mathbb{R}^n \to \mathbb{R}^{n(n-1)/2}$ where the components of v are written as

$$v = (v_{ij})_{1 \le i < j \le n} \in \mathbb{R}^{n(n-1)/2},$$

we define

$$\operatorname{curl}^* v = ((\operatorname{curl}^* v)_1, \ldots, (\operatorname{curl}^* v)_n) \in \mathbb{R}^n$$

and

$$(\operatorname{curl}^* v)_i = \sum_{j=1}^{i-1} \frac{\partial v_{ji}}{\partial x_j} - \sum_{j=i+1}^{n} \frac{\partial v_{ij}}{\partial x_j}.$$

We therefore have

$$\operatorname{div}\operatorname{curl}^* v = 0 \quad \textit{for every } v \in C^2\left(\mathbb{R}^n;\mathbb{R}^{n(n-1)/2}\right).$$

In terms of differential forms v is seen as a 2-form and the curl^* *operator is seen as the* δ *operator on 2-forms. The identity* $\operatorname{div}\operatorname{curl}^* v = 0$ *is just* $\delta\delta v = 0$.

Theorem 9.2. *Let* $r \geq 0$ *be an integer and* $0 < \alpha < 1$. *Let* $\Omega \subset \mathbb{R}^n$ *be a bounded connected open* $C^{r+2,\alpha}$ *set. The following conditions are then equivalent:*

(i) $f \in C^{r,\alpha}\left(\overline{\Omega}\right)$ *satisfies*

$$\int_\Omega f = 0.$$

(ii) *There exists* $u \in C^{r+1,\alpha}\left(\overline{\Omega};\mathbb{R}^n\right)$ *verifying*

$$\begin{cases} \operatorname{div} u = f & \text{in } \Omega, \\ u = 0 & \text{on } \partial\Omega. \end{cases} \tag{9.1}$$

Furthermore, the correspondence $f \to u$ *can be chosen linear and there exists* $C = C(r,\alpha,\Omega) > 0$ *such that*

$$\|u\|_{C^{r+1,\alpha}} \leq C\|f\|_{C^{r,\alpha}}.$$

Remark 9.3. (i) If Ω is not connected, then the condition $\int_\Omega f = 0$ has to hold on each connected component of Ω.

(ii) As we already said, this result is part of the studies on the Poincaré lemma (cf. Chapter 8). However, because of its importance in applications, it has received considerable attention and has usually been treated independently of the context of Hodge–Morrey decomposition and the Poincaré lemma. Precise references can be found in Bogovski [12], Borchers and Sohr [14], Dacorogna [27, 28], Dacorogna and Moser [33], Dautray and Lions [35], Galdi [46], Girault and Raviart [50], Kapitanskii and Pileckas [59], Ladyzhenskaya [64], Ladyzhenskaya and Solonnikov [65], Necas [79], Tartar [94] and Von Wahl [103, 104].

(iii) Similar type of results hold for $f \in L^p$, $1 < p < \infty$, finding $u \in W^{1,p}$. However, the result is false if $p = 1$ or $p = \infty$ and it is also false in $C^{0,\alpha}$ when $\alpha = 0$ or $\alpha = 1$; see Bourgain and Brézis [16], Dacorogna, Fusco and Tartar [30], Mc Mullen [73] and Preiss [83].

Proof. (ii)$\Rightarrow$(i). This implication is just the divergence theorem.

(i)$\Rightarrow$(ii). We split the proof into two steps.

Step 1. We first find $w \in C^{r+2,\alpha}$ (cf. Gilbarg and Trudinger [49] or Ladyzhenskaya and Uraltseva [66]) satisfying

$$\begin{cases} \Delta w = f & \text{in } \Omega, \\ \partial w/\partial \nu = 0 & \text{on } \partial\Omega, \end{cases}$$

where ν is the exterior unit normal to $\partial\Omega$. In order to get uniqueness, we assume that

$$\int_{\Omega} w = 0.$$

The uniqueness ensures that the correspondence $f \to w$ is linear. Moreover, there exists $c_1 = c_1(r, \alpha, \Omega) > 0$ such that

$$\|w\|_{C^{r+2,\alpha}} \leq c_1 \|f\|_{C^{r,\alpha}}. \tag{9.2}$$

Step 2. We then make the ansatz

$$u = \operatorname{curl}^* v + \operatorname{grad} w. \tag{9.3}$$

Since $\operatorname{div}\operatorname{curl}^* v = 0$ for any v, it remains to find $v \in C^{r+2,\alpha}$ such that

$$\operatorname{curl}^* v = -\operatorname{grad} w \quad \text{on } \partial\Omega.$$

An easy computation (using the fact that $\partial w/\partial\nu = 0$) shows that a solution of this problem is given by

$$\operatorname{grad} v_{ij} = \left(\frac{\partial w}{\partial x_i}\nu_j - \frac{\partial w}{\partial x_j}\nu_i\right)\nu \quad \text{on } \partial\Omega,$$

whose solvability is ensured by Lemma 8.8 and, moreover, there exists $c_2 = c_2(r, \alpha, \Omega) > 0$ such that

$$\|v\|_{C^{r+2,\alpha}} \leq c_2 \|w\|_{C^{r+2,\alpha}}. \tag{9.4}$$

The combination of (9.2)–(9.4) leads to the proof of the first part of the theorem.

Step 3. Since the constructions of Steps 1 and 2 are linear, so is the correspondence $f \to u$. The inequality follows from the previous steps. □

9.2 Regularity of Divergence-Free Vector Fields

The next result uses in a more direct way the results of Chapter 8, namely Theorem 8.4.

Theorem 9.4. *Let $r \geq 0$ be an integer and $0 < \alpha < 1$. Let $\Omega \subset \mathbb{R}^n$ be a bounded open contractible smooth set. The following conditions are then equivalent:*

(i) Let $u \in C^{r,\alpha}\left(\overline{\Omega}; \mathbb{R}^n\right)$ satisfy

$$\operatorname{div} u = 0 \quad \text{in } \Omega.$$

(ii) There exists $v \in C^{r+1,\alpha}\left(\overline{\Omega}; \mathbb{R}^{n(n-1)/2}\right)$ such that

$$u = \operatorname{curl}^* v \quad \text{in } \Omega.$$

Remark 9.5. (i) If Ω is no longer contractible, we then should add another condition to (i), namely

$$\int_\Omega \langle u;\psi\rangle = 0 \quad \text{for every } \psi \in \mathscr{H}_T(\Omega;\mathbb{R}^n),$$

where

$$\mathscr{H}_T(\Omega;\mathbb{R}^n) = \left\{ \psi \in C^\infty\left(\overline{\Omega};\mathbb{R}^n\right) : \begin{bmatrix} \operatorname{curl}\psi = 0 & \text{in } \Omega \\ \operatorname{div}\psi = 0 & \text{in } \Omega \\ \nu \wedge \psi = 0 & \text{on } \partial\Omega \end{bmatrix} \right\},$$

where ν is the exterior unit normal to $\partial\Omega$. In Section 6.1, $\mathscr{H}_T(\Omega;\mathbb{R}^n)$ is denoted by $\mathscr{H}_T\left(\Omega;\Lambda^1\right)$. If Ω is contractible, then

$$\mathscr{H}_T(\Omega;\mathbb{R}^n) = \{0\}$$

and thus the condition

$$\int_\Omega \langle u;\psi\rangle = 0 \quad \text{for every } \psi \in \mathscr{H}_T(\Omega;\mathbb{R}^n)$$

is trivially fulfilled.

(ii) We recall that $\nu \wedge \psi = 0$ stands for

$$\nu_i\psi_j - \nu_j\psi_i = 0 \text{ for every } 1 \le i,j \le n.$$

(iii) When $r = 0$, $\operatorname{div} u = 0$ is understood in the sense of distributions.

(iv) The correspondence $u \to v$ can be chosen linear and continuous, as in Theorem 8.4.

Proof. (ii)$\Rightarrow$(i). The condition $\operatorname{div} u = 0$ follows at once from the fact that $\operatorname{div}\operatorname{curl}^* v = 0$. To obtain the condition in (i) of the above remark we integrate by parts, namely

$$\int_\Omega \langle u;\psi\rangle = \int_\Omega \langle \operatorname{curl}^* v;\psi\rangle = -\int_\Omega \langle v;\operatorname{curl}\psi\rangle + \int_{\partial\Omega} \langle v;\nu\wedge\psi\rangle.$$

The result then follows since $\psi \in \mathscr{H}_T(\Omega;\mathbb{R}^n)$.

(i)$\Rightarrow$(ii). This follows from Theorem 8.4. □

9.3 Some More Results

9.3.1 A First Result

In Corollary 10.8, we use the next proposition.

Proposition 9.6. *Let $r \ge 0$ be an integer and $0 < \alpha < 1$. Let $O,\Omega \subset \mathbb{R}^n$ be bounded open smooth sets such that O is contractible, Ω is connected and*

$$O \subset \overline{O} \subset \Omega.$$

Let $f \in C^{r,\alpha}(\overline{\Omega})$ be such that

$$\int_O f = \int_\Omega f = 0.$$

Then there exists $u \in C^{r+1,\alpha}(\overline{\Omega};\mathbb{R}^n)$ such that

$$\begin{cases} \operatorname{div} u = f & \textit{in } \Omega, \\ u = 0 & \textit{on } \partial O \cup \partial\Omega. \end{cases} \tag{9.5}$$

Proof. We split the proof into four steps.

Step 1. Using Theorem 9.2, there exist $w_1 \in C^{r+1,\alpha}(\overline{O};\mathbb{R}^n)$ and $v \in C^{r+1,\alpha}(\overline{\Omega};\mathbb{R}^n)$ such that

$$\begin{cases} \operatorname{div} w_1 = f & \text{in } O, \\ w_1 = 0 & \text{on } \partial O \end{cases} \tag{9.6}$$

and

$$\begin{cases} \operatorname{div} v = f & \text{in } \Omega, \\ v = 0 & \text{on } \partial\Omega. \end{cases} \tag{9.7}$$

Step 2. Let $w_2 \in C^{r+1,\alpha}(\overline{O};\mathbb{R}^n)$ be defined by

$$w_2 = w_1 - v.$$

Using (9.6) and (9.7), we obtain

$$\begin{cases} \operatorname{div} w_2 = 0 & \text{in } O, \\ w_2 = -v & \text{on } \partial O. \end{cases} \tag{9.8}$$

Since $\operatorname{div} w_2 = 0$, there exists, invoking Theorem 9.4, $h \in C^{r+2,\alpha}(\overline{O};\mathbb{R}^{n(n-1)/2})$ such that

$$\operatorname{curl}^* h = w_2\,.$$

Step 3. Using Theorem 16.11 componentwise, there exists

$$\widetilde{h} \in C^{r+2,\alpha}(\overline{\Omega};\mathbb{R}^{n(n-1)/2})$$

such that

$$\widetilde{h} = h \quad \text{in } \overline{O}.$$

Let $\phi \in C^\infty(\Omega;[0,1])$ be such that

$$\phi \equiv 1 \quad \text{in } \overline{O} \quad \text{and} \quad \operatorname{supp}\phi \subset \Omega.$$

Finally, let $w \in C^{r+1,\alpha}(\overline{\Omega};\mathbb{R}^n)$ be defined by

$$w = \operatorname{curl}^*(\phi \widetilde{h}).$$

Step 4. Let us show that $u \in C^{r+1,\alpha}(\overline{\Omega};\mathbb{R}^n)$ defined by

$$u = v + w$$

verifies (9.5). Using (9.7), we have

$$\operatorname{div} u = \operatorname{div} v + \operatorname{div} w = f + \operatorname{div} \operatorname{curl}^*(\phi \widetilde{h}) = f \quad \text{in } \Omega.$$

Using the definition of ϕ, we have $w = 0$ on $\partial\Omega$ and therefore, appealing to (9.7),

$$u = v + w = 0 \text{ on } \partial\Omega.$$

Using again the definition of ϕ, we obtain

$$w = \operatorname{curl}^*(\phi \widetilde{h}) = \operatorname{curl}^*(h) = w_2 \quad \text{in } \overline{O}.$$

Combining the above result with (9.6) and (9.8), we have

$$u = v + w = v + w_2 = w_1 = 0 \quad \text{on } \partial O.$$

This concludes the proof of the lemma. □

9.3.2 A Second Result

The following proposition is used in Theorem 10.11. It is a weaker version of Theorem 9.2 from the point of view of regularity, but it gives an additional information on the support of the solution.

Proposition 9.7. *Let $r \geq 1$ be an integer and Ω be a bounded connected open set in $\mathbb{R}^n$. Then for every $h \in C_0^r(\Omega)$ such that*

$$\int_\Omega h = 0,$$

there exists $u \in C_0^r(\Omega;\mathbb{R}^n)$ such that

$$\operatorname{div} u = h \quad \text{in } \Omega.$$

The proof of the above proposition will rely on two lemmas. The first lemma gives an explicit solution when $\Omega = (0,1)^n$ and the second one will allow us, decomposing the domain Ω, to suppose that $\Omega = (0,1)^n$. In the sequel, we let, for any integer $s \geq 1$,

$$Q^s = (0,1)^s.$$

Lemma 9.8. *Let $r \geq 1$ be an integer. Then for every $h \in C_0^r(Q^n)$ such that*

$$\int_{Q^n} h = 0,$$

there exists $u \in C_0^r(Q^n; \mathbb{R}^n)$ such that

$$\operatorname{div} u = h \quad \textit{in } Q^n.$$

Proof. We proceed by induction on n. The case $n = 1$ is immediate; just define

$$u(x) = \int_0^x h(y)dy.$$

Suppose now that the lemma holds true for $n-1$ and let us prove it for n. Define $g \in C_0^r\left(Q^{n-1}\right)$ by

$$g(x_2, \ldots, x_n) = \int_0^1 h(y, x_2, \ldots, x_n)dy.$$

Thus, there exists $v \in C_0^r\left(Q^{n-1}; \mathbb{R}^{n-1}\right)$ satisfying

$$\operatorname{div} v = g \quad \text{in } Q^{n-1}.$$

We then choose $\xi \in C_0^\infty(0,1)$ satisfying

$$\int_0^1 \xi(y)dy = 1.$$

We now define u^1, the first component of u, by

$$u^1 = \int_0^{x_1} h(y, x_2, \ldots, x_n)dy - \int_0^{x_1} \xi(y)dy \int_0^1 h(y, x_2, \ldots, x_n)dy,$$

and u is defined by

$$u(x) = \left(u^1(x), \xi(x_1)v(x_2, \ldots, x_n)\right).$$

It is straightforward to see that u has all of the desired properties. □

We next turn to our second lemma (cf. [78] and also [33]).

Lemma 9.9. *Let $\Omega_1, \Omega \subset \mathbb{R}^n$ be bounded and connected open sets. Let $U_0, \ldots, U_N$ be $N+1$ bounded open sets in $\mathbb{R}^n$ such that*

$$\overline{\Omega}_1 \subset \bigcup_{j=0}^{N} U_j \subset \Omega \tag{9.9}$$

and, for every $0 \leq j \leq N$,

$$U_j \cap \Omega_1 \neq \emptyset. \tag{9.10}$$

Then for every $h \in C^r(\overline{\Omega})$ *with*

$$\operatorname{supp} h \subset \Omega_1 \quad \textit{and} \quad \int_\Omega h = 0,$$

there exist $h_0, \ldots, h_N \in C^r(\overline{\Omega})$ *with the following properties:*

$$\sum_{s=0}^{N} h_s = h \quad \textit{in } \overline{\Omega}, \quad \operatorname{supp} h_j \subset U_j \cap \Omega_1 \quad \textit{and} \quad \int_{U_j} h_j = 0, \quad 0 \le j \le N.$$

Proof. We divide the proof into two steps.

Step 1. We start with some preliminaries.

(i) Using (9.9) we let $\psi_0, \ldots, \psi_N \in C^\infty(\mathbb{R}^n; [0,1])$ be a partition of unity of $\overline{\Omega}_1$ subordinate to $\{U_j\}$, meaning that

$$\sum_{s=0}^{N} \psi_s = 1 \quad \text{in } \overline{\Omega}_1 \quad \text{and} \quad \operatorname{supp} \psi_j \subset U_j \quad \text{for every } 0 \le j \le N. \tag{9.11}$$

(ii) Since Ω_1 is connected and (9.10) holds, the U_j can be ordered in such a manner that for every $1 \le k \le N$,

$$\left(U_k \cap \bigcup_{j<k} U_j\right) \cap \Omega_1 \neq \emptyset.$$

This is easily shown by induction on k. Suppose first that $k = 1$. If it were not possible to choose one of the U_l, $1 \le l \le N$, and rename it U_1, such that the above equation holds true, then we would have

$$(U_l \cap U_0) \cap \Omega_1 = \emptyset \quad \text{for } 1 \le l \le N.$$

This would imply that the two open sets, both nonempty due to (9.10),

$$\bigcup_{l=1}^{N} U_l \cap \Omega_1 \quad \text{and} \quad U_0 \cap \Omega_1$$

form a disjoint partition of Ω_1, contradicting the connectivity. The choice of U_k for $2 \le k \le N$ is made by the same argument. Therefore, for every $1 \le k \le N$, we can find an integer $\rho(k) < k$ such that

$$U_k \cap U_{\rho(k)} \cap \Omega_1 \neq \emptyset. \tag{9.12}$$

We define the matrix $A \in \mathbb{R}^{(N+1) \times N}$ by

$$A_k^j = \begin{cases} 1 & \text{if } j = k \\ -1 & \text{if } j = \rho(k) \\ 0 & \text{otherwise for } 0 \le j \le N,\ 1 \le k \le N. \end{cases}$$

Since each of the N columns of the matrix A contains exactly one pair $(+1,-1)$, we have that for every $1 \le k \le N$,

$$\sum_{j=0}^{N} A_k^j = 0. \tag{9.13}$$

Observe that if we cancel the first line of A, the resulting $N \times N$ matrix is then invertible.

(iii) Invoking (9.12), we can easily construct $\eta_1, \dots, \eta_N \in C^\infty(\mathbb{R}^n; [0,1])$ such that for every $1 \le k \le N$,

$$\operatorname{supp} \eta_k \subset U_k \cap U_{\rho(k)} \cap \Omega_1 \quad \text{and} \quad \int_{\Omega_1} \eta_k = 1.$$

Step 2. For every $0 \le j \le N$, we define $h_j \in C^r(\overline{\Omega})$ by

$$h_j = h\psi_j - \sum_{k=1}^{N} \lambda_k A_k^j \eta_k, \tag{9.14}$$

where the λ_k are real numbers and will be chosen appropriately. First, we notice that

$$\operatorname{supp}(h\psi_j) \subset U_j \cap \Omega_1 \quad \text{for every } 0 \le j \le N.$$

Moreover, for $0 \le j \le N$, we have that $A_k^j \neq 0$ only if $k = j$ or $\rho(k) = j$ and for those k, the support of η_k lies in

$$U_k \cap U_{\rho(k)} \cap \Omega_1 \subset U_j \cap \Omega_1 .$$

Thus, we have for every $0 \le j \le N$ that (for any λ_k),

$$\operatorname{supp} h_j \subset U_j \cap \Omega_1 .$$

Then (again for any λ_k) we have, using (9.11) and (9.13), that

$$\sum_{j=0}^{N} h_j = h \quad \text{in } \Omega_1,$$

and since the left- and right-hand sides of the above equation are both zero outside of Ω_1, the last equality holds in all Ω. In order to have

$$\int_\Omega h_j = 0, \quad \text{for every } 0 \le j \le N,$$

the $\lambda_1, \dots, \lambda_N$ have to satisfy the $N+1$ following equations (integrating (9.14))

$$\sum_{k=1}^{N} \lambda_k A_k^j = \int_\Omega h\psi_j , \quad \text{for every } 0 \le j \le N. \tag{9.15}$$

Using (9.13) and

$$0 = \int_\Omega h = \sum_{j=0}^{N} \int_\Omega h\psi_j,$$

we observe that if (9.15) is true for $j = 1, \dots, N$, then (9.15) is automatically verified for $j = 0$. From the properties of A, the N remaining equations are uniquely solvable. The proof of the lemma is then complete. □

We can now conclude with the proof of Proposition 9.7.

Proof. We divide the proof into three steps.

Step 1. (i) Let Ω_1 be a connected open set in $\mathbb{R}^n$ such that

$$\operatorname{supp} h \subset \Omega_1 \subset \overline{\Omega}_1 \subset \Omega.$$

Then choose $N \in \mathbb{N}$, $a_0, \dots, a_N \in \Omega$ and $\eta_0, \dots, \eta_N > 0$, such that

$$\overline{\Omega}_1 \subset \bigcup_{j=0}^{N} [a_j + \eta_j Q^n] \subset \Omega$$

and

$$[a_j + \eta_j Q^n] \cap \Omega_1 \neq \emptyset \quad \text{for every } 0 \leq j \leq N.$$

(ii) Applying Lemma 9.9 to h, there exist $h_0, \dots, h_N \in C^r(\overline{\Omega})$ such that

$$\int_\Omega h_j = 0 \quad \text{and} \quad \operatorname{supp} h_j \subset a_j + \eta_j Q^n \quad \text{for every } 0 \leq j \leq N, \tag{9.16}$$

$$\sum_{j=0}^{N} h_j = h \quad \text{in } \overline{\Omega}. \tag{9.17}$$

Step 2. Fix $0 \leq j \leq N$. It is obvious, by a simple change of variables, that the result of Lemma 9.8 remains true if we change Q^n into $a + \eta Q^n$ for any $a \in \mathbb{R}^n$ and $\eta > 0$. We therefore apply Lemma 9.8 to h_j (this is possible in view of (9.16)) and we get $u_j \in C_0^r(a_j + \eta_j Q^n; \mathbb{R}^n)$ and

$$\operatorname{div} u_j = h_j \quad \text{in } a_j + \eta_j Q^n.$$

Step 3. Extending all the u_j by 0 to the whole of $\overline{\Omega}$, we have, using (9.17), that u defined by

$$u = \sum_{j=0}^{n} u_j$$

has all of the desired properties. □

Part III
The Case $k = n$

Chapter 10
The Case $f \cdot g > 0$

10.1 The Main Theorem

The main theorem of this chapter has been established by Dacorogna and Moser [33].

Theorem 10.1 (Dacorogna–Moser theorem). *Let $r \geq 0$ be an integer and $0 < \alpha < 1$. Let $\Omega \subset \mathbb{R}^n$ be a bounded connected open $C^{r+2,\alpha}$ set. Let $f, g \in C^{r,\alpha}\left(\overline{\Omega}\right)$ be such that $f \cdot g > 0$ in $\overline{\Omega}$ and*

$$\int_\Omega f = \int_\Omega g. \tag{10.1}$$

Then there exists $\varphi \in \operatorname{Diff}^{r+1,\alpha}\left(\overline{\Omega};\overline{\Omega}\right)$ satisfying

$$\begin{cases} g(\varphi(x)) \det \nabla \varphi(x) = f(x), & x \in \Omega, \\ \varphi(x) = x, & x \in \partial\Omega. \end{cases} \tag{10.2}$$

Moreover, if $c > 0$ is such that

$$\left\|\frac{1}{f}\right\|_{C^0}, \left\|\frac{1}{g}\right\|_{C^0}, \|f\|_{C^{0,\alpha}}, \|g\|_{C^{0,\alpha}} \leq c,$$

then there exists a constant $C = C(c, r, \alpha, \Omega) > 0$ such that

$$\|\varphi\|_{C^{r+1,\alpha}} \leq C\left[1 + \|f\|_{C^{r,\alpha}} + \|g\|_{C^{r,\alpha}}\right].$$

Remark 10.2. (i) Recall that $\operatorname{Diff}^{r,\alpha}\left(\overline{\Omega};\overline{\Omega}\right)$ denotes the set of diffeomorphisms φ so that $\varphi\left(\overline{\Omega}\right) = \overline{\Omega}$, $\varphi \in C^{r,\alpha}\left(\overline{\Omega};\mathbb{R}^n\right)$ and $\varphi^{-1} \in C^{r,\alpha}\left(\overline{\Omega};\mathbb{R}^n\right)$.

(ii) Identifying functions with n-forms and depending on the context, we sometimes prefer to write (10.2) as

$$\begin{cases} \varphi^*(g) = f & \text{in } \Omega, \\ \varphi = \text{id} & \text{on } \partial\Omega. \end{cases}$$

(iii) If Ω is not connected, then condition (10.1) has to hold on each connected component.

(iv) The sufficient conditions are also necessary. More precisely, if φ satisfies (10.2), then necessarily, for nonvanishing f and g, we have $f \cdot g > 0$ in $\overline{\Omega}$ (cf. Corollary 19.10) and (10.1) holds (cf. (19.3)). Moreover, the function

$$\frac{f}{g \circ \varphi} \in C^{r,\alpha}\left(\overline{\Omega}\right);$$

hence, if one of the functions f or g is in $C^{r,\alpha}$, then so is the other one.

(v) The study of this problem originated in the seminal work of Moser [78]. This result has generated a considerable amount of work, notably by Banyaga [10], Dacorogna [26], Reimann [84], Tartar [95] and Zehnder [107]. Posterior contributions to [33] can be found in Rivière and Ye [85] and Ye [106]. Burago and Kleiner [19] and Mc Mullen [73], independently, proved that the result is false if $r = \alpha = 0$, suggesting that the gain of regularity is to be expected only when $0 < \alpha < 1$.

The estimate in the theorem has a sharper form when $g \equiv 1$.

Theorem 10.3 (Dacorogna–Moser theorem). *Let $r \geq 0$ be an integer and $0 < \alpha < 1$. Let $\Omega \subset \mathbb{R}^n$ be a bounded connected open $C^{r+2,\alpha}$ set and $f : \overline{\Omega} \to \mathbb{R}$. Then the two following statements are equivalent:*

(i) The function $f \in C^{r,\alpha}\left(\overline{\Omega}\right)$, $f > 0$ in $\overline{\Omega}$ and satisfies

$$\int_{\Omega} f = \operatorname{meas} \Omega. \tag{10.3}$$

(ii) There exists $\varphi \in \operatorname{Diff}^{r+1,\alpha}\left(\overline{\Omega};\overline{\Omega}\right)$ satisfying

$$\begin{cases} \det \nabla \varphi(x) = f(x), & x \in \Omega, \\ \varphi(x) = x, & x \in \partial\Omega. \end{cases} \tag{10.4}$$

Moreover, if $c > 0$ is such that

$$\left\| \frac{1}{f} \right\|_{C^0}, \|f\|_{C^{0,\alpha}} \leq c,$$

then there exists a constant $C = C(c, r, \alpha, \Omega) > 0$ such that

$$\|\varphi - \mathrm{id}\|_{C^{r+1,\alpha}} \leq C \|f - 1\|_{C^{r,\alpha}}.$$

We will give in Section 10.4 two proofs of Theorem 10.1 (one of them relying on Theorem 10.3). However, before that, we give two intermediate results.

The first one in Section 10.2 presents the celebrated flow method introduced by Moser [78]. It is a very simple and elegant method for solving our problem; however, it fails to give the expected gain in regularity.

The second result discussed in Section 10.3 is based on a fixed point argument and gives the main theorem under a smallness condition.

In Section 10.5, we give a more constructive way to find solutions of (10.2). It has the advantage to require less smoothness of the domain and, more importantly, we are also able to obtain results such as

$$\operatorname{supp}(\varphi - \mathrm{id}) \subset \Omega,$$

provided $\operatorname{supp}(f-g) \subset \Omega$. However, its main drawback is that it does not provide the expected gain of regularity.

10.2 The Flow Method

All over the present section when dealing with maps

$$g : \mathbb{R} \times \mathbb{R}^n \to \mathbb{R}^N$$

we write, depending on the context,

$$g = g(t,x) = g_t(x), \quad t \in \mathbb{R},\, x \in \mathbb{R}^n.$$

The flow method introduced by Moser easily generalizes to the case of k-forms; see Theorem 12.7. Before stating the main theorem of the present section, we start with a lemma, which is a particular case of Theorem 12.7.

Lemma 10.4. *Let $r \geq 1$ be an integer, $0 \leq \alpha \leq 1$, $T > 0$ and $\Omega \subset \mathbb{R}^n$ be a bounded open Lipschitz set. Let*

$$u \in C^{r,\alpha}([0,T] \times \overline{\Omega}; \mathbb{R}^n) \quad \text{and} \quad f \in C^{r,\alpha}([0,T] \times \overline{\Omega})$$

and, moreover, for every $0 \leq t \leq T$,

$$\begin{cases} \operatorname{div}(f_t u_t) = -\dfrac{d}{dt} f_t & \text{in } \Omega, \\ u_t = 0 & \text{on } \partial\Omega. \end{cases} \tag{10.5}$$

Then for every $0 \leq t \leq T$, $\varphi_t : \overline{\Omega} \to \mathbb{R}^n$ defined by

$$\begin{cases} \dfrac{d}{dt} \varphi_t = u_t \circ \varphi_t, & 0 \leq t \leq T, \\ \varphi_0 = \mathrm{id} \end{cases} \tag{10.6}$$

belongs to $\operatorname{Diff}^{r,\alpha}\left(\overline{\Omega}; \overline{\Omega}\right)$ *and verifies*

$$\begin{cases} \varphi_t^*(f_t) = f_0 & \text{in } \Omega, \\ \varphi_t = \mathrm{id} & \text{on } \partial\Omega. \end{cases} \tag{10.7}$$

Moreover, for each $x \in \overline{\Omega}$ *such that* $u_t(x) = 0$ *for every* $0 \le t \le T$*, then*

$$\varphi_t(x) = x \quad \text{for every } 0 \le t \le T.$$

Furthermore, if

$$\|u_t\|_{C^1(\overline{\Omega})} \le c \quad \text{for every } 0 \le t \le T,$$

then there exists a constant $C = C(c, r, T, \Omega) > 0$ *such that for every* $t \in [0, T]$,

$$\|\varphi_t - \mathrm{id}\|_{C^{r,\alpha}(\overline{\Omega})} \le C \int_0^t \|u_s\|_{C^{r,\alpha}(\overline{\Omega})}\, ds.$$

Before proving the lemma we need the following elementary result.

Proposition 10.5. *Let* $\Omega_1, \Omega_2 \subset \mathbb{R}^n$ *be bounded open Lipschitz sets and* $T > 0$*. Let*

$$u \in C^1([0,T] \times \overline{\Omega}_2; \mathbb{R}^n) \quad \text{and} \quad \varphi \in C^1([0,T] \times \overline{\Omega}_1; \overline{\Omega}_2)$$

such that, in Ω_1,

$$\frac{d}{dt}\varphi_t = u_t \circ \varphi_t \quad \text{for every } 0 \le t \le T. \tag{10.8}$$

Then for every $f \in C^1([0,T] \times \overline{\Omega}_2)$*, the following equality holds in* Ω_1 *and for* $0 \le t \le T$*:*

$$\frac{d}{dt}[\varphi_t^*(f_t)] = \varphi_t^*\left(\frac{d}{dt} f_t + \operatorname{div}(f_t u_t)\right).$$

Remark 10.6. Although this proposition is a simple consequence of Theorem 12.5, we give here an elementary proof without using any tool of differential geometry.

Proof. We start by recalling a well-known fact (cf., e.g., Theorem 7.2 in Chapter 1 of Coddington and Levinson [22]). The solution of (10.8) satisfies

$$\det \nabla \varphi_t(x) = [\det \nabla \varphi_0(x)] \exp\left[\int_0^t (\operatorname{div} u_s)(\varphi_s(x))\, ds\right].$$

Since the right-hand side of the above identity is C^1 in t, we get

$$\frac{d}{dt}[\det \nabla \varphi_t(x)] = \det \nabla \varphi_t(x) \cdot (\operatorname{div} u_t)(\varphi_t(x)). \tag{10.9}$$

We also obtain

$$\begin{aligned}
\frac{d}{dt}[\varphi_t^*(f_t)] &= \frac{d}{dt}[\det \nabla \varphi_t \cdot f_t(\varphi_t)] \\
&= \frac{d}{dt}[\det \nabla \varphi_t]\, f_t(\varphi_t) + \det \nabla \varphi_t \left[\left(\frac{d}{dt} f_t\right)(\varphi_t) + \left\langle \nabla f_t(\varphi_t); \frac{d}{dt}\varphi_t \right\rangle\right]
\end{aligned}$$

and thus, appealing to (10.8) and (10.9), we find

$$\begin{aligned}\frac{d}{dt}[\varphi_t^*(f_t)] &= \det\nabla\varphi_t\left[(\operatorname{div}u_t)(\varphi_t)\cdot f_t(\varphi_t)+\left(\frac{d}{dt}f_t\right)(\varphi_t)+\langle\nabla f_t(\varphi_t);u_t(\varphi_t)\rangle\right]\\ &= \det\nabla\varphi_t\left[\left(\frac{d}{dt}f_t\right)(\varphi_t)+\operatorname{div}(f_tu_t)(\varphi_t)\right]=\varphi_t^*\left(\frac{d}{dt}f_t+\operatorname{div}(f_tu_t)\right),\end{aligned}$$

which concludes the proof. □

We now prove Lemma 10.4.

Proof. We split the proof into two steps.

Step 1. Using Theorem 12.1, for every $0\le t\le T$, the solution φ_t of (10.6) belongs to $\operatorname{Diff}^{r,\alpha}(\overline{\Omega};\overline{\Omega})$ and verifies $\varphi_t=\mathrm{id}$ on $\partial\Omega$. Moreover, for each $x\in\overline{\Omega}$ such that $u_t(x)=0$ for every $0\le t\le T$, then

$$\varphi_t(x)=x\quad\text{for every } 0\le t\le T.$$

Furthermore, if

$$\|u_t\|_{C^1}\le c\quad\text{for every } 0\le t\le T,$$

then, using (12.3), we immediately deduce that for every $0\le t\le T$,

$$\|\varphi_t-\mathrm{id}\|_{C^{r,\alpha}}\le C\int_0^t\|u_s\|_{C^{r,\alpha}}\,ds,$$

where $C=C(c,r,T,\Omega)>0$. Finally, we have

$$\varphi\in C^{r,\alpha}([0,T]\times\overline{\Omega};\overline{\Omega}).$$

Step 2. Using Proposition 10.5 and the hypotheses on u_t and f_t, we get that, in Ω,

$$\frac{d}{dt}[\varphi_t^*(f_t)]=\varphi_t^*\left(\frac{d}{dt}f_t+\operatorname{div}(f_tu_t)\right)=0,$$

which implies the result since $\varphi_0=\mathrm{id}$. □

We now turn to the Moser theorem [78], which did not however consider the boundary condition.

Theorem 10.7. *Let $r\ge1$ be a integer, $0\le\alpha\le1$ and $\Omega\subset\mathbb{R}^n$ be a bounded connected open $C^{r+2,\alpha}$ set. Let also $f,g\in C^{r,\alpha}(\overline{\Omega})$ be such that $f\cdot g>0$ in $\overline{\Omega}$ and*

$$\int_\Omega f=\int_\Omega g.$$

Then there exists $\varphi\in\operatorname{Diff}^{r,\alpha}(\overline{\Omega};\overline{\Omega})$ satisfying

$$\begin{cases} g(\varphi(x))\det\nabla\varphi(x)=f(x), & x\in\Omega,\\ \varphi(x)=x, & x\in\partial\Omega.\end{cases}\tag{10.10}$$

Furthermore, if $0 < \gamma \leq \alpha < 1$ *and if* $c > 0$ *is such that*

$$\left\| \frac{1}{f} \right\|_{C^0}, \left\| \frac{1}{g} \right\|_{C^0}, \|f\|_{C^1}, \|g\|_{C^1} \leq c,$$

then there exists a constant $C = C(c, r, \alpha, \gamma, \Omega) > 0$ *such that*

$$\|\varphi - \mathrm{id}\|_{C^{r,\alpha}} \leq C\left[\|f\|_{C^{r,\alpha}} + \|g\|_{C^{r,\alpha}}\right] \|f - g\|_{C^{0,\gamma}} + C\|f - g\|_{C^{r-1,\alpha}}.$$

Proof. Step 1. Define, for $0 \leq t \leq 1$, $x \in \overline{\Omega}$,

$$f_t(x) = (1-t)f(x) + tg(x)$$

and

$$u_t(x) = \frac{u(x)}{f_t(x)}, \tag{10.11}$$

where $u \in C^{r,\alpha}\left(\overline{\Omega}; \mathbb{R}^n\right)$ (if $0 < \alpha < 1$, then $u \in C^{r+1,\alpha}\left(\overline{\Omega}; \mathbb{R}^n\right)$) satisfies

$$\begin{cases} \operatorname{div} u = f - g & \text{in } \Omega, \\ u = 0 & \text{on } \partial\Omega. \end{cases} \tag{10.12}$$

Such a u exists by Theorem 9.2. Note, however, that u_t (see (10.11)) is only in $C^{r,\alpha}$ (even if $0 < \alpha < 1$), since f and g are only in $C^{r,\alpha}$. Since (10.11) and (10.12) hold, we have

$$\begin{cases} \operatorname{div}(u_t f_t) = -\dfrac{d}{dt} f_t = f - g & \text{in } \Omega, \\ u_t = 0 & \text{on } \partial\Omega. \end{cases} \tag{10.13}$$

We can then apply Lemma 10.4 and have, defining $\phi_t : \overline{\Omega} \to \mathbb{R}^n$ for every $t \in [0,1]$ as the solution of

$$\begin{cases} \dfrac{d}{dt}\phi_t = u_t \circ \phi_t, & 0 \leq t \leq 1, \\ \phi_0 = \mathrm{id}, \end{cases}$$

that

$$\varphi = \phi_1$$

has all of the desired properties.

Step 2. Let us now show the estimate (recall that in the present step, $0 < \alpha < 1$). We have that the solution of (10.13), found in Theorem 9.2, satisfies

$$\|u_t f_t\|_{C^{r,\alpha}} \leq C_1 \|f - g\|_{C^{r-1,\alpha}} \quad \text{and} \quad \|u_t f_t\|_{C^{1,\gamma}} \leq C_1 \|f - g\|_{C^{0,\gamma}},$$

and thus, invoking Theorem 16.28 and Proposition 16.29, we have

$$\begin{aligned}\|u_t\|_{C^{r,\alpha}} = \left\|u_t \frac{f_t}{f_t}\right\|_{C^{r,\alpha}} &\le C_2 \left[\|u_t f_t\|_{C^{r,\alpha}} \left\|\frac{1}{f_t}\right\|_{C^0} + \|u_t f_t\|_{C^0} \left\|\frac{1}{f_t}\right\|_{C^{r,\alpha}}\right] \\ &\le C_3 \left[\|f-g\|_{C^{r-1,\alpha}} + \|f-g\|_{C^{0,\gamma}} \|f_t\|_{C^{r,\alpha}}\right] \\ &\le C_3 \|f-g\|_{C^{r-1,\alpha}} + C_3 \|f-g\|_{C^{0,\gamma}} \left[\|f\|_{C^{r,\alpha}} + \|g\|_{C^{r,\alpha}}\right].\end{aligned}$$

Similarly, we also have

$$\|u_t\|_{C^1} \le C_4 .$$

Combining the above estimates with the one in Lemma 10.4, we obtain the claim.

□

In Section 11.1, we will need a slight improvement of the above theorem.

Corollary 10.8. *Let $r \ge 1$ be an integer. Let $O, \Omega \subset \mathbb{R}^n$ be bounded open smooth sets and such that O is contractible, Ω is connected and*

$$O \subset \overline{O} \subset \Omega .$$

Let also $f, g \in C^r(\overline{\Omega})$ be such that $f \cdot g > 0$ in $\overline{\Omega}$ with

$$\int_O f = \int_O g \quad \text{and} \quad \int_\Omega f = \int_\Omega g .$$

Then there exists $\varphi \in \operatorname{Diff}^r(\overline{\Omega};\overline{\Omega})$ such that

$$\begin{cases} g(\varphi(x)) \det \nabla\varphi(x) = f, & x \in \Omega, \\ \varphi(x) = x, & x \in \partial O \cup \partial\Omega . \end{cases} \tag{10.14}$$

Proof. We decompose the proof into two steps.

Step 1. Since $f - g \in C^r(\overline{\Omega})$, then, for example,

$$f - g \in C^{r-1,1/2}(\overline{\Omega}).$$

Therefore, using Proposition 9.6, there exists $u \in C^{r,1/2}(\overline{\Omega};\mathbb{R}^n)$ (in particular, in $C^r(\overline{\Omega};\mathbb{R}^n)$) such that

$$\begin{cases} \operatorname{div} u = f - g & \text{in } \Omega, \\ u = 0 & \text{on } \partial O \cup \partial\Omega . \end{cases}$$

Step 2. Define, for $0 \le t \le 1$, $x \in \overline{\Omega}$,

$$f_t(x) = (1-t) f(x) + t g(x)$$

and

$$u_t(x) = \frac{u(x)}{f_t(x)} .$$

Note that

$$\operatorname{div}(u_t f_t) = \operatorname{div} u = f - g = -\frac{d}{dt} f_t \quad \text{in } \Omega$$

and $u_t = 0$ on $\partial\Omega \cup \partial O$. We can then apply Lemma 10.4 and have, defining $\phi_t : \overline{\Omega} \to \mathbb{R}^n$ for every $t \in [0,1]$ as the solution of

$$\begin{cases} \dfrac{d}{dt}\phi_t = u_t \circ \phi_t\,, & 0 \le t \le 1, \\ \phi_0 = \mathrm{id}, \end{cases}$$

that

$$\varphi = \phi_1$$

has all of the desired properties. □

10.3 The Fixed Point Method

We now prove Theorem 10.1 when $g \equiv 1$ and under a smallness assumption on the $C^{0,\gamma}$ norm of $f - 1$. The following result is in Dacorogna and Moser [33] and follows earlier considerations by Zehnder [107].

Theorem 10.9. *Let $r \ge 0$ be an integer and $0 < \alpha, \gamma < 1$ with $\gamma \le r + \alpha$. Let $\Omega \subset \mathbb{R}^n$ be a bounded connected open $C^{r+2,\alpha}$ set. Let $f \in C^{r,\alpha}\left(\overline{\Omega}\right)$, $f > 0$ in $\overline{\Omega}$ and*

$$\int_\Omega f = \operatorname{meas} \Omega.$$

Then there exists $\varepsilon = \varepsilon(r, \alpha, \gamma, \Omega) > 0$ such that if $\|f - 1\|_{C^{0,\gamma}} \le \varepsilon$, then there exists $\varphi \in \operatorname{Diff}^{r+1,\alpha}\left(\overline{\Omega}; \overline{\Omega}\right)$ satisfying

$$\begin{cases} \det \nabla \varphi(x) = f(x), & x \in \Omega, \\ \varphi(x) = x, & x \in \partial\Omega. \end{cases} \tag{10.15}$$

Moreover, there exists a constant $c = c(r, \alpha, \gamma, \Omega) > 0$ such that if $\|f - 1\|_{C^{0,\gamma}} \le \varepsilon$, then φ satisfies

$$\|\varphi - \mathrm{id}\|_{C^{r+1,\alpha}} \le c\,\|f - 1\|_{C^{r,\alpha}} \quad \text{and} \quad \|\varphi - \mathrm{id}\|_{C^{1,\gamma}} \le c\,\|f - 1\|_{C^{0,\gamma}}\,.$$

Proof. For the convenience of the reader we will not use the abstract fixed point theorem (cf. Theorem 18.1), but we will redo the proof. We divide the proof into two steps.

Step 1. We start by introducing some notations.

(i) Let

$$X = \left\{a \in C^{r+1,\alpha}\left(\overline{\Omega}; \mathbb{R}^n\right) : a = 0 \text{ on } \partial\Omega\right\},$$
$$Y = \left\{b \in C^{r,\alpha}\left(\overline{\Omega}\right) : \textstyle\int_\Omega b = 0\right\}.$$

Define $L : X \to Y$ by $La = \operatorname{div} a$. Note that L is well defined by the divergence theorem. As seen in Theorem 9.2, there exist a bounded linear operator $L^{-1} : Y \to X$ and a constant $K_1 = K_1(r, \alpha, \gamma, \Omega) > 0$, such that

$$LL^{-1} = \mathrm{id} \quad \text{in } Y,$$

$$\left\|L^{-1}b\right\|_{C^{1,\gamma}} \leq K_1 \left\|b\right\|_{C^{0,\gamma}}, \tag{10.16}$$

$$\left\|L^{-1}b\right\|_{C^{r+1,\alpha}} \leq K_1 \left\|b\right\|_{C^{r,\alpha}}. \tag{10.17}$$

(ii) Let for ξ, any $n \times n$ matrix,

$$Q(\xi) = \det(I + \xi) - 1 - \operatorname{trace}(\xi), \tag{10.18}$$

where I stands for the identity matrix. Note that Q is a sum of monomials of degree t, $2 \leq t \leq n$. Hence, there exists a constant $k > 0$ such that for every $\xi, \eta \in \mathbb{R}^{n \times n}$,

$$|Q(\xi) - Q(\eta)| \leq k \left(|\xi| + |\eta| + |\xi|^{n-1} + |\eta|^{n-1}\right) |\xi - \eta|.$$

With the same method, we can find (cf. Theorem 16.28) a constant $K_2 = K_2(r, \Omega) > 0$ such that if $v, w \in C^{r+1,\alpha}$ with $\|v\|_{C^{1,\gamma}}, \|w\|_{C^{1,\gamma}} \leq 1$, then

$$\begin{gathered} \|Q(\nabla v) - Q(\nabla w)\|_{C^{0,\gamma}} \leq K_2 \left(\|v\|_{C^{1,\gamma}} + \|w\|_{C^{1,\gamma}}\right) \|v - w\|_{C^{1,\gamma}}, \\ \|Q(\nabla v)\|_{C^{r,\alpha}} \leq K_2 \|v\|_{C^1} \|v\|_{C^{r+1,\alpha}}. \end{gathered} \tag{10.19}$$

Step 2. In order to solve (10.15), we set $v(x) = \varphi(x) - x$ and we rewrite it as

$$\begin{cases} \operatorname{div} v = f - 1 - Q(\nabla v) & \text{in } \Omega, \\ v = 0 & \text{on } \partial\Omega. \end{cases} \tag{10.20}$$

If we set

$$N(v) = f - 1 - Q(\nabla v),$$

then (10.20) is satisfied for any $v \in X$ with

$$v = L^{-1} N(v). \tag{10.21}$$

Note first that the equation is well defined (i.e., $N : X \to Y$), since if $v = 0$ on $\partial\Omega$, then $\int_\Omega N(v(x))\,dx = 0$. Indeed, from (10.18) we have that

$$\begin{aligned} \int_\Omega N(v(x))\,dx &= \int_\Omega [f(x) - 1 - Q(\nabla v(x))]\,dx \\ &= \int_\Omega [f(x) + \operatorname{div} v(x) - \det(I + \nabla v(x))]\,dx; \end{aligned}$$

since $v = 0$ on $\partial\Omega$ and $\int_\Omega f = \operatorname{meas} \Omega$, it follows immediately (cf. Corollary 19.10 and the divergence theorem) that the right-hand side of the above identity is 0.

We now solve (10.21) by the contraction principle. We first let

$$B = \left\{ u \in C^{r+1,\alpha}\left(\overline{\Omega};\mathbb{R}^n\right) : \left[\begin{array}{l} u = 0 \text{ on } \partial\Omega \\ \|u\|_{C^{1,\gamma}} \leq 2K_1 \|f-1\|_{C^{0,\gamma}} \\ \|u\|_{C^{r+1,\alpha}} \leq 2K_1 \|f-1\|_{C^{r,\alpha}} \end{array} \right. \right\}.$$

We endow B with the $C^{1,\gamma}$ norm. We observe that B is complete (cf. Proposition 16.23) and we will show that by choosing $\|f-1\|_{C^{0,\gamma}}$ small enough, then $L^{-1}N : B \to B$ is a contraction mapping. The contraction principle will then immediately lead to a solution $v \in B$ and, hence, in $C^{r+1,\alpha}$ of (10.21). Indeed, let

$$\|f-1\|_{C^{0,\gamma}} \leq \min\left\{\frac{1}{8K_1^2K_2}, \frac{1}{2K_1}\right\}. \tag{10.22}$$

If $v, w \in B$ (note that by construction $2K_1 \|f-1\|_{C^{0,\gamma}} \leq 1$), we will show that

$$\left\|L^{-1}N(v) - L^{-1}N(w)\right\|_{C^{1,\gamma}} \leq \frac{1}{2}\|v-w\|_{C^{1,\gamma}}, \tag{10.23}$$

$$\left\|L^{-1}N(v)\right\|_{C^{1,\gamma}} \leq 2K_1 \|f-1\|_{C^{0,\gamma}}, \; \left\|L^{-1}N(v)\right\|_{C^{r+1,\alpha}} \leq 2K_1 \|f-1\|_{C^{r,\alpha}}. \tag{10.24}$$

Inequality (10.23) follows from (10.16), (10.19) and (10.22) through

$$\begin{aligned} \left\|L^{-1}N(v) - L^{-1}N(w)\right\|_{C^{1,\gamma}} &\leq K_1 \|N(v) - N(w)\|_{C^{0,\gamma}} \\ &= K_1 \|Q(\nabla v) - Q(\nabla w)\|_{C^{0,\gamma}} \\ &\leq K_1K_2 \left(\|v\|_{C^{1,\gamma}} + \|w\|_{C^{1,\gamma}}\right) \|v-w\|_{C^{1,\gamma}} \\ &\leq 4K_1^2K_2 \|f-1\|_{C^{0,\gamma}} \|v-w\|_{C^{1,\gamma}} \\ &\leq \frac{1}{2}\|v-w\|_{C^{1,\gamma}}. \end{aligned}$$

To obtain the first inequality in (10.24), we observe that

$$\left\|L^{-1}N(0)\right\|_{C^{1,\gamma}} \leq K_1 \|N(0)\|_{C^{0,\gamma}} = K_1 \|f-1\|_{C^{0,\gamma}},$$

and, hence, combining (10.23) with the above inequality, we have immediately the first inequality in (10.24). To obtain the second one, we just have to observe that

$$\left\|L^{-1}N(v)\right\|_{C^{r+1,\alpha}} \leq K_1 \|N(v)\|_{C^{r,\alpha}} \leq K_1 \|f-1\|_{C^{r,\alpha}} + K_1 \|Q(\nabla v)\|_{C^{r,\alpha}} \tag{10.25}$$

and use the second inequality in (10.19) to get, recalling that $v \in B$,

$$\begin{aligned} \|Q(\nabla v)\|_{C^{r,\alpha}} &\leq K_2 \|v\|_{C^1} \|v\|_{C^{r+1,\alpha}} \leq K_2 \|v\|_{C^{1,\gamma}} \|v\|_{C^{r+1,\alpha}} \\ &\leq 2K_1K_2 \|f-1\|_{C^{0,\gamma}} \|v\|_{C^{r+1,\alpha}}. \end{aligned}$$

The above inequality combined with (10.22) gives

$$\|Q(\nabla v)\|_{C^{r,\alpha}} \leq \frac{1}{4K_1}\|v\|_{C^{r+1,\alpha}}.$$

Combining this last inequality, (10.25) and the fact that $v \in B$, we deduce that

$$\left\| L^{-1} N(v) \right\|_{C^{r+1,\alpha}} \leq 2K_1 \| f - 1 \|_{C^{r,\alpha}} .$$

Thus, the contraction principle gives immediately the existence of a $C^{r+1,\alpha}$ solution.

It now remains to show that $\varphi(x) = v(x) + x$ is a diffeomorphism. This is a consequence of the fact that $\det \nabla \varphi = f > 0$ and $\varphi(x) = x$ on $\partial \Omega$ (see Theorem 19.12). The estimates in the statement of the theorem follow by construction, since $v \in B$.

□

10.4 Two Proofs of the Main Theorem

10.4.1 First Proof

We start by proving Theorem 10.3, following the original proof of Dacorogna and Moser [33].

Proof. We divide the proof into four steps. Let $r \geq 0$ be an integer and $0 < \alpha < 1$. The first step is to prove that (ii) ⇒ (i) and the three other steps to prove the reverse implication.

Step 1. Assume that $\varphi \in \mathrm{Diff}^{r+1,\alpha}\left(\overline{\Omega}; \overline{\Omega}\right)$ satisfies

$$\begin{cases} \det \nabla \varphi(x) = f(x), & x \in \Omega, \\ \varphi(x) = x, & x \in \partial \Omega. \end{cases}$$

Then, clearly, $f \in C^{r,\alpha}\left(\overline{\Omega}\right)$. We also have, from Corollary 19.10, that $f > 0$ in $\overline{\Omega}$. Finally,

$$\int_{\Omega} f = \operatorname{meas} \Omega$$

in view of (19.3).

Step 2 (approximation). We first approximate $f \in C^{r,\alpha}$ by functions $f_\eta \in C^\infty$ in an appropriate way. Let $r \geq t \geq 0$ be an integer and $0 \leq \gamma \leq \alpha$. Let

$$\left\| \frac{1}{f} \right\|_{C^0}, \| f \|_{C^{0,\alpha}} \leq c.$$

Then for every $\eta > 0$ small, we can find (see Proposition 16.46) $f_\eta \in C^\infty\left(\overline{\Omega}\right)$ with $f_\eta > 0$ in $\overline{\Omega}$ and a constant $C = C(c, r, \Omega) > 0$ so that

$$\int_{\Omega} \frac{f}{f_\eta} = \operatorname{meas} \Omega,$$

$$\|f_\eta\|_{C^{t,\gamma}} \leq C\|f\|_{C^{t,\gamma}}, \; \|f_\eta\|_{C^{r+1,\alpha}} \leq \frac{C}{\eta}\|f\|_{C^{r,\alpha}}, \; \|f_\eta\|_{C^{1,\alpha}} \leq \frac{C}{\eta}\|f\|_{C^{0,\alpha}},$$

$$\|f_\eta - 1\|_{C^{t,\gamma}} \leq C\|f-1\|_{C^{t,\gamma}}, \; \|f_\eta - 1\|_{C^{1,\alpha}} \leq \frac{C}{\eta}\|f-1\|_{C^{0,\alpha}},$$

$$\left\|\frac{f}{f_\eta} - 1\right\|_{C^{t,\gamma}} \leq C\|f-1\|_{C^{t,\gamma}} \quad \text{and} \quad \left\|\frac{f}{f_\eta} - 1\right\|_{C^{0,\gamma}} \leq C\eta^{\alpha-\gamma}.$$

Step 3 (existence and regularity). We now prove the existence of a solution with appropriate regularity.

Step 3.1 (choice of an appropriate η). Let $\varepsilon = \varepsilon(r,\alpha,\Omega)$ be the ε in the statement of Theorem 10.9 with $\gamma = \alpha/2$. Then choose $\eta_0 > 0$ small enough so that

$$C\eta_0^{\alpha/2} \leq \varepsilon.$$

Note that η_0 only depends on c, r, α and Ω. Next, define $h = f_{\eta_0}$. In particular, by definition of η_0 and by the last inequality in Step 2, we have

$$\left\|\frac{f}{h} - 1\right\|_{C^{0,\alpha/2}} \leq \varepsilon. \tag{10.26}$$

Step 3.2 (conclusion). Using (10.26) and Theorem 10.9 (with $\gamma = \alpha/2$), we can find $\varphi_1 \in \mathrm{Diff}^{r+1,\alpha}\left(\overline{\Omega};\overline{\Omega}\right)$, a solution of

$$\begin{cases} \det\nabla\varphi_1(x) = \dfrac{f(x)}{h(x)}, & x \in \Omega, \\ \varphi_1(x) = x, & x \in \partial\Omega. \end{cases}$$

We further let $\varphi_2 \in \mathrm{Diff}^{r+1,\alpha}\left(\overline{\Omega};\overline{\Omega}\right)$ be a solution of

$$\begin{cases} \det\nabla\varphi_2(y) = h\left(\varphi_1^{-1}(y)\right), & y \in \Omega, \\ \varphi_2(y) = y, & y \in \partial\Omega. \end{cases}$$

Such a solution exists by Theorem 10.7, since $h \circ \varphi_1^{-1} \in C^{r+1,\alpha}\left(\overline{\Omega}\right)$ (cf. Theorem 16.31) and

$$\int_\Omega h\left(\varphi_1^{-1}(y)\right) dy = \int_\Omega h(x)\det\nabla\varphi_1(x)\,dx = \int_\Omega f(x)\,dx = \operatorname{meas}\Omega.$$

Finally, observe that the function $\varphi = \varphi_2 \circ \varphi_1$ has all of the claimed properties.

Step 4 (estimate). We now prove the estimate, first showing estimates for φ_1, then for φ_2 and, finally, for φ. We recall that

$$\left\|\frac{1}{f}\right\|_{C^0}, \quad \|f\|_{C^{0,\alpha}} \leq c. \tag{10.27}$$

In what follows, $C_1, C_2, \ldots$ will be generic constants depending only on c, r, α and Ω.

Step 4.1. Since (10.26) holds and, by construction, we deduce that, using Theorem 10.9,

$$\|\varphi_1 - \mathrm{id}\|_{C^{r+1,\alpha}} \leq C_1 \left\| \frac{f}{h} - 1 \right\|_{C^{r,\alpha}} \quad \text{and} \quad \|\varphi_1 - \mathrm{id}\|_{C^{1,\alpha/2}} \leq C_1 \left\| \frac{f}{h} - 1 \right\|_{C^{0,\alpha/2}}.$$

Invoking Step 2 with $\gamma = \alpha/2$ and $\eta = \eta_0$, we hence find

$$\|\varphi_1 - \mathrm{id}\|_{C^{r+1,\alpha}} \leq C_2 \|f - 1\|_{C^{r,\alpha}} \tag{10.28}$$

and

$$\|\varphi_1 - \mathrm{id}\|_{C^{1,\alpha/2}} \leq C_2 \|f - 1\|_{C^{0,\alpha/2}}. \tag{10.29}$$

From the last estimate, from (10.27) and from Theorem 19.12, we deduce that

$$\|\varphi_1\|_{C^1}, \left\|\varphi_1^{-1}\right\|_{C^1} \leq C_3. \tag{10.30}$$

The next inequality is obtained, combining Theorem 16.31, (10.30) and Theorem 16.32:

$$\begin{aligned}
\left\|\varphi_1^{-1} - \mathrm{id}\right\|_{C^{r+1,\alpha}} &= \left\|(\varphi_1 - \mathrm{id}) \circ \varphi_1^{-1}\right\|_{C^{r+1,\alpha}} \\
&\leq C_4 \|\varphi_1 - \mathrm{id}\|_{C^{r+1,\alpha}} + C_4 \|\varphi_1 - \mathrm{id}\|_{C^1} \|\varphi_1\|_{C^{r+1,\alpha}} \\
&\leq C_4 \|\varphi_1 - \mathrm{id}\|_{C^{r+1,\alpha}} + C_5 \|\varphi_1 - \mathrm{id}\|_{C^1} \left[1 + \|\varphi_1 - \mathrm{id}\|_{C^{r+1,\alpha}}\right] \\
&\leq C_6 \|\varphi_1 - \mathrm{id}\|_{C^{r+1,\alpha}}
\end{aligned}$$

and hence, using (10.28),

$$\left\|\varphi_1^{-1} - \mathrm{id}\right\|_{C^{r+1,\alpha}} \leq C_7 \|f - 1\|_{C^{r,\alpha}}. \tag{10.31}$$

We also find, in a similar way,

$$\left\|\varphi_1^{-1} - \mathrm{id}\right\|_{C^{1,\alpha/2}} \leq C_7 \|f - 1\|_{C^{0,\alpha/2}}. \tag{10.32}$$

We now use Theorem 16.31, Step 2 (recalling that $1/\eta_0$ only depends on c, r, α and Ω), (10.30), Theorem 16.32 and (10.28), to find

$$\begin{aligned}
\left\|h \circ \varphi_1^{-1}\right\|_{C^{r+1,\alpha}} &\leq C_8 \|h\|_{C^{r+1,\alpha}} + C_8 \|h\|_{C^1} \left\|\varphi_1^{-1}\right\|_{C^{r+1,\alpha}} \\
&\leq C_9 \|f\|_{C^{r,\alpha}} + C_9 \|f\|_{C^{0,\alpha}} \|\varphi_1\|_{C^{r+1,\alpha}} \\
&\leq C_9 \|f\|_{C^{r,\alpha}} + C_{10} \|f\|_{C^{0,\alpha}} \left[1 + \|f\|_{C^{r,\alpha}}\right]
\end{aligned}$$

and thus, since $\|f\|_{C^{0,\alpha}} \leq c$, we get

$$\left\|h \circ \varphi_1^{-1}\right\|_{C^{r+1,\alpha}} \leq C_{11} \|f\|_{C^{r,\alpha}}. \tag{10.33}$$

Similarly, we obtain (using also Proposition 16.29)

$$\left\|h \circ \varphi_1^{-1}\right\|_{C^1}, \left\|\frac{1}{h \circ \varphi_1^{-1}}\right\|_{C^1} \leq C_{12}. \tag{10.34}$$

Finally, we have the last estimate, appealing to Theorem 16.31, (10.30) and (10.28):

$$\begin{aligned}\left\|h \circ \varphi_1^{-1} - 1\right\|_{C^{r,\alpha}} &= \left\|(h-1) \circ \varphi_1^{-1}\right\|_{C^{r,\alpha}} \\ &\leq C_{13} \|h-1\|_{C^{r,\alpha}} + C_{13} \|h-1\|_{C^1} \left\|\varphi_1^{-1}\right\|_{C^{r,\alpha}} \\ &\leq C_{14} \|f-1\|_{C^{r,\alpha}} + C_{14} \|f-1\|_{C^{0,\alpha}} \|\varphi_1\|_{C^{r+1,\alpha}} \\ &\leq C_{14} \|f-1\|_{C^{r,\alpha}} + C_{15} \|f-1\|_{C^{0,\alpha}} \left[1 + \|f-1\|_{C^{r,\alpha}}\right] \\ &\leq C_{16} \|f-1\|_{C^{r,\alpha}} + C_{15} \|f-1\|_{C^{0,\alpha}} \|f-1\|_{C^{r,\alpha}}\end{aligned}$$

and thus, since $\|f\|_{C^{0,\alpha}} \leq c$, we find

$$\left\|h \circ \varphi_1^{-1} - 1\right\|_{C^{r,\alpha}} \leq C_{17} \|f-1\|_{C^{r,\alpha}} . \tag{10.35}$$

We analogously obtain

$$\left\|h \circ \varphi_1^{-1} - 1\right\|_{C^{1,\alpha/2}} \leq C_{17} \|f-1\|_{C^{0,\alpha/2}} . \tag{10.36}$$

Step 4.2. We now turn to estimate φ_2. We have, according to Theorem 10.7 and (10.34),

$$\begin{aligned}&\|\varphi_2 - \mathrm{id}\|_{C^{r+1,\alpha}} \\ &\quad \leq C_{18} \left[1 + \left\|h \circ \varphi_1^{-1}\right\|_{C^{r+1,\alpha}}\right] \left\|h \circ \varphi_1^{-1} - 1\right\|_{C^{0,\alpha/2}} + C_{18} \left\|h \circ \varphi_1^{-1} - 1\right\|_{C^{r,\alpha}} .\end{aligned}$$

Using (10.33), (10.36) and (10.35), we find

$$\begin{aligned}\|\varphi_2 - \mathrm{id}\|_{C^{r+1,\alpha}} &\leq C_{19} \left[1 + \|f\|_{C^{r,\alpha}}\right] \|f-1\|_{C^{0,\alpha/2}} + C_{19} \|f-1\|_{C^{r,\alpha}} \\ &\leq C_{19} \left[2 + \|f-1\|_{C^{r,\alpha}}\right] \|f-1\|_{C^{0,\alpha/2}} + C_{19} \|f-1\|_{C^{r,\alpha}} \\ &\leq C_{20} \|f-1\|_{C^{r,\alpha}}\end{aligned}$$

and, similarly,

$$\|\varphi_2 - \mathrm{id}\|_{C^{1,\alpha/2}} \leq C_{20} \|f-1\|_{C^{0,\alpha/2}} . \tag{10.37}$$

Step 4.3. We are now in a position to conclude with the estimate on φ. Combining Step 4.2 and (10.31), we find

$$\begin{aligned}\left\|\varphi_2 - \varphi_1^{-1}\right\|_{C^{r+1,\alpha}} &\leq \|\varphi_2 - \mathrm{id}\|_{C^{r+1,\alpha}} + \left\|\varphi_1^{-1} - \mathrm{id}\right\|_{C^{r+1,\alpha}} \\ &\leq C_{21} \|f-1\|_{C^{r,\alpha}} .\end{aligned}$$

Moreover, by (10.32) and (10.37), we get

$$\left\|\varphi_2 - \varphi_1^{-1}\right\|_{C^{1,\alpha/2}} \leq C_{22} .$$

Since $\varphi = \varphi_2 \circ \varphi_1$, we have, invoking Theorem 16.31, (10.30), (10.28) and the two previous estimates, that

$$\begin{aligned}
\|\varphi - \mathrm{id}\|_{C^{r+1,\alpha}} &= \left\|\varphi_2 \circ \varphi_1 - \varphi_1^{-1} \circ \varphi_1\right\|_{C^{r+1,\alpha}} = \left\|\left(\varphi_2 - \varphi_1^{-1}\right) \circ \varphi_1\right\|_{C^{r+1,\alpha}} \\
&\leq C_{23}\left\|\varphi_2 - \varphi_1^{-1}\right\|_{C^{r+1,\alpha}} + C_{23}\left\|\varphi_2 - \varphi_1^{-1}\right\|_{C^1}\|\varphi_1\|_{C^{r+1,\alpha}} \\
&\leq C_{23}\left\|\varphi_2 - \varphi_1^{-1}\right\|_{C^{r+1,\alpha}} \\
&\quad + C_{24}\left\|\varphi_2 - \varphi_1^{-1}\right\|_{C^1}\left[1 + \|\varphi_1 - \mathrm{id}\|_{C^{r+1,\alpha}}\right] \\
&\leq C_{25}\left\|\varphi_2 - \varphi_1^{-1}\right\|_{C^{r+1,\alpha}} + C_{24}\left\|\varphi_2 - \varphi_1^{-1}\right\|_{C^1}\|\varphi_1 - \mathrm{id}\|_{C^{r+1,\alpha}} \\
&\leq C_{26}\|f - 1\|_{C^{r,\alpha}}.
\end{aligned}$$

This achieves the proof of the theorem. □

We may now turn to the first proof of Theorem 10.1.

Proof. We divide the proof into two steps.

Step 1. First find, by Theorem 10.3,

$$\psi_1, \psi_2 \in \mathrm{Diff}^{r+1,\alpha}(\overline{\Omega}; \overline{\Omega})$$

satisfying

$$\begin{cases}
\det \nabla \psi_2(x) = \dfrac{f(x) \operatorname{meas} \Omega}{\int_\Omega f(x)\,dx}, & x \in \Omega, \\
\det \nabla \psi_1(x) = \dfrac{g(x) \operatorname{meas} \Omega}{\int_\Omega g(x)\,dx}, & x \in \Omega, \\
\psi_1(x) = \psi_2(x) = x, & x \in \partial\Omega.
\end{cases}$$

It is then easy to see that $\varphi = \psi_1^{-1} \circ \psi_2$ satisfies (10.2).

Step 2. From Theorem 16.31 we have

$$\begin{aligned}
\|\varphi\|_{C^{r+1,\alpha}} &= \left\|\psi_1^{-1} \circ \psi_2\right\|_{C^{r+1,\alpha}} \\
&\leq C_1\left(\left\|\psi_1^{-1}\right\|_{C^{r+1,\alpha}}\|\psi_2\|_{C^1}^{r+1+\alpha} + \left\|\psi_1^{-1}\right\|_{C^1}\|\psi_2\|_{C^{r+1,\alpha}} + \left\|\psi_1^{-1}\right\|_{C^0}\right).
\end{aligned}$$

From the fact that

$$\left\|\frac{1}{f}\right\|_{C^0}, \quad \left\|\frac{1}{g}\right\|_{C^0}, \quad \|f\|_{C^{0,\alpha}}, \quad \|g\|_{C^{0,\alpha}} \leq c,$$

we get, from Theorems 10.3 and 19.12,

$$\|\psi_1\|_{C^1}, \left\|\psi_1^{-1}\right\|_{C^1}, \|\psi_2\|_{C^1} \leq C_2,$$
$$\|\psi_1\|_{C^{r+1,\alpha}} \leq C_2\left[1 + \|g\|_{C^{r,\alpha}}\right] \quad \text{and} \quad \|\psi_2\|_{C^{r+1,\alpha}} \leq C_2\left[1 + \|f\|_{C^{r,\alpha}}\right].$$

We then combine all these estimates and Theorem 16.32 to get the claim. □

10.4.2 Second Proof

We now turn to the second proof of Theorem 10.1, following the ideas of Rivière and Ye [85]. We prove here the theorem, but without the estimates, under the additional hypotheses $r \geq 1$ and Ω a smooth set. But before that we prove an intermediate result.

Theorem 10.10. *Let $r \geq 1$ be an integer, $0 < \alpha < 1$, $\Omega \subset \mathbb{R}^n$ be a bounded connected open smooth set and $f \in C^{r,\alpha}(\overline{\Omega})$ with $f > 0$ in $\overline{\Omega}$. Then, for every ε small, there exist $f_\varepsilon \in C^\infty(\overline{\Omega})$ and $\varphi_\varepsilon \in \operatorname{Diff}^{r+1,\alpha}(\overline{\Omega};\overline{\Omega})$ satisfying*

$$\begin{cases} f_\varepsilon(\varphi_\varepsilon(x)) \det \nabla \varphi_\varepsilon(x) = f(x), & x \in \Omega, \\ \varphi_\varepsilon(x) = x, & x \in \partial\Omega, \end{cases}$$

$$\lim_{\varepsilon \to 0} \|f_\varepsilon - f\|_{C^r} = 0.$$

Proof. We divide the proof into four steps.

Step 1 (definition of f_ε). We apply Proposition 16.47 to f and let $s \geq r \geq t \geq 0$ be integers and $0 \leq \alpha, \beta, \gamma \leq 1$ be such that

$$t + \gamma \leq r + \alpha \leq s + \beta.$$

Therefore, for every $\varepsilon > 0$ small, there exist $f_\varepsilon \in C^\infty(\overline{\Omega})$ with $f_\varepsilon > 0$ in $\overline{\Omega}$ and a constant $C_1 = C_1(s, \Omega, \|f\|_{C^0}, \|1/f\|_{C^0})$ such that

$$\int_\Omega f_\varepsilon = \int_\Omega f, \tag{10.38}$$

$$\|f_\varepsilon\|_{C^{s,\beta}} \leq \frac{C_1}{\varepsilon^{(s+\beta)-(r+\alpha)}} \|f\|_{C^{r,\alpha}}, \tag{10.39}$$

$$\|f_\varepsilon - f\|_{C^{t,\gamma}} \leq C_1 \varepsilon^{(r+\alpha)-(t+\gamma)} [\|f\|_{C^{r,\alpha}} + \|f\|_{C^{r,\alpha}}^2], \tag{10.40}$$

$$\left\| \frac{d}{d\varepsilon} f_\varepsilon \right\|_{C^{s,\beta}} \leq \frac{C_1}{\varepsilon^{(s+\beta)-(r+\alpha)+1}} [\|f\|_{C^{r,\alpha}} + \|f\|_{C^{r,\alpha}}^2], \tag{10.41}$$

$$\left\| \frac{d}{d\varepsilon} f_\varepsilon \right\|_{C^{t,\gamma}} \leq C_1 \varepsilon^{(r+\alpha)-(t+\gamma)-1} [\|f\|_{C^{r,\alpha}} + \|f\|_{C^{r,\alpha}}^2]. \tag{10.42}$$

Moreover, defining, for some $\overline{\varepsilon} \leq 1$ small enough, $F : (0, \overline{\varepsilon}] \times \overline{\Omega} \to \mathbb{R}$ by $F(\varepsilon, x) = f_\varepsilon(x)$, we have

$$F \in C^\infty((0, \overline{\varepsilon}] \times \overline{\Omega}). \tag{10.43}$$

Using (10.40) and choosing $\overline{\varepsilon}$ even smaller, we can assume that for every $\varepsilon \in (0, \overline{\varepsilon}]$,

$$\|f_\varepsilon\|_{C^0} \leq 2\|f\|_{C^0} \quad \text{and} \quad \left\| \frac{1}{f_\varepsilon} \right\|_{C^0} \leq 2 \left\| \frac{1}{f} \right\|_{C^0}. \tag{10.44}$$

Combining (10.44) and Proposition 16.29, we get for every $\varepsilon \in (0,\overline{\varepsilon}]$,

$$\left\| \frac{1}{f_\varepsilon} \right\|_{C^{s,\beta}} \leq C_2 \|f_\varepsilon\|_{C^{s,\beta}}, \tag{10.45}$$

where $C_2 = C_2(s,\Omega,\|f\|_{C^0},\|1/f\|_{C^0})$.

Step 2. Choose $\delta > 0$ small enough so that $[\alpha-\delta,\alpha+\delta] \subset (0,1)$. We show that for every $\varepsilon \in (0,\overline{\varepsilon}]$, there exist $u_\varepsilon \in C^\infty\left(\overline{\Omega};\mathbb{R}^n\right)$ and a constant

$$C_3 = C_3(r,\alpha,\delta,\Omega,\|f\|_{C^0},\|1/f\|_{C^0})$$

such that

$$\operatorname{div}(f_\varepsilon u_\varepsilon) = -\frac{d}{d\varepsilon} f_\varepsilon \quad \text{in } \Omega, \tag{10.46}$$

$$\|u_\varepsilon\|_{C^{r+1,\gamma}} \leq \frac{C_3}{\varepsilon^{1+\gamma-\alpha}} [\|f\|_{C^{r,\alpha}} + \|f\|^3_{C^{r,\alpha}}], \quad \gamma \in [\alpha-\delta,\alpha+\delta]. \tag{10.47}$$

Moreover, defining $u : (0,\overline{\varepsilon}] \times \overline{\Omega} \to \mathbb{R}^n$ by $u(\varepsilon,x) = u_\varepsilon(x)$, we will show that $u \in C^\infty((0,\overline{\varepsilon}] \times \overline{\Omega};\mathbb{R}^n)$.

Step 2.1. According to (10.38) we have for every $\varepsilon \in (0,\overline{\varepsilon}]$,

$$\int_\Omega \frac{d}{d\varepsilon} f_\varepsilon = 0.$$

We can therefore find, by Theorem 9.2, $w_\varepsilon \in C^\infty(\overline{\Omega};\mathbb{R}^n)$ and a constant $C_4 = C_4(r,\alpha,\delta,\Omega)$ such that

$$\begin{cases} \operatorname{div}(w_\varepsilon) = -\dfrac{d}{d\varepsilon} f_\varepsilon & \text{in } \Omega, \\ w_\varepsilon = 0 & \text{on } \partial\Omega, \end{cases}$$

and for every integer $q \leq r$ and every $\gamma \in [\alpha-\delta,\alpha+\delta]$,

$$\|w_\varepsilon\|_{C^{q+1,\gamma}} \leq C_4 \left\| \frac{d}{d\varepsilon} f_\varepsilon \right\|_{C^{q,\gamma}}. \tag{10.48}$$

Moreover, using (10.43) and defining $w : (0,\overline{\varepsilon}] \times \overline{\Omega} \to \mathbb{R}^n$ by $w(\varepsilon,x) = w_\varepsilon(x)$, we have $w \in C^\infty((0,\overline{\varepsilon}] \times \overline{\Omega};\mathbb{R}^n)$.

Step 2.2. Since $f_\varepsilon > 0$ in $\overline{\Omega}$, we can define for every $\varepsilon \in (0,\overline{\varepsilon}]$,

$$u_\varepsilon = \frac{w_\varepsilon}{f_\varepsilon}.$$

First, note that for every $\varepsilon \in (0,\overline{\varepsilon}]$, (10.46) holds,

$$u_\varepsilon \in C^\infty\left(\overline{\Omega};\mathbb{R}^n\right), \quad \text{and} \quad u_\varepsilon = 0 \quad \text{on } \partial\Omega.$$

Moreover, defining $u : (0,\overline{\varepsilon}] \times \overline{\Omega} \to \mathbb{R}^n$ by $u(\varepsilon,x) = u_\varepsilon(x)$, we have $u \in C^\infty((0,\overline{\varepsilon}] \times \overline{\Omega};\mathbb{R}^n)$.

Step 2.3. To conclude Step 2, it only remains to prove (10.47). Using Theorem 16.28, (10.44), (10.45) and (10.48), we obtain

$$\begin{aligned}
\|u_\varepsilon\|_{C^{r+1,\gamma}} &\leq C_5 \|w_\varepsilon\|_{C^{r+1,\gamma}} \left\| \frac{1}{f_\varepsilon} \right\|_{C^0} + C_5 \|w_\varepsilon\|_{C^0} \left\| \frac{1}{f_\varepsilon} \right\|_{C^{r+1,\gamma}} \\
&\leq C_6 \|w_\varepsilon\|_{C^{r+1,\gamma}} + C_6 \|w_\varepsilon\|_{C^{1,\alpha}} \|f_\varepsilon\|_{C^{r+1,\gamma}} \\
&\leq C_7 \left\| \frac{d}{d\varepsilon} f_\varepsilon \right\|_{C^{r,\gamma}} + C_7 \left\| \frac{d}{d\varepsilon} f_\varepsilon \right\|_{C^{0,\alpha}} \|f_\varepsilon\|_{C^{r+1,\gamma}}
\end{aligned}$$

and hence, appealing to (10.39), (10.41), (10.42) and recalling that $\varepsilon \leq 1$ and that $r \geq 1$, we find

$$\begin{aligned}
\|u_\varepsilon\|_{C^{r+1,\gamma}} &\leq \frac{C_8}{\varepsilon^{1+\gamma-\alpha}} [\|f\|_{C^{r,\alpha}} + \|f\|^2_{C^{r,\alpha}}] \\
&\quad + C_8 \frac{\varepsilon^{r-1}}{\varepsilon^{1+\gamma-\alpha}} [\|f\|_{C^{r,\alpha}} + \|f\|^2_{C^{r,\alpha}}] \|f\|_{C^{r,\alpha}} \\
&\leq \frac{C_9}{\varepsilon^{1+\gamma-\alpha}} [\|f\|_{C^{r,\alpha}} + \|f\|^3_{C^{r,\alpha}}],
\end{aligned}$$

where $C_i = C_i(r, \alpha, \delta, \Omega, \|f\|_{C^0}, \|1/f\|_{C^0})$. Therefore, the claim is proven.

Step 3. We can now finish the proof.

Step 3.1. Since $u \in C^\infty((0, \overline{\varepsilon}] \times \overline{\Omega}; \mathbb{R}^n)$, $u_\varepsilon = 0$ on $\partial\Omega$ and (10.47) holds, we deduce, using Theorem 12.4, that the solution $\varphi : [0, \overline{\varepsilon}] \times \overline{\Omega} \to \overline{\Omega}$, $\varphi(\varepsilon, x) = \varphi_\varepsilon(x)$, of

$$\begin{cases}
\dfrac{d}{d\varepsilon} \varphi_\varepsilon = u_\varepsilon \circ \varphi_\varepsilon, & 0 < \varepsilon \leq \overline{\varepsilon}, \\
\varphi_0 = \mathrm{id}
\end{cases}$$

verifies

$$\varphi \in C^{r+1}([0, \overline{\varepsilon}] \times \overline{\Omega}; \overline{\Omega}) \tag{10.49}$$

and that for every $\varepsilon \in [0, \overline{\varepsilon}]$,

$$\varphi_\varepsilon \in \mathrm{Diff}^{r+1,\alpha}(\overline{\Omega}; \overline{\Omega}) \quad \text{and} \quad \varphi_\varepsilon = \mathrm{id} \text{ on } \partial\Omega.$$

Step 3.2. Since (10.46) holds, we have that, using Proposition 10.5, for every $0 < \varepsilon_1 \leq \varepsilon_2 \leq \overline{\varepsilon}$,

$$\varphi^*_{\varepsilon_2}(f_{\varepsilon_2}) = \varphi^*_{\varepsilon_1}(f_{\varepsilon_1}) \quad \text{in } \Omega.$$

Since using (10.40) and (10.49),

$$\lim_{\varepsilon \to 0} \|f_\varepsilon - f\|_{C^0} = 0 \quad \text{and} \quad \lim_{\varepsilon \to 0} \|\varphi_\varepsilon - \varphi_0\|_{C^1} = 0,$$

we immediately deduce that for every $\varepsilon \in [0, \overline{\varepsilon}]$,

$$\varphi^*_\varepsilon(f_\varepsilon) = \varphi^*_0(f) = f.$$

Finally, using again (10.40), we deduce that

$$\lim_{\varepsilon \to 0} \|f_\varepsilon - f\|_{C^r} = 0,$$

which concludes the proof. □

We can now deal with the second proof of Theorem 10.1.

Proof. Step 1. By Theorem 10.10 applied to f and g there exist $f_1, g_1 \in C^\infty(\overline{\Omega})$ and $\varphi_1, \varphi_2 \in \mathrm{Diff}^{r+1,\alpha}(\overline{\Omega};\overline{\Omega})$ such that

$$\begin{cases} \varphi_1^*(f_1) = f & \text{in } \Omega, \\ \varphi_1 = \mathrm{id} & \text{on } \partial\Omega \end{cases} \quad \text{and} \quad \begin{cases} \varphi_2^*(g_1) = g & \text{in } \Omega, \\ \varphi_2 = \mathrm{id} & \text{on } \partial\Omega. \end{cases} \tag{10.50}$$

Step 2. Since (10.50) holds, we have (cf. Corollary 19.10), $f_1, g_1 > 0$ in $\overline{\Omega}$ and (cf. (19.3))

$$\int_\Omega f_1 = \int_\Omega f = \int_\Omega g = \int_\Omega g_1 .$$

Therefore, using Theorem 10.7, there exists $\varphi_3 \in \mathrm{Diff}^\infty(\overline{\Omega};\overline{\Omega})$ such that

$$\begin{cases} \varphi_3^*(g_1) = f_1 & \text{in } \Omega, \\ \varphi_3 = \mathrm{id} & \text{on } \partial\Omega. \end{cases}$$

Step 3. Using the above steps, we find that

$$\varphi = \varphi_2^{-1} \circ \varphi_3 \circ \varphi_1$$

has all of the desired properties. □

10.5 A Constructive Method

In this section we present a constructive method (cf. Theorem 5 in Dacorogna and Moser [33]) to solve the problem

$$\begin{cases} \varphi^*(g) = f & \text{in } \Omega, \\ \varphi = \mathrm{id} & \text{on } \partial\Omega. \end{cases}$$

The following theorem is only a particular case of the results in [33]. The proof that we provide here is substantially different from the original proof.

Theorem 10.11. *Let $r \geq 1$ be an integer, Ω be a bounded connected open set in $\mathbb{R}^n$ and $f, g \in C^r(\overline{\Omega})$ such that*

$$f \cdot g > 0 \text{ in } \Omega, \quad \int_\Omega f = \int_\Omega g \quad \text{and} \quad \mathrm{supp}(f - g) \subset \Omega.$$

Then there exists $\varphi \in \operatorname{Diff}^r(\overline{\Omega};\overline{\Omega})$ *such that*

$$\begin{cases} \varphi^*(g) = f & \text{in } \Omega, \\ \operatorname{supp}(\varphi - \mathrm{id}) \subset \Omega. \end{cases}$$

Proof. We divide the proof into two steps, but before that, we note that we can assume, by choosing Ω smaller if necessary, that $f \cdot g > 0$ in $\overline{\Omega}$ and that Ω is smooth, since $\operatorname{supp}(f-g) \subset \Omega$.

Step 1. Using Proposition 9.7, we can find $u \in C^r(\overline{\Omega};\mathbb{R}^n)$ such that

$$\operatorname{div} u = f - g \quad \text{in } \Omega \quad \text{and} \quad \operatorname{supp} u \subset \Omega.$$

Step 2. Define for $0 \le t \le 1$, $x \in \overline{\Omega}$,

$$f_t(x) = (1-t)f(x) + tg(x)$$

and

$$u_t(x) = \frac{u(x)}{f_t(x)}.$$

Note that

$$\operatorname{div}(u_t f_t) = \operatorname{div} u = f - g = -\frac{d}{dt} f_t \quad \text{in } \Omega$$

and $\operatorname{supp} u_t = \operatorname{supp} u \subset \Omega$. We can then apply Lemma 10.4 and have, defining $\phi_t : \overline{\Omega} \to \mathbb{R}^n$ for every $t \in [0,1]$, as the solution of

$$\begin{cases} \dfrac{d}{dt}\phi_t = u_t \circ \phi_t, & 0 \le t \le 1, \\ \phi_0 = \mathrm{id}, \end{cases}$$

that

$$\varphi = \phi_1$$

has all of the desired properties. □

Chapter 11
The Case Without Sign Hypothesis on f

11.1 Main Result

The aim of this chapter is to solve the problem

$$\begin{cases} g(\varphi(x)) \det \nabla \varphi(x) = f(x), & x \in \Omega, \\ \varphi(x) = x, & x \in \partial\Omega, \end{cases}$$

equivalently written as

$$\begin{cases} \varphi^*(g) = f & \text{in } \Omega, \\ \varphi = \text{id} & \text{on } \partial\Omega, \end{cases} \tag{11.1}$$

with $g > 0$ in $\mathbb{R}^n$ but with no sign restriction on f. Of course, the solution cannot be a diffeomorphism; nevertheless, if $f \geq 0$ and under further restrictions, it can be a homeomorphism (see Theorem 11.1(iii)).

The main result of this chapter, established by Cupini, Dacorogna and Kneuss [25], is the following. In the sequel, we denote by B_R the open ball of radius R centered at the origin.

Theorem 11.1. *Let $n \geq 2$ and $r \geq 1$ be integers and Ω a bounded open set in $\mathbb{R}^n$ such that $\overline{\Omega}$ is C^{r+1}-diffeomorphic to $\overline{B}_1$. Let $g \in C^r(\mathbb{R}^n)$ with $g > 0$ and $f \in C^r(\overline{\Omega})$ be such that*

$$\int_\Omega g = \int_\Omega f.$$

Then for every $\varepsilon > 0$, there exists $\varphi = \varphi_\varepsilon \in C^r(\overline{\Omega}; \mathbb{R}^n)$ satisfying (11.1), namely

$$\begin{cases} \varphi^*(g) = f & \text{in } \Omega, \\ \varphi = \text{id} & \text{on } \partial\Omega \end{cases}$$

and

$$\overline{\Omega} \subset \varphi\left(\overline{\Omega}\right) \subset \overline{\Omega} + B_\varepsilon.$$

Moreover, the three following properties hold:

(i) If either $f > 0$ on $\partial\Omega$ or $f \geq 0$ in $\overline{\Omega}$, then ε can be taken to be 0. In other words, there exists $\varphi \in C^r(\overline{\Omega};\overline{\Omega})$ satisfying (11.1).

(ii) If $\operatorname{supp}(g-f) \subset \Omega$, then φ can be chosen such that

$$\varphi \in C^r(\overline{\Omega};\overline{\Omega}) \quad \textit{and} \quad \operatorname{supp}(\varphi - \mathrm{id}) \subset \Omega.$$

(iii) If $f \geq 0$ in $\overline{\Omega}$ and $f^{-1}(0) \cap \Omega$ is countable, then φ can be chosen such that

$$\varphi \in C^r(\overline{\Omega};\overline{\Omega}) \cap \operatorname{Hom}(\overline{\Omega};\overline{\Omega}).$$

Remark 11.2. (i) Note that, in view of (19.2), we always have $\overline{\Omega} \subset \varphi(\overline{\Omega})$ as soon as $\varphi = \mathrm{id}$ on $\partial\Omega$.

(ii) In general, without further hypothesis on f as the extra statement (i), it is not possible to find a solution that remains in $\overline{\Omega}$. In fact, if f is negative in some part of $\partial\Omega$, then any solution must go out of $\overline{\Omega}$ (cf. Proposition 11.3).

(iii) The above theorem is also valid in Hölder spaces.

The proof of the theorem will be discussed in Section 11.3, but we want to explain the two main steps. First, observe that the fact that f is not strictly positive precludes the use of either the flow method or the fixed point method developed in Chapter 10; the proof will be more constructive. Here are the main steps for Ω the unit ball. The idea is to look for radial solutions of the problem; however, to achieve this, we have to rearrange f in an appropriate way. We therefore will look for solutions of the form

$$\varphi = \psi \circ \chi^{-1}$$

with $\psi = \chi = \mathrm{id}$ on $\partial\Omega$.

— First, we rearrange f with a diffeomorphism χ, so that

$$f_1 = \chi^*(f)$$

satisfies $f_1(0) > 0$ and has nice symmetry properties, for instance, among others,

$$\int_0^r s^{n-1} f_1\left(s\frac{x}{|x|}\right) ds > 0 \quad \text{for every } x \neq 0 \text{ and } r \in (0,1].$$

This will be the most difficult part of our proof and will be achieved in Section 11.6 (with the help of Section 11.5). Note that in view of Proposition 11.6 the function f_1 cannot therefore be strictly positive if f is not strictly positive.

— We then find a map ψ so that

$$\psi^*(g) = f_1 .$$

This will be achieved using Section 11.4 and Chapter 10. Note that the map ψ cannot be a diffeomorphism if f_1 vanishes even at a single point.

11.2 Remarks and Related Results

In this section Ω will be a bounded open set in $\mathbb{R}^n$. We start by showing that if $f < 0$ in some parts of $\partial\Omega$, then any solution of

$$\begin{cases} \varphi^*(g) = f & \text{in } \Omega, \\ \varphi = \text{id} & \text{on } \partial\Omega \end{cases} \tag{11.2}$$

must go out of $\overline{\Omega}$—more precisely,

$$\overline{\Omega} \underset{\neq}{\subset} \varphi\left(\overline{\Omega}\right).$$

We recall, using (19.2), that we necessarily have

$$\overline{\Omega} \subset \varphi(\overline{\Omega}).$$

Proposition 11.3. *Let Ω be a bounded open C^1 set in $\mathbb{R}^n$ and $\varphi \in C^1(\overline{\Omega};\mathbb{R}^n)$ with $\varphi = \text{id}$ on $\partial\Omega$. If there exists $\overline{x} \in \partial\Omega$ such that* $\det\nabla\varphi(\overline{x}) < 0$, *then*

$$\overline{\Omega} \underset{\neq}{\subset} \varphi(\overline{\Omega}). \tag{11.3}$$

Proof. We divide the proof into two steps.

Step 1 (simplification). Since Ω is C^1 (cf. Definition 16.5), there exists $\psi \in \text{Diff}^1(\overline{B}_1;\psi(\overline{B}_1))$ with $\psi(0) = \overline{x}$ and

$$\psi(\overline{B}_1 \cap \{x_n = 0\}) \subset \partial\Omega,$$
$$\psi(\overline{B}_1 \cap \{x_n > 0\}) \subset \Omega,$$
$$\psi(\overline{B}_1 \cap \{x_n < 0\}) \subset (\overline{\Omega})^c.$$

Therefore, using that $\varphi(\overline{x}) = \overline{x}$, we can choose $\varepsilon > 0$ small enough so that

$$\widetilde{\varphi} : \overline{B}_\varepsilon \cap \{x_n \geq 0\} \to \mathbb{R}^n \quad \text{with} \quad \widetilde{\varphi}(x) = \psi^{-1}(\varphi(\psi(x)))$$

is well defined. We observe that $\widetilde{\varphi}$ satisfies

$$\widetilde{\varphi} = \text{id on } \overline{B}_\varepsilon \cap \{x_n = 0\} \quad \text{and} \quad \det\nabla\widetilde{\varphi}(0) = \det\nabla\varphi(\overline{x}) < 0. \tag{11.4}$$

To prove (11.3) it is enough to show that

$$\widetilde{\varphi}(\overline{B}_{\varepsilon'} \cap \{x_n > 0\}) \subset \{x_n < 0\} \tag{11.5}$$

for a certain $0 < \varepsilon' \leq \varepsilon$.

Step 2. We finally show (11.5). Using (11.4), we immediately obtain

$$\frac{\partial \widetilde{\varphi}_n}{\partial x_n}(0) = \det \nabla \widetilde{\varphi}(0) < 0,$$

and therefore, by continuity, there exists $0 < \varepsilon' \leq \varepsilon$ such that

$$\frac{\partial \widetilde{\varphi}_n}{\partial x_n} < 0 \quad \text{in } B_{\varepsilon'} \cap \{x_n > 0\}. \tag{11.6}$$

Combining (11.6) and the fact that $\widetilde{\varphi}_n(0) = 0$ (by (11.4)), we get (11.5). □

We now discuss the special case $n = 1$ in the context $g > 0$ and with no sign restriction on f.

Proposition 11.4. *Let $n = 1$, $r \geq 0$, $\Omega = (a,b)$, $g \in C^r(\mathbb{R})$ with $g > 0$ and $f \in C^r([a,b])$. Let*

$$F(x) = \int_a^x f(t)\,dt \quad \text{and} \quad G(x) = \int_a^x g(t)\,dt.$$

Then there exists $\varphi \in C^{r+1}([a,b];\mathbb{R})$ a solution of (11.2) if and only if

$$F(b) = G(b) \quad \text{and} \quad F([a,b]) \subset G(\mathbb{R}).$$

Remark 11.5. Let F and G be as in the proposition with $F(b) = G(b)$. Then the following statements are verified:

(i) We always have

$$G([a,b]) \subset F([a,b]).$$

Moreover, when $f \geq 0$, the previous inclusion is an equality.

(ii) In general,

$$F([a,b]) \underset{\neq}{\subset} G([a,b]).$$

This is for example always the case when $f(a) < 0$ or $f(b) < 0$.

(iii) The inclusion

$$F([a,b]) \subset G(\mathbb{R})$$

is not always fulfilled.

Proof. Step 1. First, note that Problem (11.2) becomes

$$\begin{cases} G(\varphi(x)) = F(x), & x \in (a,b), \\ G(b) = F(b). \end{cases}$$

Indeed, (11.2) is equivalent to

$$\begin{cases} [G(\varphi(x))]' = F'(x) & \text{if } x \in (a,b), \\ \varphi(a) = a \quad \text{and} \quad \varphi(b) = b. \end{cases}$$

We therefore get

$$G(\varphi(x)) = F(x) + c.$$

Since $\varphi(a) = a$ and $G(a) = F(a)$, we deduce that $c = 0$ and thus our claim.

Step 2. Since G is strictly monotone (because $g > 0$), the solution φ (if it exists) is given by

$$\varphi(x) = G^{-1}(F(x)).$$

Therefore, the conclusion easily follows. □

We now show that Problem (11.2) is not symmetric in g and f.

Proposition 11.6. *Let $g \in C^0(\mathbb{R}^n)$ with $g^{-1}(0) \cap \overline{\Omega} \neq \emptyset$ and $f \in C^0(\overline{\Omega})$ with $f > 0$ in $\overline{\Omega}$. Then no $\varphi \in C^1(\overline{\Omega};\mathbb{R}^n)$ can satisfy (11.2).*

Proof. We proceed by contradiction. Assume that $\varphi \in C^1(\overline{\Omega};\mathbb{R}^n)$ is a solution of (11.2). Since $\varphi = \mathrm{id}$ on $\partial\Omega$, then (see (19.2) below)

$$\varphi(\overline{\Omega}) \supset \overline{\Omega}.$$

Thus, there exists $z \in \overline{\Omega}$ such that $\varphi(z) \in \overline{\Omega}$ and $g(\varphi(z)) = 0$, which is the desired contradiction, since

$$g(\varphi(z)) \det \nabla\varphi(z) = f(z) > 0.$$

The proposition is therefore proved. □

In the following proposition, we state a necessary condition (see (11.7) below) for the existence of a one-to-one solution of (11.2). Moreover, we show that not all solutions of (11.2), verifying (11.7), are one-to-one.

Proposition 11.7. *Let*

$$g \in C^0(\mathbb{R}^n), \quad g > 0 \text{ in } \mathbb{R}^n, \quad f \in C^0(\overline{\Omega}) \quad \text{and} \quad \int_\Omega f = \int_\Omega g.$$

Then the following claims hold true:

(i) If $\varphi \in C^1(\overline{\Omega};\mathbb{R}^n)$ is a one-to-one solution of (11.2), then $\varphi \in \mathrm{Hom}(\overline{\Omega};\overline{\Omega})$ and

$$f \geq 0 \quad \text{and} \quad \mathrm{int}(f^{-1}(0)) = \emptyset. \tag{11.7}$$

(ii) There exists f satisfying (11.7) such that not all solutions $\varphi \in C^1(\overline{\Omega};\mathbb{R}^n)$ of (11.2) are one-to-one.

Proof. (i) By Lemma 19.11, we have that $\varphi \in \mathrm{Hom}(\overline{\Omega};\overline{\Omega})$. Applying Proposition 19.14, we have the claim.

(ii) We provide a counterexample in two dimensions. Let $f \in C^1(\overline{B}_1)$ be such that $f \geq 0$,

$$f^{-1}(0) = \{(t,0) : t \in [1/2, 3/4]\}, \quad f \equiv 1 \text{ in a neighborhood of } 0$$

and, for every $x \neq 0$,

$$\int_0^1 s f\left(s\frac{x}{|x|}\right) ds = \frac{1}{2}.$$

Define next $\alpha : \overline{B}_1 \to [0,1]$, through $\alpha(0) = 0$ and, for $0 < |x| \leq 1$,

$$\frac{\alpha(x)^2}{2} = \int_0^{\alpha(x)} s\,ds = \int_0^{|x|} s f\left(s\frac{x}{|x|}\right) ds.$$

As in Step 2 of the proof of Lemma 11.11 (with $g = 1$), the map

$$\varphi(x) = \alpha(x)\frac{x}{|x|}$$

is in $C^1(\overline{B}_1; \overline{B}_1)$, with

$$\varphi^*(1) = f \quad \text{and} \quad \varphi = \text{id on } \partial B_1 .$$

Since $\varphi(1/2, 0) = \varphi(3/4, 0)$, φ is not one-to-one. □

The next proposition can be proved with the same technique as the one developed in this chapter and we refer to [60] for details.

Proposition 11.8. *Let $r \geq 1$ and $n \geq 2$ be integers. Let $g \in C^r(\mathbb{R}^n)$ with $g > 0$ in $\mathbb{R}^n$, $f \in C^r(\overline{B}_1)$ satisfying*

$$\int_{B_1} g = \int_{B_1} f .$$

Then there exist $\delta = \delta(n, r, g, f)$ and $\gamma = \gamma(n, r, g, f)$ such that for every $g_1, g_2 \in C^r(\mathbb{R}^n)$, $f_1, f_2 \in C^r(\overline{B}_1)$ satisfying, for $i = 1, 2$,

$$\int_{B_1} g_i = \int_{B_1} f_i, \quad \|f_i - f\|_{C^r(B_1)} \leq \delta \quad \text{and} \quad \|g_i - g\|_{C^r(B_2)} \leq \delta,$$

there exist $\varphi_i \in C^r(\overline{B}_1; B_2)$, $i = 1, 2$, such that for every $0 \leq k \leq r-1$,

$$\varphi_i^*(g_i) = f_i \text{ in } B_1, \quad \varphi_i = \text{id on } \partial B_1,$$

$$\|\varphi_1 - \varphi_2\|_{C^k(\overline{B}_1)} \leq \gamma(\|f_1 - f_2\|_{C^k(\overline{B}_1)} + \|g_1 - g_2\|_{C^k(\overline{B}_2)}),$$

$$\|\varphi_i\|_{C^r(\overline{B}_1)} \leq \gamma.$$

Remark 11.9. We can make the conclusion of the proposition more precise. In the sense that for every $\varepsilon > 0$, by letting δ and γ depending of ε we can replace B_2 above by $B_{1+\varepsilon}$.

11.3 Proof of the Main Result

We can now discuss the proof of the main theorem. For the sake of simplicity, we will split it into two proofs. First, we establish the main statement of the theorem and then we show its three extra statements.

Proof. We divide the proof into five steps and we fix $\varepsilon > 0$.

Step 1 (transfer of the problem into the ball). Since $\overline{\Omega}$ is C^{r+1}-diffeomorphic to $\overline{B}_1$, there exists $\varphi_1 \in \operatorname{Diff}^{r+1}(\overline{B}_1;\overline{\Omega})$. With no loss of generality we can assume that $\det\nabla\varphi_1 > 0$. Indeed, if $\det\nabla\varphi_1 < 0$ (note that since φ_1 is a diffeomorphism, then $\det\nabla\varphi_1 \neq 0$ everywhere), then replace $\varphi_1(x)$ by $\varphi_1(-x_1,x_2,\ldots,x_n)$. Using Corollary 16.15, we extend φ_1 and choose $\varepsilon_1 > 0$ small enough so that $\varphi_1 \in \operatorname{Diff}^{r+1}(\overline{B}_{1+\varepsilon_1};\varphi_1(\overline{B}_{1+\varepsilon_1}))$ with

$$\varphi_1(\overline{B}_{1+\varepsilon_1}) \subset \overline{\Omega} + B_\varepsilon .$$

Define

$$f_1 = \varphi_1^*(f) \in C^r(\overline{B}_1) \quad \text{and} \quad g_1 = \varphi_1^*(g) \in C^r(\overline{B}_{1+\varepsilon_1}).$$

By the change of variables formula, we have that

$$\int_{B_1} f_1 = \int_\Omega f = \int_\Omega g = \int_{B_1} g_1 > 0. \tag{11.8}$$

Step 2 (positive radial integration). Since (11.8) holds, we may apply Lemma 11.21 to f_1. Therefore, there exists $\varphi_2 \in \operatorname{Diff}^\infty(\overline{B}_1;\overline{B}_1)$ with

$$\operatorname{supp}(\varphi_2 - \mathrm{id}) \subset B_1$$

such that, letting $f_2 = \varphi_2^*(f_1) \in C^r(\overline{B}_1)$, we have $f_2(0) > 0$ and

$$\int_0^r s^{n-1} f_2\left(s\frac{x}{|x|}\right) ds > 0 \quad \text{for every } x \neq 0 \text{ and } r \in (0,1]$$

$$\int_r^1 s^{n-1} f_2\left(s\frac{x}{|x|}\right) ds > -\frac{\displaystyle\int_{B_{1+\varepsilon_1}} g_1 - \int_{B_1} g_1}{n \operatorname{meas}(B_1)} \quad \text{for every } x \neq 0 \text{ and } r \in [0,1].$$

The change of variables formula and (11.8) lead to

$$\int_{B_1} f_2 = \int_{B_1} \varphi_2^*(f_1) = \int_{B_1} f_1 = \int_{B_1} g_1 . \tag{11.9}$$

Step 3 (radial solution). By the previous step, f_2 satisfies all of the hypotheses of Lemma 11.10 (with $m = \int_{B_{1+\varepsilon_1}} g_1$). Therefore, there exist $g_2 \in C^r(\mathbb{R}^n)$ with $g_2 > 0$ in $\mathbb{R}^n$ and

$$\int_{B_{1+\varepsilon_1}} g_2 = \int_{B_{1+\varepsilon_1}} g_1$$

and $\varphi_3 \in C^r(\overline{B}_1;\overline{B}_{1+\varepsilon_1})$ verifying

$$\begin{cases} \varphi_3^*(g_2) = f_2 & \text{in } B_1, \\ \varphi_3 = \text{id} & \text{on } \partial B_1 . \end{cases}$$

Note that, using (19.3),

$$\int_{B_1} g_2 = \int_{B_1} f_2$$

and therefore, by (11.9),

$$\int_{B_1} g_2 = \int_{B_1} g_1 .$$

Step 4 (positive resolution). Since $g_1, g_2 \in C^r(\overline{B}_{1+\varepsilon_1})$, $g_1, g_2 > 0$ in $\overline{B}_{1+\varepsilon_1}$,

$$\int_{B_1} g_1 = \int_{B_1} g_2 \quad \text{and} \quad \int_{B_{1+\varepsilon_1}} g_1 = \int_{B_{1+\varepsilon_1}} g_2 ,$$

there exists, using Corollary 10.8, $\varphi_4 \in \text{Diff}^r(\overline{B}_{1+\varepsilon_1};\overline{B}_{1+\varepsilon_1})$ such that

$$\begin{cases} \varphi_4^*(g_1) = g_2 & \text{in } B_{1+\varepsilon_1}, \\ \varphi_4 = \text{id} & \text{on } \partial B_1 \cup \partial B_{1+\varepsilon_1} . \end{cases}$$

Step 5 (conclusion). By the above steps, we have that

$$\varphi = \varphi_1 \circ \varphi_4 \circ \varphi_3 \circ \varphi_2^{-1} \circ \varphi_1^{-1} \in C^r(\overline{\Omega};\mathbb{R}^n)$$

satisfies

$$\overline{\Omega} \subset \varphi(\overline{\Omega}) \subset \overline{\Omega} + B_\varepsilon,$$

$$\begin{cases} \varphi^*(g) = f & \text{in } \Omega, \\ \varphi = \text{id} & \text{on } \partial\Omega. \end{cases}$$

Indeed, for $x \in \partial\Omega$, since $\varphi_1(\partial B_1) = \partial\Omega$ (see Theorem 19.6) and $\varphi_i = \text{id}$ on ∂B_1, $i = 2, 3, 4$, we have

$$\begin{aligned} \varphi(x) &= \varphi_1 \circ \varphi_4 \circ \varphi_3 \circ \varphi_2^{-1} \circ \varphi_1^{-1}(x) \\ &= \varphi_1(\varphi_1^{-1}(x)) = x. \end{aligned}$$

Thus, using (19.2), we have that $\overline{\Omega} \subset \varphi(\overline{\Omega})$. Noticing that

$$\varphi_1^{-1}(\overline{\Omega}) = \overline{B}_1, \quad \varphi_2^{-1}(\overline{B}_1) = \overline{B}_1, \quad \varphi_3(\overline{B}_1) \subset \overline{B}_{1+\varepsilon_1},$$

$$\varphi_4(\overline{B}_{1+\varepsilon_1}) = \overline{B}_{1+\varepsilon_1} \quad \text{and} \quad \varphi_1(\overline{B}_{1+\varepsilon_1}) \subset \overline{\Omega} + B_\varepsilon,$$

we have

$$\varphi(\overline{\Omega}) \subset \overline{\Omega} + B_\varepsilon .$$

Eventually, using several times the third statement in Theorem 3.10,

$$\begin{aligned}\varphi^*(g) &= \left(\varphi_1\circ\varphi_4\circ\varphi_3\circ\varphi_2^{-1}\circ\varphi_1^{-1}\right)^*(g)\\ &= (\varphi_1^{-1})^*\left((\varphi_2^{-1})^*\left(\varphi_3^*\left(\varphi_4^*\left(\varphi_1^*(g)\right)\right)\right)\right)\\ &= (\varphi_1^{-1})^*\left((\varphi_2^{-1})^*\left(\varphi_3^*\left(\varphi_4^*(g_1)\right)\right)\right)\\ &= (\varphi_1^{-1})^*\left((\varphi_2^{-1})^*\left(\varphi_3^*(g_2)\right)\right)\\ &= (\varphi_1^{-1})^*\left((\varphi_2^{-1})^*(f_2)\right)\\ &= (\varphi_1^{-1})^*(f_1) = f,\end{aligned}$$

which concludes the proof. □

We now prove the three extra statements of Theorem 11.1.

Proof. We divide the proof into seven steps.

Step 1 (transfer of the problem into the ball). Since $\overline{\Omega}$ is C^{r+1}-diffeomorphic to $\overline{B}_1$, there exists $\varphi_1\in\mathrm{Diff}^{r+1}(\overline{B}_1;\overline{\Omega})$. With no loss of generality we can assume that $\det\nabla\varphi_1>0$. Indeed, if $\det\nabla\varphi_1<0$, then replace $\varphi_1(x)$ by $\varphi_1(-x_1,x_2,\ldots,x_n)$. Define

$$f_1=\varphi_1^*(f)\in C^r(\overline{B}_1)\quad\text{and}\quad g_1=\varphi_1^*(g)\in C^r(\overline{B}_1).$$

From the change of variables formula, we get

$$\int_{B_1} f_1=\int_\Omega f=\int_\Omega g=\int_{B_1} g_1>0. \tag{11.10}$$

We notice the following facts:

(i) If $f>0$ on $\partial\Omega$, then

$$f_1>0\quad\text{on }\partial B_1 \tag{11.11}$$

since $\varphi_1(\partial B_1)=\partial\Omega$ by the invariance of domain theorem (see Theorem 19.6).

(ii) If $\mathrm{supp}(g-f)\subset\Omega$, then

$$\mathrm{supp}(g_1-f_1)\subset B_1. \tag{11.12}$$

(iii) If $f\geq 0$ in $\overline{\Omega}$, then

$$f_1\geq 0\quad\text{in }\overline{B}_1 \tag{11.13}$$

since $\det\nabla\varphi_1>0$ in $\overline{B}_1$.

(iv) If $f\geq 0$ in $\overline{\Omega}$ and $f^{-1}(0)\cap\Omega$ is countable, then

$$f_1\geq 0\text{ in }\overline{B}_1\quad\text{and}\quad f_1^{-1}(0)\cap B_1\quad\text{is countable.} \tag{11.14}$$

Step 2 (positive radial integration). Applying Corollary 11.23 to f_1, which is justified by (11.10) and (11.11) if $f>0$ on $\partial\Omega$ and by (11.10) and (11.13) if $f\geq 0$ in $\overline{\Omega}$, we can find $\varphi_2\in\mathrm{Diff}^\infty(\overline{B}_1;\overline{B}_1)$ with

$$\mathrm{supp}(\varphi_2-\mathrm{id})\subset B_1$$

such that, letting $f_2 = \varphi_2^*(f_1) \in C^r(\overline{B}_1)$, we have $f_2(0) > 0$ and

$$\int_0^r s^{n-1} f_2\left(s\frac{x}{|x|}\right) ds > 0 \quad \text{for every } x \neq 0 \text{ and } r \in (0,1],$$

$$\int_r^1 s^{n-1} f_2\left(s\frac{x}{|x|}\right) ds \geq 0, \quad \text{for every } x \neq 0 \text{ and } r \in [0,1].$$

Moreover, using the change of variables formula and (11.10), we obtain

$$\int_{B_1} f_2 = \int_{B_1} \varphi_2^*(f_1) = \int_{B_1} f_1 = \int_{B_1} g_1 . \tag{11.15}$$

Finally, we notice the two following facts:

(i) If $\operatorname{supp}(g-f) \subset \Omega$, then by (11.12) and since $\operatorname{supp}(\varphi_2 - \mathrm{id}) \subset B_1$, we have

$$\operatorname{supp}(g_1 - f_2) \subset B_1 . \tag{11.16}$$

(ii) If $f \geq 0$ in $\overline{\Omega}$ and $f^{-1}(0) \cap \Omega$ is countable, then by (11.14), we get that

$$f_2 \geq 0 \text{ in } \overline{B}_1 \quad \text{and} \quad f_2^{-1}(0) \cap B_1 \quad \text{is countable.} \tag{11.17}$$

Step 3 (radial solution). Since f_2 satisfies all the hypotheses of Lemma 11.11, there exist $g_2 \in C^r(\overline{B}_1)$ with $g_2 > 0$ in $\overline{B}_1$ and $\varphi_3 \in C^r(\overline{B}_1; \overline{B}_1)$ verifying

$$\begin{cases} \varphi_3^*(g_2) = f_2 & \text{in } B_1, \\ \varphi_3 = \mathrm{id} & \text{on } \partial B_1 . \end{cases}$$

Note that using (19.3),

$$\int_{B_1} g_2 = \int_{B_1} f_2$$

and therefore, using (11.15),

$$\int_{B_1} g_2 = \int_{B_1} g_1 .$$

We, moreover, have the two following facts:

(i) If $\operatorname{supp}(g-f) \subset \Omega$ (which implies, in particular, by (11.16) that $f_2 > 0$ on ∂B_1), the first extra statement of Lemma 11.11 implies that g_2 and φ_3 can be chosen so that

$$\operatorname{supp}(g_2 - f_2) \subset B_1 \quad \text{and} \quad \operatorname{supp}(\varphi_3 - \mathrm{id}) \subset B_1 . \tag{11.18}$$

(ii) If $f \geq 0$ in $\overline{\Omega}$ and $f^{-1}(0) \cap \Omega$ is countable (which implies by (11.17) that $f_2 \geq 0$ in $\overline{B}_1$ and $f_2^{-1}(0) \cap B_1$ is countable), the second extra statement of Lemma 11.11 implies that φ_3 can be chosen so that

$$\varphi_3 \in \operatorname{Hom}(\overline{B}_1; \overline{B}_1). \tag{11.19}$$

Step 4 (positive resolution). Since $g_1, g_2 \in C^r(\overline{B}_1)$, $g_1, g_2 > 0$ in $\overline{B}_1$ and

$$\int_{B_1} g_1 = \int_{B_1} g_2 \,,$$

using Theorem 10.7, we can find $\varphi_4 \in \operatorname{Diff}^r(\overline{B}_1;\overline{B}_1)$ such that

$$\begin{cases} \varphi_4^*(g_1) = g_2 & \text{in } B_1, \\ \varphi_4 = \mathrm{id} & \text{on } \partial B_1 . \end{cases}$$

We, moreover, have the following fact: If $\operatorname{supp}(g-f) \subset \Omega$, then by (11.16) and (11.18) we get that $\operatorname{supp}(g_1 - g_2) \subset B_1$. Therefore, using Theorem 10.11 instead of Theorem 10.7, we can furthermore assume that

$$\operatorname{supp}(\varphi_4 - \mathrm{id}) \subset B_1 . \tag{11.20}$$

Step 5 (conclusion). Using the above steps, we have that

$$\varphi = \varphi_1 \circ \varphi_4 \circ \varphi_3 \circ \varphi_2^{-1} \circ \varphi_1^{-1} \in C^r(\overline{\Omega};\overline{\Omega})$$

satisfies

$$\begin{cases} \varphi^*(g) = f & \text{in } \Omega, \\ \varphi = \mathrm{id} & \text{on } \partial\Omega. \end{cases}$$

Indeed, for $x \in \partial\Omega$, since $\varphi_1(\partial B_1) = \partial\Omega$ (see Theorem 19.6) and $\varphi_i = \mathrm{id}$ on ∂B_1, $i = 2,3,4$, we have

$$\begin{aligned} \varphi(x) &= \varphi_1 \circ \varphi_4 \circ \varphi_3 \circ \varphi_2^{-1} \circ \varphi_1^{-1}(x) \\ &= \varphi_1(\varphi_1^{-1}(x)) = x. \end{aligned}$$

Since $\varphi_1^{-1}(\overline{\Omega}) = \overline{B}_1$, $(\varphi_2)^{-1}(\overline{B}_1) = \overline{B}_1$, $\varphi_4(\overline{B}_1) = \overline{B}_1$, $\varphi_3(\overline{B}_1) = \overline{B}_1$ (by (19.2)) and $\varphi_1(\overline{B}_1) = \overline{\Omega}$, we have that

$$\varphi(\overline{\Omega}) = \overline{\Omega}.$$

Finally, exactly as in Step 5 of the previous proof, we prove that

$$\varphi^*(g) = f \quad \text{in } \Omega,$$

which shows the first extra statement.

Step 6. We show the second extra assertion. If $\operatorname{supp}(g-f) \subset \Omega$, then (11.18) and (11.20) imply the result, since

$$\operatorname{supp}(\varphi_2 - \mathrm{id}),\ \operatorname{supp}(\varphi_3 - \mathrm{id}),\ \operatorname{supp}(\varphi_4 - \mathrm{id}) \subset B_1 .$$

Step 7. Finally, we show the third extra assertion. If $f \geq 0$ in $\overline{\Omega}$ and $f^{-1}(0) \cap \Omega$ is countable, then (11.19) implies the assertion since φ_1, φ_2 and φ_4 are diffeomorphisms. □

11.4 Radial Solution

In this section we give sufficient conditions on f in order to have a positive g and a radial solution φ of (11.2) in the unit ball (i.e., a solution of the form $\alpha(x)x/|x|$ with $\alpha:\overline{B}_1\to\mathbb{R}$). For the sake of simplicity, we split the discussion into two lemmas.

Lemma 11.10. *Let $r\geq 1$ be an integer, $m>0$ and $f\in C^r(\overline{B}_1)$ be such that*

$$f(0)>0,\quad m>\int_{B_1} f,$$

$$\int_0^r s^{n-1} f\left(s\frac{x}{|x|}\right) ds>0 \quad \textit{for every } x\neq 0 \textit{ and } r\in(0,1], \tag{11.21}$$

$$\int_r^1 s^{n-1} f\left(s\frac{x}{|x|}\right) ds>-\frac{m-\int_{B_1} f}{n\operatorname{meas}(B_1)} \quad \textit{for every } x\neq 0 \textit{ and } r\in[0,1]. \tag{11.22}$$

Then for every $\varepsilon>0$, there exist $g=g_{m,\varepsilon}\in C^r(\mathbb{R}^n)$ with $g>0$ in $\mathbb{R}^n$ and

$$\int_{B_{1+\varepsilon}} g=m$$

and $\varphi=\varphi_{m,\varepsilon}\in C^r(\overline{B}_1;B_{1+\varepsilon})$ such that

$$\begin{cases}\varphi^*(g)=f & \textit{in } B_1,\\ \varphi=\mathrm{id} & \textit{on } \partial B_1.\end{cases}$$

Proof. We split the proof into two steps. Fix $\varepsilon>0$.

Step 1 (construction of g). In this step we construct a function $g\in C^r(\mathbb{R}^n)$ with the following properties:

$$g>0 \text{ in } \mathbb{R}^n,\quad g=f \text{ in a neighborhood of } 0,\quad \int_{B_{1+\varepsilon}} g=m,$$

$$\int_0^1 s^{n-1} g\left(s\frac{x}{|x|}\right) ds=\int_0^1 s^{n-1} f\left(s\frac{x}{|x|}\right) ds \quad \text{for every } x\neq 0, \tag{11.23}$$

$$\int_0^{1+\varepsilon} s^{n-1} g\left(s\frac{x}{|x|}\right) ds>\int_0^r s^{n-1} f\left(s\frac{x}{|x|}\right) ds \quad \text{for every } x\neq 0 \text{ and } r\in[0,1]. \tag{11.24}$$

Step 1.1 (preliminaries). Since $f(0)>0$ and (11.21) and (11.22) hold, there exists $\delta>0$ small enough such that

$$f>0 \quad \text{in } B_\delta,\quad \min_{x\neq 0}\int_\delta^1 s^{n-1} f\left(s\frac{x}{|x|}\right) ds>0, \tag{11.25}$$

$$\int_r^1 s^{n-1} f\left(s\frac{x}{|x|}\right) ds > -\frac{m - \int_{B_1} f}{n \operatorname{meas}(B_1)} + \delta \quad \text{for every } x \neq 0 \text{ and } r \in [0,1]. \quad (11.26)$$

Let $\eta \in C^\infty([0,\infty);[0,1])$ be such that

$$\eta(s) = \begin{cases} 1 & \text{if } 0 \le s \le \delta/2 \\ 0 & \text{if } \delta \le s. \end{cases}$$

Define then $\overline{h} : \mathbb{R}^n \setminus \{0\} \to \mathbb{R}$ by

$$\overline{h}(x) = \frac{\int_0^1 s^{n-1}(1-\eta(s)) f\left(s\frac{x}{|x|}\right) ds}{\int_0^1 s^{n-1}(1-\eta(s))\, ds}.$$

It is easily seen that $\overline{h} \in C^r(\mathbb{R}^n \setminus \{0\})$,

$$\overline{h}(x) = \overline{h}(\lambda x) \text{ for every } \lambda > 0,$$

and, using (11.25),

$$\overline{h} > 0, \quad \text{in } \mathbb{R}^n \setminus \{0\}.$$

Now define, for $x \in \mathbb{R}^n$,

$$h(x) = \eta(|x|) f(x) + (1 - \eta(|x|))\overline{h}(x).$$

Using the definition of $\overline{h}$ and η, we have that

$$\begin{cases} h \in C^r(\mathbb{R}^n), \quad h > 0 \text{ in } \mathbb{R}^n, \quad h = f \text{ in } B_{\delta/2}, \\ \int_0^1 s^{n-1} h\left(s\frac{x}{|x|}\right) ds = \int_0^1 s^{n-1} f\left(s\frac{x}{|x|}\right) ds \text{ for every } x \neq 0. \end{cases} \quad (11.27)$$

For every $0 < \mu < \varepsilon$, let $\rho_\mu \in C^\infty(\mathbb{R}^n;[0,1])$ be such that

$$\rho_\mu = \begin{cases} 1 & \text{in } \overline{B}_1, \\ 0 & \text{in } (B_{1+\mu})^c \end{cases}$$

and define

$$c_\mu = \frac{m - \int_{B_{1+\varepsilon}} \rho_\mu h}{\int_{B_{1+\varepsilon}} (1-\rho_\mu)}.$$

Integrating the last equation of (11.27) on the unit sphere, we obtain that

$$\int_{B_1} h = \int_{B_1} f$$

and, thus, we get

$$\lim_{\mu\to 0} c_\mu = \frac{m - \int_{B_1} h}{\operatorname{meas}(B_{1+\varepsilon}\setminus B_1)} = \frac{m - \int_{B_1} f}{\operatorname{meas}(B_{1+\varepsilon}\setminus B_1)} = \frac{m - \int_{B_1} f}{[(1+\varepsilon)^n - 1]\operatorname{meas}(B_1)}.$$

This implies

$$\lim_{\mu\to 0} \frac{(1+\varepsilon)^n - (1+\mu)^n}{n} c_\mu = \frac{(1+\varepsilon)^n - 1}{n} \lim_{\mu\to 0} c_\mu = \frac{m - \int_{B_1} f}{n \operatorname{meas}(B_1)}$$

and therefore, by (11.26) we can choose μ_1 small enough such that $c_{\mu_1} > 0$ and

$$\int_r^1 s^{n-1} f\left(s\frac{x}{|x|}\right) ds > -\frac{(1+\varepsilon)^n - (1+\mu_1)^n}{n} c_{\mu_1}, \quad \text{for every } x \neq 0 \text{ and } r \in [0,1]. \tag{11.28}$$

Step 1.2 (conclusion). Let us show that the function

$$g = \rho_{\mu_1} h + (1-\rho_{\mu_1}) c_{\mu_1} \in C^r(\mathbb{R}^n)$$

has all of the desired properties. Indeed, since $h > 0$ in $\mathbb{R}^n$ and $c_{\mu_1} > 0$, we have that $g > 0$ in $\mathbb{R}^n$. By definition of c_{μ_1}, we see that

$$\int_{B_{1+\varepsilon}} g = m.$$

Using the last equation of (11.27) and the fact that $g = h$ in $\overline{B}_1$, we get (11.23). We finally show (11.24). Using (11.23), this is equivalent to showing

$$\int_1^{1+\varepsilon} s^{n-1} g\left(s\frac{x}{|x|}\right) ds > -\int_r^1 s^{n-1} f\left(s\frac{x}{|x|}\right) ds \quad \text{for every } x \neq 0 \text{ and } r \in [0,1].$$

Let $x \neq 0$ and $r \in [0,1]$. We have, since $g = c_{\mu_1}$ in $\overline{B}_{1+\varepsilon} \setminus B_{1+\mu_1}$ and (11.28) holds,

$$\begin{aligned}\int_1^{1+\varepsilon} s^{n-1} g\left(s\frac{x}{|x|}\right) ds &> \int_{1+\mu_1}^{1+\varepsilon} s^{n-1} g\left(s\frac{x}{|x|}\right) ds = \int_{1+\mu_1}^{1+\varepsilon} s^{n-1} c_{\mu_1} ds \\ &= \frac{(1+\varepsilon)^n - (1+\mu_1)^n}{n} c_{\mu_1} > -\int_r^1 s^{n-1} f\left(s\frac{x}{|x|}\right) ds\end{aligned}$$

and therefore the assertion.

Step 2 (construction of φ). We will construct a solution φ of the form

$$\varphi(x) = \alpha(x)\frac{x}{|x|},$$

where $\alpha : \overline{B}_1 \to \mathbb{R}$.

Step 2.1 (definition of α). Let $\alpha : \overline{B}_1 \to \mathbb{R}$ be such that $\alpha(0) = 0$ and, for $0 < |x| \le 1$,

$$\int_0^{\alpha(x)} s^{n-1} g\left(s\frac{x}{|x|}\right) ds = \int_0^{|x|} s^{n-1} f\left(s\frac{x}{|x|}\right) ds. \tag{11.29}$$

Since $g > 0$, using (11.21) and (11.24), we get, for every $x \in \overline{B}_1 \setminus \{0\}$, that $\alpha(x)$ is well defined and verifies $0 < \alpha(x) < 1 + \varepsilon$. Since $g = f$ in a neighborhood of 0, we obtain that

$$\alpha(x) = |x| \quad \text{in the same neighborhood of } 0.$$

By (11.23), we immediately have

$$\alpha(x) = 1 \quad \text{on } \partial B_1 .$$

Therefore, by the implicit function theorem, which can be used since $\alpha > 0$ and $g > 0$, we have that $\alpha \in C^r(\overline{B}_1 \setminus \{0\})$. Moreover, since $\alpha(x) = |x|$ in a neighborhood of 0, the function $x \to \alpha(x)/|x|$ is $C^r(\overline{B}_1)$.

Step 2.2 (conclusion). We finally show that

$$\varphi(x) = \frac{\alpha(x)}{|x|}\, x$$

is in $C^r(\overline{B}_1; B_{1+\varepsilon})$ and verifies

$$\begin{cases} \varphi^*(g) = f & \text{in } B_1, \\ \varphi = \text{id} & \text{on } \partial B_1 . \end{cases}$$

In fact, by the properties of α, it is obvious that $\varphi \in C^r(\overline{B}_1; B_{1+\varepsilon})$ and that $\varphi = \text{id}$ on ∂B_1 . Using Lemma 11.12, we obtain

$$\det \nabla \varphi(x) = \frac{\alpha^{n-1}(x)}{|x|^n} \sum_{i=1}^n x_i \frac{\partial \alpha}{\partial x_i}(x). \tag{11.30}$$

Computing the derivative of (11.29) with respect to x_i , we get

$$\begin{aligned} &\alpha^{n-1}(x) g(\varphi(x)) \frac{\partial \alpha}{\partial x_i}(x) + \sum_{j=1}^n \int_0^{\alpha(x)} s^n \frac{\partial g}{\partial x_j}\left(s\frac{x}{|x|}\right) \left(\frac{|x|\delta_{ij} - \frac{x_i x_j}{|x|}}{|x|^2}\right) ds \\ &= |x|^{n-1} f(x) \frac{x_i}{|x|} + \sum_{j=1}^n \int_0^{|x|} s^n \frac{\partial f}{\partial x_j}\left(s\frac{x}{|x|}\right) \left(\frac{|x|\delta_{ij} - \frac{x_i x_j}{|x|}}{|x|^2}\right) ds, \end{aligned}$$

where $\delta_{ij} = 1$ if $i = j$ and $\delta_{ij} = 0$ otherwise. Multiplying the above equality by x_i, adding up the terms with respect to i and using

$$\sum_{i=1}^n x_i \left(\frac{|x|\delta_{ij} - \frac{x_i x_j}{|x|}}{|x|^2}\right) = 0, \quad 1 \le j \le n,$$

we obtain

$$\alpha^{n-1}(x)g(\varphi(x))\sum_{i=1}^{n}x_i\frac{\partial\alpha}{\partial x_i}(x)=|x|^n f(x).$$

This equality, together with (11.30), implies $\varphi^*(g)=f$, which shows the assertion. □

Lemma 11.11. *Let $r\geq 1$ be an integer, $f\in C^r(\overline{B}_1)$ be such that $f(0)>0$ and*

$$\int_0^r s^{n-1}f\left(s\frac{x}{|x|}\right)ds>0 \quad \textit{for every } x\neq 0 \textit{ and } r\in(0,1], \tag{11.31}$$

$$\int_r^1 s^{n-1}f\left(s\frac{x}{|x|}\right)ds\geq 0 \quad \textit{for every } x\neq 0 \textit{ and } r\in[0,1]. \tag{11.32}$$

Then there exists $g\in C^r(\overline{B}_1)$ with $g>0$ in $\overline{B}_1$ and $\varphi\in C^r(\overline{B}_1;\overline{B}_1)$ such that

$$\begin{cases}\varphi^*(g)=f & \textit{in } B_1,\\ \varphi=\mathrm{id} & \textit{on } \partial B_1\,.\end{cases}$$

Furthermore, the following two extra properties hold:

(i) If $f>0$ on ∂B_1, then g and φ can be chosen so that

$$\operatorname{supp}(g-f)\subset B_1 \quad \textit{and} \quad \operatorname{supp}(\varphi-\mathrm{id})\subset B_1\,.$$

(ii) If $f\geq 0$ in $\overline{B}_1$ and

$$f^{-1}(0)\cap B_1 \quad \textit{is countable,}$$

then φ can be chosen in $\mathrm{Hom}(\overline{B}_1;\overline{B}_1)$.

Proof. The proof is essentially the same as the previous one. We split the proof into two steps.

Step 1 (construction of g). In this step we construct a function $g\in C^r(\overline{B}_1)$ with the following properties: $g>0$ in $\overline{B}_1$, $g=f$ in a neighborhood of 0 (and also $\operatorname{supp}(g-f)\subset B_1$ if $f>0$ on ∂B_1),

$$\int_0^1 s^{n-1}g\left(s\frac{x}{|x|}\right)ds=\int_0^1 s^{n-1}f\left(s\frac{x}{|x|}\right)ds \quad \text{for every } x\neq 0, \tag{11.33}$$

$$\int_0^1 s^{n-1}g\left(s\frac{x}{|x|}\right)ds\geq\int_0^r s^{n-1}f\left(s\frac{x}{|x|}\right)ds \quad \text{for every } x\neq 0 \text{ and } r\in[0,1]. \tag{11.34}$$

Step 1.1 (preliminaries). Since $f(0)>0$ and (11.31) holds, there exists $\delta>0$ small enough such that

$$f>0 \quad \text{in } B_\delta \quad \text{and} \quad \min_{x\neq 0}\int_\delta^1 s^{n-1}f\left(s\frac{x}{|x|}\right)ds>0. \tag{11.35}$$

Let $\eta \in C^{\infty}([0,\infty);[0,1])$ be such that

$$\eta(s)=\begin{cases}1 & \text{if } 0\le s\le \delta/2\\ 0 & \text{if } \delta\le s.\end{cases}$$

If $f>0$ on ∂B_1, we modify the definition of δ and η as follows. We assume that

$$\eta(s)=\begin{cases}1 & \text{if } 0\le s\le \delta/2 \text{ or } 1-\delta/2\le s\le 1\\ 0 & \text{if } \delta\le s\le 1-\delta,\end{cases}$$

where $\delta>0$ small enough is such that

$$f>0 \quad \text{in } B_\delta\cup(\overline{B}_1\setminus B_{1-\delta}) \quad \text{and} \quad \min_{x\neq 0}\int_\delta^{1-\delta} s^{n-1} f\left(s\frac{x}{|x|}\right)ds>0. \tag{11.36}$$

Define next $\overline{h}:\mathbb{R}^n\setminus\{0\}\to\mathbb{R}$ by

$$\overline{h}(x)=\frac{\displaystyle\int_0^1 s^{n-1}(1-\eta(s))f(s\tfrac{x}{|x|})\,ds}{\displaystyle\int_0^1 s^{n-1}(1-\eta(s))\,ds}.$$

It is easily seen that $\overline{h}\in C^r(\mathbb{R}^n\setminus\{0\})$, that

$$\overline{h}(x)=\overline{h}(\lambda x) \text{ for every } \lambda>0,$$

and, using (11.35) or (11.36), that

$$\overline{h}>0.$$

Step 1.2 (conclusion). Let us show that g defined by

$$g(x)=\eta(|x|)f(x)+(1-\eta(|x|))\overline{h}(x), \quad x\in\overline{B}_1,$$

has all of the required properties. Using the definition of $\overline{h}$ and η, we see that $g\in C^r(\overline{B}_1)$ satisfies $g>0$ in $\overline{B}_1$, (11.33) and $g=f$ in $B_{\delta/2}$ (if, moreover, $f>0$ on ∂B_1, then $\operatorname{supp}(g-f)\subset B_1$). Finally, we show (11.34). Let $x\neq 0$ and $r\in[0,1]$. Using (11.32) and (11.33), we get

$$\begin{aligned}\int_0^1 s^{n-1}g\left(s\frac{x}{|x|}\right)ds &= \int_0^1 s^{n-1}f\left(s\frac{x}{|x|}\right)ds\\ &= \int_0^r s^{n-1}f\left(s\frac{x}{|x|}\right)ds+\int_r^1 s^{n-1}f\left(s\frac{x}{|x|}\right)ds\\ &\ge \int_0^r s^{n-1}f\left(s\frac{x}{|x|}\right)ds,\end{aligned}$$

which ends the construction of g.

Step 2 (construction of φ). We will construct, as before, a solution φ of the form

$$\varphi(x) = \alpha(x)\,\frac{x}{|x|},$$

where $\alpha : \overline{B}_1 \to \mathbb{R}$.

Step 2.1 (definition of α). Let $\alpha : \overline{B}_1 \to \mathbb{R}$ be such that $\alpha(0) = 0$ and, for $0 < |x| \leq 1$,

$$\int_0^{\alpha(x)} s^{n-1} g\left(s\frac{x}{|x|}\right) ds = \int_0^{|x|} s^{n-1} f\left(s\frac{x}{|x|}\right) ds.$$

Since $g > 0$, using (11.31) and (11.34), we get for every $x \in \overline{B}_1 \setminus \{0\}$ that $\alpha(x)$ is well defined and verifies $0 < \alpha(x) \leq 1$. Since $g = f$ in a neighborhood of 0, we obtain that

$$\alpha(x) = |x| \quad \text{in the same neighborhood of } 0.$$

By (11.33), we immediately have

$$\alpha(x) = 1 \quad \text{on } \partial B_1\,.$$

Moreover, if $\operatorname{supp}(g - f) \subset B_1$, then α also verifies

$$\alpha(x) = |x| \quad \text{in a neighborhood of } \partial B_1\,. \tag{11.37}$$

Therefore, by the implicit function theorem, which can be used since $\alpha > 0$ and $g > 0$, we have that $\alpha \in C^r(\overline{B}_1 \setminus \{0\})$. Moreover, since $\alpha(x) = |x|$ in a neighborhood of 0, the map $x \to \alpha(x)x/|x|$ is $C^r(\overline{B}_1)$.

Step 2.2 (conclusion). We show that

$$\varphi(x) = \frac{\alpha(x)}{|x|}\,x$$

is in $C^r(\overline{B}_1;\overline{B}_1)$ and verifies

$$\begin{cases} \varphi^*(g) = f & \text{in } B_1, \\ \varphi = \text{id} & \text{on } \partial B_1\,. \end{cases}$$

In fact, by the properties of α, it is obvious that $\varphi \in C^r(\overline{B}_1;\overline{B}_1)$ and that $\varphi = \text{id}$ on ∂B_1. Finally, proceeding exactly as in Step 2.2 of the proof of Lemma 11.10, we obtain that

$$\varphi^*(g) = f \quad \text{in } B_1,$$

which concludes the proof of the main statement.

It remains to show the two extra statements.

(i) If $f > 0$ on ∂B_1, then we have $\operatorname{supp}(g - f) \subset B_1$. Hence, it follows from (11.37) that

$$\operatorname{supp}(\varphi - \mathrm{id}) \subset B_1,$$

which proves the first extra statement.

(ii) If $f \geq 0$ and

$$f^{-1}(0) \cap B_1 \quad \text{is countable,}$$

we immediately obtain

$$\alpha(x) \neq \alpha(rx) \quad \text{for every } x \in \overline{B}_1 \setminus \{0\} \text{ and } r \in [0,1),$$

which implies that $\varphi \in \operatorname{Hom}(\overline{B}_1; \overline{B}_1)$ and establishes the second statement and ends the proof. □

In the proof of Lemmas 11.10 and 11.11, we used the following elementary result.

Lemma 11.12. *Let $\lambda \in C^1(\overline{B}_1)$ and $\varphi \in C^1(\overline{B}_1; \mathbb{R}^n)$ be such that $\varphi(x) = \lambda(x)x$. Then*

$$\det \nabla \varphi(x) = \lambda^n(x) + \lambda^{n-1}(x) \sum_{i=1}^n x_i \frac{\partial \lambda}{\partial x_i}(x).$$

In particular, if $\lambda(x) = \alpha(x)/|x|$ for some α, then

$$\det \nabla \varphi(x) = \frac{\alpha^{n-1}(x)}{|x|^n} \sum_{i=1}^n x_i \frac{\partial \alpha}{\partial x_i}(x).$$

Proof. Since $\nabla \varphi = \lambda \,\mathrm{Id} + x \otimes \nabla \lambda$ and $x \otimes \nabla \lambda$ is a rank-1 matrix, the first equality holds true. The second one easily follows. □

11.5 Concentration of Mass

We start with an elementary lemma.

Lemma 11.13. *Let $c \in C^0([0,1]; B_1)$. Then for every $\varepsilon > 0$ such that*

$$c([0,1]) + B_\varepsilon \subset B_1,$$

there exists $\varphi_\varepsilon \in \operatorname{Diff}^\infty(\overline{B}_1; \overline{B}_1)$ satisfying

$$\varphi_\varepsilon(c(0)) = c(1) \quad \text{and} \quad \operatorname{supp}(\varphi_\varepsilon - \mathrm{id}) \subset c([0,1]) + B_\varepsilon.$$

Proof. Define $\eta_\varepsilon \in C^\infty(\mathbb{R}^n; [0,1])$ such that

$$\eta_\varepsilon = \begin{cases} 1 & \text{in } B_{\varepsilon/4} \\ 0 & \text{in } \left(B_{\varepsilon/2}\right)^c. \end{cases}$$

Set, for $a \in \mathbb{R}^n$,

$$\eta_{a,\varepsilon}(x) = \eta_\varepsilon(x-a).$$

We then have

$$\delta \|\nabla \eta_{a,\varepsilon}\|_{C^0} = \delta \|\nabla \eta_\varepsilon\|_{C^0} < 1/(2n) \tag{11.38}$$

for a suitable $\delta > 0$. Let $x_i \in B_1$, $1 \le i \le N$, with $x_1 = c(0)$ and $x_N = c(1)$, be such that

$$x_i \in c([0,1]) \text{ for } 1 \le i \le N \quad \text{and} \quad |x_{i+1} - x_i| < \delta \text{ for } 1 \le i \le N-1$$

and define

$$\varphi_i(x) = x + \eta_{x_i,\varepsilon}(x)(x_{i+1} - x_i), \quad 1 \le i \le N-1.$$

Since (11.38) holds and $\operatorname{supp}(\varphi_i - \operatorname{id}) \subset c([0,1]) + B_\varepsilon \subset B_1$, we have

$$\det \nabla \varphi_i > 0 \quad \text{and} \quad \varphi_i = \operatorname{id} \text{ on } \partial B_1.$$

Therefore, $\varphi_i \in \operatorname{Diff}^\infty(\overline{B}_1; \overline{B}_1)$ by Theorem 19.12. Moreover, $\varphi_i(x_i) = x_{i+1}$. Then the diffeomorphism

$$\varphi_\varepsilon = \varphi_{N-1} \circ \cdots \circ \varphi_1$$

has all of the required properties. □

Before stating the main result of this section, we need some notations and elementary properties of pullbacks and connected components.

Notation 11.14. *(i) Let $\Omega \subset \mathbb{R}^n$ be a bounded open set. For $f \in C^0(\overline{\Omega})$, we adopt the following notations:*

$$F^+ = f^{-1}((0,\infty)) \quad \text{and} \quad F^- = f^{-1}((-\infty,0)).$$

Moreover, if $x \in F^\pm$, then

$$F_x^\pm \text{ denotes the connected component of } F^\pm \text{ containing } x.$$

(ii) Given a set $A \subset \mathbb{R}^n$, we let

$$1_A(x) = \begin{cases} 1 & \text{if } x \in A \\ 0 & \text{otherwise.} \end{cases}$$

In the following lemma we state an easy property of pullbacks.

Lemma 11.15. *Let $\Omega \subset \mathbb{R}^n$ be open and bounded and $f \in C^0(\overline{\Omega})$,*

$$\varphi \in \operatorname{Diff}^1(\overline{\Omega}; \overline{\Omega}) \quad \text{with} \quad \det \nabla \varphi > 0,$$

$x \in F^+$, $y \in F^-$. If $\widetilde{f} = \varphi^*(f)$, then $\varphi^{-1}(F^+) = \widetilde{F}^+$, $\varphi^{-1}(F^-) = \widetilde{F}^-$,

$$\varphi^{-1}(F_x^+) = \widetilde{F}^+_{\varphi^{-1}(x)} \quad \text{and} \quad \varphi^{-1}(F_y^-) = \widetilde{F}^-_{\varphi^{-1}(y)}.$$

The following lemma is a trivial result about the cardinality of the connected components of super (sub)-level sets of continuous functions and we state it for the sake of completeness.

Lemma 11.16. *Let $f \in C^0(\overline{B}_1)$. Let $\{F^+_{x_i}\}_{i\in I^+}$ and $\{F^-_{y_j}\}_{j\in I^-}$ be the connected components of F^+, respectively of F^-. Then I^+ and I^- are at most countable. Moreover, if $|I^+| = \infty$, respectively $|I^-| = \infty$, then*

$$\lim_{k\to\infty} \operatorname{meas}\left(F^+ \setminus \bigcup_{i=1}^{k} F^+_{x_i}\right) = 0, \quad \textit{respectively} \quad \lim_{k\to\infty} \operatorname{meas}\left(F^- \setminus \bigcup_{j=1}^{k} F^-_{y_j}\right) = 0.$$

We now give the first main result of the present section.

Lemma 11.17 (Concentration of the positive mass). *Let $r \geq 1$ be an integer, $f \in C^r(\overline{B}_1)$ and $z \in F^+$. Let also A_i, $1 \leq i \leq M$, be M closed sets pairwise disjoint of positive measure such that*

$$A_i \subset F^+_z \cap B_1, \quad 1 \leq i \leq M.$$

Then for every $\varepsilon > 0$ small enough, there exists $\varphi_{\varepsilon,f,\{A_i\}} \in \operatorname{Diff}^r(\overline{B}_1;\overline{B}_1)$ (which will be simply denoted φ_ε) satisfying the following properties:

$$\operatorname{supp}(\varphi_\varepsilon - \operatorname{id}) \subset F^+_z \cap B_1,$$

$$\varphi^*_\varepsilon(f) \geq \frac{\int_{F^+_z} f}{M \operatorname{meas}(A_i)} - \varepsilon \quad \textit{in } A_i, \quad 1 \leq i \leq M. \tag{11.39}$$

Remark 11.18. Indeed, the above lemma allows one to concentrate the positive mass of the connected component containing z into the union of the A_i. The conclusion of the lemma immediately implies that

$$\int_{F^+_z} f = \int_{F^+_z} \varphi^*_\varepsilon(f) \geq \sum_{i=1}^{M} \int_{A_i} \varphi^*_\varepsilon(f) \geq \int_{F^+_z} f - \varepsilon \sum_{i=1}^{M} \operatorname{meas}(A_i).$$

Proof. We split the proof into three steps.

Step 1 (simplification). Using Theorem 10.11, it is sufficient to prove the existence of $f_\varepsilon \in C^r(\overline{B}_1)$, such that

$$f_\varepsilon > 0 \quad \text{in } F^+_z, \quad \operatorname{supp}(f - f_\varepsilon) \subset F^+_z \cap B_1 \quad \text{and} \quad \int_{F^+_z} f_\varepsilon = \int_{F^+_z} f$$

satisfying also (11.39) with $\varphi^*_\varepsilon(f)$ replaced by f_ε.

Step 2 (definition of f_ε and conclusion). Let $K \subset F^+_z \cap B_1$ be a closed set with

$$\bigcup_{i=1}^{M} A_i \subset \operatorname{int} K \subset K \subset F^+_z \cap B_1$$

and let, for every $\varepsilon > 0$ and $1 \le i \le M$,

$$\eta_{i,\varepsilon} \in C^\infty(\overline{B}_1;[0,1]) \quad \text{and} \quad \xi_\varepsilon \in C^\infty(\overline{B}_1;[0,1])$$

be such that

$$\operatorname{supp}(\eta_{i,\varepsilon}) \cap \operatorname{supp}(\eta_{j,\varepsilon}) = \emptyset \quad \text{for } i \neq j, \tag{11.40}$$

$$A_i \subset \{x \in \overline{B}_1 : \eta_{i,\varepsilon}(x) = 1\} \subset \operatorname{supp} \eta_{i,\varepsilon} \subset \operatorname{int} K, \tag{11.41}$$

$$K \subset \{x \in \overline{B}_1 : \xi_\varepsilon(x) = 1\} \subset \operatorname{supp}(\xi_\varepsilon) \subset F_z^+ \cap B_1, \tag{11.42}$$

$$\lim_{\varepsilon \to 0} \xi_\varepsilon = 1_{F_z^+ \cap B_1} \quad \text{and} \quad \lim_{\varepsilon \to 0} \eta_{i,\varepsilon} = 1_{A_i}. \tag{11.43}$$

Define f_ε, ε small, as

$$f_\varepsilon = \begin{cases} \sum_{i=1}^M \eta_{i,\varepsilon} C_{i,\varepsilon}^+ + (1 - \sum_{i=1}^M \eta_{i,\varepsilon}) \cdot \varepsilon & \text{in } K \\ \xi_\varepsilon \cdot \varepsilon + (1 - \xi_\varepsilon) f & \text{elsewhere,} \end{cases}$$

where

$$C_{i,\varepsilon}^+ = \frac{\int_{F_z^+} f}{M \operatorname{meas}(A_i)}, \quad 1 \le i \le M-1, \tag{11.44}$$

and $C_{M,\varepsilon}^+$ is the unique constant defined implicitly by the equation

$$\int_{F_z^+} f_\varepsilon = \int_{F_z^+} f.$$

We claim that f_ε has, up to rescaling ε, all of the required properties. Using (11.41) and (11.42), we get that

$$f_\varepsilon \in C^r(\overline{B}_1), \quad \operatorname{supp}(f_\varepsilon - f) \subset F_z^+ \cap B_1, \quad \int_{F_z^+} f_\varepsilon = \int_{F_z^+} f.$$

We claim that

$$\lim_{\varepsilon \to 0} C_{i,\varepsilon}^+ = \frac{\int_{F_z^+} f}{M \operatorname{meas}(A_i)}, \quad 1 \le i \le M. \tag{11.45}$$

By (11.44), it is obviously sufficient to prove the assertion for $i = M$. Using (11.43), (11.44), and the dominated convergence theorem, we get

$$\begin{aligned} \int_{F_z^+} f &= \lim_{\varepsilon \to 0} \int_{F_z^+} f_\varepsilon = \sum_{i=1}^{M-1} \int_{F_z^+} 1_{A_i} \frac{\int_{F_z^+} f}{M \operatorname{meas}(A_i)} + \int_{F_z^+} 1_{A_M} \lim_{\varepsilon \to 0} C_{M,\varepsilon}^+ \\ &= \frac{M-1}{M} \int_{F_z^+} f + \operatorname{meas}(A_M) \lim_{\varepsilon \to 0} C_{M,\varepsilon}^+ \end{aligned}$$

and thus the assertion holds. By the definition of f_ε, (11.40) and (11.45), we get that, for ε small,

$$f_\varepsilon > 0 \quad \text{in } F_z^+ .$$

Finally, since, by (11.41),

$$f_\varepsilon = C_{i,\varepsilon}^+ \quad \text{in } A_i, \quad 1 \le i \le M,$$

(11.45) directly implies, up to rescaling ε, (11.39), which ends the proof. □

We now give a similar result for the negative mass.

Lemma 11.19 (Concentration of the negative mass). *Let $r \ge 1$ be an integer, $f \in C^r(\overline{B}_1)$ and $y \in F^-$. Let also A_i, $1 \le i \le M$, be M closed sets pairwise disjoint of positive measure such that*

$$A_i \subset F_y^- \cap B_1 \quad \text{and} \quad \operatorname{meas}(\partial A_i) = 0, \quad 1 \le i \le M. \tag{11.46}$$

Then for every $\varepsilon > 0$ small enough, there exists $\varphi_{\varepsilon,f,\{A_i\}} \in \operatorname{Diff}^r(\overline{B}_1;\overline{B}_1)$ (simply denoted φ_ε) satisfying the following properties:

$$\operatorname{supp}(\varphi_\varepsilon - \operatorname{id}) \subset F_y^- \cap B_1,$$

$$\frac{\int_{F_y^-} f}{M \operatorname{meas}(A_i)} - \varepsilon \le \varphi_\varepsilon^*(f) < 0 \quad \text{in } A_i, \quad 1 \le i \le M, \tag{11.47}$$

$$\int_0^1 s^{n-1} (1_{F_y^- \setminus (\cup_{i=1}^M A_i)} \varphi_\varepsilon^*(f)) \left(s\frac{x}{|x|} \right) ds \ge -\varepsilon, \quad x \ne 0. \tag{11.48}$$

Remark 11.20. Integrating the last inequality over the unit sphere, we indeed obtain that the negative mass of the connected component containing y is concentrated into the union of the A_i.

Proof. We split the proof into three steps.

Step 1 (simplification). Using Theorem 10.11, it is sufficient to prove the existence of $f_\varepsilon \in C^r(\overline{B}_1)$, such that

$$f_\varepsilon < 0 \quad \text{in } F_y^-, \quad \operatorname{supp}(f - f_\varepsilon) \subset F_y^- \cap B_1 \quad \text{and} \quad \int_{F_y^-} f_\varepsilon = \int_{F_y^-} f$$

satisfying also (11.47) and (11.48) with $\varphi_\varepsilon^*(f)$ replaced by f_ε.

Step 2 (preliminaries). It is easily seen that the family of closed sets K_ε, ε small, defined by

$$K_\varepsilon = \{x \in \overline{F_y^- \cap B_{1-\varepsilon}} : f(x) \le -\varepsilon\}$$

has the following properties:

$$K_\varepsilon \subset K_{\varepsilon'} \quad \text{if } \varepsilon' < \varepsilon \quad \text{and} \quad \bigcup_{\varepsilon > 0} K_\varepsilon = F_y^- \cap B_1, \tag{11.49}$$

$$f\,|_{(F_y^- \cap B_{1-\varepsilon}) \setminus K_\varepsilon} > -\varepsilon. \tag{11.50}$$

Let $\xi_\varepsilon \in C^\infty(\overline{B}_1;[0,1])$ be such that

$$\xi_\varepsilon = 1 \text{ in } K_\varepsilon \quad \text{and} \quad \operatorname{supp} \xi_\varepsilon \subset F_y^- \cap B_1 \,. \tag{11.51}$$

Using (11.49) and (11.51), we immediately deduce that

$$\lim_{\varepsilon\to 0} \xi_\varepsilon = 1_{F_y^- \cap B_1} \quad \text{and} \quad \bigcup_{i=1}^{M} A_i \subset \operatorname{int}(K_\varepsilon) \quad \text{for } \varepsilon \text{ small.} \tag{11.52}$$

Finally, for every $1 \le i \le M$ and ε small enough, let $\eta_{i,\varepsilon} \in C^\infty(\overline{B}_1;[0,1])$ be such that

$$\operatorname{supp}(\eta_{i,\varepsilon}) \subset \operatorname{int}(A_i) \quad \text{and} \quad \lim_{\varepsilon\to 0} \eta_{i,\varepsilon} = 1_{\operatorname{int}(A_i)}, \quad 1 \le i \le M. \tag{11.53}$$

Step 3 (definition of f_ε and conclusion). Define f_ε, ε small, as

$$f_\varepsilon = \begin{cases} \sum_{i=1}^{M} \eta_{i,\varepsilon} C_{i,\varepsilon}^- + (1 - \sum_{i=1}^{M} \eta_{i,\varepsilon}) \cdot (-\varepsilon) & \text{in } \bigcup_{i=1}^{M} A_i \\ \xi_\varepsilon \cdot (-\varepsilon) + (1-\xi_\varepsilon) f & \text{elsewhere,} \end{cases}$$

where

$$C_{i,\varepsilon}^- = \frac{\int_{F_y^-} f}{M \operatorname{meas}(A_i)}, \quad 1 \le i \le M-1, \tag{11.54}$$

and $C_{M,\varepsilon}^-$ is the unique constant defined implicitly by the equation

$$\int_{F_y^-} f_\varepsilon = \int_{F_y^-} f \,.$$

We claim that, up to rescaling ε, f_ε has all the required properties. Using (11.51)–(11.53), we obtain that

$$f_\varepsilon \in C^r(\overline{B}_1), \quad \operatorname{supp}(f - f_\varepsilon) \subset F_y^- \cap B_1 \,, \quad \int_{F_y^-} f_\varepsilon = \int_{F_y^-} f.$$

We assert that

$$\lim_{\varepsilon\to 0} C_{i,\varepsilon}^- = \frac{\int_{F_y^-} f}{M \operatorname{meas}(A_i)}, \quad 1 \le i \le M. \tag{11.55}$$

By (11.54), it is obviously sufficient to prove (11.55) for $i = M$. Using (11.52) and (11.53) and noticing (using (11.46))

$$\operatorname{meas}(A_i) = \operatorname{meas}(\operatorname{int}(A_i)),$$

we get, by the dominated convergence theorem,

$$\int_{F_y^-} f = \lim_{\varepsilon \to 0} \int_{F_y^-} f_\varepsilon = \sum_{i=1}^{M-1} \int_{F_y^-} 1_{\mathrm{int}(A_i)} \frac{\int_{F_y^-} f}{M \operatorname{meas}(A_i)} + \int_{F_y^-} 1_{\mathrm{int}(A_M)} \lim_{\varepsilon \to 0} C_{M,\varepsilon}^-$$
$$= \frac{M-1}{M} \int_{F_y^-} f + \operatorname{meas}(A_M) \lim_{\varepsilon \to 0} C_{M,\varepsilon}^-$$

and, thus, the assertion is verified. Equation (11.55) immediately implies $f_\varepsilon < 0$ in F_y^- for ε small and also, rescaling ε if necessary, (11.47).

It remains to prove (11.48). First, we claim that

$$f_\varepsilon \,|_{(F_y^- \cap B_{1-\varepsilon}) \setminus (\cup_{i=1}^M A_i)} \geq -\varepsilon, \tag{11.56}$$

$$f_\varepsilon \geq -D \tag{11.57}$$

for some $D > 0$ independent of ε. In fact, (11.56) is obtained combining the fact that (by definition of f_ε)

$$f_\varepsilon = -\varepsilon \quad \text{in} \quad K_\varepsilon \setminus \left(\cup_{i=1}^M A_i\right)$$

and, by (11.50) and the definition of f_ε,

$$f_\varepsilon \,|_{(F_y^- \cap B_{1-\varepsilon}) \setminus K_\varepsilon} \geq -\varepsilon.$$

Equation (11.57) is an immediate consequence of (11.55) and the definition of f_ε. Using (11.56) and (11.57) we get, for ε small and every $x \neq 0$,

$$\int_0^1 s^{n-1} (1_{F_y^- \setminus (\cup_{i=1}^M A_i)} f_\varepsilon) \left(s \frac{x}{|x|}\right) ds \geq \int_0^1 (1_{F_y^- \setminus (\cup_{i=1}^M A_i)} f_\varepsilon) \left(s \frac{x}{|x|}\right) ds$$
$$= \int_0^{1-\varepsilon} (1_{F_y^- \setminus (\cup_{i=1}^M A_i)} f_\varepsilon) \left(s \frac{x}{|x|}\right) ds + \int_{1-\varepsilon}^1 (1_{F_y^- \setminus (\cup_{i=1}^M A_i)} f_\varepsilon) \left(s \frac{x}{|x|}\right) ds$$
$$\geq \int_0^{1-\varepsilon} -\varepsilon ds + \int_{1-\varepsilon}^1 (-D) ds \geq -\varepsilon - \varepsilon D = -(D+1)\varepsilon.$$

Replacing ε by $\varepsilon/(D+1)$, we have shown (11.48) while still conserving the inequality (11.47). This ends the proof. □

11.6 Positive Radial Integration

Lemma 11.21 is the central part of the proof of Theorem 11.1. We show how to modify the mass distribution $f \in C^0(\overline{B}_1)$ satisfying $\int_{B_1} f > 0$, in order to have strictly positive integrals on every radius starting from 0 and almost positive integrals on every radius starting from any point of the boundary (see Lemma 11.21). Moreover, if f is strictly positive on the boundary or if $f \geq 0$ in $\overline{B}_1$, we will be able to modify the mass of f in order to have strictly positive integrals on every radius starting either from 0 or from any point of the boundary (see Corollary 11.23).

Lemma 11.21 (Positive radial integration). *Let $f \in C^0(\overline{B}_1)$ be such that*

$$\int_{B_1} f > 0. \tag{11.58}$$

Then for every $\sigma > 0$, there exists $\varphi = \varphi_\sigma \in \mathrm{Diff}^\infty(\overline{B}_1;\overline{B}_1)$ such that

$$\operatorname{supp}(\varphi - \mathrm{id}) \subset B_1\,, \quad \varphi^*(f)(0) > 0,$$

$$\int_0^r s^{n-1}\varphi^*(f)\left(s\frac{x}{|x|}\right) ds > 0 \quad \textit{for every } x \neq 0 \textit{ and } r \in (0,1], \tag{11.59}$$

$$\int_r^1 s^{n-1}\varphi^*(f)\left(s\frac{x}{|x|}\right) ds > -\sigma \quad \textit{for every } x \neq 0 \textit{ and } r \in [0,1]. \tag{11.60}$$

Remark 11.22. (i) If $f \geq 0$, the proof is straightforward (see Corollary 11.23).

(ii) If f_1 satisfies $\varphi^*(f_1)(0) > 0$, (11.59) and (11.60), for a certain φ as in the lemma, then every $f \geq f_1$ also satisfies $\varphi^*(f)(0) > 0$, (11.59) and (11.60) with the same φ. Indeed, we clearly have

$$\varphi^*(f_1)(x) = f_1(\varphi(x))\underbrace{\det\nabla\varphi(x)}_{>0} \leq f(\varphi(x))\det\nabla\varphi(x) = \varphi^*(f)(x).$$

(iii) Integrating (11.59) over the sphere with $r = 1$, we get $\int_{B_1}\varphi^*(f) > 0$ and, therefore, (11.58) is necessary using the change of variables formula.

(iv) In general, (11.60) cannot be assumed to be positive or 0 for every x and r. This is, for example, always the case when $f(\overline{x}) < 0$ for some $\overline{x} \in \partial B_1$. Indeed, noting that

$$\varphi^*(f)(\overline{x}) = f(\overline{x})\det\nabla\varphi(\overline{x}) < 0,$$

we have that (11.60) will be strictly negative for $x = \overline{x}$ and r sufficiently close to 1.

(v) We could have replaced, without any changes, the unit ball by any ball centered at 0.

As a corollary, we have the following result.

Corollary 11.23. *Let $f \in C^0(\overline{B}_1)$ be such that*

$$\int_{B_1} f > 0 \tag{11.61}$$

and

$$\textit{either} \quad f > 0 \textit{ on } \partial B_1 \quad \textit{or} \quad f \geq 0 \textit{ in } \overline{B}_1\,.$$

Then there exists $\varphi \in \operatorname{Diff}^{\infty}(\overline{B}_1;\overline{B}_1)$ *such that*

$$\operatorname{supp}(\varphi - \mathrm{id}) \subset B_1, \quad \varphi^*(f)(0) > 0,$$

$$\int_0^r s^{n-1}\varphi^*(f)\left(s\frac{x}{|x|}\right)ds > 0 \quad \textit{for every } x \neq 0 \textit{ and } r \in (0,1], \tag{11.62}$$

$$\int_r^1 s^{n-1}\varphi^*(f)\left(s\frac{x}{|x|}\right)ds \geq 0 \quad \textit{for every } x \neq 0 \textit{ and } r \in [0,1]. \tag{11.63}$$

Proof (Corollary 11.23). We split the proof into two parts.

Part 1. We prove the corollary when $f \geq 0$ in $\overline{B}_1$. By (11.61) there exists $a \in B_1$ with $f(a) > 0$. Using Lemma 11.13, there exists $\varphi \in \operatorname{Diff}^{\infty}(\overline{B}_1;\overline{B}_1)$ such that

$$\operatorname{supp}(\varphi - \mathrm{id}) \subset B_1 \quad \text{and} \quad \varphi(0) = a.$$

Since $\varphi^*(f)(0) = f(\varphi(0))\det\nabla\varphi(0) > 0$ and $\varphi^*(f) \geq 0$ in $\overline{B}_1$, it is immediate that φ has all of the required properties.

Part 2. We prove the corollary when $f > 0$ on ∂B_1.

Part 2.1. By (11.61), there exist $0 < \eta < 1$ and $\varepsilon > 0$ such that

$$\int_{B_\eta} f > 0 \quad \text{and} \quad f > \varepsilon \quad \text{on } \overline{B}_1 \setminus B_\eta .$$

Using Lemma 11.21 with B_η instead of B_1, there exists $\varphi \in \operatorname{Diff}^{\infty}(\overline{B}_\eta;\overline{B}_\eta)$ verifying

$$\operatorname{supp}(\varphi - \mathrm{id}) \subset B_\eta, \quad \varphi^*(f)(0) > 0,$$

$$\int_0^r s^{n-1}\varphi^*(f)\left(s\frac{x}{|x|}\right)ds > 0 \quad \text{for every } x \neq 0 \text{ and } r \in (0,\eta], \tag{11.64}$$

$$\int_r^\eta s^{n-1}\varphi^*(f)\left(s\frac{x}{|x|}\right)ds > -\frac{\varepsilon(1-\eta^n)}{n} \quad \text{for every } x \neq 0 \text{ and } r \in [0,\eta]. \tag{11.65}$$

Part 2.2. Let us show that φ (extended by the identity to $\overline{B}_1$) has all of the required properties. Trivially, $\varphi \in \operatorname{Diff}^{\infty}(\overline{B}_1;\overline{B}_1)$,

$$\varphi^*(f)(0) > 0 \quad \text{and} \quad \operatorname{supp}(\varphi - \mathrm{id}) \subset B_\eta \subset B_1 .$$

Since $\varphi^*(f) = f > 0$ in $\overline{B}_1 \setminus B_\eta$, (11.64) directly implies (11.62). Finally, we show (11.63). Using again that $\varphi^*(f) = f > 0$ in $\overline{B}_1 \setminus B_\eta$, it is obvious that (11.63) is verified for every $r \subset [\eta, 1]$. Suppose that $r \in [0, \eta)$. Combining the fact that $\varphi^*(f) = f > \varepsilon$ in $\overline{B}_1 \setminus B_\eta$ and (11.65), we obtain for every $x \neq 0$,

$$\begin{aligned}\int_r^1 s^{n-1}\varphi^*(f)\left(s\frac{x}{|x|}\right)ds &= \int_\eta^1 s^{n-1}\varphi^*(f)\left(s\frac{x}{|x|}\right)ds + \int_r^\eta s^{n-1}\varphi^*(f)\left(s\frac{x}{|x|}\right)ds \\ &> \int_\eta^1 s^{n-1}\varepsilon\, ds - \frac{\varepsilon(1-\eta^n)}{n} = 0.\end{aligned}$$

The proof is therefore complete. □

Finally, we give the proof of Lemma 11.21.

Proof. Since the proof is rather long, we divide it into five steps. The three following facts will be crucial.

(i) For fixed $a,b \in B_1$, there exists, from Lemma 11.13, $\varphi \in \operatorname{Diff}^\infty(\overline{B}_1;\overline{B}_1)$ such that $\varphi(a)=b$. This will be used in Step 1.3 and Step 3.1.

(ii) From Lemmas 11.17 and 11.19, we concentrate the mass contained in connected components of F^+ and F^- in sectors of cones. This will be achieved in Step 4.

(iii) From Remark 11.22(ii), it is sufficient to prove the result for a function $f_1 \leq f$. This will be used in Steps 1.1, 1.2 and 1.4.

Step 1. We show that we can, without loss of generality, assume that

$$f \in C^\infty(\overline{B}_1), \quad F^- \text{ connected}, \quad f(0)>0 \quad \text{and} \quad \int_{B_1\setminus F_0^+} f>0, \tag{11.66}$$

recalling that F_0^+ is the connected component of $F^+ = f^{-1}((0,\infty))$ containing 0.

Step 1.1. We start by showing that we can assume $f \in C^\infty(\overline{B}_1)$. First, using Theorem 16.11, we extend f so that $f \in C^0(\mathbb{R}^n)$. Then we choose $\delta>0$ small enough such that

$$\int_{B_1} f > \delta \operatorname{meas}(B_1).$$

By continuity of f, there exists $f_\delta \in C^\infty(\mathbb{R}^n)$ such that

$$f_\delta(x) < f(x) < f_\delta(x)+\delta \quad \text{for every } x \in \overline{B}_1.$$

Note that

$$\int_{B_1} f_\delta > \int_{B_1} f - \delta \operatorname{meas}(B_1) > 0.$$

Using Remark 11.22(ii), we have the assertion. From now on, we write f instead of f_δ and we can therefore assume that $f \in C^\infty(\overline{B}_1)$.

Step 1.2. We show that we can, without loss of generality, assume that F^- is connected.

Step 1.2.1 (preliminaries). For every $\varepsilon>0$ there exist $M \in \mathbb{N}$, $a_1,\ldots,a_M \in B_1$ and $\delta_1,\ldots,\delta_M>0$ (depending all on ε) such that

$$\bigcup_{i=1}^M \overline{B}_{\delta_i}(a_i) \subset F^+ \cap B_1$$

$$\overline{B}_{\delta_i}(a_i) \cap \overline{B}_{\delta_j}(a_j) = \emptyset \quad \text{for every } i \neq j,$$

$$\operatorname{meas}\left(F^+ \setminus \left(\cup_{i=1}^M B_{\delta_i}(a_i)\right)\right) < \varepsilon.$$

Using the last equation and since

$$\int_{B_1} f = \int_{F^+} f + \int_{F^-} f > 0,$$

we can choose $\varepsilon > 0$ (and, therefore, also M, a_i and δ_i) small enough so that

$$\int_{\cup_{i=1}^M B_{\delta_i}(a_i)} f + \int_{F^-} f > 0.$$

We then choose $\delta > 0$ small enough such that

$$\bigcup_{i=1}^M \overline{B}_{\delta_i+4\delta}(a_i) \subset F^+ \cap B_1,$$

$$\overline{B}_{\delta_i+4\delta}(a_i) \cap \overline{B}_{\delta_j+4\delta}(a_j) = \emptyset \quad \text{for every } i \neq j,$$

$$\int_{\cup_{i=1}^M B_{\delta_i}(a_i)} f + \int_{F^-} f > \delta \operatorname{meas}(B_1). \tag{11.67}$$

Let $\xi \in C^\infty(\overline{B}_1;[0,1])$ be such that

$$\xi = 1 \quad \text{in } \bigcup_{i=1}^M \left(\overline{B}_{\delta_i+3\delta}(a_i) \setminus B_{\delta_i+\delta}(a_i)\right),$$

$$\operatorname{supp} \xi \subset \bigcup_{i=1}^M \left(B_{\delta_i+4\delta}(a_i) \setminus \overline{B}_{\delta_i}(a_i)\right),$$

$$\{x \in \overline{B}_1 \setminus \left(\cup_{i=1}^M B_{\delta_i+2\delta}(a_i)\right) : \xi(x) < 1\} \quad \text{is connected.} \tag{11.68}$$

Using Theorem 16.11, we extend f so that $f \in C^\infty(\mathbb{R}^n)$. Define $\tilde{f} : \mathbb{R}^n \to \mathbb{R}$ by

$$\tilde{f}(x) = \min\{f(x), 0\}.$$

By continuity of $\widetilde{f}$, there exists $h_\delta \in C^\infty(\mathbb{R}^n)$ such that

$$h_\delta(x) < \widetilde{f}(x) < h_\delta(x) + \delta \quad \text{for every } x \in \overline{B}_1. \tag{11.69}$$

In particular, note that

$$h_\delta < 0 \quad \text{in } \overline{B}_1.$$

Step 1.2.2 (conclusion). Let $f_\delta : \overline{B}_1 \to \mathbb{R}$ be defined by

$$f_\delta = \begin{cases} (1-\xi)f & \text{in } \bigcup_{i=1}^M B_{\delta_i+2\delta}(a_i) \\ (1-\xi)h_\delta & \text{in } \overline{B}_1 \setminus \bigcup_{i=1}^M B_{\delta_i+2\delta}(a_i). \end{cases} \tag{11.70}$$

It is easily seen that f_δ is of class C^∞ and satisfies the following properties:

$$f_\delta(x)\begin{cases} = h_\delta(x) < \min\{f(x),0\} \le f(x) & \text{if } x \in \overline{B}_1 \setminus \bigcup_{i=1}^M B_{\delta_i+4\delta}(a_i) \\ \le 0 < f(x) & \text{if } x \in \bigcup_{i=1}^M (B_{\delta_i+4\delta}(a_i) \setminus B_{\delta_i+3\delta}(a_i)) \\ = 0 < f(x) & \text{if } x \in \bigcup_{i=1}^M (B_{\delta_i+3\delta}(a_i) \setminus B_{\delta_i+\delta}(a_i)) \\ \le f(x) & \text{if } x \in \bigcup_{i=1}^M (B_{\delta_i+\delta}(a_i) \setminus B_{\delta_i}(a_i)) \\ = f(x) & \text{if } x \in \bigcup_{i=1}^M B_{\delta_i}(a_i). \end{cases}$$

In particular, $f_\delta \le f$. We, moreover, have, since $h_\delta < 0$ and

$$f_\delta \ge 0 \quad \text{in} \bigcup_{i=1}^M B_{\delta_i+2\delta}(a_i),$$

that

$$\begin{aligned} F_\delta^- &= \{x \in \overline{B}_1 : f_\delta(x) < 0\} = \{x \in \overline{B}_1 \setminus \cup_{i=1}^M B_{\delta_i+2\delta}(a_i) : f_\delta(x) < 0\} \\ &= \{x \in \overline{B}_1 \setminus \cup_{i=1}^M B_{\delta_i+2\delta}(a_i) : (1-\xi(x))h_\delta(x) < 0\} \\ &= \{x \in \overline{B}_1 \setminus \cup_{i=1}^M B_{\delta_i+2\delta}(a_i) : \xi(x) < 1\}, \end{aligned}$$

which is a connected set by (11.68). We thus have that

$$F_\delta^- \subset \overline{B}_1 \setminus \bigcup_{i=1}^M B_{\delta_i+2\delta}(a_i) \quad \text{and} \quad F_\delta^- \text{ is connected.}$$

Observe next that

$$\begin{aligned} \int_{F_\delta^-} f_\delta &= \int_{F_\delta^-} (1-\xi)h_\delta \ge \int_{F_\delta^-} h_\delta > \int_{F_\delta^-} \left(\widetilde{f} - \delta\right) \ge \int_{F_\delta^-} \widetilde{f} - \delta\,\text{meas}(B_1) \\ &= \int_{F_\delta^- \cap F^-} \widetilde{f} + \int_{F_\delta^- \setminus F^-} \widetilde{f} - \delta\,\text{meas}(B_1) = \int_{F_\delta^- \cap F^-} \widetilde{f} - \delta\,\text{meas}(B_1) \\ &= \int_{F_\delta^- \cap F^-} f - \delta\,\text{meas}(B_1) \ge \int_{F^-} f - \delta\,\text{meas}(B_1). \end{aligned}$$

This leads to

$$\begin{aligned} \int_{B_1} f_\delta &= \int_{F_\delta^+} f_\delta + \int_{F_\delta^-} f_\delta \ge \int_{\cup_{i=1}^M B_{\delta_i}(a_i)} f_\delta + \int_{F_\delta^-} f_\delta = \int_{\cup_{i=1}^M B_{\delta_i}(a_i)} f + \int_{F_\delta^-} f_\delta \\ &> \int_{\cup_{i=1}^M B_{\delta_i}(a_i)} f + \int_{F^-} f - \delta\,\text{meas}(B_1) > 0, \end{aligned}$$

where we have used (11.67) in the last inequality. From now on, we write f in place of f_δ, since $f_\delta \le f$ and Remark 11.22(ii) holds. We may therefore assume, in the remaining part of the proof, that $f \in C^\infty\left(\overline{B}_1\right)$ and F^- is connected.

Step 1.3. Let us prove that we can assume that $f(0) > 0$. In fact, suppose that $f(0) \leq 0$. We prove that there exists a diffeomorphism φ_1 such that $\varphi_1^*(f)(0) > 0$. Indeed, since $\int_{B_1} f > 0$, there exists $a \in B_1$ such that $f(a) > 0$. By Lemma 11.13, there exists $\varphi_1 \in \operatorname{Diff}^\infty(\overline{B}_1;\overline{B}_1)$ such that

$$\operatorname{supp}(\varphi_1 - \mathrm{id}) \subset B_1 \quad \text{and} \quad \varphi_1(0) = a.$$

Since $\varphi_1^*(f)(0) = f(a)\det\nabla\varphi_1(0) > 0$, we have the result. Note that, using the change of variables formula,

$$\int_{B_1} \varphi_1^*(f) = \int_{B_1} f > 0.$$

Note also that $\varphi_1^*(f) \in C^\infty\left(\overline{B}_1\right)$ and, using Lemma 11.15,

$$\left(\varphi_1^*(f)\right)^{-1}((-\infty,0)) = \varphi_1^{-1}(F^-) \quad \text{is connected.}$$

From now on, we write f in place of $\varphi_1^*(f)$ and thus we can assume, without loss of generality, that $f \in C^\infty\left(\overline{B}_1\right)$, F^- is connected and $f(0) > 0$.

Step 1.4. We finally show that we can assume that

$$\int_{B_1 \setminus F_0^+} f > 0.$$

In fact, since $f(0) > 0$ and $\int_{B_1} f > 0$, if $\delta_1 > 0$ is small enough, we have that $B_{4\delta_1} \subset F_0^+$ and

$$\int_{B_1 \setminus B_{4\delta_1}} f > 0. \tag{11.71}$$

Let $\eta \in C^\infty([0,1];[0,1])$ be such that

$$\eta(r) = \begin{cases} 1 & \text{if } r \leq \delta_1 \text{ or } 4\delta_1 \leq r \leq 1 \\ 0 & \text{if } 2\delta_1 \leq r \leq 3\delta_1 . \end{cases}$$

Let $h \in C^\infty(\overline{B}_1)$ defined by $h(x) = \eta(|x|)f(x)$. We then have

$$h(0) > 0, \quad H^- = F^- \text{ connected} \quad \text{and} \quad B_{\delta_1} \subset H_0^+ \subset B_{2\delta_1} .$$

Using (11.71), we get

$$\int_{B_1 \setminus H_0^+} h \geq \int_{B_1 \setminus B_{4\delta_1}} h = \int_{B_1 \setminus B_{4\delta_1}} f > 0.$$

Since $h \leq f$, we may, according to Remark 11.22(ii), proceed replacing f with $h = \eta f$. The proof of Step 1 is therefore complete.

Step 2. In this step we start by selecting N connected components of $F^+ \setminus F_0^+$. Then we select an appropriate amount of points in each of them.

Step 2.1 (selection of N connected components of $F^+ \setminus F_0^+$). Let $F_{x_i}^+$, $i \in I^+$, $x_i \in B_1 \setminus F_0^+$, be the pairwise disjoint connected components of $F^+ \setminus F_0^+$. Notice that I^+ is not empty by Step 1.4 and is at most countable; see Lemma 11.16. We claim that there exists $N \in \mathbb{N}$ such that

$$\int_{\cup_{i=1}^N F_{x_i}^+} f + \int_{F^-} f > 0. \tag{11.72}$$

In fact, suppose that I^+ is infinite (otherwise the assertion is trivial because of the fourth statement in (11.66)). Since, by the fourth statement in (11.66),

$$\int_{F^+ \setminus F_0^+} f + \int_{F^-} f > 0$$

and since, using Lemma 11.16,

$$\lim_{N \to \infty} \int_{\cup_{i=1}^N F_{x_i}^+} f = \int_{F^+ \setminus F_0^+} f,$$

we have (11.72) for N large enough.

Step 2.2 (selection of M_i points in $F_{x_i}^+$, $1 \le i \le N$ and of $M_1 + \cdots + M_N - 1$ points in F^-). We start by defining the integers M_i. We claim that there exist $M_1, \ldots, M_N \in \mathbb{N}$ such that

$$\frac{\int_{F_{x_i}^+} f}{M_i} + \frac{\int_{F^-} f}{(\sum_{j=1}^N M_j) - 1} > 0 \quad \text{for every } 1 \le i \le N. \tag{11.73}$$

In order to simplify the notations, let

$$m_i^+ = \int_{F_{x_i}^+} f, \quad 1 \le i \le N \quad \text{and} \quad m^- = \int_{F^-} f.$$

We claim that for an integer ν large enough,

$$M_1 = \nu \quad \text{and} \quad M_i = \left[\frac{m_i^+}{m_1^+}\nu\right], \quad 2 \le i \le N,$$

where $[x]$ stands for the integer part of x, satisfy (11.73). Indeed, let $1 \le i \le N$; then since

$$\frac{m_i^+}{m_1^+}\nu - 1 < M_i < \frac{m_i^+}{m_1^+}\nu + 1, \quad 1 \le i \le N,$$

we deduce

$$\frac{m_i^+}{M_i} + \frac{m^-}{(\sum_{j=1}^N M_j) - 1} \ge \frac{m_i^+}{\frac{m_i^+}{m_1^+}\nu + 1} + \frac{m^-}{\frac{\sum_{j=1}^N m_j^+}{m_1^+}\nu - N - 1}.$$

Therefore, since, by (11.72),

$$\sum_{j=1}^{N} m_j^+ + m^- > 0,$$

we get

$$\lim_{\nu\to\infty}\left[\nu\left(\frac{m_i^+}{M_i}+\frac{m^-}{(\sum_{j=1}^{N} M_j)-1}\right)\right]\geq m_1^+\left(1+\frac{m^-}{\sum_{j=1}^{N} m_j^+}\right)>0.$$

This proves the assertion.

Finally, choose M_1 distinct points

$$z_1,\dots,z_{M_1}\in F_{x_1}^+.$$

Then choose M_2 distinct points

$$z_{M_1+1},\dots,z_{M_1+M_2}\in F_{x_2}^+$$

and so on, and finally choose M_N distinct points

$$z_{M_1+\cdots+M_{N-1}+1},\dots,z_{M_1+\cdots+M_N}\in F_{x_N}^+.$$

Similarly, choose $M_1+\cdots+M_N-1$ distinct points

$$y_1,\dots,y_{M_1+\cdots+M_N-1}\in F^-.$$

We define

$$M=M_1+\cdots+M_N.$$

In particular, we have

$$f(z_k)>0,\quad 1\leq k\leq M \quad\text{and}\quad f(y_j)<0,\quad 1\leq j\leq M-1.$$

Step 3. In this step we move the $2M-1$ points selected in the above step so that they are on the same radial axis and well ordered; moreover, we define some cone sectors.

Step 3.1 (displacement of the points z_k and y_j). Choose $(2M-1)$ points, $\widetilde{z}_1,\dots,\widetilde{z}_M$ and $\widetilde{y}_1,\dots,\widetilde{y}_{M-1}$ such that

$$0<|\widetilde{z}_1|<|\widetilde{y}_1|<|\widetilde{z}_2|<|\widetilde{y}_2|<\cdots<|\widetilde{z}_{M-1}|<|\widetilde{y}_{M-1}|<|\widetilde{z}_M|<1,$$

$$\frac{\widetilde{z}_k}{|\widetilde{z}_k|}=\frac{\widetilde{y}_j}{|\widetilde{y}_j|}\quad\text{for every } 1\leq k\leq M,\ 1\leq j\leq M-1.$$

Then choose ε_1 small enough and

$$c_l\in C^0([0,1];B_1),\ 1\leq l\leq 2M-1,$$

such that the sets

$$c_l([0,1]) + B_{\varepsilon_1} \quad \text{are pairwise disjoint and contained in } B_1 \setminus \{0\},$$
$$\begin{cases} c_l(0) = \widetilde{z}_l & c_l(1) = z_l & \text{if } 1 \le l \le M, \\ c_l(0) = \widetilde{y}_{l-M} & c_l(1) = y_{l-M} & \text{if } M+1 \le l \le 2M-1. \end{cases}$$

Applying, for $1 \le l \le 2M-1$, Lemma 11.13 with $\varepsilon = \varepsilon_1$ and $c = c_l$, we get $\psi_l \in \mathrm{Diff}^\infty(\overline{B}_1; \overline{B}_1)$ with

$$\psi_l(c_l(0)) = c_l(1) \quad \text{and} \quad \operatorname{supp}(\psi_l - \mathrm{id}) \subset c_l([0,1]) + B_{\varepsilon_1} \subset B_1 \setminus \{0\}.$$

Thus, defining $\varphi_2 = \psi_1 \circ \cdots \circ \psi_{2M-1}$, we get that $\operatorname{supp}(\varphi_2 - \mathrm{id}) \subset B_1 \setminus \{0\}$ (and thus, in particular $\varphi_2(0) = 0$) and

$$\varphi_2(\widetilde{z}_k) = z_k, \quad 1 \le k \le M \quad \text{and} \quad \varphi_2(\widetilde{y}_j) = y_j, \quad 1 \le j \le M-1.$$

To complete, we also define

$$\widetilde{x}_i = \varphi_2^{-1}(x_i) \quad 1 \le i \le N.$$

Step 3.2 (definition of cone sectors). For $0 < \delta < 1$, let K_δ be the closed cone having vertex 0 and axis $\mathbb{R}_+ \widetilde{y}_1$ and such that

$$\operatorname{meas}(K_\delta \cap B_1) = \delta \operatorname{meas} B_1 .$$

This immediately implies that

$$\operatorname{meas}(K_\delta \cap B_r) = \delta \operatorname{meas} B_r \quad \text{for every } r > 0. \tag{11.74}$$

Define

$$\widetilde{f} = \varphi_2^*(f).$$

By the properties of f and φ_2 we get that

$$\widetilde{f}(0) > 0, \quad \widetilde{f}(\widetilde{z}_k) > 0,\ 1 \le k \le M \quad \text{and} \quad \widetilde{f}(\widetilde{y}_j) < 0,\ 1 \le j \le M-1.$$

Therefore, there exists $\delta > 0$ small enough such that

$$\begin{cases} \widetilde{f} > \delta \quad \text{in } B_\delta, \\ K_\delta \cap \left(\overline{B}_{|\widetilde{z}_k|+\delta} \setminus B_{|\widetilde{z}_k|-\delta}\right) \subset \widetilde{F}^+ \cap B_1, & 1 \le k \le M, \\ K_\delta \cap \left(\overline{B}_{|\widetilde{y}_j|+\delta} \setminus B_{|\widetilde{y}_j|-\delta}\right) \subset \widetilde{F}^- \cap B_1, & 1 \le j \le M-1; \end{cases}$$

in particular,

$$\delta < |\widetilde{z}_1| - \delta < |\widetilde{z}_1| + \delta < |\widetilde{y}_1| - \delta < |\widetilde{y}_1| + \delta < |\widetilde{z}_2| - \delta < |\widetilde{z}_2| + \delta$$
$$< \cdots < |\widetilde{y}_{M-1}| - \delta < |\widetilde{y}_{M-1}| + \delta < |\widetilde{z}_M| - \delta < |\widetilde{z}_M| + \delta < 1.$$

Using Lemma 11.15 and (11.72), we get that $\widetilde{f} \in C^{\infty}\left(\overline{B}_1\right)$ is such that $\widetilde{F}^-$ is connected and

$$\int_{\cup_{i=1}^{N} \widetilde{F}_{\widetilde{x}_i}^{+}} \widetilde{f} + \int_{\widetilde{F}^-} \widetilde{f} > 0.$$

From now on, we write f, x_i, z_k and y_j instead of $\widetilde{f} = \varphi_2^*(f)$, $\widetilde{x}_i$, $\widetilde{z}_k$ and $\widetilde{y}_j$, respectively. Define

$$\begin{cases} S_k^+ = K_\delta \cap \left(\overline{B}_{|z_k|+\delta} \setminus B_{|z_k|-\delta}\right), & 1 \leq k \leq M, \\ S_j^- = K_\delta \cap \left(\overline{B}_{|y_j|+\delta} \setminus B_{|y_j|-\delta}\right), & 1 \leq j \leq M-1, \end{cases}$$

in particular,

$$\begin{aligned} \delta < |z_1| - \delta < |z_1| + \delta < |y_1| - \delta < |y_1| + \delta < |z_2| - \delta < |z_2| + \delta \\ < \cdots < |y_{M-1}| - \delta < |y_{M-1}| + \delta < |z_M| - \delta < |z_M| + \delta < 1. \end{aligned}$$

Choosing δ even smaller, we can assume, without loss of generality, that

$$\frac{\delta^{n+1}}{n} < \sigma, \tag{11.75}$$

where σ is the σ in the statement of the lemma. Note that f has the following properties:

$$S_k^+ \subset F_{x_{t(k)}}^+,$$

where $t(k)$ is defined by

$$t(k) = \begin{cases} 1 & \text{if } 1 \leq k \leq M_1 \\ \vdots & \vdots \\ N & \text{if } M_1 + \cdots + M_{N-1} + 1 \leq k \leq M, \end{cases}$$

$f > \delta$ in $B_\delta \subset F_0^+$, F^- is connected and

$$\int_{\cup_{i=1}^{N} F_{x_i}^{+}} f + \int_{F^-} f > 0. \tag{11.76}$$

Step 4. In this step we concentrate the positive and the negative mass in the cone sectors defined in the previous step.

Step 4.1 (concentration of the positive mass in S_k^+, $1 \leq k \leq M$). Using (11.73), we can find ϵ_1 small enough such that

$$\frac{\int_{F_{x_i}^+} f}{M_i} - 2\epsilon_1 \operatorname{meas} B_1 + \frac{\int_{F^-} f}{M-1} > 0, \quad 1 \leq i \leq N. \tag{11.77}$$

Applying, for $1 \le i \le N$, Lemma 11.17 to f, $z = x_i$, $\varepsilon = \varepsilon_1$ and

$$A_1 = S^+_{1+\sum_{j=1}^{i-1} M_j}, \dots, A_{M_i} = S^+_{\sum_{j=1}^{i} M_j},$$

we get $\psi_i \in \operatorname{Diff}^\infty(\overline{B}_1;\overline{B}_1)$ with $\operatorname{supp}(\psi_i - \mathrm{id}) \subset F^+_{x_i} \cap B_1$ and

$$(\psi_i)^*(f) \ge \frac{\int_{F^+_{x_i}} f}{M_i \operatorname{meas} S^+_k} - \varepsilon_1 \text{ in } S^+_k, \quad 1 + \sum_{j=1}^{i-1} M_j \le k \le \sum_{j=1}^{i} M_j.$$

Letting $\varphi_3 = \psi_1 \circ \cdots \circ \psi_N \in \operatorname{Diff}^\infty(\overline{B}_1;\overline{B}_1)$, we obtain that

$$\operatorname{supp}(\varphi_3 - \mathrm{id}) \subset \bigcup_{j=1}^{N} (F^+_{x_j} \cap B_1) \subset B_1 \setminus F^+_0,$$

$$\varphi_3^*(f) = f > \delta \quad \text{in } B_\delta,$$

and, for every $1 \le i \le N$,

$$\varphi_3^*(f) \ge \frac{\int_{F^+_{x_i}} f}{M_i \operatorname{meas} S^+_k} - \varepsilon_1 \text{ in } S^+_k, \quad 1 + \sum_{j=1}^{i-1} M_j \le k \le \sum_{j=1}^{i} M_j.$$

We define, for $1 \le k \le M$,

$$C^+_k = \frac{\int_{F^+_{x_i}} f}{M_i \operatorname{meas} S^+_k} - \varepsilon_1 \quad \text{if } 1 + \sum_{j=1}^{i-1} M_j \le k \le \sum_{j=1}^{i} M_j$$

and we replace $\varphi_3^*(f)$ by f. We therefore have, using (11.77) and the fact that $\operatorname{meas}(S^+_k) \le \operatorname{meas}(B_1)$,

$$\begin{cases} f \ge C^+_k \quad \text{in } S^+_k, \quad 1 \le k \le M, & f > \delta \quad \text{in } B_\delta, \\ C^+_k \operatorname{meas}(S^+_k) + \dfrac{\int_{F^-} f}{M-1} - \varepsilon_1 \operatorname{meas}(B_1) > 0, & 1 \le k \le M. \end{cases} \tag{11.78}$$

We also have

$$\bigcup_{k=1}^{M} S^+_k \subset F^+ \setminus F^+_0.$$

Step 4.2 (concentration of the negative mass in S^-_j, $1 \le j \le M-1$). Using Lemma 11.19, recalling that F^- is connected, with $A_j = S^-_j$, $1 \le j \le M-1$, and

$$\varepsilon = \min\{\varepsilon_1, \delta^{n+1}/n\},$$

where δ has been defined in Step 3.2, we get $\varphi_4 \in \operatorname{Diff}^\infty(\overline{B}_1;\overline{B}_1)$ with $\operatorname{supp}(\varphi_4 - \mathrm{id}) \subset F^- \cap B_1$ and

$$\begin{cases} \dfrac{\int_{F^-} f}{(M-1)\operatorname{meas} S_j^-} - \varepsilon_1 \le \varphi_4^*(f) < 0 \quad \text{in } S_j^-, & 1 \le j \le M-1, \\ \displaystyle\int_0^1 s^{n-1}\left(1_{F^- \setminus \left(\cup_{j=1}^{M-1} S_j^-\right)} \varphi_4^*(f)\right)\left(s\frac{x}{|x|}\right) ds \ge -\frac{\delta^{n+1}}{n}, & x \ne 0. \end{cases}$$

Defining

$$C_j^- = \frac{\int_{F^-} f}{(M-1)\operatorname{meas} S_j^-} - \varepsilon_1, \quad 1 \le j \le M-1,$$

we thus get, using the second inequality of (11.78),

$$\begin{cases} C_j^- \le \varphi_4^*(f) < 0 \quad \text{in } S_j^-, & 1 \le j \le M-1, \\ C_k^+ \operatorname{meas} S_k^+ + C_j^- \operatorname{meas} S_j^- > 0, & 1 \le j \le M-1,\ 1 \le k \le M, \\ \displaystyle\int_0^1 s^{n-1}\left(1_{F^- \setminus \left(\cup_{j=1}^{M-1} S_j^-\right)} \varphi_4^*(f)\right)\left(s\frac{x}{|x|}\right) ds \ge -\frac{\delta^{n+1}}{n}, & x \ne 0. \end{cases}$$

Note that $\varphi_4^*(f) = f$ in F^+. Finally, as usual, we replace $\varphi_4^*(f)$ by f. We therefore obtain, using (11.78) and recalling (by (11.75)) that $\frac{\delta^{n+1}}{n} < \sigma$,

$$\begin{cases} f > \delta \quad \text{in } B_\delta \subset F_0^+, & \\ f \ge C_k^+ \quad \text{in } S_k^+ \subset F^+ \setminus F_0^+, & 1 \le k \le M, \\ f \ge C_j^- \quad \text{in } S_j^- \subset F^-, & 1 \le j \le M-1, \\ C_k^+ \operatorname{meas} S_k^+ + C_j^- \operatorname{meas} S_j^- > 0, & 1 \le k \le M,\ 1 \le j \le M-1, \\ \displaystyle\int_0^1 s^{n-1}\left(1_{F^- \setminus \left(\cup_{j=1}^{M-1} S_j^-\right)} f\right)\left(s\frac{x}{|x|}\right) ds \ge -\frac{\delta^{n+1}}{n} > -\sigma, & x \ne 0. \end{cases} \tag{11.79}$$

Step 4.3 (summary of the properties of f). We claim that f has the following properties:

$$f > \delta \quad \text{in } B_\delta \subset F_0^+, \tag{11.80}$$

$$\bigcup_{k=1}^{M} S_k^+ \subset F^+ \setminus F_0^+, \quad \bigcup_{j=1}^{M-1} S_j^- \subset F^-, \tag{11.81}$$

$$\int_0^1 s^{n-1}\left(1_{F^- \setminus \left(\cup_{j=1}^{M-1} S_j^-\right)} f\right)\left(s\frac{x}{|x|}\right) ds \ge -\frac{\delta^{n+1}}{n} > -\sigma \quad \text{if } x \ne 0 \tag{11.82}$$

and for every $x \ne 0$ and $1 \le k \le M$, $1 \le j \le M-1$,

$$\int_0^1 s^{n-1}(1_{S_k^+} f)\left(s\frac{x}{|x|}\right) ds + \int_0^1 s^{n-1}(1_{S_j^-} f)\left(s\frac{x}{|x|}\right) ds > 0. \tag{11.83}$$

In fact, (11.80)–(11.82) are just the first, second and fifth inequalities of (11.79), respectively. Let us show (11.83). Fix $1 \le k \le M$ and $1 \le j \le M-1$. Recall that

$$S_k^+ = K_\delta \cap (\overline{B}_{|z_k|+\delta} \setminus B_{|z_k|-\delta}) \quad \text{and} \quad S_j^- = K_\delta \cap (\overline{B}_{|y_j|+\delta} \setminus B_{|y_j|-\delta}),$$

where K_δ is a cone with vertex 0 and aperture δ. Thus, according to (11.74),

$$\begin{aligned}\operatorname{meas} S_k^+ &= \delta\,[(|z_k|+\delta)^n - (|z_k|-\delta)^n] \operatorname{meas} B_1,\\ \operatorname{meas} S_j^- &= \delta\,[(|y_j|+\delta)^n - (|y_j|-\delta)^n] \operatorname{meas} B_1.\end{aligned}$$

Then, using (11.79), we get

$$\begin{aligned}&\int_0^1 s^{n-1}(1_{S_k^+} f)\left(s\frac{x}{|x|}\right) ds + \int_0^1 s^{n-1}(1_{S_j^-} f)\left(s\frac{x}{|x|}\right) ds\\ &\quad\ge \int_{|z_k|-\delta}^{|z_k|+\delta} s^{n-1} C_k^+\, ds + \int_{|y_j|-\delta}^{|y_j|+\delta} s^{n-1} C_j^-\, ds\\ &\quad= C_k^+ \frac{(|z_k|+\delta)^n - (|z_k|-\delta)^n}{n} + C_j^- \frac{(|y_j|+\delta)^n - (|y_j|-\delta)^n}{n}\\ &\quad= C_k^+ \frac{\operatorname{meas} S_k^+}{n\,\delta \operatorname{meas} B_1} + C_j^- \frac{\operatorname{meas} S_j^-}{n\,\delta \operatorname{meas} B_1} > 0,\end{aligned}$$

which is the claim.

Step 5 (conclusion). Let

$$\varphi = \varphi_1 \circ \varphi_2 \circ \varphi_3 \circ \varphi_4 .$$

Note that, by construction, $\operatorname{supp}(\varphi - \mathrm{id}) \subset B_1$. Because of all of the successive replacements of f in Steps 1–4 by a new f, the lemma has to be proved for $\varphi = \mathrm{id}$.

Step 5.1. First, note that $f(0) > 0$ by (11.80).

Step 5.2. We now show (11.59). We divide the discussion into three steps.

Step 5.2.1. If $r \le \delta$, (11.80) directly implies the assertion.

Step 5.2.2. We now suppose that either $x \notin K_\delta$ and $r \in (\delta, 1]$ or $x \in K_\delta$ and $r \in (\delta, |y_1| - \delta)$ and thus, in particular,

$$\left[0, r\frac{x}{|x|}\right] \cap \left(\cup_{j=1}^{M-1} S_j^-\right) = \emptyset.$$

Observe that (11.80) and (11.82) then imply

$$\begin{aligned}\int_0^r s^{n-1} f\left(s\frac{x}{|x|}\right) ds &\ge \int_0^r s^{n-1}(1_{F_0^+} f)\left(s\frac{x}{|x|}\right) ds + \int_0^r s^{n-1}(1_{F^-} f)\left(s\frac{x}{|x|}\right) ds\\ &= \int_0^r s^{n-1}(1_{F_0^+} f)\left(s\frac{x}{|x|}\right) ds + \int_0^r s^{n-1}(1_{F^- \setminus \cup_{j=1}^{M-1} S_j^-} f)\left(s\frac{x}{|x|}\right) ds\end{aligned}$$

$$> \int_0^\delta s^{n-1}\delta\,ds + \int_0^r s^{n-1}(1_{F^- \setminus \cup_{j=1}^{M-1} S_j^-} f)\left(s\frac{x}{|x|}\right) ds$$
$$\geq \int_0^\delta s^{n-1}\delta\,ds + \int_0^1 s^{n-1}(1_{F^- \setminus \cup_{j=1}^{M-1} S_j^-} f)\left(s\frac{x}{|x|}\right) ds \geq 0$$

and the assertion is proved.

Step 5.2.3. It only remains to show the assertion when $x \in K_\delta$ and $r \in [\,|y_1| - \delta, 1]$. We get

$$\int_0^r s^{n-1} f\left(s\frac{x}{|x|}\right) ds$$
$$= \int_0^r s^{n-1}(1_{F_0^+} f)\left(s\frac{x}{|x|}\right) ds + \int_0^r s^{n-1}(1_{F^+ \setminus F_0^+} f)\left(s\frac{x}{|x|}\right) ds$$
$$+ \int_0^r s^{n-1}(1_{F^-} f)\left(s\frac{x}{|x|}\right) ds$$

and thus

$$\int_0^r s^{n-1} f\left(s\frac{x}{|x|}\right) ds$$
$$= \int_0^r s^{n-1}(1_{F_0^+} f)\left(s\frac{x}{|x|}\right) ds + \int_0^r s^{n-1}(1_{F^- \setminus \cup_{j=1}^{M-1} S_j^-} f)\left(s\frac{x}{|x|}\right) ds$$
$$+ \int_0^r s^{n-1}(1_{F^+ \setminus F_0^+} f)\left(s\frac{x}{|x|}\right) ds + \int_0^r s^{n-1}(1_{\cup_{j=1}^{M-1} S_j^-} f)\left(s\frac{x}{|x|}\right) ds.$$

Since $r \geq |y_1| - \delta \geq |z_1| + \delta \geq \delta$, (11.80) holds, and $f < 0$ in $F^- \setminus \cup_{j=1}^{M-1} S_j^-$, we obtain

$$\int_0^r s^{n-1}(1_{F_0^+} f)\left(s\frac{x}{|x|}\right) ds + \int_0^r s^{n-1}(1_{F^- \setminus \cup_{j=1}^{M-1} S_j^-} f)\left(s\frac{x}{|x|}\right) ds$$
$$\geq \int_0^\delta s^{n-1}\delta\,ds + \int_0^1 s^{n-1}(1_{F^- \setminus \cup_{j=1}^{M-1} S_j^-} f)\left(s\frac{x}{|x|}\right) ds.$$

and hence, according to (11.82),

$$\int_0^r s^{n-1}(1_{F_0^+} f)\left(s\frac{x}{|x|}\right) ds + \int_0^r s^{n-1}(1_{F^- \setminus \cup_{j=1}^{M-1} S_j^-} f)\left(s\frac{x}{|x|}\right) ds \geq 0.$$

We therefore find, using (11.81), that

$$\int_0^r s^{n-1} f\left(s\frac{x}{|x|}\right) ds \geq \int_0^r s^{n-1}(1_{F^+ \setminus F_0^+} f)\left(s\frac{x}{|x|}\right) ds + \int_0^r s^{n-1}(1_{\cup_{j=1}^{M-1} S_j^-} f)\left(s\frac{x}{|x|}\right) ds$$
$$\geq \sum_{k=1}^{M-1} \left\{ \int_0^r s^{n-1}(1_{S_k^+} f)\left(s\frac{x}{|x|}\right) ds + \int_0^r s^{n-1}(1_{S_k^-} f)\left(s\frac{x}{|x|}\right) ds \right\}.$$

Define

$$A=\sum_{k=1}^{M-1}\left\{\int_0^r s^{n-1}(1_{S_k^+}f)\left(s\frac{x}{|x|}\right)ds+\int_0^r s^{n-1}(1_{S_k^-}f)\left(s\frac{x}{|x|}\right)ds\right\}.$$

In order to conclude the proof of Step 5.2.3 and thus of Step 5.2, it is sufficient to show that $A>0$. We consider several cases.

Case 1: $r\in[|y_1|-\delta,|z_2|+\delta)$. We then have

$$\begin{aligned}A=&\int_0^r s^{n-1}(1_{S_2^+}f)\left(s\frac{x}{|x|}\right)ds+\int_0^r s^{n-1}(1_{S_1^+}f)\left(s\frac{x}{|x|}\right)ds\\&+\int_0^r s^{n-1}(1_{S_1^-}f)\left(s\frac{x}{|x|}\right)ds\end{aligned}$$

and thus, recalling that $r\geq|y_1|-\delta>|z_1|+\delta$,

$$\begin{aligned}A&\geq\int_0^r s^{n-1}(1_{S_1^+}f)\left(s\frac{x}{|x|}\right)ds+\int_0^r s^{n-1}(1_{S_1^-}f)\left(s\frac{x}{|x|}\right)ds\\&\geq\int_0^1 s^{n-1}(1_{S_1^+}f)\left(s\frac{x}{|x|}\right)ds+\int_0^1 s^{n-1}(1_{S_1^-}f)\left(s\frac{x}{|x|}\right)ds,\end{aligned}$$

which is positive, according to (11.83).

Case 2: $r\in[|z_i|+\delta,|z_{i+1}|+\delta)$, $2\leq i\leq M-1$. We therefore find

$$\begin{aligned}A&=\sum_{k=1}^{i+1}\int_0^r s^{n-1}(1_{S_k^+}f)\left(s\frac{x}{|x|}\right)ds+\sum_{k=1}^{i}\int_0^r s^{n-1}(1_{S_k^-}f)\left(s\frac{x}{|x|}\right)ds\\&\geq\sum_{k=1}^{i}\left\{\int_0^r s^{n-1}(1_{S_k^+}f)\left(s\frac{x}{|x|}\right)ds+\int_0^r s^{n-1}(1_{S_k^-}f)\left(s\frac{x}{|x|}\right)ds\right\}\\&\geq\sum_{k=1}^{i}\left\{\int_0^1 s^{n-1}(1_{S_k^+}f)\left(s\frac{x}{|x|}\right)ds+\int_0^1 s^{n-1}(1_{S_k^-}f)\left(s\frac{x}{|x|}\right)ds\right\}\end{aligned}$$

which is positive, in view of (11.83).

Case 3: $r\in[|z_M|+\delta,1]$. We now have

$$A=\sum_{k=1}^{M-1}\left\{\int_0^1 s^{n-1}(1_{S_k^+}f)\left(s\frac{x}{|x|}\right)ds+\int_0^1 s^{n-1}(1_{S_k^-}f)\left(s\frac{x}{|x|}\right)ds\right\},$$

which is positive, according to (11.83).

Step 5.3. We finally prove (11.60) and we divide the proof into two steps.

Step 5.3.1. First, suppose that either $x\notin K_\delta$ or

$$x\in K_\delta\quad\text{and}\quad r\in(|y_{M-1}|+\delta,1]$$

and thus, in particular,

$$\left[r\frac{x}{|x|}, \frac{x}{|x|}\right] \bigcap \left(\bigcup_{j=1}^{M-1} S_j^-\right) = \emptyset.$$

Inequality (11.82) then implies

$$\begin{aligned}\int_r^1 s^{n-1} f\left(s\frac{x}{|x|}\right) ds &\geq \int_r^1 s^{n-1}(1_{F^-} f)\left(s\frac{x}{|x|}\right) ds\\ &= \int_r^1 s^{n-1}(1_{F^-\setminus\cup_{j=1}^{M-1} S_j^-} f)\left(s\frac{x}{|x|}\right) ds\\ &\geq \int_0^1 s^{n-1}(1_{F^-\setminus\cup_{j=1}^{M-1} S_j^-} f)\left(s\frac{x}{|x|}\right) ds > -\sigma,\end{aligned}$$

which proves the assertion.

Step 5.3.2. It only remains to show the assertion when $x \in K_\delta$ and $r \in [0, |y_{M-1}| + \delta]$. We get, using the fact that $f < 0$ in F^-, (11.81) and $f > 0$ in F_0^+, that

$$\begin{aligned}\int_r^1 s^{n-1} f\left(s\frac{x}{|x|}\right) ds &= \int_r^1 s^{n-1}(1_{F^-} f)\left(s\frac{x}{|x|}\right) ds + \int_r^1 s^{n-1}(1_{F^+} f)\left(s\frac{x}{|x|}\right) ds\\ &\geq \int_r^1 s^{n-1}(1_{F^-\setminus\cup_{j=1}^{M-1} S_j^-} f)\left(s\frac{x}{|x|}\right) ds\\ &\quad + \int_r^1 s^{n-1}(1_{\cup_{j=1}^{M-1} S_j^-} f)\left(s\frac{x}{|x|}\right) ds\\ &\quad + \int_r^1 s^{n-1}(1_{F^+\setminus F_0^+} f)\left(s\frac{x}{|x|}\right) ds\end{aligned}$$

and hence, appealing to (11.82) and since $f > 0$ in S_1^+,

$$\begin{aligned}\int_r^1 s^{n-1} f\left(s\frac{x}{|x|}\right) ds &> -\sigma + \int_r^1 s^{n-1}(1_{\cup_{j=1}^{M-1} S_j^-} f)\left(s\frac{x}{|x|}\right) ds\\ &\quad + \int_r^1 s^{n-1}(1_{F^+\setminus F_0^+} f)\left(s\frac{x}{|x|}\right) ds\\ &\geq -\sigma + \sum_{k=2}^{M} \int_r^1 s^{n-1}(1_{S_{k-1}^-} f)\left(s\frac{x}{|x|}\right) ds\\ &\quad + \sum_{k=2}^{M} \int_r^1 s^{n-1}(1_{S_k^+} f)\left(s\frac{x}{|x|}\right) ds.\end{aligned}$$

Define

$$B = \sum_{k=2}^{M} \left\{\int_r^1 s^{n-1}(1_{S_k^+} f)\left(s\frac{x}{|x|}\right) ds + \int_r^1 s^{n-1}(1_{S_{k-1}^-} f)\left(s\frac{x}{|x|}\right) ds\right\}.$$

In order to obtain the claim, it remains to prove that $B > 0$. This is obtained exactly as in Step 5.2.3.

Case 1: $r \in [|z_{M-1}| - \delta, |y_{M-1}| + \delta]$. We then have

$$B = \int_r^1 s^{n-1}(1_{S_M^+} f)\left(s\frac{x}{|x|}\right) ds + \int_r^1 s^{n-1}(1_{S_{M-1}^+} f)\left(s\frac{x}{|x|}\right) ds$$
$$+ \int_r^1 s^{n-1}(1_{S_{M-1}^-} f)\left(s\frac{x}{|x|}\right) ds$$

and thus, recalling that $r \leq |y_{M-1}| + \delta < |z_M| - \delta$,

$$B \geq \int_r^1 s^{n-1}(1_{S_M^+} f)\left(s\frac{x}{|x|}\right) ds + \int_r^1 s^{n-1}(1_{S_{M-1}^-} f)\left(s\frac{x}{|x|}\right) ds$$
$$\geq \int_0^1 s^{n-1}(1_{S_M^+} f)\left(s\frac{x}{|x|}\right) ds + \int_0^1 s^{n-1}(1_{S_{M-1}^-} f)\left(s\frac{x}{|x|}\right) ds,$$

which leads to $B > 0$, in view of (11.83).

Case 2: $r \in [|z_{i-1}| - \delta, |z_i| - \delta)$, $2 \leq i \leq M-1$. We thus deduce

$$B = \sum_{k=i-1}^{M} \int_r^1 s^{n-1}(1_{S_k^+} f)\left(s\frac{x}{|x|}\right) ds + \sum_{k=i}^{M} \int_r^1 s^{n-1}(1_{S_{k-1}^-} f)\left(s\frac{x}{|x|}\right) ds$$
$$\geq \sum_{k=i}^{M} \left\{ \int_r^1 s^{n-1}(1_{S_k^+} f)\left(s\frac{x}{|x|}\right) ds + \int_r^1 s^{n-1}(1_{S_{k-1}^-} f)\left(s\frac{x}{|x|}\right) ds \right\}$$
$$\geq \sum_{k=i}^{M} \left\{ \int_0^1 s^{n-1}(1_{S_k^+} f)\left(s\frac{x}{|x|}\right) ds + \int_0^1 s^{n-1}(1_{S_{k-1}^-} f)\left(s\frac{x}{|x|}\right) ds \right\}$$

and, using (11.83), we get that $B > 0$.

Case 3: $r \in [0, |z_1| - \delta)$. We therefore find

$$B = \sum_{k=2}^{M} \left\{ \int_0^1 s^{n-1}(1_{S_k^+} f)\left(s\frac{x}{|x|}\right) ds + \int_0^1 s^{n-1}(1_{S_{k-1}^-} f)\left(s\frac{x}{|x|}\right) ds \right\};$$

using once more (11.83), we get that $B > 0$. This concludes the proof of the lemma. □

Part IV
The Case $0 \leq k \leq n-1$

Chapter 12
General Considerations on the Flow Method

Let $T > 0$, $\Omega \subset \mathbb{R}^n$ be an open set and

$$g : [0,T] \times \overline{\Omega} \to \mathbb{R}^N.$$

Throughout the present chapter, when dealing with such maps, we write, depending on the context,

$$g = g(t,x) = g_t(x), \quad t \in [0,T], \, x \in \overline{\Omega}.$$

Moreover, unless specified otherwise, we write $\|g_t\|_{C^{r,\alpha}}$ instead of $\|g_t\|_{C^{r,\alpha}(\overline{\Omega})}$ when t is fixed.

On several occasions we will use the fact that for bounded Lipschitz sets (cf. Corollary 16.13), the $\|.\|_{C^{0,1}}$ and the $\|.\|_{C^1}$ norms are equivalent.

12.1 Basic Properties of the Flow

We start with a global result.

Theorem 12.1. *Let $r \geq 1$ be an integer, $0 \leq \alpha \leq 1$, $T > 0$ and $\Omega \subset \mathbb{R}^n$ be a bounded open Lipschitz set. Let $u \in C^{r,\alpha}((0,T] \times \overline{\Omega}; \mathbb{R}^n)$ be such that $u_t = 0$ on $\partial\Omega$ for every $t \in (0,T]$ and*

$$\int_0^T \|u_t\|_{C^{r,\alpha}} dt < \infty. \tag{12.1}$$

Then there exists a unique solution $\varphi \in C^{r,\alpha}([0,T] \times \overline{\Omega}; \overline{\Omega})$ of

$$\begin{cases} \dfrac{d}{dt}\varphi_t = u_t \circ \varphi_t, & 0 < t \leq T, \\ \varphi_0 = \mathrm{id}. \end{cases} \tag{12.2}$$

Furthermore, for every $t \in [0,T]$, $\varphi_t \in \operatorname{Diff}^{r,\alpha}(\overline{\Omega};\overline{\Omega})$, $\varphi_t = \operatorname{id}$ *on* $\partial\Omega$ *and*

$$\frac{d}{dt}\nabla^k \varphi_t = \nabla^k \frac{d}{dt}\varphi_t = \nabla^k (u_t \circ \varphi_t), \quad 1 \leq k \leq r.$$

Moreover, there exists a constant $C = C(r,\Omega)$ *such that*

$$\|\varphi_t - \operatorname{id}\|_{C^{r,\alpha}} \leq C \exp\left[C \int_0^t \|u_s\|_{C^1}\, ds\right] \int_0^t \|u_s\|_{C^{r,\alpha}}\, ds. \tag{12.3}$$

Finally, if $x \in \overline{\Omega}$ *is such that* $u_t(x) = 0$ *for every* $0 < t \leq T$, *then*

$$\varphi_t(x) = x \quad \text{for every } 0 \leq t \leq T.$$

Proof. We only show (12.3), the other properties being well known (see, e.g., [22]). We split the proof into two steps. In what follows we will always suppose $t \in [0,T]$ and $C_1, C_2, \ldots$ will denote generic constants depending only on r and Ω.

Step 1. We start by showing that

$$\|\varphi_t\|_{C^1} \leq C_1 \exp\left[C_1 \int_0^t \|u_s\|_{C^1}\, ds\right]. \tag{12.4}$$

First, integrating (12.2), we get for $x, y \in \overline{\Omega}$,

$$\begin{aligned}|\varphi_t(x) - \varphi_t(y)| &= \left| x - y + \int_0^t (u_s(\varphi_s(x)) - u_s(\varphi_s(y)))\, ds \right| \\ &\leq |x-y| + C_2 \int_0^t \|u_s\|_{C^1} |\varphi_s(x) - \varphi_s(y)|\, ds.\end{aligned}$$

Applying Lemma 12.3, we obtain

$$|\varphi_t(x) - \varphi_t(y)| \leq |x-y| \exp\left[C_2 \int_0^t \|u_s\|_{C^1}\, ds\right]$$

and thus

$$[\varphi_t]_{C^{0,1}} \leq \exp\left[C_2 \int_0^t \|u_s\|_{C^1}\, ds\right].$$

Combining the last equation with the fact that $\varphi_t(\overline{\Omega}) = \overline{\Omega}$, we immediately get that

$$\|\varphi_t\|_{C^1} \leq C_3 + \exp\left[C_2 \int_0^t \|u_s\|_{C^1}\, ds\right] \leq C_4 \exp\left[C_4 \int_0^t \|u_s\|_{C^1}\, ds\right],$$

which proves the claim.

Step 2 (conclusion). We now show (12.3). Integrating (12.2) and using Theorem 16.31, we get

$$\begin{aligned}
\|\varphi_t - \mathrm{id}\,\|_{C^{r,\alpha}} &= \left\| \int_0^t u_s \circ \varphi_s \, ds \right\|_{C^{r,\alpha}} \le \int_0^t \|u_s \circ \varphi_s\|_{C^{r,\alpha}} \, ds \\
&\le C_5 \int_0^t \left[\|u_s\|_{C^{r,\alpha}} \|\varphi_s\|_{C^1}^{r+\alpha} + \|u_s\|_{C^1} \|\varphi_s\|_{C^{r,\alpha}} + \|u_s\|_{C^0} \right] ds \\
&\le C_6 \int_0^t \|u_s\|_{C^{r,\alpha}} (1 + \|\varphi_s\|_{C^1}^{r+\alpha}) \, ds \\
&\quad + C_5 \int_0^t \|u_s\|_{C^1} \|\varphi_s - \mathrm{id}\,\|_{C^{r,\alpha}} \, ds
\end{aligned}$$

and hence, since (12.4) holds,

$$\begin{aligned}
\|\varphi_t - \mathrm{id}\,\|_{C^{r,\alpha}} \le C_7 \exp\left[C_7 \int_0^t \|u_s\|_{C^1} \, ds \right] \int_0^t \|u_s\|_{C^{r,\alpha}} \, ds \\
+ C_5 \int_0^t \|u_s\|_{C^1} \|\varphi_s - \mathrm{id}\,\|_{C^{r,\alpha}} \, ds.
\end{aligned}$$

Noticing that

$$C_7 \exp\left[C_7 \int_0^t \|u_s\|_{C^1} ds \right] \int_0^t \|u_s\|_{C^{r,\alpha}} \, ds$$

is increasing in t, we get, using Lemma 12.3,

$$\begin{aligned}
&\|\varphi_t - \mathrm{id}\,\|_{C^{r,\alpha}} \\
&\quad \le C_7 \exp\left[C_7 \int_0^t \|u_s\|_{C^1} \, ds \right] \int_0^t \|u_s\|_{C^{r,\alpha}} ds \cdot \exp\left[C_5 \int_0^t \|u_s\|_{C^1} \, ds \right],
\end{aligned}$$

which concludes the proof. □

We also have a local version of the above theorem.

Theorem 12.2. *Let $r \ge 1$ be an integer, $0 \le \alpha \le 1$, $T > 0$ and V be a neighborhood of $x_0 \in \mathbb{R}^n$. Let*

$$u \in C^{r,\alpha}((0,T] \times \overline{V}; \mathbb{R}^n)$$

be such that $u_t(x_0) = 0$ for every $t \in (0,T]$ and

$$\int_0^T \|u_t\|_{C^{r,\alpha}} \, dt < \infty. \tag{12.5}$$

Then there exist a neighborhood $U \subset V$ of x_0 and a unique solution $\varphi \in C^{r,\alpha}([0,T] \times U; \mathbb{R}^n)$ of

$$\begin{cases} \dfrac{d}{dt} \varphi_t = u_t \circ \varphi_t, & 0 < t \le T, \\ \varphi_0 = \mathrm{id}. \end{cases}$$

Moreover, for every $t \in [0,T]$, $\varphi_t \in \mathrm{Diff}^{r,\alpha}(U; \varphi_t(U))$ and verifies $\varphi_t(x_0) = x_0$.

We finally recall the classical Grönwall lemma.

Lemma 12.3 (Grönwall lemma). *Let $T>0$, $h:[0,T]\to[0,\infty)$ increasing, $f\in C^0([0,T];[0,\infty))$ and $g\in C^0((0,T];[0,\infty))\cap L^1(0,T)$ be such that*

$$f(t)\leq h(t)+\int_0^t g(s)f(s)\,ds\quad \text{for every } t\in[0,T].$$

Then

$$f(t)\leq h(t)\exp\left[\int_0^t g(s)\,ds\right]\quad \text{for every } t\in[0,T].$$

12.2 A Regularity Result

The next result is essentially in Rivière and Ye [85].

Theorem 12.4. *Let $r\geq 1$ and*

$$0<\delta<\alpha<\alpha+\delta<1.$$

Let $\Omega\subset\mathbb{R}^n$ be a bounded open Lipschitz set, $c,T>0$ and

$$u\in C^{r,\alpha+\delta}((0,T]\times\overline{\Omega};\mathbb{R}^n)$$

be such that for every $t\in(0,T]$, $u_t=0$ on $\partial\Omega$ and

$$\|u_t\|_{C^{r,\gamma}}\leq\frac{c}{t^{1+\gamma-\alpha}}\quad\text{for every } \gamma\in[\alpha-\delta,\alpha+\delta]. \tag{12.6}$$

Then, for every $\gamma<\alpha$, there exists a unique solution $\varphi\in C^{r,\gamma}([0,T]\times\overline{\Omega};\overline{\Omega})$ of

$$\begin{cases}\dfrac{d}{dt}\varphi_t=u_t\circ\varphi_t, & 0<t\leq T,\\ \varphi_0=\mathrm{id}.\end{cases} \tag{12.7}$$

Moreover, for every $0\leq t\leq T$, $\varphi_t\in\mathrm{Diff}^{r,\alpha}(\overline{\Omega};\overline{\Omega})$ and verifies $\varphi_t=\mathrm{id}$ on $\partial\Omega$.

Proof. All of the results, except the fact that $\varphi_t\in C^{r,\alpha}$, are easy consequences of Theorem 12.1. We split the proof into three steps. In the proof we will always assume that $t\in[0,T]$ and $C_1,C_2,\ldots$ will be generic constants depending only on c,r,α,δ,Ω and T.

Step 1. Using (12.6), we have for every $\gamma\in[\alpha-\delta,\alpha)$,

$$\int_0^T\|u_s\|_{C^{r,\gamma}}ds\leq\int_0^T\frac{c}{s^{1+\gamma-\alpha}}\,ds=\frac{c\,T^{\alpha-\gamma}}{\alpha-\gamma}<\infty. \tag{12.8}$$

Therefore, using Theorem 12.1, there exists $\varphi:[0,T]\times\overline{\Omega}\to\overline{\Omega}$, a unique solution of (12.7) such that for every $\gamma<\alpha$,

$$\varphi\in C^{r,\gamma}([0,T]\times\overline{\Omega};\overline{\Omega}),$$

with

$$\varphi_t \in \operatorname{Diff}^{r,\gamma}(\overline{\Omega};\overline{\Omega}), \quad \varphi_t = \operatorname{id} \text{ on } \partial\Omega,$$

and

$$\frac{d}{dt}\nabla^r \varphi_t = \nabla^r (u_t \circ \varphi_t), \tag{12.9}$$

$$\|\varphi_t - \operatorname{id}\|_{C^{r,\gamma}} \le C_1 \exp\left[C_1 \int_0^t \|u_s\|_{C^1}\, ds\right] \int_0^t \|u_s\|_{C^{r,\gamma}}\, ds. \tag{12.10}$$

Therefore, it only remains to prove that $\varphi_t \in C^{r,\alpha}(\overline{\Omega};\overline{\Omega})$ to establish the theorem. This will be done in Steps 2 and 3.

Step 2. If we show (cf. Step 3) that for every $\gamma \in [\alpha-\delta,\alpha)$,

$$\|\varphi_t - \operatorname{id}\|_{C^{r,\gamma}} \le C_2 + C_2 \int_0^t \|\nabla u_s\|_{C^0} \|\varphi_s - \operatorname{id}\|_{C^{r,\gamma}}\, ds, \tag{12.11}$$

then, using Lemma 12.3 and noticing that $\int_0^T \|\nabla u_s\|_{C^0} ds < \infty$ by (12.8), we deduce that

$$\|\varphi_t - \operatorname{id}\|_{C^{r,\gamma}} \le C_2 \exp\left[C_2 \int_0^t \|\nabla u_s\|_{C^0}\, ds\right] < \infty.$$

Hence, letting γ tend to α and recalling that C_2 is independent of γ, we obtain that $\|\varphi_t - \operatorname{id}\|_{C^{r,\alpha}} < \infty$, which concludes the theorem.

Step 3. We show (12.11). We start by noticing that (12.8) implies

$$\int_0^t \|u_s\|_{C^1}\, ds \le \int_0^t \|u_s\|_{C^{r,\alpha-\delta}}\, ds \le \frac{ct^\delta}{\delta}.$$

Thus, combining the previous inequality with (12.10) we deduce that

$$\|\varphi_t - \operatorname{id}\|_{C^{r,\alpha-\delta}} \le C_3 t^\delta. \tag{12.12}$$

Let $\gamma \in [\alpha-\delta,\alpha)$. Since

$$\|\varphi_t - \operatorname{id}\|_{C^r} \le C_3 T^\delta \quad \text{and} \quad [\nabla^r(\varphi_t - \operatorname{id})]_{C^{0,\gamma}} = [\nabla^r \varphi_t]_{C^{0,\gamma}},$$

to obtain (12.11), it is enough to prove that for every $x, y \in \overline{\Omega}$,

$$\begin{aligned}&|\nabla^r \varphi_t(x) - \nabla^r \varphi_t(y)| \\ &\quad \le |x-y|^\gamma C_4 \left(1 + \int_0^t \|\nabla u_s\|_{C^0} \|\varphi_s - \operatorname{id}\|_{C^{r,\gamma}} ds\right).\end{aligned} \tag{12.13}$$

Integrating (12.9) and using Lemma 16.33, we obtain

$$\begin{aligned}&|\nabla^r \varphi_t(x) - \nabla^r \varphi_t(y)| \\ &\quad = \left|\int_0^t [\nabla^r(u_s \circ \varphi_s)(x) - \nabla^r(u_s \circ \varphi_s)(y)]\, ds\right|\end{aligned}$$

$$\le C_5|x-y|^\gamma \int_0^t \|\nabla u_s\|_{C^0} \|\varphi_s - \mathrm{id}\|_{C^{r,\gamma}} ds$$
$$+ C_5|x-y|^\gamma \int_0^t \|u_s\|_{C^{r,\gamma}} \|\varphi_s - \mathrm{id}\|_{C^1} (1+\|\varphi_s\|_{C^1}^{r+\gamma})\, ds$$
$$+ C_5 \int_0^t |\nabla^r u_s(\varphi_s(x)) - \nabla^r u_s(\varphi_s(y))|\, ds.$$

We now estimate the three terms of the last equation separately and show that they are of the form of the right-hand side of (12.13). We start by noticing that the first term already has the desired form.

Estimate of the second term. Using (12.6) and (12.12), we obtain

$$\int_0^t \|u_s\|_{C^{r,\gamma}} \|\varphi_s - \mathrm{id}\|_{C^1} (1+\|\varphi_s\|_{C^1}^{r+\gamma})\, ds$$
$$\le C_6 \int_0^t \frac{c}{s^{1+\gamma-\alpha}} s^\delta ds = C_6\, c\, \frac{t^{\alpha+\delta-\gamma}}{\alpha+\delta-\gamma} \le C_7,$$

where in the last inequality, we have used that $\alpha+\delta-\gamma \in (\delta, 2\delta]$ since $\gamma \in [\alpha-\delta, \alpha)$.

Estimate of the third term. First, note that since $\gamma < \alpha$, it is sufficient to prove that

$$\int_0^t |\nabla^r u_s(\varphi_s(x)) - \nabla^r u_s(\varphi_s(y))|\, ds \le C_8|x-y|^\alpha.$$

Next, observe that the previous inequality will be verified if we show that

$$\int_0^{|x-y|} |\nabla^r u_s(\varphi_s(x)) - \nabla^r u_s(\varphi_s(y))|\, ds \le C_9|x-y|^\alpha$$

and that if $|x-y| < t$,

$$\int_{|x-y|}^t |\nabla^r u_s(\varphi_s(x)) - \nabla^r u_s(\varphi_s(y))|\, ds \le C_{10}|x-y|^\alpha.$$

Let us show these last two inequalities. Using (12.12), we obtain

$$\|\varphi_t\|_{C^{0,1}} \le C_{11}\,.$$

Appealing to (12.6), we get

$$\int_0^{|x-y|} |\nabla^r u_s(\varphi_s(x)) - \nabla^r u_s(\varphi_s(y))|\, ds$$
$$\le \int_0^{|x-y|} \|u_s\|_{C^{r,\alpha-\delta}} |\varphi_s(x) - \varphi_s(y)|^{\alpha-\delta} ds$$
$$\le C_{12}|x-y|^{\alpha-\delta} \int_0^{|x-y|} \frac{c}{s^{1-\delta}} ds = \frac{C_{12}\, c}{\delta} |x-y|^\alpha.$$

Similarly, if $|x-y|<t$,

$$\begin{aligned}
&\int_{|x-y|}^{t} |\nabla^r u_s(\varphi_s(x)) - \nabla^r u_s(\varphi_s(y))|\, ds \\
&\quad \le \int_{|x-y|}^{t} \|u_s\|_{C^{r,\alpha+\delta}} |\varphi_s(x) - \varphi_s(y)|^{\alpha+\delta}\, ds \\
&\quad \le C_{13}|x-y|^{\alpha+\delta} \int_{|x-y|}^{t} \frac{c}{s^{1+\delta}}\, ds \le C_{13}|x-y|^{\alpha+\delta} \int_{|x-y|}^{\infty} \frac{c}{s^{1+\delta}}\, ds \\
&\quad = \frac{C_{13}\, c}{\delta} |x-y|^{\alpha}
\end{aligned}$$

which ends the proof of the theorem. □

12.3 The Flow Method

We start by recalling a well-known result of differential geometry.

Theorem 12.5. *Let Ω_1 and $\Omega_2 \subset \mathbb{R}^n$ be open sets, $T>0$ and $0 \le k \le n$ be an integer. Let*

$$u \in C^1([0,T] \times \Omega_2; \mathbb{R}^n) \quad \text{and} \quad \varphi \in C^1([0,T] \times \Omega_1; \Omega_2)$$

be such that in Ω_1,

$$\frac{d}{dt}\varphi_t = u_t \circ \varphi_t \quad \text{for every } 0 \le t \le T. \tag{12.14}$$

Then for every $f \in C^1([0,T] \times \Omega_2; \Lambda^k)$, the following equality holds in Ω_1 and for $0 \le t \le T$:

$$\frac{d}{dt}[\varphi_t^*(f_t)] = \varphi_t^*\left(\frac{d}{dt} f_t + d(u_t \lrcorner f_t) + u_t \lrcorner (df_t)\right), \tag{12.15}$$

where u_t has been identified with a 1-form.

Remark 12.6. (i) Let $a \in C^1(U;\mathbb{R}^n)$ and $\omega \in C^1(U;\Lambda^k)$. The Lie derivative is defined as

$$\mathscr{L}_a \omega = \left.\frac{d}{dt}\right|_{t=0} \varphi_t^*(\omega),$$

where

$$\begin{cases} \dfrac{d}{dt}\varphi_t = a \circ \varphi_t, \\ \varphi_0 = \mathrm{id}. \end{cases}$$

The Cartan formula states that

$$\mathscr{L}_a \omega = a \lrcorner d\,\omega + d(a \lrcorner \omega).$$

The formula follows at once from Theorem 12.5.

(ii) Note that when $k = n$, then necessarily $d f_t = 0$ and (identifying as usual functions with n-forms and 1-forms with vector fields)

$$d(u_t \lrcorner f_t) = \operatorname{div}(f_t u_t),$$

therefore recovering Proposition 10.5.

Proof. Since both sides of (12.15) are linear in f_t, we can assume with no loss of generality that

$$f_t = a_t\, dx^{i_1} \wedge \cdots \wedge dx^{i_k}$$

for some $1 \le i_1 < \cdots < i_k \le n$ and some $a \in C^1([0,T] \times \Omega_2)$. We split the proof into three steps.

Step 1. We compute the right-hand side of (12.15). First,

$$\frac{d}{dt} f_t = \left[\frac{d}{dt} a_t\right] dx^{i_1} \wedge \cdots \wedge dx^{i_k}.$$

Since

$$u_t \lrcorner f_t = \sum_{l=1}^{k} (-1)^{l+1} a_t\, u_t^{i_l}\, dx^{i_1} \wedge \cdots \wedge dx^{i_{l-1}} \wedge dx^{i_{l+1}} \wedge \cdots \wedge dx^{i_k},$$

we deduce that

$$\begin{aligned} d(u_t \lrcorner f_t) &= \sum_{l=1}^{k} (-1)^{l+1} \sum_{j=1}^{n} \frac{\partial a_t}{\partial x_j} u_t^{i_l}\, dx^j \wedge dx^{i_1} \wedge \cdots \wedge dx^{i_{l-1}} \wedge dx^{i_{l+1}} \wedge \cdots \wedge dx^{i_k} \\ &\quad + \sum_{l=1}^{k} (-1)^{l+1} \sum_{j=1}^{n} a_t \frac{\partial u_t^{i_l}}{\partial x_j}\, dx^j \wedge dx^{i_1} \wedge \cdots \wedge dx^{i_{l-1}} \wedge dx^{i_{l+1}} \wedge \cdots \wedge dx^{i_k}. \end{aligned}$$

Next, from

$$d f_t = \sum_{j=1}^{n} \frac{\partial a_t}{\partial x_j}\, dx^j \wedge dx^{i_1} \wedge \cdots \wedge dx^{i_k},$$

we get

$$\begin{aligned} u_t \lrcorner d f_t &= \sum_{j=1}^{n} \frac{\partial a_t}{\partial x_j} u_t^j\, dx^{i_1} \wedge \cdots \wedge dx^{i_k} \\ &\quad + \sum_{j=1}^{n} \frac{\partial a_t}{\partial x_j} \sum_{l=1}^{k} (-1)^l u_t^{i_l}\, dx^j \wedge dx^{i_1} \wedge \cdots \wedge dx^{i_{l-1}} \wedge dx^{i_{l+1}} \wedge \cdots \wedge dx^{i_k}. \end{aligned}$$

Therefore, combining the previous three equations, it follows that

$$\begin{aligned}
&\frac{d}{dt}f_t + d(u_t \lrcorner f_t) + u_t \lrcorner df_t \\
&= \left[\frac{d}{dt}a_t\right] dx^{i_1} \wedge \cdots \wedge dx^{i_k} + \sum_{j=1}^{n} \frac{\partial a_t}{\partial x_j} u_t^j \, dx^{i_1} \wedge \cdots \wedge dx^{i_k} \\
&\quad + \sum_{l=1}^{k} \sum_{j=1}^{n} a_t \frac{\partial u_t^{i_l}}{\partial x_j} dx^{i_1} \wedge \cdots \wedge dx^{i_{l-1}} \wedge dx^j \wedge dx^{i_{l+1}} \wedge \cdots \wedge dx^{i_k}.
\end{aligned}$$

Step 2. We now compute the left-hand side of (12.15).

Step 2.1. Standard results on ordinary differential equations (cf. (7.13) in the proof of Theorem 7.2 in Chapter 1 of Coddington and Levinson [22]) give that $\nabla \varphi_t$ is differentiable in t and satisfies

$$\frac{d}{dt}[\nabla \varphi_t] = \nabla u_t\,(\varphi_t)\,\nabla \varphi_t\,.$$

This is indeed what is immediately obtained by formal differentiation of (12.14). In particular, for every $1 \leq i \leq n$, we have

$$\frac{d}{dt}[d\varphi_t^i] = \sum_{m=1}^{n} \left[\sum_{j=1}^{n} \frac{\partial u_t^i}{\partial x_j}(\varphi_t) \frac{\partial \varphi_t^j}{\partial x_m}\right] dx^m = \sum_{j=1}^{n} \frac{\partial u_t^i}{\partial x_j}(\varphi_t)\, d\varphi_t^j\,.$$

Step 2.2. Using Step 2.1 we can differentiate $\varphi_t^*(f_t)$ with respect to t. Since

$$\varphi_t^*(f_t) = a_t\,(\varphi_t)\, d\varphi_t^{i_1} \wedge \cdots \wedge d\varphi_t^{i_k}\,,$$

we deduce that

$$\begin{aligned}
\frac{d}{dt}[\varphi_t^*(f_t)] &= \left[\frac{d}{dt}a_t\right](\varphi_t)\, d\varphi_t^{i_1} \wedge \cdots \wedge d\varphi_t^{i_k} \\
&\quad + \sum_{j=1}^{n} \left[\frac{\partial a_t}{\partial x_j}(\varphi_t)\right]\left[\frac{d}{dt}\varphi_t^j\right] d\varphi_t^{i_1} \wedge \cdots \wedge d\varphi_t^{i_k} \\
&\quad + \sum_{l=1}^{k} a_t(\varphi_t)\, d\varphi_t^{i_1} \wedge \cdots \wedge d\varphi_t^{i_{l-1}} \wedge \frac{d}{dt}\left(d\varphi_t^{i_l}\right) \wedge d\varphi_t^{i_{l+1}} \wedge \cdots \wedge d\varphi_t^{i_k}.
\end{aligned}$$

Using (12.14) and Step 2.1, we have

$$\begin{aligned}
\frac{d}{dt}[\varphi_t^*(f_t)] &= \left[\frac{d}{dt}a_t\right](\varphi_t)\, d\varphi_t^{i_1} \wedge \cdots \wedge d\varphi_t^{i_k} \\
&\quad + \sum_{j=1}^{n} \left[\frac{\partial a_t}{\partial x_j}(\varphi_t)\right]\left[u_t^j(\varphi_t)\right] d\varphi_t^{i_1} \wedge \cdots \wedge d\varphi_t^{i_k} \\
&\quad + \sum_{l=1}^{k} \sum_{j=1}^{n} a_t(\varphi_t) \frac{\partial u_t^{i_l}}{\partial x_j}(\varphi_t)\, d\varphi_t^{i_1} \wedge \cdots \wedge d\varphi_t^{i_{l-1}} \wedge d\varphi_t^j \wedge d\varphi_t^{i_{l+1}} \wedge \cdots \wedge d\varphi_t^{i_k}.
\end{aligned}$$

Step 3 (conclusion). Since we trivially have

$$\varphi_t^*\left(\left[\frac{d}{dt}a_t\right]dx^{i_1}\wedge\cdots\wedge dx^{i_k}\right)=\left[\frac{d}{dt}a_t(\varphi_t)\right]d\varphi_t^{i_1}\wedge\cdots\wedge d\varphi_t^{i_k},$$

$$\varphi_t^*\left(\sum_{j=1}^{n}\frac{\partial a_t}{\partial x_j}u_t^j dx^{i_1}\wedge\cdots\wedge dx^{i_k}\right)=\sum_{j=1}^{n}\left[\frac{\partial a_t}{\partial x_j}(\varphi_t)\right]\left[u_t^j(\varphi_t)\right]d\varphi_t^{i_1}\wedge\cdots\wedge d\varphi_t^{i_k},$$

$$\varphi_t^*\left(\sum_{l=1}^{k}\sum_{j=1}^{n}a_t\frac{\partial u_t^{i_l}}{\partial x_j}dx^{i_1}\wedge\cdots\wedge dx^{i_{l-1}}\wedge dx^j\wedge dx^{i_{l+1}}\wedge\cdots\wedge dx^{i_k}\right)$$
$$=\sum_{l=1}^{k}\sum_{j=1}^{n}a_t(\varphi_t)\frac{\partial u_t^{i_l}}{\partial x_j}(\varphi_t)\,d\varphi_t^{i_1}\wedge\cdots\wedge d\varphi_t^{i_{l-1}}\wedge d\varphi_t^j\wedge d\varphi_t^{i_{l+1}}\wedge\cdots\wedge d\varphi_t^{i_k},$$

we have indeed shown the theorem in view of Steps 1 and 2. □

As a consequence, we have the following result essentially established by Moser [78].

Theorem 12.7. *Let $r\geq 1$ and $0\leq k\leq n$ be integers, $0\leq\alpha\leq 1$, $T>0$ and $\Omega\subset\mathbb{R}^n$ be a bounded open Lipschitz set. Let*

$$u\in C^{r,\alpha}\left([0,T]\times\overline{\Omega};\mathbb{R}^n\right)\quad\textit{and}\quad f\in C^{r,\alpha}\left([0,T]\times\overline{\Omega};\Lambda^k\right)$$

be such that for every $t\in[0,T]$,

$$u_t=0\quad\textit{on }\partial\Omega,\quad df_t=0\quad\textit{in }\Omega,$$

$$d(u_t\lrcorner f_t)=-\frac{d}{dt}f_t\quad\textit{in }\Omega.$$

Then for every $t\in[0,T]$, the solution φ_t of

$$\begin{cases}\dfrac{d}{dt}\varphi_t=u_t\circ\varphi_t, & 0\leq t\leq T,\\ \varphi_0=\mathrm{id}\end{cases}\tag{12.16}$$

belongs to $\mathrm{Diff}^{r,\alpha}\left(\overline{\Omega};\overline{\Omega}\right)$, *satisfies* $\varphi_t=\mathrm{id}$ *on* $\partial\Omega$ *and*

$$\varphi_t^*(f_t)=f_0\quad\textit{in }\Omega.$$

Proof. We split the proof into two steps.

Step 1. Using Theorem 12.1, for every $0\leq t\leq T$, the solution φ_t of (12.16) belongs to $\mathrm{Diff}^{r,\alpha}(\overline{\Omega};\overline{\Omega})$ and verifies $\varphi_t=\mathrm{id}$ on $\partial\Omega$. Moreover, defining $\varphi:[0,T]\times\overline{\Omega}\to\overline{\Omega}$ by $\varphi(t,x)=\varphi_t(x)$, we have

$$\varphi\in C^{r,\alpha}([0,T]\times\overline{\Omega};\overline{\Omega}).$$

Step 2. Using Theorem 12.5 and the hypotheses on u_t and f_t, we find that in Ω,

$$\frac{d}{dt}[\varphi_t^*(f_t)] = \varphi_t^*\left(\frac{d}{dt}f_t + d(u_t \lrcorner f_t) + u_t \lrcorner (df_t)\right) = 0,$$

which implies the result since $\varphi_0 = \mathrm{id}$. □

We also have the local version of the above theorem.

Theorem 12.8. *Let $r \geq 1$ and $0 \leq k \leq n$ be integers, $0 \leq \alpha \leq 1$, $T > 0$ and $x_0 \in \mathbb{R}^n$. Let*

$$u \in C^{r,\alpha}([0,T] \times \mathbb{R}^n; \mathbb{R}^n) \quad \textit{and} \quad f \in C^{r,\alpha}([0,T] \times \mathbb{R}^n; \Lambda^k)$$

be such that for every $t \in [0,T]$, $u_t(x_0) = 0$ and

$$df_t = 0 \quad \textit{and} \quad d(u_t \lrcorner f_t) = -\frac{d}{dt}f_t \quad \textit{in a neighborhood of } x_0.$$

Then there exists a neighborhood U of x_0 in which for every $t \in [0,T]$, the solution φ_t of

$$\begin{cases} \dfrac{d}{dt}\varphi_t = u_t \circ \varphi_t, & 0 \leq t \leq T, \\ \varphi_0 = \mathrm{id} \end{cases} \tag{12.17}$$

belongs to $\mathrm{Diff}^{r,\alpha}(U; \varphi_t(U))$, *satisfies $\varphi_t(x_0) = x_0$ and*

$$\varphi_t^*(f_t) = f_0 \quad \textit{in } U.$$

Proof. We split the proof in two steps.

Step 1. Using Theorem 12.2, there exists a neighborhood U of x_0 in which for every $0 \leq t \leq T$, the solution φ_t of (12.17) belongs to $\mathrm{Diff}^{r,\alpha}(U; \varphi_t(U))$ and $\varphi_t(x_0) = x_0$. Moreover, defining $\varphi : [0,T] \times U \to \mathbb{R}^n$ by $\varphi(t,x) = \varphi_t(x)$, then

$$\varphi \in C^{r,\alpha}([0,T] \times U; \mathbb{R}^n).$$

Step 2. Since $\varphi_t(x_0) = x_0$ for every $t \in [0,T]$ and since φ is C^0, we can, choosing if necessary U smaller, assume that for every $t \in [0,T]$,

$$df_t = 0 \quad \text{and} \quad d(u_t \lrcorner f_t) = -\frac{d}{dt}f_t \quad \text{in } \varphi_t(U).$$

Using Theorem 12.5 and the hypotheses on u_t and f_t, we know that in U,

$$\frac{d}{dt}[\varphi_t^*(f_t)] = \varphi_t^*\left(\frac{d}{dt}f_t + d(u_t \lrcorner f_t) + u_t \lrcorner (df_t)\right) = 0,$$

which implies the result since $\varphi_0 = \mathrm{id}$. □

Chapter 13
The Cases $k = 0$ and $k = 1$

13.1 The Case of 0-Forms and of Closed 1-Forms

13.1.1 The Case of 0-Forms

We start with 0-forms. We begin our study with a local existence theorem.

Theorem 13.1. *Let $r \geq 1$ be an integer, $x_0 \in \mathbb{R}^n$ and f and g be C^r functions in a neighborhood of x_0 such that $f(x_0) = g(x_0)$,*

$$\nabla f(x_0) \neq 0 \quad \text{and} \quad \nabla g(x_0) \neq 0.$$

Then there exist a neighborhood U of x_0 and $\varphi \in \operatorname{Diff}^r(U; \varphi(U))$ such that $\varphi(x_0) = x_0$ and

$$\varphi^*(g)(x) = g(\varphi(x)) = f(x).$$

Furthermore, if

$$\frac{\partial f}{\partial x_i}(x_0) \cdot \frac{\partial g}{\partial x_i}(x_0) \neq 0$$

for a certain $1 \leq i \leq n$, then φ can be chosen of the form

$$\varphi(x) = (x_1, \ldots, x_{i-1}, \varphi^i(x), x_{i+1}, \ldots, x_n).$$

Proof. Without loss of generality we may assume that $x_0 = 0$. We split the proof into two steps.

Step 1. We prove the main statement. Since $\nabla f(0) \neq 0$ and $\nabla g(0) \neq 0$, we can find

$$A_2, \ldots, A_n, B_2, \ldots, B_n \in \mathbb{R}^n$$

such that letting

$$F(x) = (f(x), \langle A_2; x\rangle, \ldots, \langle A_n; x\rangle) \quad \text{and} \quad G(x) = (g(x), \langle B_2; x\rangle, \ldots, \langle B_n; x\rangle),$$

then

$$\det \nabla F(0) \neq 0 \quad \text{and} \quad \det \nabla G(0) \neq 0.$$

Hence, since $F(0) = G(0)$, we deduce that

$$F \in \mathrm{Diff}^r(U; F(U)), \quad G \in \mathrm{Diff}^r(U; G(U)) \quad \text{and} \quad G^{-1} \circ F \in \mathrm{Diff}^r(U; (G^{-1} \circ F)(U))$$

for a neighborhood U of 0 small enough. Therefore, $\varphi = G^{-1} \circ F$ has all of the desired properties.

Step 2. We now prove the extra property. Define

$$F(x) = (x_1, \ldots, x_{i-1}, f(x), x_{i+1}, \ldots, x_n),$$

$$G(x) = (x_1, \ldots, x_{i-1}, g(x), x_{i+1}, \ldots, x_n)$$

and note that $\varphi = G^{-1} \circ F$ has all of the required properties. The proof is therefore complete. □

We now have the following global result.

Theorem 13.2. *Let $\Omega \subset \mathbb{R}^n$ be a bounded open Lipschitz set. Let $r \geq 1$ be an integer and f and $g \in C^r(\overline{\Omega})$ with $f = g$ on $\partial\Omega$ and*

$$\frac{\partial f}{\partial x_i} \cdot \frac{\partial g}{\partial x_i} > 0 \text{ in } \overline{\Omega} \tag{13.1}$$

for a certain $1 \leq i \leq n$. Then there exists a diffeomorphism $\varphi \in \mathrm{Diff}^r(\overline{\Omega}; \overline{\Omega})$ of the form

$$\varphi(x) = (x_1, \ldots, x_{i-1}, \varphi^i(x), x_{i+1}, \ldots, x_n)$$

satisfying

$$\begin{cases} \varphi^*(g) = g \circ \varphi = f & \text{in } \Omega, \\ \varphi = \mathrm{id} & \text{on } \partial\Omega. \end{cases}$$

Proof. Let e_i be the ith vector of the Euclidean basis of $\mathbb{R}^n$. We will find φ of the form $\varphi(x) = x + u(x)e_i$, where $u : \overline{\Omega} \to \mathbb{R}$. Since Ω is Lipschitz, we can extend (according to Theorem 16.11) f and g to $C^r(\mathbb{R}^n)$ functions. We therefore also have

$$\frac{\partial f}{\partial x_i} \cdot \frac{\partial g}{\partial x_i} > 0 \quad \text{in a neighborhood of } \overline{\Omega}. \tag{13.2}$$

By compactness, for every $x \in \Omega$, there exist $s_x, t_x \in \mathbb{R}$ with $s_x < 0 < t_x$ such that

$$x + s_x e_i, x + t_x e_i \in \partial\Omega \quad \text{and} \quad (x + s_x e_i, x + t_x e_i) \subset \Omega.$$

Define $h : \mathbb{R}^n \times \mathbb{R} \to \mathbb{R}$ by

$$h(x, v) = g(x + v e_i) - f(x).$$

We claim that there exists $u \in C^r(\overline{\Omega})$ such that

$$h(x,u(x)) = 0, \text{ for } x \in \overline{\Omega}, \quad u = 0 \text{ on } \partial\Omega \quad \text{and} \quad 1 + \frac{\partial u}{\partial x_i}(x) > 0 \text{ for } x \in \overline{\Omega}.$$

(i) For every $x \in \partial\Omega$, let $u(x) = 0$ and note that $h(x,u(x)) = h(x,0) = g(x) - f(x) = 0$ since $f = g$ on $\partial\Omega$.

(ii) Let $x \in \Omega$. Since $f = g$ on $\partial\Omega$, we have

$$h(x,s_x) = f(x + s_x e_i) - f(x) \quad \text{and} \quad h(x,t_x) = f(x + t_x e_i) - f(x).$$

Hence, recalling that $\partial f(x)/\partial x_i \neq 0$ for every $x \in \overline{\Omega}$, we get

$$h(x,s_x) \cdot h(x,t_x) < 0.$$

Note that $v \to h(x,v)$ is monotone. Therefore, there exists a unique $u(x) \in (s_x, t_x)$ verifying $h(x,u(x)) = 0$.

(iii) Using the implicit function theorem and (13.2), we immediately deduce that $u \in C^r(\overline{\Omega})$ and that

$$1 + \frac{\partial u}{\partial x_i}(x) = \frac{\partial f}{\partial x_i}(x) \left(\frac{\partial g}{\partial x_i}(x + u(x)) \right)^{-1} > 0 \quad \text{for every } x \in \overline{\Omega}.$$

This proves the claim. Finally, letting $\varphi(x) = x + u(x)e_i$, we get that $g \circ \varphi = f$ in $\overline{\Omega}$, $\varphi \in C^r(\overline{\Omega};\mathbb{R}^n)$, $\det \nabla \varphi > 0$ in $\overline{\Omega}$ and $\varphi = \text{id}$ on $\partial\Omega$. Hence, using Theorem 19.12, we have $\varphi \in \text{Diff}^r(\overline{\Omega};\overline{\Omega})$, which concludes the proof. □

13.1.2 The Case of Closed 1-Forms

We get as immediate corollaries similar results for closed 1-forms. Recall that 1-forms are written as

$$f = \sum_{i=1}^{n} f_i dx^i \quad \text{and} \quad g = \sum_{i=1}^{n} g_i dx^i.$$

We start first with the local version.

Corollary 13.3. *Let $r \geq 0$ be an integer, $x_0 \in \mathbb{R}^n$ and f and g be C^r closed 1-forms in a neighborhood of x_0 such that*

$$f(x_0) \neq 0 \quad \textit{and} \quad g(x_0) \neq 0.$$

Then there exist a neighborhood U of x_0 and $\varphi \in \text{Diff}^{r+1}(U;\varphi(U))$ such that $\varphi(x_0) = x_0$ and

$$\varphi^*(g) = f \quad \textit{in } U.$$

Furthermore, if

$$f_i(x_0)\cdot g_i(x_0)\neq 0$$

for a certain $1\leq i\leq n$, *then* φ *can be chosen of the form*

$$\varphi(x)=(x_1,\ldots,x_{i-1},\varphi^i(x),x_{i+1},\ldots,x_n).$$

Remark 13.4. When $r=0$, the fact that a 1-form ω is closed has to be understood in the sense of distributions.

Proof. Using Corollary 8.6, there exist a small ball V centered at x_0 and $F,G\in C^{r+1}(V)$ such that

$$dF=f \quad \text{and} \quad dG=g \quad \text{in } V.$$

Adding, if necessary, a constant, we can also assume that $F(x_0)=G(x_0)$. Note that if $f_i(x_0)\cdot g_i(x_0)\neq 0$ for a certain $1\leq i\leq n$, then

$$\frac{\partial F}{\partial x_i}(x_0)\cdot\frac{\partial G}{\partial x_i}(x_0)\neq 0.$$

We are then in a position to apply Theorem 13.1 to get $U\subset V$, a neighborhood of x_0 and $\varphi\in\operatorname{Diff}^{r+1}(U;\varphi(U))$ such that $\varphi(x_0)=x_0$ and

$$\varphi^*(G)=F,$$

which implies

$$\varphi^*(dG)=dF$$

and concludes the proof. □

We now conclude with the global version obtained in Bandyopadhyay and Dacorogna [8].

Corollary 13.5. *Let* $\Omega\subset\mathbb{R}^n$ *be a bounded simply connected smooth open set. Let* $r\geq 0$ *be an integer and* $f,g\in C^r\left(\overline{\Omega};\Lambda^1\right)$ *be closed and such that*

$$\nu\wedge f=\nu\wedge g \text{ on } \partial\Omega \quad \text{and} \quad f_i\cdot g_i>0 \text{ in } \overline{\Omega}$$

for a certain $1\leq i\leq n$. *Then there exists a diffeomorphism* $\varphi\in\operatorname{Diff}^{r+1}\left(\overline{\Omega};\overline{\Omega}\right)$ *of the form*

$$\varphi(x)=(x_1,\ldots,x_{i-1},\varphi^i(x),x_{i+1},\ldots,x_n)$$

satisfying

$$\begin{cases}\varphi^*(g)=f & \text{in } \Omega,\\ \varphi=\operatorname{id} & \text{on } \partial\Omega.\end{cases}$$

Proof. We first claim that there exist $F,G\in C^{r+1}(\overline{\Omega})$ such that $F,G\in C^{r+1}\left(\overline{\Omega}\right)$ and

$$dF=f,\ dG=g \text{ in } \Omega \quad \text{and} \quad F=G \text{ on } \partial\Omega.$$

Indeed, by Theorem 8.16 and the remark following it and recalling that $\mathscr{H}_T(\Omega,\Lambda^1) = \{0\}$ since Ω is simply connected (see Remark 6.6), there exists $H \in C^{r+1}(\overline{\Omega};\Lambda^1)$ such that

$$dH = f - g \text{ in } \Omega \quad \text{and} \quad H = 0 \text{ on } \partial\Omega.$$

Then, using Corollary 8.6, there exists $G \in C^{r+1}(\overline{\Omega})$ such that $dG = g$ in Ω. Letting $F = H + G$, we have the claim. In particular, note that

$$\frac{\partial F}{\partial x_i} \cdot \frac{\partial G}{\partial x_i} > 0 \text{ in } \overline{\Omega}.$$

Finally, apply Theorem 13.2 to get $\varphi \in \operatorname{Diff}^{r+1}(\overline{\Omega};\overline{\Omega})$ of the desired form so that

$$\varphi^*(G) = F \quad \text{in } \Omega,$$

which implies

$$\varphi^*(dG) = dF \quad \text{in } \Omega.$$

The proof is therefore complete. □

13.2 Darboux Theorem for 1-Forms

13.2.1 Main Results

The following result is classical and due to Darboux [34]; see, for example Bryant et al. [18], Olver [80], or Sternberg [93]. This result is equivalent to the Darboux theorem (cf. the remark below) for closed 2-forms.

Theorem 13.6. *Let $r \geq 3$ and $2 \leq 2m \leq n$ be integers. Let $0 < \alpha < 1$, $x_0 \in \mathbb{R}^n$ and w be a $C^{r,\alpha}$ 1-form such that*

$$\operatorname{rank}[dw] = 2m \quad \textit{in a neighborhood of } x_0 .$$

Then there exist a neighborhood U of x_0 and

$$\varphi \in \begin{cases} \operatorname{Diff}^{r,\alpha}(U;\varphi(U)) & \text{if } 2m = n \\ \operatorname{Diff}^{r-1,\alpha}(U;\varphi(U)) & \text{if } 2m < n \end{cases}$$

such that $\varphi(x_0) = x_0$ and

$$\varphi^*(w) = \sum_{i=1}^{m} x_{2i-1} dx^{2i} + dS \quad \textit{in } U,$$

with

$$S \in \begin{cases} C^{r,\alpha}(U) & \text{if } 2m = n \\ C^{r-1,\alpha}(U) & \text{if } 2m < n. \end{cases}$$

Remark 13.7. (i) The above result is equivalent to the Darboux theorem for closed 2-forms. This last theorem reads (see Theorems 14.1 and 14.3) as follows. Let $n \geq 2m$, $x_0 \in \mathbb{R}^n$ and f be a $C^{r,\alpha}$ closed 2-form satisfying

$$\operatorname{rank}[f] = 2m \quad \text{in a neighborhood of } x_0 .$$

Then there exist a neighborhood U of x_0 and

$$\varphi \in \begin{cases} \operatorname{Diff}^{r+1,\alpha}(U;\varphi(U)) & \text{if } n = 2m \\ \operatorname{Diff}^{r,\alpha}(U;\varphi(U)) & \text{if } n > 2m \end{cases}$$

such that $\varphi(x_0) = x_0$ and

$$\varphi^*(f) = \sum_{i=1}^{m} dx^{2i-1} \wedge dx^{2i} \quad \text{in } U.$$

The fact that the Darboux theorem for 2-forms implies the one for 1-forms is precisely the proof of Theorem 13.6 below. The other implication follows immediately, once it has been observed that we can choose, for example, U to be a ball so that, f being closed in U, we can find (cf. Theorem 8.3) $w \in C^{r+1,\alpha}(U;\Lambda^1)$ such that $f = dw$. We then apply the theorem to w, getting

$$\varphi^*(f) = \varphi^*(dw) = d\varphi^*(w) = \sum_{i=1}^{m} dx^{2i-1} \wedge dx^{2i}.$$

(ii) The hypothesis $r \geq 3$ can be weakened if we use a weak version of the fourth statement of Theorem 3.10. Indeed, it is enough to assume $r \geq 1$ if $n = 2m$ and $r \geq 2$ if $n > 2m$ (cf. Csató [23]).

Proof. Using Theorem 14.1 if $2m = n$ or Theorem 14.3 if $2m < n$, there exist a neighborhood U of x_0 and

$$\varphi \in \begin{cases} \operatorname{Diff}^{r,\alpha}(U;\varphi(U)) & \text{if } 2m = n \\ \operatorname{Diff}^{r-1,\alpha}(U;\varphi(U)) & \text{if } 2m < n \end{cases}$$

such that $\varphi(x_0) = x_0$ and

$$\varphi^*(dw) = \sum_{i=1}^{m} dx^{2i-1} \wedge dx^{2i} \quad \text{in } U.$$

Note that

$$d\left[\varphi^*(w) - \sum_{i=1}^{m} x_{2i-1} dx^{2i}\right] = 0 \quad \text{in } U$$

and

$$\varphi^*(w) - \sum_{i=1}^{m} x_{2i-1} dx^{2i} \in \begin{cases} C^{r-1,\alpha}(U;\Lambda^1) & \text{if } 2m = n \\ C^{r-2,\alpha}(U;\Lambda^1) & \text{if } 2m < n. \end{cases}$$

Thus, by Theorem 8.3, restricting U if necessary, there exists

$$S \in \begin{cases} C^{r,\alpha}(U) & \text{if } 2m = n \\ C^{r-1,\alpha}(U) & \text{if } 2m < n \end{cases}$$

such that

$$dS = \varphi^*(w) - \sum_{i=1}^{m} x_{2i-1} dx^{2i},$$

which concludes the proof. □

The next two theorems refine the above result (cf. Bryant et al. [18] or Sternberg [93]). In particular, the second one gives a sufficient condition ensuring that $S = 0$.

Theorem 13.8. *Let $r \geq 3$ and $2 \leq 2m \leq n$ be integers. Let $0 < \alpha < 1$, $x_0 \in \mathbb{R}^n$ and w be a $C^{r,\alpha}$ 1-form such that*

$$\operatorname{rank}[dw] = 2m \quad \textit{in a neighborhood of } x_0$$

and

$$w \wedge \underbrace{dw \wedge \cdots \wedge dw}_{m \text{ times}}(x_0) \neq 0.$$

Then there exist a neighborhood U of x_0 and

$$\varphi \in \operatorname{Diff}^{r-1,\alpha}(U;\varphi(U))$$

such that $\varphi(x_0) = x_0$ and

$$\varphi^*(w) = \sum_{i=1}^{m} x_{2i-1}\, dx^{2i} + dx^{2m+1} \quad \textit{in } U.$$

Remark 13.9. Since $w \wedge (dw)^m$ is a $(2m+1)$-form and since

$$w \wedge (dw)^m(x_0) \neq 0,$$

we necessarily have $2m < n$.

Proof. With no loss of generality, we can assume $x_0 = 0$. Since (according to Remark 13.9) we necessarily have $2m < n$, we get, using Theorem 13.6, a neighborhood V of 0 and $\varphi_1 \in \operatorname{Diff}^{r-1,\alpha}(V;\varphi_1(V))$ such that $\varphi_1(0) = 0$ and

$$\varphi_1^*(w) = \sum_{i=1}^{m} x_{2i-1}\, dx^{2i} + dS \quad \text{in } V,$$

with $S \in C^{r-1,\alpha}(V)$. Since, by hypothesis,

$$w \wedge (dw)^m(0) \neq 0,$$

we get that since $\varphi_1(0) = 0$,

$$\varphi_1^*(w) \wedge (d\varphi_1^*(w))^m(0) \neq 0,$$

which is equivalent to

$$dS \wedge dx^1 \wedge \cdots \wedge dx^{2m}(0) \neq 0.$$

Permuting, if necessary, the coordinates $x_{2m+1}, \ldots, x_n$, we can therefore assume with no loss of generality that

$$\frac{\partial S}{\partial x_{2m+1}}(0) \neq 0.$$

Now, define, for $x \in V$,

$$\varphi_2(x) = (x_1, \ldots, x_{2m}, S(x) - S(0), x_{2m+2}, \ldots, x_n).$$

Taking V smaller, if necessary, we obtain that $\varphi_2 \in \mathrm{Diff}^{r-1,\alpha}(V; \varphi_2(V))$, $\varphi_2(0) = 0$ and

$$\varphi_2^*(dx^i) = \begin{cases} dx^i & \text{if } i \neq 2m+1 \\ dS & \text{if } i = 2m+1. \end{cases}$$

Finally, letting $U = \varphi_2(V)$ and $\varphi = \varphi_1 \circ (\varphi_2)^{-1}$, we easily obtain that $\varphi \in \mathrm{Diff}^{r-1,\alpha}(U; \varphi(U))$, $\varphi(0) = 0$ and

$$\varphi^*(w) = \sum_{i=1}^m x_{2i-1} dx^{2i} + dx^{2m+1} \quad \text{in } U,$$

which ends the proof. □

Theorem 13.10. *Let $2 \leq 2m \leq n$ be an integer, $x_0 \in \mathbb{R}^n$ and w a C^∞ 1-form such that*

$$\operatorname{rank}[dw] = 2m \quad \textit{in a neighborhood of } x_0,$$

$w(x_0) \neq 0$ and

$$w \wedge \underbrace{dw \wedge \cdots \wedge dw}_{m \text{ times}} = 0 \quad \textit{in a neighborhood of } x_0.$$

Then there exist an open set U and

$$\varphi \in \mathrm{Diff}^\infty(U; \varphi(U))$$

such that $\varphi(U)$ is a neighborhood of x_0 and

$$\varphi^*(w) = \sum_{i=1}^m x_{2i-1}\, dx^{2i} \quad \textit{in } U.$$

Remark 13.11. (i) If $w \in C^r$, the following proof shows in fact that $\varphi \in C^{r-2m+1}$ if $2m = n$ and $\varphi \in C^{r-2m}$ if $2m < n$.

(ii) If we, moreover, want $\varphi(x_0) = x_0$, then the conclusion becomes

$$\varphi^*(w) = \sum_{i=1}^{m} (x_{2i-1} - c_{2i-1})\, dx^{2i} \quad \text{in } U$$

for some $c_{2i-1} \in \mathbb{R}$, $1 \le i \le m$. Note that the c_{2i-1} cannot be arbitrary. For example, the c_{2i-1} can never verify $c_{2i-1} = (x_0)_{2i-1}$ for every $1 \le i \le m$. Indeed,

$$\varphi^*(w)(x_0) = \sum_{i=1}^{m} ((x_0)_{2i-1} - c_{2i-1})\, dx^{2i}$$

and thus we have the assertion since, recalling that $\varphi(x_0) = x_0$,

$$\varphi^*(w)(x_0) \neq 0 \Leftrightarrow w(x_0) \neq 0.$$

Proof. We split the proof into two steps. With no loss of generality, we can assume that $x_0 = 0$.

Step 1 (simplification). Let us first prove that we can assume that $n = 2m$. Applying Theorem 14.3 to dw, we can find a neighborhood U of 0 and $\psi \in \mathrm{Diff}^\infty(U; \psi(U))$ such that $\psi(0) = 0$ and

$$\psi^*(dw) = \sum_{i=1}^{m} dx^{2i-1} \wedge dx^{2i} \quad \text{in } U. \tag{13.3}$$

Note that since $\psi(0) = 0$, we have, restricting U if necessary,

$$\psi^*(w)(0) \neq 0 \quad \text{and} \quad \psi^*(w) \wedge (d(\psi^*(w)))^m = 0 \quad \text{in } U. \tag{13.4}$$

The last equation being equivalent to

$$\psi^*(w) \wedge dx^1 \wedge \cdots \wedge dx^{2m} = 0 \quad \text{in } U,$$

we immediately deduce

$$[\psi^*(w)]_i = 0 \quad \text{in } U \quad \text{for every } 2m+1 \le i \le n$$

and, hence,

$$\psi^*(w)(x) = \sum_{i=1}^{2m} b_i(x)\, dx^i \quad \text{in } U.$$

Combining the previous equation with (13.3), we get that

$$b_i(x) = b_i(x_1, \ldots, x_{2m}) \quad \text{in a neighborhood of } 0, \quad \text{for every } 1 \le i \le 2m.$$

We thus have the claim, replacing $\psi^*(w)$ by w.

Step 2 (conclusion). Applying Lemma 13.12 to w, we can find a neighborhood V of 0 in $\mathbb{R}^{2m}$ and $\varphi_1 \in \mathrm{Diff}^\infty(V;\varphi_1(V))$ such that $\varphi_1(0)=0$ and

$$\varphi_1^*(w) = \sum_{i=1}^{m} b_{2i-1}(x_1,\ldots,x_{2m})\,dx^{2i-1} \quad \text{in } V \tag{13.5}$$

for some $b_{2i-1} \in C^\infty(V)$. Since

$$\operatorname{rank}[d(\varphi_1^*(w))(0)] = \operatorname{rank}[dw(0)] = 2m,$$

we know that $(d\varphi_1^*(w))^m(0) \neq 0$, which is equivalent to, using (13.5),

$$dx^1 \wedge dx^3 \wedge \cdots \wedge dx^{2m-1} \wedge db_1 \wedge db_3 \wedge \cdots \wedge db_{2m-1}(0) \neq 0. \tag{13.6}$$

Now, define, for $x \in V$,

$$\varphi_2(x) = (b_1(x), x_1, b_3(x), x_3, \ldots, b_{2m-1}(x), x_{2m-1}).$$

Using (13.6), we obtain that $\varphi_2 \in \mathrm{Diff}^\infty(V;\varphi_2(V))$, taking V smaller if necessary. Finally, letting $U=\varphi_2(V)$ and $\varphi = \varphi_1 \circ (\varphi_2)^{-1}$, we easily obtain that $\varphi \in \mathrm{Diff}^\infty(U;\varphi(U))$ and

$$\varphi^*(w) = \sum_{i=1}^{m} x_{2i-1}\,dx^{2i} \quad \text{in } U,$$

which ends the proof. □

13.2.2 A Technical Result

We still need to prove the following lemma.

Lemma 13.12. *Let $m \geq 1$ be an integer, $x_0 \in \mathbb{R}^{2m}$ and w be a C^∞ 1-form defined in a neighborhood of x_0 such that $w(x_0) \neq 0$ and*

$$\operatorname{rank}[dw(x_0)] = 2m.$$

Then there exist a neighborhood U of x_0 and

$$\varphi \in \mathrm{Diff}^\infty(U;\varphi(U))$$

such that $\varphi(x_0)=x_0$ and

$$[\varphi^*(w)]_{2i} = 0 \quad \textit{in } U \textit{ for every } 1 \leq i \leq m. \tag{13.7}$$

Remark 13.13. If $w \in C^r$, then the following proof gives $\varphi \in C^{r-2(m-1)}$.

For the proof of the lemma we will need the two following elementary results, the first of which is purely algebraic.

Lemma 13.14. *Let* $f \in \Lambda^2(\mathbb{R}^{2m})$ *with* $\operatorname{rank}[f] = 2m$ *and*

$$a = \sum_{i=1}^{2m-1} a_i e^i \in \Lambda^1(\mathbb{R}^{2m})$$

with $a \neq 0$. *Then there exists* $A \in \mathrm{GL}(2m)$ *of the form*

$$A = \begin{pmatrix} & & & 0 \\ & B & & \vdots \\ & & & 0 \\ 0 & \cdots & 0 & 1 \end{pmatrix},$$

where $B \in \mathrm{GL}(2m-1)$ *and such that*

$$\sum_{1 \le i < j \le 2m-1} (A^*(f))_{ij} e^i \wedge e^j = \sum_{i=1}^{m-1} e^{2i-1} \wedge e^{2i} \quad \textit{and} \quad \sum_{i=1}^{2m-2} (A^*(a))_i e^i \neq 0.$$

Proof. Step 1. Using Proposition 2.24(ii), there exists $\widetilde{A} \in \mathrm{GL}(2m)$ such that

$$\widetilde{A}^*(f) = \sum_{i=1}^{m} e^{2i-1} \wedge e^{2i} \quad \text{and} \quad \widetilde{A}^*(e^{2m}) = e^{2m}.$$

Note that the condition $\widetilde{A}^*(e^{2m}) = e^{2m}$ is equivalent to

$$\widetilde{A} = \begin{pmatrix} & & & \widetilde{A}^1_{2m} \\ & \widetilde{B} & & \vdots \\ & & & \widetilde{A}^{2m-1}_{2m} \\ 0 & \cdots & 0 & 1 \end{pmatrix},$$

where $\widetilde{B} \in \mathrm{GL}(2m-1)$ is given by $\widetilde{B}^i_j = \widetilde{A}^i_j$. Define

$$A = \begin{pmatrix} & & & 0 \\ & \widetilde{B} & & \vdots \\ & & & 0 \\ 0 & \cdots & 0 & 1 \end{pmatrix}$$

and observe that for $1 \le i < j \le 2m-1$,

$$\begin{aligned}(A^*(f))_{ij} &= \left(\sum_{1 \le p < q \le 2m} f_{pq} A^p \wedge A^q \right)_{ij} = \sum_{1 \le p < q \le 2m} f_{pq} \left(A^p_i A^q_j - A^p_j A^q_i \right) \\ &= \sum_{1 \le p < q \le 2m} f_{pq} \left(\widetilde{A}^p_i \widetilde{A}^q_j - \widetilde{A}^p_j \widetilde{A}^q_i \right) = (\widetilde{A}^*(f))_{ij}.\end{aligned}$$

We therefore have

$$\sum_{1\le i<j\le 2m-1}(A^*(f))_{ij}e^i\wedge e^j=\sum_{1\le i<j\le 2m-1}(\widetilde{A}^*(f))_{ij}e^i\wedge e^j=\sum_{i=1}^{m-1}e^{2i-1}\wedge e^{2i}.$$

Note that the previous equation is equivalent to

$$A^*(f)=\sum_{i=1}^{m-1}e^{2i-1}\wedge e^{2i}+h\wedge e^{2m} \tag{13.8}$$

for a certain $h=\sum_{i=1}^{2m-1}h_ie^i\in\Lambda^1(\mathbb{R}^{2m})$.

Step 2. Since $a\neq 0$, we have $A^*(a)=\sum_{i=1}^{2m-1}A^*(a)_ie^i\neq 0$ and thus there exists $1\le i\le 2m-1$ such that $A^*(a)_i\neq 0$. If $1\le i\le 2m-2$, the matrix A has all of the required properties. If $A^*(a)_i=0$ for $1\le i\le 2m-2$, we proceed as follows. Define

$$P=\begin{pmatrix}1&0&\cdots&0\\ 0&\ddots&0&\vdots\\ 1&0&1&0\\ 0&\cdots&0&1\end{pmatrix}\Leftrightarrow P_j^i=\begin{cases}1&\text{if } i=j\\ 1&\text{if } i=2m-1 \text{ and } j=1\\ 0&\text{otherwise}\end{cases}$$

and let us show that $AP\in\mathrm{GL}(2m)$ has all the claimed properties. Indeed, first note that AP has the desired form. Since

$$P^*(e^i)=\begin{cases}e^i&\text{if } i\neq 2m-1\\ e^1+e^{2m-1}&\text{if } i=2m-1,\end{cases}$$

we deduce that, using (13.8),

$$(AP)^*(f)=P^*(A^*(f))=\sum_{i=1}^{m-1}e^{2i-1}\wedge e^{2i}+P^*(h)\wedge e^{2m}.$$

We therefore get

$$\sum_{1\le i<j\le 2m-1}((AP)^*(f))_{ij}e^i\wedge e^j=\sum_{i=1}^{m-1}e^{2i-i}\wedge e^{2i}.$$

Note also that

$$((AP)^*(a))_1=(P^*(A^*(a)))_1=(A^*(a))_{2m-1}\neq 0.$$

The proof is therefore complete. □

We now give the second result.

Lemma 13.15. *Let $U\subset\mathbb{R}^n$ be an open set, $n\ge 2$ and $w\in C^\infty\left(U;\Lambda^1\right)$ be such that*

$$(dx^n)\lrcorner dw=w\quad\text{in } U. \tag{13.9}$$

Then

$$\begin{cases} w = e^{x_n} \sum_{i=1}^{n-1} b_{in}(x_1,\ldots,x_{n-1})\,dx^i, \\ dw = -e^{x_n} \sum_{1\le i<j\le n} b_{ij}(x_1,\ldots,x_{n-1})\,dx^i \wedge dx^j \end{cases}$$

for some $b_{ij} \in C^\infty$.

Proof. We first write

$$dw = \sum_{1\le i<j\le n} a_{ij}\,dx^i \wedge dx^j$$

and observe that, as a direct consequence of (13.9), we have

$$w = -\sum_{i=1}^{n-1} a_{in}\,dx^i. \tag{13.10}$$

We finally show that for every $1 \le i < j \le n$ and $x = (x_1,\ldots,x_n) \in U$,

$$a_{ij}(x) = -e^{x_n} b_{ij}(x_1,\ldots,x_{n-1})$$

for some $b_{ij} \in C^\infty$. For this, it is enough to prove that for every $1 \le i < j \le n$,

$$a_{ij} = \frac{\partial a_{ij}}{\partial x_n}.$$

Let $1 \le i < j \le n$. First, since $ddw = 0$ and hence, in particular, $(ddw)_{ijn} = 0$, we have (with the convention that $a_{nn} = 0$)

$$\frac{\partial a_{jn}}{\partial x_i} - \frac{\partial a_{in}}{\partial x_j} + \frac{\partial a_{ij}}{\partial x_n} = 0.$$

Using (13.10) and the previous equation, we obtain

$$a_{ij} = (dw)_{ij} = -\left(\frac{\partial a_{jn}}{\partial x_i} - \frac{\partial a_{in}}{\partial x_j}\right) = \frac{\partial a_{ij}}{\partial x_n},$$

which concludes the proof. □

Finally, we prove Lemma 13.12.

Proof. With no loss of generality we can assume $x_0 = 0$. In the sequel, U will be a generic neighborhood of 0. We prove the lemma by induction on m and we split the proof into three steps.

Step 1. We start by introducing some notations. Let

$$x = (x_1,\ldots,x_{2m-2},x_{2m-1},x_{2m}) \in \mathbb{R}^n.$$

For every $(x_{2m-1},x_{2m})\in\mathbb{R}^2$, define $i_{(x_{2m-1},x_{2m})}:\mathbb{R}^{2m-2}\to\mathbb{R}^{2m}$ by

$$i_{(x_{2m-1},x_{2m})}(x_1,\ldots,x_{2m-2})=x.$$

Let $1\le k\le n$ and

$$g=\sum_{1\le i_1<\cdots<i_k\le 2m} g_{i_1\cdots i_k}\,dx^{i_1}\wedge\cdots\wedge dx^{i_k}\in C^0\left(\mathbb{R}^{2m};\Lambda^k(\mathbb{R}^{2m})\right).$$

Then for every $(x_{2m-1},x_{2m})\in\mathbb{R}^2$, we have

$$i^*_{(x_{2m-1},x_{2m})}(g)\in C^0\left(\mathbb{R}^{2m-2};\Lambda^k(\mathbb{R}^{2m-2})\right)$$

and, explicitly,

$$i^*_{(x_{2m-1},x_{2m})}(g)(x_1,\ldots,x_{2m-2})=\sum_{1\le i_1<\cdots<i_k\le 2m-2} g_{i_1\cdots i_k}(x)dx^{i_1}\wedge\cdots\wedge dx^{i_k}.$$

Step 2 (the case $m=1$). In that case, we have

$$w(x)=w_1(x_1,x_2)\,dx^1+w_2(x_1,x_2)\,dx^2.$$

Since, by hypothesis, $(w_1(0),w_2(0))\neq(0,0)$, there exist, applying Remark 4.3(ii), a neighborhood U of 0 and $\varphi\in\mathrm{Diff}^\infty(U;\varphi(U))$ such that $\varphi(0)=0$ and

$$\frac{\partial\varphi}{\partial x_2}=(-w_2(\varphi),w_1(\varphi))\quad\text{in } U.$$

We thus get, using the above equation,

$$\begin{aligned}\varphi^*(w)&=w_1(\varphi)d\varphi^1+w_2(\varphi)d\varphi^2\\&=\left[w_1(\varphi)\frac{\partial\varphi^1}{\partial x_1}+w_2(\varphi)\frac{\partial\varphi^2}{\partial x_1}\right]dx^1+\left[w_1(\varphi)\frac{\partial\varphi^1}{\partial x_2}+w_2(\varphi)\frac{\partial\varphi^2}{\partial x_2}\right]dx^2\\&=\left[w_1(\varphi)\frac{\partial\varphi^1}{\partial x_1}+w_2(\varphi)\frac{\partial\varphi^2}{\partial x_1}\right]dx^1,\end{aligned}$$

which is the desired assertion.

Step 3 (induction). We assume that the lemma has been proved for $m-1$ and prove it for m.

Step 3.1 (preliminaries). In this step we show the existence of a neighborhood U of 0 and $\psi\in\mathrm{Diff}^\infty(U;\psi(U))$ with $\psi(0)=0$ such that for every $x=(x_1,\ldots,x_{2m-2},x_{2m-1},x_{2m})\in U$,

$$i^*_{(x_{2m-1},x_{2m})}\left(\psi^*(w)\right)(x_1,\ldots,x_{2m-2})\neq 0,\tag{13.11}$$

$$\operatorname{rank}\left[d(i^*_{(x_{2m-1},x_{2m})}(\psi^*(w)))(x_1,\ldots,x_{2m-2})\right]=2m-2,\tag{13.12}$$

$$\psi^*(w)(x) = e^{x_{2m}} \sum_{i=1}^{2m-1} c_i(x_1,\ldots,x_{2m-1})dx^i \quad \text{in } U \tag{13.13}$$

for some $c_i \in C^\infty(U)$.

(i) Since $\operatorname{rank}[dw] = 2m$ in a neighborhood of 0 and Proposition 2.50 holds, we can find a neighborhood U of 0 and a unique $v \in C^\infty(U;\Lambda^1)$ such that

$$v \lrcorner dw = w \quad \text{in } U.$$

Note that $v(0) \neq 0$ since $w(0) \neq 0$. Hence, using Remark 4.3(ii), there exist a neighborhood U of 0 and $\chi \in \operatorname{Diff}^\infty(U;\chi(U))$ such that $\chi(0) = 0$ and

$$\frac{\partial \chi}{\partial x_{2m}} = v \circ \chi \quad \text{in } U.$$

Using Theorem 3.10 and Proposition 3.11, we thus get

$$\chi^*(w) = \chi^*(v \lrcorner dw) = dx^{2m} \lrcorner d\chi^*(w) \quad \text{in } U.$$

Therefore, applying Lemma 13.15, we have

$$d\chi^*(w)(x) = -e^{x_{2m}} \sum_{1\leq i<j\leq 2m} b_{ij}(x_1,\ldots,x_{2m-1})\, dx^i \wedge dx^j \quad \text{for every } x \in U,$$

$$\chi^*(w)(x) = e^{x_{2m}} \sum_{i=1}^{2m-1} b_{i(2m)}(x_1,\ldots,x_{2m-1})\, dx^i \quad \text{for every } x \in U \tag{13.14}$$

for some $b_{ij} \in C^\infty$.

(ii) Apply Lemma 13.14 to

$$f = d\chi^*(w)(0) \in \Lambda^2(\mathbb{R}^{2m}) \quad \text{and} \quad a = \chi^*(w)(0) \in \Lambda^1(\mathbb{R}^{2m})$$

to get $A \in \mathrm{GL}(n)$ of the form

$$A = \begin{pmatrix} & & 0 \\ & B & \vdots \\ & & 0 \\ 0 & \cdots\ 0 & 1 \end{pmatrix}$$

such that

$$\begin{gathered} \sum_{1\leq i<j\leq 2m-1} (A^*(f))_{ij} e^i \wedge e^j = \sum_{i=1}^{m-1} e^{2i-1} \wedge e^{2i}, \\ \sum_{i=1}^{2m-2} (A^*(a))_i e^i \neq 0. \end{gathered} \tag{13.15}$$

(iii) Let $\theta(x) = A \cdot x$. We now prove that $\psi = \chi \circ \theta$ has all of the desired properties claimed by Step 3.1. In the following, we will frequently use (cf. Remark 3.9) that for any $\varphi \in C^1(\mathbb{R}^M;\mathbb{R}^N)$, any k-form α, and any fixed $x \in \mathbb{R}^M$,

$$\varphi^*(\alpha)(x) = (\nabla\varphi(x))^*[\alpha(\varphi(x))].$$

First, note that $\psi(0)=0$ since $\chi(0)=\theta(0)=0$. We now show (13.11). Restricting if necessary U, it is enough to show the property for $x=0$. Using the second statement in (13.15), we deduce

$$i^*_{(0,0)}(\psi^*(w))(0,\dots,0) = \sum_{i=1}^{2m-2} A^*(a)_i e^i \neq 0,$$

which proves the claim. We next prove (13.12). As before, restricting if necessary U, it is enough to prove the assertion for $x=0$. Using the first equality in (13.15), we obtain

$$\begin{aligned} di^*_{(0,0)}(\psi^*(w))(0,\dots,0) &= i^*_{(0,0)}(\psi^*(dw))(0,\dots,0) \\ &= \sum_{1\le i<j\le 2m-2} (A^*(f))_{ij} e^i \wedge e^j = \sum_{i=1}^{m-1} e^{2i-1}\wedge e^{2i}. \end{aligned}$$

This establishes the claim. Finally, using (13.14) and since

$$\theta(x) = (\theta^1(x_1,\dots,x_{2m-1}),\dots,\theta^{2m-1}(x_1,\dots,x_{2m-1}),x_{2m}),$$

we have

$$\begin{aligned} &\theta^*(\chi^*(w))(x) \\ &= e^{x_{2m}} \sum_{i=1}^{2m-1} b_{i(2m)}\left[\theta^1(x_1,\dots,x_{2m-1}),\dots,\theta^{2m-1}(x_1,\dots,x_{2m-1})\right] d\theta^i \\ &= e^{x_{2m}} \sum_{i=1}^{2m-1} c_i(x_1,\dots,x_{2m-1})\, dx^i, \quad x \in U, \end{aligned}$$

for some $c_i \in C^\infty$; therefore, (13.13) is fulfilled.

Step 3.2 (conclusion). Using (13.11) and (13.12), we get that

$$i^*_{(x_{2m-1},x_{2m})}(\psi^*(w))$$

satisfies the induction hypothesis for $m-1$, for x_{2m-1},x_{2m} small. Moreover, note that using (13.13),

$$i^*_{(x_{2m-1},x_{2m})}(\psi^*(w))(x_1,\dots,x_{2m-2}) = e^{x_{2m}} \sum_{i=1}^{2m-2} c_i(x_i,\dots,x_{2m-1})\, dx^i.$$

Hence, by the induction hypothesis and thanks to the special form of the coefficients of

$$i^*_{(x_{2m-1},x_{2m})}(\psi^*(w))$$

with respect to x_{2m}, there exist a neighborhood $\widehat{U}$ of 0 in $\mathbb{R}^{2m-2}$ and, for every x_{2m-1} small, $\phi_{x_{2m-1}} \in \operatorname{Diff}^\infty(\widehat{U};\phi_{x_{2m-1}}(\widehat{U}))$, verifying

$$\left((\phi_{x_{2m-1}})^*(i^*_{(x_{2m-1},x_{2m})}(\psi^*(w))\right)_{2i} = 0 \quad \text{in } U, \quad 1 \le i \le m-1. \tag{13.16}$$

Furthermore, since the construction is smooth in the parameters, we have in fact

$$(x_1,\ldots,x_{2m-1}) \to \phi_{x_{2m-1}}(x_1,\ldots,x_{2m-2}) \quad \text{is } C^\infty.$$

Define, for a neighborhood U of 0 small enough, $\phi \in \operatorname{Diff}^\infty(U;\phi(U))$ by

$$\phi(x) = \phi(x_1,\ldots,x_{2m}) = (\phi_{x_{2m-1}}(x_1,\ldots,x_{2m-2}),x_{2m-1},x_{2m}).$$

Since $\phi \circ i_{(x_{2m-1},x_{2m})} = i_{(x_{2m-1},x_{2m})} \circ \phi_{x_{2m-1}}$, we obtain

$$(\phi_{x_{2m-1}})^*(i^*_{(x_{2m-1},x_{2m})}(\psi^*(w))) = i^*_{(x_{2m-1},x_{2m})}(\phi^*(\psi^*(w))).$$

Note also that for every $1 \le s \le 2m-2$ and for every 1-form g,

$$\left[i^*_{(x_{2m-1},x_{2m})}(g)(x_1,\ldots,x_{2m-2})\right]_s = [g(x_1,\ldots,x_{2m-2},x_{2m-1},x_{2m})]_s\,.$$

Therefore, combining (13.16) with the above two equations, one gets

$$[\phi^*(\psi^*(w))]_{2i} = 0 \quad \text{in } U, \quad 1 \le i \le m-1.$$

Moreover, since the first $(2m-1)$ components of ϕ do not depend on x_{2m}, we obtain, using (13.13),

$$[\phi^*(\psi^*(w))]_{2m} = 0 \quad \text{in } U.$$

Finally, letting $\varphi = \psi \circ \phi$, we have indeed found the desired diffeomorphism. □

Chapter 14
The Case $k = 2$

14.1 Notations

We recall, from Chapter 2, some notations that we will use throughout the present chapter. As usual, when necessary, we identify in a natural way 1-forms with vectors in $\mathbb{R}^n$.

(i) If $u \in \Lambda^1(\mathbb{R}^n)$ and $f \in \Lambda^2(\mathbb{R}^n)$, then (cf. Proposition 2.12)

$$u \lrcorner f = \sum_{j=1}^{n} \left[\sum_{i=1}^{n} f_{ij} u_i \right] dx^j \in \Lambda^1(\mathbb{R}^n).$$

(ii) Given $f \in \Lambda^2(\mathbb{R}^n)$, the matrix $\overline{f} \in \mathbb{R}^{n \times n}$ (denoted in Notation 2.30 as $\overline{f}_{\lrcorner,1}$) is defined, by abuse of notations, as

$$\overline{f} u = u \lrcorner f \quad \text{for every } u \in \Lambda^1(\mathbb{R}^n) \approx \mathbb{R}^n.$$

(iii) The *rank* of $f \in \Lambda^2(\mathbb{R}^n)$ is defined (cf. Proposition 2.32(i)) by

$$\operatorname{rank}[f] = \operatorname{rank}\left(\overline{f}\right).$$

We also recall that in the present chapter we denote by rank what was denoted by rank_1 in Chapter 2. In particular, if $\operatorname{rank}[f] = n$, then $\overline{f}$ is invertible and

$$v = u \lrcorner f \Leftrightarrow u = \left(\overline{f}\right)^{-1} v.$$

(iv) When n is even, identifying n-forms with 0-forms, we have (cf. Proposition 2.37(iii))

$$\left|\det \overline{f}\right|^{1/2} = \frac{1}{(n/2)!} \left| f^{n/2} \right|,$$

where $f^m = \underbrace{f \wedge \cdots \wedge f}_{m \text{ times}}$.

(v) Let $r \geq 0$ be an integer and $0 \leq \alpha \leq 1$. Let $f \in \Lambda^2(\mathbb{R}^n)$ with $\operatorname{rank}[f] = n$ (thus, in particular, n is even). In view of Corollary 16.30 and of the previous point, if $c > 0$ is such that

$$\left\| \frac{1}{f^{n/2}} \right\|_{C^0}, \quad \|f\|_{C^0} \leq c,$$

then there exists a constant $C = C(c, r, \Omega) > 0$ such that

$$\| (\overline{f})^{-1} \|_{C^{r,\alpha}} \leq C \|f\|_{C^{r,\alpha}}.$$

(vi) Finally, we recall the notion of harmonic fields with a vanishing tangential part (cf. Section 6.1). If $\Omega \subset \mathbb{R}^n$ is a bounded open smooth set, then

$$\mathscr{H}_T(\Omega;\Lambda^2) = \{\omega \in C^\infty(\overline{\Omega};\Lambda^2) : d\omega = 0,\ \delta\omega = 0 \text{ in } \Omega \text{ and } \nu \wedge \omega = 0 \text{ on } \partial\Omega\}.$$

Recall that if Ω is contractible, then

$$\mathscr{H}_T(\Omega;\Lambda^2) = \{0\} \quad \text{if } n \geq 3.$$

In terms of the components of

$$\omega = \sum_{1 \leq i < j \leq n} \omega_{ij}\, dx^i \wedge dx^j,$$

we have

$$d\omega = 0 \Leftrightarrow \frac{\partial \omega_{ij}}{\partial x_k} - \frac{\partial \omega_{ik}}{\partial x_j} + \frac{\partial \omega_{jk}}{\partial x_i} = 0,\ \forall 1 \leq i < j < k \leq n,$$

$$\delta\omega = 0 \Leftrightarrow \sum_{j=1}^{n} \frac{\partial \omega_{ij}}{\partial x_j} = 0,\ \forall 1 \leq i \leq n,$$

$$\nu \wedge \omega = 0 \Leftrightarrow \omega_{ij}\nu_k - \omega_{ik}\nu_j + \omega_{jk}\nu_i = 0,\ \forall 1 \leq i < j < k \leq n.$$

14.2 Local Result for Forms with Maximal Rank

The following result is the classical Darboux theorem for closed 2-forms but with optimal regularity. This is a delicate point and it has been obtained by Bandyopadhyay and Dacorogna [8]. The other existing results provide solutions that are only in $C^{r,\alpha}$, whereas in the theorem below we find a solution which belongs to $C^{r+1,\alpha}$.

Theorem 14.1 (Darboux theorem with optimal regularity). *Let $r \geq 0$ and $n = 2m \geq 4$ be integers. Let $0 < \alpha < 1$ and $x_0 \in \mathbb{R}^n$. Let ω_m be the standard symplectic form of rank $2m$,*

$$\omega_m = \sum_{i=1}^{m} dx^{2i-1} \wedge dx^{2i}.$$

Let ω be a 2-form. The two following statements are then equivalent:

(i) The 2-form ω is closed, is in $C^{r,\alpha}$ in a neighborhood of x_0, and verifies

$$\operatorname{rank}\left[\omega\left(x_0\right)\right] = n.$$

(ii) There exist a neighborhood U of x_0 and $\varphi \in \operatorname{Diff}^{r+1,\alpha}(U;\varphi(U))$ such that

$$\varphi^*\left(\omega_m\right) = \omega \quad \text{in } U \quad \text{and} \quad \varphi\left(x_0\right) = x_0.$$

Remark 14.2. (i) When $r = 0$, the hypothesis $d\omega = 0$ is to be understood in the sense of distributions.

(ii) The theorem is still valid when $n = 2$, but it is then the result of Dacorogna and Moser [33] (cf. Theorem 10.3).

Proof. The necessary part is obvious and we discuss only the sufficient part. We divide the proof into four steps.

Step 1. Without loss of generality we take $x_0 = 0$. We can, according to Proposition 2.24(ii), also always assume that

$$\omega(0) = \omega_m.$$

Step 2. Our theorem will follow from Theorem 18.1. So we need to define the spaces and the operators and then check all of the hypotheses.

1) We choose V a sufficiently small ball centered at 0 and we define the sets

$$X_1 = C^{1,\alpha}(\overline{V};\mathbb{R}^n) \quad \text{and} \quad Y_1 = C^{0,\alpha}(\overline{V};\Lambda^2),$$

$$X_2 = C^{r+1,\alpha}(\overline{V};\mathbb{R}^n) \quad \text{and} \quad Y_2 = \{b \in C^{r,\alpha}(\overline{V};\Lambda^2) : db = 0 \text{ in } V\}.$$

Using Proposition 16.23, we immediately deduce that (H_{XY}) of Theorem 18.1 is fulfilled.

2) Define $L : X_2 \to Y_2$ by

$$La = d[a \lrcorner \omega_m] = b.$$

We will show that there exists $L^{-1} : Y_2 \to X_2$ a linear right inverse of L and a constant $C_1 = C_1(r,\alpha,V)$ such that

$$\|L^{-1}b\|_{X_i} \le C_1 \|b\|_{Y_i} \quad \text{for every } b \in Y_2 \text{ and } i = 1,2.$$

Once shown this, (H_L) of Theorem 18.1 will be satisfied. First, using Theorem 8.3, find $w \in C^{r+1,\alpha}(\overline{V};\Lambda^1)$ and $C_1 = C_1(r,\alpha,V) > 0$ such that

$$dw = b \quad \text{in } V,$$

$$\|w\|_{C^{r+1,\alpha}} \le C_1 \|b\|_{C^{r,\alpha}} \quad \text{and} \quad \|w\|_{C^{1,\alpha}} \le C_1 \|b\|_{C^{0,\alpha}}.$$

Moreover, the correspondence $b \to w$ can be chosen to be linear. Next, define $a \in C^{r+1,\alpha}(\overline{V};\mathbb{R}^n)$ by

$$a_{2i-1} = w_{2i} \quad \text{and} \quad a_{2i} = -w_{2i-1}, \quad 1 \le i \le m,$$

and note that

$$a \lrcorner \omega_m = w.$$

Finally, defining $L^{-1} : Y_2 \to X_2$ by $L^{-1}(b) = a$, we easily check that L^{-1} is linear,

$$LL^{-1} = \mathrm{id} \quad \text{on } Y_2$$

and

$$\|L^{-1}b\|_{X_i} \le C_1 \|b\|_{Y_i} \text{ for every } b \in Y_2 \text{ and } i = 1,2.$$

So (H_L) of Theorem 18.1 is satisfied.

3) We then let Q be defined by

$$Q(u) = \omega_m - (\mathrm{id}+u)^* \omega_m + d\left[u \lrcorner \omega_m\right].$$

Since

$$d\left[u \lrcorner \omega_m\right] = \sum_{i=1}^m \left[du^{2i-1} \wedge dx^{2i} + dx^{2i-1} \wedge du^{2i}\right],$$

$$\omega_m - (\mathrm{id}+u)^* \omega_m = \sum_{i=1}^m \left[dx^{2i-1} \wedge dx^{2i} - \left(dx^{2i-1} + du^{2i-1}\right) \wedge \left(dx^{2i} + du^{2i}\right)\right],$$

we get

$$Q(u) = -\sum_{i=1}^m du^{2i-1} \wedge du^{2i}.$$

4) Note that $Q(0) = 0$ and $dQ(u) = 0$ in V. Appealing to Theorem 16.28 (a similar but more involved estimate can be found in Lemma 14.8), there exists a constant $C_2 = C_2(r,V)$ such that for every $u, v \in C^{r+1,\alpha}(\overline{\Omega};\mathbb{R}^n)$, the following estimates hold:

$$\begin{aligned}
\|Q(u) - Q(v)\|_{C^{0,\alpha}} &\le \sum_{i=1}^m \|du^{2i-1} \wedge du^{2i} - dv^{2i-1} \wedge dv^{2i}\|_{C^{0,\alpha}} \\
&\le \sum_{i=1}^m \|du^{2i-1} \wedge \left(du^{2i} - dv^{2i}\right)\|_{C^{0,\alpha}} \\
&\quad + \sum_{i=1}^m \|\left(dv^{2i-1} - du^{2i-1}\right) \wedge dv^{2i}\|_{C^{0,\alpha}} \\
&\le C_2(\|u\|_{C^{1,\alpha}} + \|v\|_{C^{1,\alpha}})\|u - v\|_{C^{1,\alpha}}
\end{aligned}$$

and

$$\begin{aligned}\|Q(u)\|_{C^{r,\alpha}} &\le \sum_{i=1}^{m} \|du^{2i-1} \wedge du^{2i}\|_{C^{r,\alpha}} \\ &\le C \sum_{i=1}^{m} \left[\|du^{2i-1}\|_{C^{r,\alpha}} \| du^{2i}\|_{C^0} + \|du^{2i}\|_{C^{r,\alpha}} \| du^{2i-1}\|_{C^0} \right] \\ &\le C_2 \|u\|_{C^{1,\alpha}} \|u\|_{C^{r+1,\alpha}} .\end{aligned}$$

We therefore see that property (H_Q) is valid for every ρ and we choose $\rho = 1/(2n)$,

$$c_1(r,s) = C_2(r+s) \quad \text{and} \quad c_2(r,s) = C_2 rs.$$

5) Setting $\varphi = \mathrm{id} + u$, we can rewrite the equation $\varphi^*(\omega_m) = \omega$ as

$$\begin{aligned}Lu = d[u \lrcorner \omega_m] &= \omega - (\mathrm{id}+u)^* \omega_m + d[u \lrcorner \omega_m] \\ &= \omega - \omega_m + [\omega_m - (\mathrm{id}+u)^* \omega_m + d[u \lrcorner \omega_m]] \\ &= \omega - \omega_m + Q(u).\end{aligned}$$

Step 3. We may now apply Theorem 18.1 and get that there exists $\psi \in C^{r+1,\alpha}(\overline{V};\mathbb{R}^n)$ such that $\psi^*(\omega_m) = \omega$ in V with $\|\nabla\psi - I\|_{C^0} \le 1/(2n)$, provided

$$\|\omega - \omega_m\|_{C^{0,\alpha}} \le \frac{1}{2C_1 \max\{4C_1C_2, 1\}}. \tag{14.1}$$

Setting $\varphi(x) = \psi(x) - \psi(0)$, we have indeed proved that there exists $\varphi \in C^{r+1,\alpha}(\overline{V};\mathbb{R}^n)$ satisfying

$$\varphi^*(\omega_m) = \omega \text{ in } V, \quad \|\nabla\varphi - I\|_{C^0} \le \frac{1}{2n} \quad \text{and} \quad \varphi(0) = 0.$$

Step 4. We may now conclude the proof of the theorem.

Step 4.1. Let $0 < \varepsilon < 1$ and define

$$\omega^\varepsilon(x) = \omega(\varepsilon x).$$

Observe that $\omega^\varepsilon \in C^{r,\alpha}(\overline{V};\Lambda^2)$, $d\omega^\varepsilon = 0$, $\omega^\varepsilon(0) = \omega_m$ and

$$\|\omega^\varepsilon - \omega_m\|_{C^{0,\alpha}(\overline{V})} \to 0 \quad \text{as } \varepsilon \to 0.$$

Choose ε sufficiently small so that

$$\|\omega^\varepsilon - \omega_m\|_{C^{0,\alpha}(\overline{V})} \le \frac{1}{2C_1 \max\{4C_1C_2, 1\}}.$$

Apply Step 3 to find $\psi_\varepsilon \in C^{r+1,\alpha}(\overline{V};\mathbb{R}^n)$ satisfying

$$\psi_\varepsilon^*(\omega_m) = \omega^\varepsilon \text{ in } V, \quad \|\nabla\psi_\varepsilon - I\|_{C^0} \le \frac{1}{2n} \quad \text{and} \quad \psi_\varepsilon(0) = 0.$$

Step 4.2. Let

$$\chi_\varepsilon(x) = \frac{x}{\varepsilon}$$

and define

$$\varphi = \varepsilon\, \psi_\varepsilon \circ \chi_\varepsilon .$$

Define $U = \varepsilon V$. It is easily seen that $\varphi \in C^{r+1,\alpha}(\overline{U};\mathbb{R}^n)$,

$$\varphi^*(\omega_m) = \omega \text{ in } U \quad \text{and} \quad \varphi(0) = 0.$$

Note in particular that

$$\|\nabla\varphi - I\|_{C^0(\overline{U})} = \|\nabla\psi_\varepsilon - I\|_{C^0(\overline{V})} \leq \frac{1}{2n}$$

and therefore $\det\nabla\varphi > 0$ in $\overline{U}$. Hence, restricting U, if necessary, we can assume that $\varphi \in \mathrm{Diff}^{r+1,\alpha}(\overline{U};\varphi(\overline{U}))$. This concludes the proof of the theorem. □

14.3 Local Result for Forms of Nonmaximal Rank

The main result of the present section is to obtain the Darboux theorem for degenerate closed 2-forms. We will provide, following Bandyopadhyay, Dacorogna and Kneuss [9], two proofs of the theorem. The standard proof uses the Frobenius theorem to reduce the dimension so that the forms have maximal rank and then apply the classical Darboux theorem. We will follow this path but using the more sophisticated Theorem 14.1. Our theorem will provide a solution in $C^{r,\alpha}$, whereas in the existing literature solutions are found only in $C^{r-1,\alpha}$.

We will also give a completely different proof; it will use an argument based on the flow method. Still a different proof can be found in [8] when $n = 2m+1$.

14.3.1 The Theorem and a First Proof

Theorem 14.3. *Let $n \geq 3$, $r,m \geq 1$ be integers and $0 < \alpha < 1$. Let $x_0 \in \mathbb{R}^n$ and ω_m be the standard symplectic form with* $\operatorname{rank}[\omega_m] = 2m < n$*, namely*

$$\omega_m = \sum_{i=1}^{m} dx^{2i-1} \wedge dx^{2i}.$$

Let ω be a $C^{r,\alpha}$ closed 2-form such that

$$\operatorname{rank}[\omega] = 2m \text{ in a neighborhood of } x_0 .$$

Then there exist a neighborhood U of x_0 and $\varphi \in \mathrm{Diff}^{r,\alpha}(U;\varphi(U))$ such that

$$\varphi^*(\omega_m) = \omega \text{ in } U \quad \text{and} \quad \varphi(x_0) = x_0 .$$

Remark 14.4. The theorem is standard in the C^∞ case. In all proofs that we have seen, the regularity that is established is, at best, that if $\omega \in C^{r,\alpha}$, then $\varphi \in C^{r-1,\alpha}$. However, our result asserts that ω and φ have the same regularity in Hölder spaces. This is, of course, better but still not optimal, as in the nondegenerate case of Theorem 14.1.

Proof. Step 1. Without loss of generality, we can assume $x_0 = 0$. We first find, appealing to Theorem 4.5, a neighborhood $V \subset \mathbb{R}^n$ of 0 and $\psi \in \operatorname{Diff}^{r,\alpha}(V;\psi(V))$ with $\psi(0) = 0$ and

$$\psi^*(\omega)(x_1,\ldots,x_n) = \widetilde{\omega}(x_1,\ldots,x_{2m}) = \sum_{1\le i<j\le 2m} \widetilde{\omega}_{ij}(x_1,\ldots,x_{2m})\,dx^i \wedge dx^j.$$

Therefore, $\psi^*(\omega) = \widetilde{\omega} \in C^{r-1,\alpha}$ in a neighborhood of 0 in $\mathbb{R}^{2m}$ and $\operatorname{rank}\widetilde{\omega} = 2m$ in a neighborhood of 0.

Step 2. We then apply Theorem 14.1 to $\widetilde{\omega}$ find a neighborhood $W \subset \mathbb{R}^{2m}$ of 0 and $\chi \in \operatorname{Diff}^{r,\alpha}(W;\chi(W))$, with $\chi(0) = 0$, such that

$$\chi^*(\omega_m) = \widetilde{\omega} \text{ in } W.$$

We set

$$\widetilde{\chi}(x) = \widetilde{\chi}(x_1,\ldots,x_{2m},x_{2m+1},\ldots,x_n) = (\chi(x_1,\ldots,x_{2m}),x_{2m+1},\ldots,x_n).$$

We then choose V smaller, if necessary, so that

$$V \subset W \times \mathbb{R}^{n-2m}.$$

We finally have that $U = \psi(V)$ and $\varphi = \widetilde{\chi} \circ \psi^{-1}$ have all of the desired properties. □

14.3.2 A Second Proof

We now provide a second proof of Theorem 14.3 under the extra assumption that ω is in C^∞. It seems that the present proof is more appropriate if one wants to look for global results.

Proof. As usual, we consider, without loss of generality, that $x_0 = 0$.

Step 1. Define, for a sufficiently small neighborhood U_1 of 0,

$$h(t,x) = h_t(x) = \omega(tx).$$

Then the homotopy h is such that $h \in C^\infty([0,1] \times U_1;\Lambda^2)$ and for every $t \in [0,1]$, the following identities hold in U_1:

$$dh_t = 0, \quad h_t^m \neq 0 \quad \text{and} \quad h_t^{m+1} = 0 \tag{14.2}$$

(recall that the last two conditions are equivalent to rank $[h_t] = 2m$) and

$$h_0 = \omega(0) \quad \text{and} \quad h_1 = \omega.$$

Step 2. Since (14.2) holds and

$$h_t^m \wedge \frac{\partial h_t}{\partial t} = \frac{1}{m+1} \frac{\partial h_t^{m+1}}{\partial t} = 0,$$

we can apply Theorem 8.22. We can therefore find a neighborhood $U_2 \subset U_1$ of 0 and $w \in C^\infty([0,1] \times U_2; \mathbb{R}^n)$, $w(t,x) = w_t(x)$, satisfying, for every $t \in [0,1]$, $w_t(0) = 0$ and

$$dw_t = -\frac{\partial h_t}{\partial t} \quad \text{and} \quad w_t \wedge h_t^m = 0 \quad \text{in } U_2.$$

We then apply Proposition 2.50 to find $u \in C^\infty([0,1] \times U_2; \mathbb{R}^n)$, $u(t,x) = u_t(x)$, with

$$u_t \lrcorner h_t = w_t \quad \text{and} \quad u_t(0) = 0.$$

Step 3. We next find the flow, associated to the vector field u_t,

$$\begin{cases} \dfrac{d}{dt}\varphi_t = u_t \circ \varphi_t, & 0 \le t \le 1, \\ \varphi_0 = \text{id}. \end{cases}$$

Theorem 12.8 gives that φ_1 is a diffeomorphism in a neighborhood $U_3 \subset U_2$ of 0 such that

$$\varphi_1^*(h_1) = h_0 \text{ in } U_3 \quad \text{and} \quad \varphi_1(0) = 0.$$

Step 4. Since h_0 is constant, we can use Proposition 2.24(ii) to find a diffeomorphism ψ of the form $\psi(x) = Ax$ with $A \in \mathrm{GL}(n)$ so that

$$\psi^*(h_0) = \omega_m = \sum_{i=1}^m dx^{2i-1} \wedge dx^{2i}.$$

Letting $\varphi = \psi^{-1} \circ \varphi_1^{-1}$, we have the claim. □

14.4 Global Result with Dirichlet Data

14.4.1 The Main Result

We now state our main theorem. It has been obtained under slightly more restrictive hypotheses by Bandyopadhyay and Dacorogna [8]; as stated, it is due to Dacorogna and Kneuss [32]. We will provide two proofs of the theorem in Sections 14.4.5 and 14.4.6.

Theorem 14.5. *Let $n > 2$ be even and $\Omega \subset \mathbb{R}^n$ be a bounded open smooth set with exterior unit normal ν. Let $0 < \alpha < 1$ and $r \geq 1$ be an integer. Let $f, g \in C^{r,\alpha}(\overline{\Omega};\Lambda^2)$ satisfying $df = dg = 0$ in Ω,*

$$\nu \wedge f, \nu \wedge g \in C^{r+1,\alpha}(\partial\Omega;\Lambda^3) \quad \textit{and} \quad \nu \wedge f = \nu \wedge g \textit{ on } \partial\Omega,$$

$$\int_\Omega \langle f;\psi\rangle\, dx = \int_\Omega \langle g;\psi\rangle\, dx \quad \textit{for every } \psi \in \mathscr{H}_T(\Omega;\Lambda^2) \tag{14.3}$$

and, for every $t \in [0,1]$,

$$\operatorname{rank}[tg + (1-t)f] = n \textit{ in } \overline{\Omega}.$$

Then there exists $\varphi \in \mathrm{Diff}^{r+1,\alpha}(\overline{\Omega};\overline{\Omega})$ such that

$$\varphi^*(g) = f \textit{ in } \Omega \quad \textit{and} \quad \varphi = \mathrm{id} \textit{ on } \partial\Omega.$$

Remark 14.6. (i) As already mentioned, we can consider, in a similar way, a general homotopy f_t with $f_0 = f$, $f_1 = g$,

$$df_t = 0, \quad \nu \wedge f_t = \nu \wedge f_0 \text{ on } \partial\Omega \quad \text{and} \quad \operatorname{rank}[f_t] = n \text{ in } \overline{\Omega},$$

$$\int_\Omega \langle f_t;\psi\rangle\, dx = \int_\Omega \langle f_0;\psi\rangle\, dx \quad \text{for every } \psi \in \mathscr{H}_T(\Omega;\Lambda^2).$$

Note that the nondegeneracy condition $\operatorname{rank}[f_t] = n$ implies (identifying, as usual, volume forms with functions)

$$f^{n/2} \cdot g^{n/2} > 0 \text{ in } \overline{\Omega}.$$

(ii) The nondegeneracy condition

$$\operatorname{rank}[tg + (1-t)f] = n \quad \text{for every } t \in [0,1]$$

is equivalent to the condition that the matrix $(\overline{g})(\overline{f})^{-1}$ has no negative eigenvalues.

(iii) If Ω is contractible, then $\mathscr{H}_T(\Omega;\Lambda^2) = \{0\}$ and, therefore, (14.3) is automatically satisfied.

(iv) Note that the extra regularity on f and g holds only on the boundary and only for their tangential parts. More precisely, recall that for $x \in \partial\Omega$, we denote by $\nu = \nu(x)$ the exterior unit normal to Ω. By

$$\nu \wedge f \in C^{r+1,\alpha}(\partial\Omega;\Lambda^3)$$

we mean that the tangential part of f is in $C^{r+1,\alpha}$, namely the 3-form F defined by

$$F(x) = \nu(x) \wedge f(x)$$

is such that

$$F \in C^{r+1,\alpha}(\partial\Omega;\Lambda^3).$$

14.4.2 The Flow Method

We now state and prove a weaker version, from the point of view of regularity, of Theorem 14.5. It has, however, the advantage of having a simple proof. It has been obtained by Bandyopadhyay and Dacorogna [8].

Theorem 14.7. *Let $n > 2$ be even and $\Omega \subset \mathbb{R}^n$ be a bounded open smooth set with exterior unit normal ν. Let $r \geq 1$ be an integer, $0 < \alpha < 1$ and $f, g \in C^{r,\alpha}(\overline{\Omega}; \Lambda^2)$ satisfy*

$$df = dg = 0 \text{ in } \Omega, \quad \nu \wedge f = \nu \wedge g \text{ on } \partial\Omega,$$

$$\int_\Omega \langle f; \psi \rangle\, dx = \int_\Omega \langle g; \psi \rangle\, dx \quad \text{for every } \psi \in \mathscr{H}_T(\Omega; \Lambda^2),$$

$$\operatorname{rank}[tg + (1-t) f] = n \text{ in } \overline{\Omega} \text{ and for every } t \in [0,1].$$

Then there exists $\varphi \in \operatorname{Diff}^{r,\alpha}(\overline{\Omega}; \overline{\Omega})$ such that

$$\varphi^*(g) = f \text{ in } \Omega \quad \text{and} \quad \varphi = \operatorname{id} \text{ on } \partial\Omega.$$

Furthermore, if $0 < \beta \leq \alpha < 1$ and if $c > 0$ is such that

$$\|f\|_{C^1}, \|g\|_{C^1}, \left\| \frac{1}{[tg + (1-t) f]^{n/2}} \right\|_{C^0} \leq c \quad \text{for every } t \in [0,1],$$

then there exists a constant $C = C(c, r, \alpha, \beta, \Omega) > 0$ such that

$$\|\varphi - \operatorname{id}\|_{C^{r,\alpha}} \leq C[\|f\|_{C^{r,\alpha}} + \|g\|_{C^{r,\alpha}}] \|f - g\|_{C^{0,\beta}} + C \|f - g\|_{C^{r-1,\alpha}}.$$

Proof. We solve (cf. Theorem 8.16)

$$\begin{cases} dw = f - g & \text{in } \Omega, \\ w = 0 & \text{on } \partial\Omega \end{cases}$$

and find $w \in C^{r+1,\alpha}(\overline{\Omega}; \Lambda^1)$ and $C_1 = C_1(r, \alpha, \beta, \Omega) > 0$ such that

$$\|w\|_{C^{r,\alpha}} \leq C_1 \|f - g\|_{C^{r-1,\alpha}} \quad \text{and} \quad \|w\|_{C^{1,\beta}} \leq C_1 \|f - g\|_{C^{0,\beta}}.$$

Since $\operatorname{rank}[tg + (1-t) f] = n$, we can find $u_t \in C^{r,\alpha}(\overline{\Omega}; \mathbb{R}^n)$ so that

$$u_t \lrcorner [tg + (1-t) f] = w \Leftrightarrow u_t = \left[t\overline{g} + (1-t)\overline{f}\right]^{-1} w.$$

Moreover (in view of Notation (v) in Section 14.1, Theorem 16.28 and Corollary 16.30), we can find constants $C_i = C_i(c, r, \alpha, \beta, \Omega) > 0$, $i = 2, 3$, such that

$$\begin{aligned} \|u_t\|_{C^{r,\alpha}} &\leq C_2 [\|f\|_{C^{r,\alpha}} + \|g\|_{C^{r,\alpha}}] \|w\|_{C^0} + C_2 \|w\|_{C^{r,\alpha}} \\ &\leq C_3 [\|f\|_{C^{r,\alpha}} + \|g\|_{C^{r,\alpha}}] \|f - g\|_{C^{0,\beta}} + C_3 \|f - g\|_{C^{r-1,\alpha}} \end{aligned}$$

and $\|u_t\|_{C^1} \leq C_3$. We then apply Theorem 12.7 to u_t and $f_t = tg + (1-t)f$ to find φ satisfying

$$\varphi^*(g) = f \text{ in } \Omega \quad \text{and} \quad \varphi = \text{id on } \partial\Omega.$$

The estimate follows from Theorem 12.1. The proof is therefore complete. □

14.4.3 The Key Estimate for Regularity

The following estimate will play a crucial role in getting the optimal regularity in Theorem 14.10. We have encountered a result of the same type in the much simpler case of volume forms (see Theorem 10.9) or in the local case (see Theorem 14.1). We will state the theorem for k-forms, although we will use it only when $k = 2$.

Lemma 14.8. *Let $n \geq 2$ and $\Omega \subset \mathbb{R}^n$ be a bounded open Lipschitz set. Let $r \geq 1$, $2 \leq k \leq n$ be integers, $c > 0$ and $0 \leq \gamma \leq \alpha \leq 1$. Let $g \in C^{r+1,\alpha}(\overline{\Omega};\Lambda^k)$ be closed, $u, v \in C^{r+1,\alpha}(\overline{\Omega};\mathbb{R}^n)$ and $c > 0$ with*

$$\|u\|_{C^{1,\gamma}}, \|v\|_{C^{1,\gamma}} \leq c,$$

$$(\text{id}+tu)(\overline{\Omega}), (\text{id}+tv)(\overline{\Omega}) \subset \overline{\Omega}, \ \forall t \in [0,1].$$

Set

$$Q(u) = g - (\text{id}+u)^*(g) + d[u \lrcorner g].$$

Then there exists a constant $C = C(c, r, \Omega)$ such that the following estimates hold:

$$\|Q(u) - Q(v)\|_{C^{0,\gamma}} \leq C\|g\|_{C^{2,\gamma}}(\|u\|_{C^{1,\gamma}} + \|v\|_{C^{1,\gamma}})\|u - v\|_{C^{1,\gamma}},$$

$$\|Q(u)\|_{C^{r,\alpha}} \leq C\|g\|_{C^{r+1,\alpha}}\|u\|_{C^1} + C\|g\|_{C^1}\|u\|_{C^{r+1,\alpha}}\|u\|_{C^1}.$$

Remark 14.9. With essentially the same argument, we can replace the last estimate by the following one. In addition to the hypotheses of the lemma, let $0 \leq \alpha < \beta \leq 1$ and $g \in C^{r+1,\beta}(\overline{\Omega};\Lambda^k)$; then the last estimate takes the following form:

$$\|Q(u)\|_{C^{r,\alpha}} \leq C\|g\|_{C^{r+1,\beta}}\|u\|_{C^1}^{1+\beta-\alpha} + C\|g\|_{C^{r+1,\alpha}}\|u\|_{C^{r+1,\alpha}}\|u\|_{C^1}$$

for every $u, v \in C^{r+1,\alpha}(\overline{\Omega};\mathbb{R}^n)$ with

$$\|u\|_{C^{1,\gamma}}, \|v\|_{C^{1,\gamma}} \leq c,$$

$$(\text{id}+tu)(\overline{\Omega}), (\text{id}+tv)(\overline{\Omega}) \subset \overline{\Omega}, \ \forall t \in [0,1].$$

Proof. We divide the proof into four steps. Since we will apply the result only when $k = 2$, we will always single out the formulas for this case. We also will constantly use Theorem 16.28.

Step 1. We start with some notations. The form g will be written as

$$g = \sum_{I \in \mathcal{T}_k} g_I \, dx^I .$$

We first need to write $(\mathrm{id}+u)^*(g)$ in a different way. For this, we observe that we have, for $I \in \mathcal{T}_k$,

$$\begin{aligned} d(x+u)^I &= \left(dx^{i_1} + du^{i_1}\right) \wedge \cdots \wedge \left(dx^{i_k} + du^{i_k}\right) \\ &= dx^I + \sum_{\substack{(J,K)=I \\ 1 \le |K| \le k}} dx^J \wedge du^K \\ &= dx^I + \sum_{\substack{(J,i)=I \\ 1 \le i \le n}} dx^J \wedge du^i + \sum_{\substack{(J,K)=I \\ 2 \le |K| \le k}} dx^J \wedge du^K , \end{aligned}$$

where we have used the notation

$$\sum_{\substack{(J,i)=I \\ 1 \le i \le n}} dx^J \wedge du^i = \sum_{\gamma=1}^{k} (-1)^{k+\gamma} dx^{i_1} \wedge \cdots \wedge dx^{i_{\gamma-1}} \wedge dx^{i_{\gamma+1}} \wedge \cdots \wedge dx^{i_k} \wedge du^{i_\gamma}$$

and similarly for

$$\sum_{\substack{(J,K)=I \\ 2 \le |K| \le k}} dx^J \wedge du^K .$$

When $k = 2$, we have

$$\begin{aligned} (dx + du)^{ij} &= (dx^i + du^i) \wedge (dx^j + du^j) \\ &= dx^i \wedge dx^j + \left[du^i \wedge dx^j + dx^i \wedge du^j\right] + du^i \wedge du^j . \end{aligned}$$

We can therefore write

$$\begin{aligned} (\mathrm{id}+u)^*(g) &= \sum_{I \in \mathcal{T}_k} g_I (\mathrm{id}+u) \, dx^I + \sum_{I \in \mathcal{T}_k} g_I (\mathrm{id}+u) \sum_{\substack{(J,K)=I \\ 1 \le |K| \le k}} dx^J \wedge du^K \\ &= g(\mathrm{id}+u) + \sum_{I \in \mathcal{T}_k} \sum_{\substack{(J,i)=I \\ 1 \le i \le n}} g_I (\mathrm{id}+u) \, dx^J \wedge du^i \\ &\quad + \sum_{I \in \mathcal{T}_k} \sum_{\substack{(J,K)=I \\ 2 \le |K| \le k}} g_I (\mathrm{id}+u) \, dx^J \wedge du^K \end{aligned}$$

so that when $k = 2$, we find

$$\begin{aligned} (\mathrm{id}+u)^*(g) &= g(\mathrm{id}+u) + \sum_{1 \le i < j \le n} g_{ij}(\mathrm{id}+u) \left[du^i \wedge dx^j + dx^i \wedge du^j\right] \\ &\quad + \sum_{1 \le i < j \le n} g_{ij}(\mathrm{id}+u) \, du^i \wedge du^j . \end{aligned}$$

We will also use, for $I \in \mathcal{T}_k$,

$$d\left[u \lrcorner dx^I\right] = \sum_{\substack{(J,i)=I \\ 1\le i\le n}} dx^J \wedge du^i,$$

which reads, when $k=2$, as

$$d\left[u \lrcorner dx^{ij}\right] = d\left[u \lrcorner \left(dx^i \wedge dx^j\right)\right] = du^i \wedge dx^j + dx^i \wedge du^j.$$

Step 2. We have, since g is closed and according to Lemma 5.4, that

$$\begin{aligned} d\left[u \lrcorner g\right] &= \sum_{I\in\mathcal{T}_k} g_I\, d\left[u \lrcorner dx^I\right] + \sum_{I\in\mathcal{T}_k} \langle \operatorname{grad} g_I; u\rangle\, dx^I \\ &= \sum_{I\in\mathcal{T}_k} \sum_{\substack{(J,i)=I \\ 1\le i\le n}} g_I\, dx^J \wedge du^i + \sum_{I\in\mathcal{T}_k} \langle \operatorname{grad} g_I; u\rangle\, dx^I \end{aligned}$$

and hence, when $k=2$,

$$d\left[u \lrcorner g\right] = \sum_{1\le i<j\le n} g_{ij}\left[du^i \wedge dx^j + dx^i \wedge du^j\right] + \sum_{1\le i<j\le n} \langle \operatorname{grad} g_{ij}; u\rangle\, dx^i \wedge dx^j.$$

In order to get the right estimates, we rewrite $Q(u)$, defined by

$$Q(u) = g - (\mathrm{id}+u)^*(g) + d\left[u \lrcorner g\right],$$

in the following way:

$$\begin{aligned} Q(u) = g - g(\mathrm{id}+u) &- \sum_{I\in\mathcal{T}_k} \sum_{\substack{(J,i)=I \\ 1\le i\le n}} g_I(\mathrm{id}+u)\, dx^J \wedge du^i \\ &- \sum_{I\in\mathcal{T}_k} \sum_{\substack{(J,K)=I \\ 2\le |K|\le k}} g_I(\mathrm{id}+u)\, dx^J \wedge du^K + d\left[u \lrcorner g\right] \end{aligned}$$

and thus

$$\begin{aligned} Q(u) = g - g(\mathrm{id}+u) &- \sum_{I\in\mathcal{T}_k} \sum_{\substack{(J,i)=I \\ 1\le i\le n}} g_I(\mathrm{id}+u)\, dx^J \wedge du^i \\ &- \sum_{I\in\mathcal{T}_k} \sum_{\substack{(J,K)=I \\ 2\le |K|\le k}} g_I(\mathrm{id}+u)\, dx^J \wedge du^K \\ &+ \sum_{I\in\mathcal{T}_k} \sum_{\substack{(J,i)=I \\ 1\le i\le n}} g_I\, dx^J \wedge du^i + \sum_{I\in\mathcal{T}_k} \langle \operatorname{grad} g_I; u\rangle\, dx^I. \end{aligned}$$

We then let

$$Q_1(u) = \sum_{I\in\mathcal{T}_k} \sum_{\substack{(J,i)=I \\ 1\le i\le n}} \left[g_I - g_I(\mathrm{id}+u)\right]\left[dx^J \wedge du^i\right],$$

$$Q_2(u)=\sum_{I\in\mathcal{T}_k}\left[g_I(\mathrm{id}+u)-g_I-\langle \operatorname{grad} g_I;u\rangle\right]dx^I,$$

$$Q_3(u)=\sum_{I\in\mathcal{T}_k}\sum_{\substack{(J,K)=I\\ 2\le|K|\le k}} g_I(\mathrm{id}+u)\,dx^J\wedge du^K$$

so that

$$Q(u)=Q_1(u)-Q_2(u)-Q_3(u).$$

We therefore have, when $k=2$, that

$$Q_1(u)=\sum_{1\le i<j\le n}\left[g_{ij}-g_{ij}(\mathrm{id}+u)\right]\left[du^i\wedge dx^j+dx^i\wedge du^j\right],$$

$$Q_2(u)=\sum_{1\le i<j\le n}\left[g_{ij}(\mathrm{id}+u)-g_{ij}-\langle \operatorname{grad} g_{ij};u\rangle\right]dx^i\wedge dx^j,$$

$$Q_3(u)=\sum_{1\le i<j\le n} g_{ij}(\mathrm{id}+u)du^i\wedge du^j.$$

Step 3. We now establish the first estimate for each of the Q_p, $p=1,2,3$. So let $u,v\in C^{r+1,\alpha}(\overline{\Omega};\mathbb{R}^n)$ with

$$\|u\|_{C^{1,\gamma}},\|v\|_{C^{1,\gamma}}\le c \quad\text{and}\quad (\mathrm{id}+tu)\left(\overline{\Omega}\right),(\mathrm{id}+tv)\left(\overline{\Omega}\right)\subset\overline{\Omega},\ \forall t\in[0,1].$$

In the sequel, C_i will denote constants that depend only on c and Ω. Since in all cases we will make the estimates component by component, we immediately drop the sum signs. Before starting, we recall (cf. Theorems 16.31 and 16.36) that there exists a constant $C_1=C_1(c,\Omega)$ such that for every $f\in C^{1,\gamma}(\overline{\Omega})$ and every $w,\widetilde{w}\in C^1(\overline{\Omega};\overline{\Omega})$ with $\|w\|_{C^1},\|\widetilde{w}\|_{C^1}\le c$,

$$\|f\circ w\|_{C^{0,\gamma}}\le C_1\|f\|_{C^{0,\gamma}},$$
$$\|f\circ w-f\circ\widetilde{w}\|_{C^0}\le C_1\|f\|_{C^1}\|w-\widetilde{w}\|_{C^0},$$
$$\|f\circ w-f\circ\widetilde{w}\|_{C^{0,\gamma}}\le C_1\|f\|_{C^{1,\gamma}}\|w-\widetilde{w}\|_{C^{0,\gamma}}.$$

Estimate for Q_1. We have

$$\begin{aligned}&\|Q_1(u)-Q_1(v)\|_{C^{0,\gamma}}\\&\quad=\left\|\left[g_I(\mathrm{id})-g_I(\mathrm{id}+u)\right]\left[dx^J\wedge du^i\right]-\left[g_I(\mathrm{id})-g_I(\mathrm{id}+v)\right]\left[dx^J\wedge dv^i\right]\right\|_{C^{0,\gamma}}\\&\quad\le\left\|\left[g_I(\mathrm{id}+v)-g_I(\mathrm{id}+u)\right]\left[dx^J\wedge dv^i\right]\right\|_{C^{0,\gamma}}\\&\qquad+\left\|\left[g_I(\mathrm{id}+u)-g_I(\mathrm{id})\right]\left[dx^J\wedge\left[dv^i-du^i\right]\right]\right\|_{C^{0,\gamma}}.\end{aligned}$$

We therefore get

$$\begin{aligned}\|Q_1(u)-Q_1(v)\|_{C^{0,\gamma}}\le{}&C_2\|[g_I(\mathrm{id}+v)-g_I(\mathrm{id}+u)]\|_{C^0}\|v\|_{C^{1,\gamma}}\\&+C_2\|[g_I(\mathrm{id}+v)-g_I(\mathrm{id}+u)]\|_{C^{0,\gamma}}\|v\|_{C^1}\\&+C_2\|[g_I(\mathrm{id}+u)-g_I(\mathrm{id})]\|_{C^0}\|u-v\|_{C^{1,\gamma}}\\&+C_2\|[g_I(\mathrm{id}+u)-g_I(\mathrm{id})]\|_{C^{0,\gamma}}\|u-v\|_{C^1}.\end{aligned}$$

Hence (bearing in mind that $\|u\|_{C^{1,\gamma}}, \|v\|_{C^{1,\gamma}} \leq c$), we get

$$\begin{aligned}
&\|Q_1(u) - Q_1(v)\|_{C^{0,\gamma}} \\
&\quad \leq C_3 \|g\|_{C^1} \|v-u\|_{C^0} \|v\|_{C^{1,\gamma}} + C_3 \|g\|_{C^{1,\gamma}} \|v-u\|_{C^{0,\gamma}} \|v\|_{C^1} \\
&\qquad + C_3 \|g\|_{C^1} \|u\|_{C^0} \|u-v\|_{C^{1,\gamma}} + C_3 \|g\|_{C^{1,\gamma}} \|u\|_{C^{0,\gamma}} \|u-v\|_{C^1}.
\end{aligned}$$

We thus have

$$\|Q_1(u) - Q_1(v)\|_{C^{0,\gamma}} \leq C \|g\|_{C^{1,\gamma}} (\|u\|_{C^{1,\gamma}} + \|v\|_{C^{1,\gamma}}) \|u-v\|_{C^{1,\gamma}}.$$

Estimate for Q_2. For Q_2 we proceed in the following way. We first observe that

$$\begin{aligned}
Q_2(u) &= \int_0^1 \frac{d}{dt} \left[(g_I(\mathrm{id}+tu) - t\langle \operatorname{grad} g_I(\mathrm{id}); u \rangle) \, dx^I \right] dt \\
&= \int_0^1 \left[\langle \operatorname{grad} g_I(\mathrm{id}+tu) - \operatorname{grad} g_I(\mathrm{id}); u \rangle \, dx^I \right] dt.
\end{aligned}$$

We therefore obtain

$$\begin{aligned}
&\|Q_2(u) - Q_2(v)\|_{C^{0,\gamma}} \\
&\quad \leq \int_0^1 \| \langle \operatorname{grad} g_I(\mathrm{id}+tu) - \operatorname{grad} g_I(\mathrm{id}); u \rangle \\
&\qquad - \langle \operatorname{grad} g_I(\mathrm{id}+tv) - \operatorname{grad} g_I(\mathrm{id}); v \rangle \|_{C^{0,\gamma}} \, dt \\
&\quad \leq \int_0^1 \{ \| \langle \operatorname{grad} g_I(\mathrm{id}+tu) - \operatorname{grad} g_I(\mathrm{id}+tv); u \rangle \|_{C^{0,\gamma}} \\
&\qquad + \| \langle \operatorname{grad} g_I(\mathrm{id}+tv) - \operatorname{grad} g_I(\mathrm{id}); u-v \rangle \|_{C^{0,\gamma}} \} \, dt
\end{aligned}$$

and, hence,

$$\begin{aligned}
&\|Q_2(u) - Q_2(v)\|_{C^{0,\gamma}} \\
&\quad \leq C_2 \int_0^1 \{ \| \operatorname{grad} g_I(\mathrm{id}+tu) - \operatorname{grad} g_I(\mathrm{id}+tv) \|_{C^{0,\gamma}} \|u\|_{C^0} \\
&\qquad + \| \operatorname{grad} g_I(\mathrm{id}+tu) - \operatorname{grad} g_I(\mathrm{id}+tv) \|_{C^0} \|u\|_{C^{0,\gamma}} \\
&\qquad + \| \operatorname{grad} g_I(\mathrm{id}+tv) - \operatorname{grad} g_I(\mathrm{id}) \|_{C^{0,\gamma}} \|u-v\|_{C^0} \\
&\qquad + \| \operatorname{grad} g_I(\mathrm{id}+tv) - \operatorname{grad} g_I(\mathrm{id}) \|_{C^0} \|u-v\|_{C^{0,\gamma}} \} \, dt.
\end{aligned}$$

This leads to (recall that $\|u\|_{C^{1,\gamma}}, \|v\|_{C^{1,\gamma}} \leq c$)

$$\begin{aligned}
&\|Q_2(u) - Q_2(v)\|_{C^{0,\gamma}} \\
&\quad \leq C_3 \|g\|_{C^{2,\gamma}} \|u-v\|_{C^{0,\gamma}} \|u\|_{C^0} + C_3 \|g\|_{C^2} \|u-v\|_{C^0} \|u\|_{C^{0,\gamma}} \\
&\qquad + C_3 \|g\|_{C^{2,\gamma}} \|v\|_{C^{0,\gamma}} \|u-v\|_{C^0} + C_3 \|g\|_{C^2} \|v\|_{C^0} \|u-v\|_{C^{0,\gamma}}.
\end{aligned}$$

We therefore have the estimate

$$\|Q_2(u) - Q_2(v)\|_{C^{0,\gamma}} \leq C \|g\|_{C^{2,\gamma}} (\|u\|_{C^{0,\gamma}} + \|v\|_{C^{0,\gamma}}) \|u-v\|_{C^{0,\gamma}}.$$

Estimate for Q_3. It remains to prove the estimate for Q_3. We get

$$\begin{aligned}
&\|Q_3(u) - Q_3(v)\|_{C^{0,\gamma}} \\
&\quad = \|g_I(\mathrm{id}+v)\, dx^J \wedge dv^K - g_I(\mathrm{id}+u)\, dx^J \wedge du^K\|_{C^{0,\gamma}} \\
&\quad \leq \|g_I(\mathrm{id}+v)(dx^J \wedge (dv^K - du^K))\|_{C^{0,\gamma}} \\
&\qquad + \|(g_I(\mathrm{id}+v) - g_I(\mathrm{id}+u))\, dx^J \wedge du^K\|_{C^{0,\gamma}},
\end{aligned}$$

which leads to (recalling that $\|u\|_{C^{1,\gamma}}, \|v\|_{C^{1,\gamma}} \leq c$ and $|K| \geq 2$, just as in (10.19))

$$\begin{aligned}
\|Q_3(u) - Q_3(v)\|_{C^{0,\gamma}} \leq{}& C_3 \|g\|_{C^{0,\gamma}} (\|u\|_{C^{1,\gamma}} + \|v\|_{C^{1,\gamma}}) \|u - v\|_{C^{1,\gamma}} \\
&+ C_3 \|g\|_{C^{1,\gamma}} \|u - v\|_{C^{0,\gamma}} \|u\|_{C^{1,\gamma}}
\end{aligned}$$

and, thus,

$$\|Q_3(u) - Q_3(v)\|_{C^{0,\gamma}} \leq C \|g\|_{C^{1,\gamma}} (\|u\|_{C^{1,\gamma}} + \|v\|_{C^{1,\gamma}}) \|u - v\|_{C^{1,\gamma}},$$

proving the estimate for Q_3.

Step 4. We next establish the second estimate for each of the Q_p, $p = 1, 2, 3$. So let $u \in C^{r+1,\alpha}(\overline{\Omega}; \mathbb{R}^n)$ with

$$\|u\|_{C^{1,\gamma}} \leq c \quad \text{and} \quad (\mathrm{id}+tu)\left(\overline{\Omega}\right) \subset \overline{\Omega}, \quad \forall\, t \in [0,1].$$

As before, C_i will denote constants that depend only on c, r and Ω. Since in all cases we will make the estimates component by component, we drop the sum signs. We recall (cf. Theorem 16.31) that there exists a constant $C_1 = C_1(c, r, \Omega)$ such that for every $f \in C^{r,\alpha}(\overline{\Omega})$ and every $w \in C^{r,\alpha}(\overline{\Omega}; \overline{\Omega})$ with $\|w\|_{C^1} \leq c$,

$$\|f \circ w\|_{C^{r,\alpha}} \leq C_1 \|f\|_{C^{r,\alpha}} + C_1 \|f\|_{C^1} \|w\|_{C^{r,\alpha}}.$$

We also claim that

$$\|g \circ (\mathrm{id}+u) - g \circ \mathrm{id}\|_{C^{r,\alpha}} \leq C_1 \|g\|_{C^{r+1,\alpha}} \|u\|_{C^1} + C_1 \|g\|_{C^1} \|u\|_{C^{r+1,\alpha}}$$

for every $u \in C^{r+1,\alpha}(\overline{\Omega}; \mathbb{R}^n)$, with

$$\|u\|_{C^1} \leq c \quad \text{and} \quad (\mathrm{id}+u)\left(\overline{\Omega}\right) \subset \overline{\Omega}.$$

Indeed, from Theorem 16.36, we have

$$\begin{aligned}
\|g \circ (\mathrm{id}+u) - g \circ \mathrm{id}\|_{C^{r,\alpha}} \leq{}& C_2 \|g\|_{C^{r+1,\alpha}} \|u\|_{C^0} + C_2 \|g\|_{C^2} [1 + \|u\|_{C^{r,\alpha}}] \|u\|_{C^0} \\
&+ C_2 \|g\|_{C^1} \|u\|_{C^{r,\alpha}},
\end{aligned}$$

and from Corollary 16.27, we get

$$\|g\|_{C^2} \|u\|_{C^{r,\alpha}} \leq C_3 [\|g\|_{C^{r+1,\alpha}} \|u\|_{C^1} + \|g\|_{C^1} \|u\|_{C^{r+1,\alpha}}]. \tag{14.4}$$

Combining the two estimates, we have our claim.

Estimate for Q_1. We have

$$\begin{aligned}\|Q_1(u)\|_{C^{r,\alpha}} &= \|\,[g_I(\mathrm{id}) - g_I(\mathrm{id}+u)]\,\left[dx^J \wedge du^i\right]\|_{C^{r,\alpha}} \\ &\le C_2\|[g_I(\mathrm{id}+u) - g_I(\mathrm{id})]\|_{C^0}\|u\|_{C^{r+1,\alpha}} \\ &\quad + C_2\|[g_I(\mathrm{id}+u) - g_I(\mathrm{id})]\|_{C^{r,\alpha}}\|u\|_{C^1}\,.\end{aligned}$$

We therefore get (bearing in mind that $\|u\|_{C^{1,\gamma}} \le c$)

$$\begin{aligned}\|Q_1(u)\|_{C^{r,\alpha}} &\le C_3\|g\|_{C^1}\|u\|_{C^0}\|u\|_{C^{r+1,\alpha}} \\ &\quad + C_3\,\|g\|_{C^{r+1,\alpha}}\,\|u\|_{C^1}^2 + C_3\,\|g\|_{C^1}\,\|u\|_{C^{r+1,\alpha}}\,\|u\|_{C^1}\end{aligned}$$

and, thus,

$$\|Q_1(u)\|_{C^{r,\alpha}} \le C\,\|g\|_{C^{r+1,\alpha}}\,\|u\|_{C^1} + C\,\|g\|_{C^1}\,\|u\|_{C^{r+1,\alpha}}\,\|u\|_{C^1}\,.$$

Estimate for Q_2. As before, we have that

$$\begin{aligned}Q_2(u) &= \int_0^1 \frac{d}{dt}\left[(g_I(\mathrm{id}+tu) - t\langle \mathrm{grad}\, g_I(\mathrm{id}); u\rangle)\, dx^I\right] dt \\ &= \int_0^1 \left[\langle \mathrm{grad}\, g_I(\mathrm{id}+tu) - \mathrm{grad}\, g_I(\mathrm{id}); u\rangle dx^I\right] dt.\end{aligned}$$

We therefore obtain

$$\begin{aligned}\|Q_2(u)\|_{C^{r,\alpha}} \le C_2 \int_0^1 \{&\|\,\mathrm{grad}\, g_I(\mathrm{id}+tu) - \mathrm{grad}\, g_I(\mathrm{id})\|_{C^{r,\alpha}}\|u\|_{C^0} \\ &+ \|\,\mathrm{grad}\, g_I(\mathrm{id}+tu) - \mathrm{grad}\, g_I(\mathrm{id})\|_{C^0}\|u\|_{C^{r,\alpha}}\}dt\end{aligned}$$

and, hence,

$$\begin{aligned}\|Q_2(u)\|_{C^{r,\alpha}} \le C_2 \int_0^1 \{&[\|\,\mathrm{grad}\, g_I(\mathrm{id}+tu)\|_{C^{r,\alpha}} + \|\,\mathrm{grad}\, g_I\|_{C^{r,\alpha}}]\,\|u\|_{C^0} \\ &+ \|\,\mathrm{grad}\, g_I(\mathrm{id}+tu) - \mathrm{grad}\, g_I(\mathrm{id})\|_{C^0}\|u\|_{C^{r,\alpha}}\}\, dt.\end{aligned}$$

This leads to (recall that $\|u\|_{C^{1,\gamma}} \le c$)

$$\|Q_2(u)\|_{C^{r,\alpha}} \le C_3\,[\|g\|_{C^{r+1,\alpha}} + \|g\|_{C^2}\|u\|_{C^{r,\alpha}}]\,\|u\|_{C^0} + C_3\|g\|_{C^2}\|u\|_{C^0}\|u\|_{C^{r,\alpha}}\,.$$

From (14.4) we get

$$\|Q_2(u)\|_{C^{r,\alpha}} \le C\,\|g\|_{C^{r+1,\alpha}}\,\|u\|_{C^1} + C_3\|g\|_{C^1}\|u\|_{C^{r+1,\alpha}}\|u\|_{C^1}\,.$$

Estimate for Q_3. We immediately have

$$\begin{aligned}\|Q_3(u)\|_{C^{r,\alpha}} &= \|g_I(\mathrm{id}+u)\, dx^J \wedge du^K\|_{C^{r,\alpha}} \\ &\le C_2\|g(\mathrm{id}+u)\|_{C^{r,\alpha}}\|\, du^K\|_{C^0} + C_2\|g\|_{C^0}\|\, du^K\|_{C^{r,\alpha}}\,.\end{aligned}$$

Since $|K| \geq 2$ and $\|u\|_{C^{1,\gamma}} \leq c$, we get

$$\|Q_3(u)\|_{C^{r,\alpha}} \leq C_3 \left[\|g\|_{C^{r,\alpha}} + \|g\|_{C^1} \|u\|_{C^{r,\alpha}}\right] \|u\|_{C^1}^{|K|} + C_3 \|g\|_{C^0} \|u\|_{C^1}^{|K|-1} \|u\|_{C^{r+1,\alpha}}$$

and, thus, since $\|u\|_{C^{1,\gamma}} \leq c$,

$$\|Q_3(u)\|_{C^{r,\alpha}} \leq C\|g\|_{C^{r,\alpha}} \|u\|_{C^1} + C\|g\|_{C^1} \|u\|_{C^{r+1,\alpha}} \|u\|_{C^1} .$$

The combination of the three estimates gives the proof of the lemma. □

14.4.4 The Fixed Point Method

The first proof of Theorem 14.5 relies on the following key theorem (obtained by Bandyopadhyay and Dacorogna [8] under more restrictive hypotheses; as stated, it is due to Dacorogna and Kneuss [32]).

Theorem 14.10. *Let $n > 2$ be even and $\Omega \subset \mathbb{R}^n$ be a bounded open smooth set. Let $r \geq 1$ be an integer and $0 < \gamma \leq \alpha < 1$. Let $g \in C^{r+1,\alpha}(\overline{\Omega};\Lambda^2)$ and $f \in C^{r,\alpha}(\overline{\Omega};\Lambda^2)$ be such that*

$$df = dg = 0 \text{ in } \Omega, \quad \nu \wedge f = \nu \wedge g \text{ on } \partial\Omega,$$

$$\int_\Omega \langle f;\psi\rangle \, dx = \int_\Omega \langle g;\psi\rangle \, dx \quad \text{for every } \psi \in \mathcal{H}_T(\Omega;\Lambda^2),$$

$$\operatorname{rank}[g] = n \text{ in } \overline{\Omega}.$$

Let $c > 0$ be such that

$$\|g\|_{C^0}, \left\|\frac{1}{[g]^{n/2}}\right\|_{C^0} \leq c$$

and define

$$\theta(g) = \frac{1}{\|g\|_{C^{1,\gamma}}^2} \min\left\{\|g\|_{C^{1,\gamma}}, \frac{1}{\|g\|_{C^{2,\gamma}}}, \frac{1}{\|g\|_{C^{r+1,\alpha}}}\right\}.$$

There exists $C = C(c,r,\alpha,\gamma,\Omega) > 0$ such that if

$$\|f-g\|_{C^{0,\gamma}} \leq C\theta(g) \quad \text{and} \quad \|f-g\|_{C^{0,\gamma}} \leq C\frac{\|f-g\|_{C^{r,\alpha}}}{\|g\|_{C^{1,\gamma}}\|g\|_{C^{r+1,\alpha}}}, \tag{14.5}$$

then there exists $\varphi \in \operatorname{Diff}^{r+1,\alpha}(\overline{\Omega};\overline{\Omega})$ verifying

$$\varphi^*(g) = f \text{ in } \Omega \quad \text{and} \quad \varphi = \text{id on } \partial\Omega. \tag{14.6}$$

Furthermore, there exists $\widetilde{C} = \widetilde{C}(c, r, \alpha, \gamma, \Omega) > 0$ *such that*

$$\|\varphi - \mathrm{id}\|_{C^{r+1,\alpha}} \leq \widetilde{C}\|g\|_{C^{r+1,\alpha}}\|f-g\|_{C^{r,\alpha}}.$$

Remark 14.11. (i) Note that since $g \in C^{r+1,\alpha}(\overline{\Omega};\Lambda^2)$ and $\nu \wedge f = \nu \wedge g$ on $\partial\Omega$, then $\nu \wedge f \in C^{r+1,\alpha}(\partial\Omega;\Lambda^3)$.

(ii) With essentially the same proof, but replacing the last estimate of Lemma 14.8 by the corresponding one in Remark 14.9, we get the following result. In addition to the hypotheses of the theorem, let $0 < \gamma \leq \alpha < \beta < 1$, $g \in C^{r+1,\beta}(\overline{\Omega};\Lambda^2)$, $f \in C^{r,\alpha}(\overline{\Omega};\Lambda^2)$ and $c > 0$ be such that

$$\begin{gathered} df = dg = 0 \text{ in } \Omega, \quad \nu \wedge f = \nu \wedge g \text{ on } \partial\Omega, \\ \int_\Omega \langle f;\psi\rangle\, dx = \int_\Omega \langle g;\psi\rangle\, dx \quad \text{for every } \psi \in \mathscr{H}_T(\Omega;\Lambda^2), \\ \operatorname{rank}[g] = n \text{ in } \overline{\Omega}, \\ \|g\|_{C^0}, \quad \left\|\frac{1}{[g]^{n/2}}\right\|_{C^0} \leq c. \end{gathered}$$

Define

$$\theta(g) = \frac{1}{\|g\|_{C^{1,\gamma}}} \min \left\{ \begin{array}{c} 1, \dfrac{1}{\|g\|_{C^{1,\gamma}}\|g\|_{C^{2,\gamma}}}, \dfrac{1}{\|g\|_{C^{r+1,\alpha}}^2}, \\ \left[\dfrac{1}{\|g\|_{C^{1,\gamma}}\|g\|_{C^{r+1,\beta}}}\right]^{\frac{1}{\beta-\alpha}} \end{array} \right\}.$$

There exist $C = C(c, r, \alpha, \beta, \gamma, \Omega) > 0$ and $\widetilde{C} = \widetilde{C}(c, r, \alpha, \beta, \gamma, \Omega) > 0$ such that if (compare with (14.5))

$$\|f-g\|_{C^{0,\gamma}} \leq C\theta(g),$$

then there exists $\varphi \in \mathrm{Diff}^{r+1,\alpha}(\overline{\Omega};\overline{\Omega})$ verifying

$$\varphi^*(g) = f \text{ in } \Omega \quad \text{and} \quad \varphi = \mathrm{id} \text{ on } \partial\Omega$$

and

$$\|\varphi - \mathrm{id}\|_{C^{r+1,\alpha}} \leq \widetilde{C}\|g\|_{C^{r+1,\alpha}}\|f-g\|_{C^{r,\alpha}}.$$

Proof. The theorem will follow from Theorem 18.1. We divide the proof into five steps; the first four to verify the hypotheses of the theorem and the last one to conclude.

Step 1. We define the spaces as follows:

$$\begin{gathered} X_1 = C^{1,\gamma}(\overline{\Omega};\mathbb{R}^n) \quad \text{and} \quad Y_1 = C^{0,\gamma}(\overline{\Omega};\Lambda^2), \\ X_2 = \{a \in C^{r+1,\alpha}(\overline{\Omega};\mathbb{R}^n) : a = 0 \text{ on } \partial\Omega\}, \end{gathered}$$

$$Y_2 = \left\{ b \in C^{r,\alpha}(\overline{\Omega};\Lambda^2) : \left[\begin{array}{c} db = 0 \text{ in } \Omega, \quad \nu \wedge b = 0 \text{ on } \partial\Omega, \\ \int_\Omega \langle b;\psi\rangle\, dx = 0, \ \forall \psi \in \mathscr{H}_T(\Omega;\Lambda^2) \end{array} \right] \right\}.$$

It is easily seen that they satisfy Hypothesis (H_{XY}) of Theorem 18.1 (see Proposition 16.23).

Step 2. Define $L : X_2 \to Y_2$ by

$$La = d[a \lrcorner g] = b.$$

We will show that there exist $L^{-1} : Y_2 \to X_2$, a linear right inverse of L and a constant $K_1 = K_1(c, r, \alpha, \gamma, \Omega)$ such that, defining

$$k_1 = K_1 \|g\|_{C^{1,\gamma}} \quad \text{and} \quad k_2 = K_1 \|g\|_{C^{r+1,\alpha}},$$

we get

$$\|L^{-1}b\|_{X_i} \le k_i \|b\|_{Y_i} \text{ for every } b \in Y_2 \text{ and } i = 1, 2.$$

Once this is shown, (H_L) of Theorem 18.1 will be satisfied.

Step 2.1. Indeed, we first solve, using Theorem 8.16, the equation

$$\begin{cases} dw = b & \text{in } \Omega, \\ w = 0 & \text{on } \partial\Omega \end{cases}$$

and find $w \in C^{r+1,\alpha}(\overline{\Omega}; \Lambda^1)$ and $C_1 = C_1(r, \alpha, \gamma, \Omega) > 0$ such that

$$\|w\|_{C^{r+1,\alpha}} \le C_1 \|b\|_{C^{r,\alpha}} \quad \text{and} \quad \|w\|_{C^{1,\gamma}} \le C_1 \|b\|_{C^{0,\gamma}}.$$

Moreover, the correspondence $b \to w$ can be chosen to be linear.

Step 2.2. Since $\operatorname{rank}[g] = n$, we can find a unique $a \in C^{r+1,\alpha}(\overline{\Omega}; \mathbb{R}^n)$ so that

$$a \lrcorner g = w,$$

which is equivalent to

$$a = [\overline{g}]^{-1} w.$$

Define $L^{-1} : Y_2 \to X_2$ by $L^{-1}(b) = a$. First, note that L^{-1} is linear and that

$$LL^{-1} = \text{id} \quad \text{on } Y_2.$$

Moreover, using Theorem 16.28, Corollary 16.30 and Step 2.1, we can find constants $C_i = C_i(c, r, \alpha, \gamma, \Omega)$, $i = 2, 3, 4$, such that

$$\begin{aligned} \|a\|_{C^{r+1,\alpha}} &\le C_2 \left\|(\overline{g})^{-1}\right\|_{C^{r+1,\alpha}} \|w\|_{C^0} + C_2 \left\|(\overline{g})^{-1}\right\|_{C^0} \|w\|_{C^{r+1,\alpha}} \\ &\le C_3 \|g\|_{C^{r+1,\alpha}} \|b\|_{C^{0,\gamma}} + C_3 \|g\|_{C^0} \|b\|_{C^{r,\alpha}} \\ &\le C_4 \|g\|_{C^{r+1,\alpha}} \|b\|_{C^{r,\alpha}} \end{aligned}$$

and, similarly,

$$\|a\|_{C^{1,\gamma}} \le C_4 \|g\|_{C^{1,\gamma}} \|b\|_{C^{0,\gamma}}.$$

Thus, the claim of Step 2 is valid.

Step 3. We define

$$Q(u) = g - (\mathrm{id}+u)^*(g) + d[u \lrcorner g].$$

We will verify that Property (H_Q) of Theorem 18.1 holds with $\rho = 1/(2n)$. The fact that $Q(0) = 0$ is evident.

Step 3.1. According to Lemma 14.8, there exists a constant $K_2 = K_2(r,\Omega)$ such that the following estimates hold:

$$\|Q(u) - Q(v)\|_{C^{0,\gamma}} \le K_2 \|g\|_{C^{2,\gamma}}(\|u\|_{C^{1,\gamma}} + \|v\|_{C^{1,\gamma}})\|u - v\|_{C^{1,\gamma}},$$

$$\|Q(u)\|_{C^{r,\alpha}} \le K_2 \|g\|_{C^{r+1,\alpha}} \|u\|_{C^1} + K_2 \|g\|_{C^1} \|u\|_{C^{r+1,\alpha}} \|u\|_{C^1}$$

for every $u, v \in C^{r+1,\alpha}(\overline{\Omega};\mathbb{R}^n)$, with

$$\|u\|_{C^{1,\gamma}}, \|v\|_{C^{1,\gamma}} \le 1/(2n),$$

$$(\mathrm{id}+tu)(\overline{\Omega}), (\mathrm{id}+tv)(\overline{\Omega}) \subset \overline{\Omega},\ \forall t \in [0,1].$$

We finally let

$$c_1(t_1,t_2) = K_2\|g\|_{C^{2,\gamma}}(t_1 + t_2),$$

$$c_2(t_1,t_2) = K_2\|g\|_{C^{r+1,\alpha}} t_1 + K_2\|g\|_{C^1} t_1 t_2.$$

Note that if

$$F(t,x) = x + tu(x) \quad \text{and} \quad \|u\|_{C^1} \le 1/(2n),$$

then for every $t \in [0,1]$,

$$\det \nabla_x F(t,x) = \det(I + t\nabla u(x)) > 0, \quad x \in \overline{\Omega}.$$

Therefore, if $u = 0$ on $\partial\Omega$, then, appealing to Theorem 19.12, we get that

$$F(t,x) \in \overline{\Omega} \quad \text{for every } (t,x) \in [0,1] \times \overline{\Omega}.$$

Thus, (18.1) and (18.2) have been verified.

Step 3.2. Let us check that

$$Q : \{u \in X_2 : \|u\|_{X_1} \le 1/(2n)\} \to Y_2$$

is well defined. We have to prove that

$$dQ(u) = 0 \text{ in } \Omega, \quad \nu \wedge Q(u) = 0 \text{ on } \partial\Omega,$$

$$\int_\Omega \langle Q(u); \psi \rangle\, dx = 0,\ \forall \psi \in \mathcal{H}_T(\Omega;\Lambda^2).$$

(i) The first condition follows immediately since $dg = 0$ and

$$dQ(u) = dg - (\mathrm{id}+u)^*(dg) + dd[u \lrcorner g].$$

(ii) The second one is true since $u=0$ on $\partial\Omega$. Indeed, clearly (using the notations Q_i used in the proof of Lemma 14.8),

$$Q_1(u)=Q_2(u)=0 \text{ on } \partial\Omega.$$

Since $u=0$ on $\partial\Omega$, each of $\operatorname{grad} u^i$ and $\operatorname{grad} u^j$ is parallel to the normal ν. Thus, $du^i\wedge du^j=0$ on $\partial\Omega$ for every $i<j$, which implies that

$$Q_3(u)=0 \text{ on } \partial\Omega.$$

Thus, we have, in fact, proved that $Q(u)=0$ on $\partial\Omega$.

(iii) Choosing $F(t,x)=x+tu(x)$ in Remark 17.4, we find that there exists Φ such that

$$\begin{cases} d\Phi = g-(\mathrm{id}+u)^*(g) & \text{in } \Omega, \\ \Phi=0 & \text{on } \partial\Omega. \end{cases}$$

Since $\Psi=\Phi+u\lrcorner g$ satisfies

$$\begin{cases} d\Psi = Q(u) & \text{in } \Omega, \\ \Psi=0 & \text{on } \partial\Omega, \end{cases}$$

we have the claim, namely

$$\int_\Omega \langle Q(u);\psi\rangle\, dx=0,\ \forall\,\psi\in\mathscr{H}_T(\Omega;\Lambda^2).$$

Step 4. With the definition of L and Q in hand, we now rewrite (14.6) as follows. Setting $\varphi=\mathrm{id}+u$, the equation $\varphi^*(g)=f$ becomes

$$\begin{aligned} Lu=d[u\lrcorner g] &= f-(\mathrm{id}+u)^*(g)+d[u\lrcorner g] \\ &= f-g+[g-(\mathrm{id}+u)^*(g)+d[u\lrcorner g]] \\ &= f-g+Q(u). \end{aligned}$$

In order to apply Theorem 18.1, it remains to see how the hypotheses

$$\begin{gathered} 2k_1\|f-g\|_{C^{0,\gamma}}\le 1/(2n), \\ 2k_1c_1(2k_1\|f-g\|_{C^{0,\gamma}},2k_1\|f-g\|_{C^{0,\gamma}})\le 1, \\ c_2(2k_1\|f-g\|_{C^{0,\gamma}},2k_2\|f-g\|_{C^{r,\alpha}})\le \|f-g\|_{C^{r,\alpha}} \end{gathered} \tag{14.7}$$

translate in our context.

(i) The first one leads to

$$\|f-g\|_{C^{0,\gamma}}\le \frac{1}{4nK_1\|g\|_{C^{1,\gamma}}}.$$

(ii) The second one gives

$$\|f-g\|_{C^{0,\gamma}} \leq \frac{1}{8K_1^2K_2\|g\|_{C^{1,\gamma}}^2\|g\|_{C^{2,\gamma}}}.$$

(iii) The third condition reads as

$$\begin{aligned}
&K_2\|g\|_{C^{r+1,\alpha}}\left(2K_1\|g\|_{C^{1,\gamma}}\|f-g\|_{C^{0,\gamma}}\right) \\
&+K_2\|g\|_{C^1}\left(2K_1\|g\|_{C^{1,\gamma}}\|f-g\|_{C^{0,\gamma}}\right)\left(2K_1\|g\|_{C^{r+1,\alpha}}\|f-g\|_{C^{r,\alpha}}\right) \\
&\quad\leq \|f-g\|_{C^{r,\alpha}}.
\end{aligned}$$

Note that the third condition is verified if

$$2K_1K_2\|g\|_{C^{r+1,\alpha}}\|g\|_{C^{1,\gamma}}\|f-g\|_{C^{0,\gamma}} \leq \frac{1}{2}\|f-g\|_{C^{r,\alpha}}$$

and

$$4K_1^2K_2\|g\|_{C^1}\|g\|_{C^{1,\gamma}}\|g\|_{C^{r+1,\alpha}}\|f-g\|_{C^{0,\gamma}}\|f-g\|_{C^{r,\alpha}} \leq \frac{1}{2}\|f-g\|_{C^{r,\alpha}}.$$

The first one leads to

$$\|f-g\|_{C^{0,\gamma}} \leq \frac{\|f-g\|_{C^{r,\alpha}}}{4K_1K_2\|g\|_{C^{r+1,\alpha}}\|g\|_{C^{1,\gamma}}}$$

and the second one is verified if

$$\|f-g\|_{C^{0,\gamma}} \leq \frac{1}{8K_1^2K_2\|g\|_{C^{r+1,\alpha}}\|g\|_{C^{1,\gamma}}^2}.$$

Combining the four conditions, we have just obtained, letting

$$\theta(g) = \frac{1}{\|g\|_{C^{1,\gamma}}^2}\min\left\{\|g\|_{C^{1,\gamma}}, \frac{1}{\|g\|_{C^{2,\gamma}}}, \frac{1}{\|g\|_{C^{r+1,\alpha}}}\right\},$$

that there exists $C = C(c, r, \alpha, \gamma, \Omega) > 0$ such that the inequalities (14.7) are satisfied if

$$\|f-g\|_{C^{0,\gamma}} \leq C\theta(g) \quad \text{and} \quad \|f-g\|_{C^{0,\gamma}} \leq C\frac{\|f-g\|_{C^{r,\alpha}}}{\|g\|_{C^{1,\gamma}}\|g\|_{C^{r+1,\alpha}}}.$$

Step 5. The hypotheses of Theorem 18.1 having been verified, we conclude that there exists $u \in C^{r+1,\alpha}(\overline{\Omega};\mathbb{R}^n)$, with $\|u\|_{C^{1,\gamma}} \leq 1/(2n)$, satisfying $u = 0$ on $\partial\Omega$ and

$$Lu = d[u \lrcorner g] = f - g + Q(u) = f - (\mathrm{id}+u)^*(g) + d[u \lrcorner g].$$

Letting $\varphi = \mathrm{id}+u$, we therefore have found that

$$\varphi^*(g) = f \text{ in } \Omega.$$

Since $u=0$ on $\partial\Omega$, we have that $\varphi=\mathrm{id}$ on $\partial\Omega$. Since $\|u\|_{C^1}\leq 1/(2n)$, we deduce that

$$\det\nabla\varphi>0 \quad \text{in } \overline{\Omega},$$

and therefore, according to Theorem 19.12, we find that $\varphi\in\mathrm{Diff}^{r+1,\alpha}\left(\overline{\Omega};\overline{\Omega}\right)$. Moreover, by construction (cf. (18.5)),

$$\|u\|_{C^{r+1,\alpha}}\leq 2k_2\|f-g\|_{C^{r,\alpha}},$$

which implies the desired estimate, namely

$$\|\varphi-\mathrm{id}\|_{C^{r+1,\alpha}}\leq\widetilde{C}\|g\|_{C^{r+1,\alpha}}\|f-g\|_{C^{r,\alpha}}.$$

The proof is thus complete. □

14.4.5 A First Proof of the Main Theorem

We first prove Theorem 14.5 for special f and general g with extra regularity and under a smallness assumption.

Proposition 14.12. *Let $\Omega\subset\mathbb{R}^n$ be a bounded open smooth set, $r\geq 1$, $0<\alpha<\beta<1$ and $g\in C^{r,\beta}\left(\overline{\Omega};\Lambda^2\right)$ with*

$$\nu\wedge g\in C^{r+1,\alpha}(\partial\Omega;\Lambda^3),\quad dg=0 \quad \text{and} \quad \operatorname{rank}[g]=n \quad \text{in } \overline{\Omega}.$$

Then for every ε small, there exist $g_\varepsilon\in C^{r+1,\alpha}\left(\overline{\Omega};\Lambda^2\right)$ and $\varphi_\varepsilon\in\mathrm{Diff}^{r+1,\alpha}\left(\overline{\Omega};\overline{\Omega}\right)$ such that

$$\begin{cases}\varphi_\varepsilon^*(g_\varepsilon)=g & \text{in } \Omega,\\ \varphi_\varepsilon=\mathrm{id} & \text{on } \partial\Omega,\end{cases}$$

$$dg_\varepsilon=0,\quad \nu\wedge g_\varepsilon=\nu\wedge g \ \text{ on } \partial\Omega,$$

$$\int_\Omega\langle g_\varepsilon;\psi\rangle=\int_\Omega\langle g;\psi\rangle,\ \forall\,\psi\in\mathscr{H}_T(\Omega;\Lambda^2),$$

$$\lim_{\varepsilon\to 0}\|g_\varepsilon-g\|_{C^{r,\alpha}(\overline{\Omega})}=0.$$

Moreover, there exists $C=C(r,\alpha,\beta,\Omega,\|g\|_{C^{1,\alpha}},\|1/g^{n/2}\|_{C^0})$, such that for every ε small,

$$\|\varphi_\varepsilon-\mathrm{id}\|_{C^{r+1,\alpha}(\overline{\Omega})}\leq C\frac{\varepsilon^{\beta-\alpha}}{\beta-\alpha}\|g\|_{C^{r,\beta}(\overline{\Omega})}+C\varepsilon\|\nu\wedge g\|_{C^{r+1,\alpha}(\partial\Omega)}. \tag{14.8}$$

Proof. For the sake of alleviating the notations we will write in the present proof, for example, $\|g\|_{C^{r,\beta}}$ instead of $\|g\|_{C^{r,\beta}(\overline{\Omega})}$. However, when we will be considering norms on the boundary of Ω, we will keep the notation $\|g\|_{C^{r,\beta}(\partial\Omega)}$.

Step 1 (definition of g_ε). Apply Theorem 16.49 and Remark 16.50(v) and get, for every $\varepsilon \in (0,1]$, that there exist $g_\varepsilon \in C^{r+1,\alpha}(\overline{\Omega};\Lambda^2)$ and a constant $C_1 = C_1(r,\alpha,\beta,\Omega)$ such that

$$dg_\varepsilon = 0 \quad \text{in } \Omega, \qquad \nu \wedge g_\varepsilon = \nu \wedge g \quad \text{on } \partial\Omega, \tag{14.9}$$

$$\int_\Omega \langle g_\varepsilon;\psi\rangle = \int_\Omega \langle g;\psi\rangle \quad \text{for every } \psi \in \mathscr{H}_T(\Omega;\Lambda^2), \tag{14.10}$$

$$\|g_\varepsilon\|_{C^{r+1,\alpha}} \le \frac{C_1}{\varepsilon^{1+\alpha-\beta}}\|g\|_{C^{r,\beta}} + C_1\|\nu\wedge g\|_{C^{r+1,\alpha}(\partial\Omega)}, \tag{14.11}$$

$$\|g_\varepsilon - g\|_{C^{r,\alpha}} \le C_1\varepsilon^{\beta-\alpha}\|g\|_{C^{r,\beta}}, \tag{14.12}$$

$$\left\|\frac{d}{d\varepsilon}g_\varepsilon\right\|_{C^{0,\alpha}} \le C_1\|g\|_{C^{1,\alpha}} \quad \text{and} \quad \left\|\frac{d}{d\varepsilon}g_\varepsilon\right\|_{C^{r,\alpha}} \le C_1\varepsilon^{\beta-\alpha-1}\|g\|_{C^{r,\beta}}. \tag{14.13}$$

Moreover, defining $G:(0,1]\times\overline{\Omega}\to\Lambda^2$ by $G(\varepsilon,x) = g_\varepsilon(x)$, we have

$$G \in C^{r+1,\alpha}\left((0;1]\times\overline{\Omega};\Lambda^2\right) \quad \text{and} \quad \frac{\partial}{\partial\varepsilon}G \in C^\infty\left((0;1]\times\overline{\Omega};\Lambda^2\right). \tag{14.14}$$

Since $\operatorname{rank}[g] = n$ in $\overline{\Omega}$ (which is equivalent to $g^{n/2}(x) \neq 0$ for every $x \in \overline{\Omega}$) and since (14.12) holds, there exists $\overline{\varepsilon} \le 1$ such that for every $\varepsilon \in (0,\overline{\varepsilon}]$,

$$\begin{gathered}\|g_\varepsilon\|_{C^0} \le 2\|g\|_{C^0}, \quad \|g_\varepsilon\|_{C^1} \le 2\|g\|_{C^1}, \\ \|1/(g_\varepsilon)^{n/2}\|_{C^0} \le 2\|1/g^{n/2}\|_{C^0}.\end{gathered} \tag{14.15}$$

Hence, combining (14.15) and Notation (v) in Section 14.1, we deduce that for every $\varepsilon \in (0,\overline{\varepsilon}]$,

$$\|(\overline{g}_\varepsilon)^{-1}\|_{C^1} \le C_2\|g_\varepsilon\|_{C^1} \quad \text{and} \quad \|(\overline{g}_\varepsilon)^{-1}\|_{C^{r+1,\alpha}} \le C_2\|g_\varepsilon\|_{C^{r+1,\alpha}}, \tag{14.16}$$

where $C_2 = C_2(r,\Omega,\|g\|_{C^0},\|1/g^{n/2}\|_{C^0})$.

Step 2. In this step we will show that for every $\varepsilon \in (0,\overline{\varepsilon}]$, there exist $u_\varepsilon \in C^{r+1,\alpha}(\overline{\Omega};\Lambda^1)$ and a constant $C_3 = C_3(r,\alpha,\beta,\Omega,\|g\|_{C^{1,\alpha}},\|1/g^{n/2}\|_{C^0})$ such that $u_\varepsilon = 0$ on $\partial\Omega$ and

$$d(u_\varepsilon \lrcorner g_\varepsilon) = -\frac{d}{d\varepsilon}g_\varepsilon \quad \text{in } \Omega, \tag{14.17}$$

$$\|u_\varepsilon\|_{C^{r+1,\alpha}} \le \frac{C_3}{\varepsilon^{1+\alpha-\beta}}\|g\|_{C^{r,\beta}} + C_3\|\nu\wedge g\|_{C^{r+1,\alpha}(\partial\Omega)}, \tag{14.18}$$

$$\|u_\varepsilon\|_{C^1} \le C_3. \tag{14.19}$$

Moreover, defining $u:(0,\overline{\varepsilon}]\times\overline{\Omega}\to\Lambda^1$ by $u(\varepsilon,x) = u_\varepsilon(x)$, we will show that $u \in C^{r+1,\alpha}\left((0,\overline{\varepsilon}]\times\overline{\Omega};\Lambda^1\right)$.

Step 2.1. Since (14.9), (14.10) and (14.14) hold, using Theorem 8.16 we can find, for every $\varepsilon \in (0,\overline{\varepsilon}]$, $w_\varepsilon \in C^\infty(\overline{\Omega};\Lambda^1)$ and a constant $C_4 = C_4(r,\alpha,\Omega)$ such that

$$dw_\varepsilon = -\frac{d}{d\varepsilon}g_\varepsilon \quad \text{in } \Omega, \qquad w_\varepsilon = 0 \quad \text{on } \partial\Omega$$

and, for every integer $q \le r$,

$$\|w_\varepsilon\|_{C^{q+1,\alpha}} \le C_4 \left\| \frac{d}{d\varepsilon} g_\varepsilon \right\|_{C^{q,\alpha}} . \tag{14.20}$$

Moreover, defining $w : (0,\overline{\varepsilon}] \times \overline{\Omega} \to \Lambda^1$ by $w(\varepsilon,x) = w_\varepsilon(x)$, we have, using (14.14), $w \in C^\infty\left((0,\overline{\varepsilon}] \times \overline{\Omega};\Lambda^1\right)$.

Step 2.2. Since by (14.15), we have, for every $\varepsilon \in (0,\overline{\varepsilon}]$, $\operatorname{rank}[g_\varepsilon] = n$ in $\overline{\Omega}$, there exists a unique $u_\varepsilon : \overline{\Omega} \to \Lambda^1$ verifying

$$u_\varepsilon \lrcorner g_\varepsilon = w_\varepsilon .$$

Note that $u_\varepsilon \in C^{r+1,\alpha}\left(\overline{\Omega};\Lambda^1\right)$ and that $u_\varepsilon = 0$ on $\partial\Omega$. Moreover, defining $u : (0,\overline{\varepsilon}] \times \overline{\Omega} \to \Lambda^1$ by $u(\varepsilon,x) = u_\varepsilon(x)$, we have $u \in C^{r+1,\alpha}\left((0,\overline{\varepsilon}] \times \overline{\Omega};\Lambda^1\right)$.

Step 2.3. To show Step 2, it only remains to prove (14.18) and (14.19). Using Theorem 16.28, (14.15), (14.16) and (14.20), it follows that

$$\begin{aligned}
\|u_\varepsilon\|_{C^{r+1,\alpha}} &= \|(\overline{g}_\varepsilon)^{-1} w_\varepsilon\|_{C^{r+1,\alpha}} \\
&\le C_5 \|(\overline{g}_\varepsilon)^{-1}\|_{C^{r+1,\alpha}} \|w_\varepsilon\|_{C^0} + C_5 \|(\overline{g}_\varepsilon)^{-1}\|_{C^0} \|w_\varepsilon\|_{C^{r+1,\alpha}} \\
&\le C_6 \|g_\varepsilon\|_{C^{r+1,\alpha}} \|w_\varepsilon\|_{C^{1,\alpha}} + C_6 \|w_\varepsilon\|_{C^{r+1,\alpha}} \\
&\le C_7 \|g_\varepsilon\|_{C^{r+1,\alpha}} \left\| \frac{d}{d\varepsilon} g_\varepsilon \right\|_{C^{0,\alpha}} + C_7 \left\| \frac{d}{d\varepsilon} g_\varepsilon \right\|_{C^{r,\alpha}}
\end{aligned}$$

and thus, invoking (14.11) and (14.13),

$$\begin{aligned}
&\|u_\varepsilon\|_{C^{r+1,\alpha}} \\
&\le C_8 \left(\frac{1}{\varepsilon^{1+\alpha-\beta}} \|g\|_{C^{r,\beta}} + \|\nu \wedge g\|_{C^{r+1,\alpha}(\partial\Omega)} \right) \|g\|_{C^{1,\alpha}} + \frac{C_8}{\varepsilon^{1+\alpha-\beta}} \|g\|_{C^{r,\beta}} \\
&\le \frac{C_9}{\varepsilon^{1+\alpha-\beta}} \|g\|_{C^{r,\beta}} + C_9 \|\nu \wedge g\|_{C^{r+1,\alpha}(\partial\Omega)},
\end{aligned}$$

where $C_i = C_i(r,\alpha,\beta,\Omega,\|g\|_{C^{1,\alpha}},\|1/g^{n/2}\|_{C^0})$. We similarly obtain

$$\begin{aligned}
\|u_\varepsilon\|_{C^1} &= \|(\overline{g}_\varepsilon)^{-1} w_\varepsilon\|_{C^1} \le C_{10} \|(\overline{g}_\varepsilon)^{-1}\|_{C^1} \|w_\varepsilon\|_{C^1} \\
&\le C_{11} \|g\|_{C^1} \left\| \frac{d}{d\varepsilon} g_\varepsilon \right\|_{C^{0,\alpha}} \le C_{12} \|g\|_{C^1} \|g\|_{C^{1,\alpha}} \le C_{13},
\end{aligned}$$

where $C_i = C_i(r,\alpha,\beta,\Omega,\|g\|_{C^{1,\alpha}},\|1/g^{n/2}\|_{C^0})$. This shows the assertion.

Step 3. We can now conclude the proof.

Step 3.1. Since $u \in C^{r+1,\alpha}((0,\overline{\varepsilon}] \times \overline{\Omega} : \mathbb{R}^n)$, $u_\varepsilon = 0$ on $\partial\Omega$ and by (14.18),

$$\int_0^{\overline{\varepsilon}} \|u_\varepsilon\|_{C^{r+1,\alpha}} d\varepsilon < \infty,$$

we deduce, using Theorem 12.1, that the solution $\varphi : [0,\overline{\varepsilon}] \times \overline{\Omega} \to \overline{\Omega}$, $\varphi(\varepsilon,x) = \varphi_\varepsilon(x)$, of

$$\begin{cases} \dfrac{d}{d\varepsilon}\varphi_\varepsilon = u_\varepsilon \circ \varphi_\varepsilon, & 0 < \varepsilon \le \overline{\varepsilon}, \\ \varphi_0 = \mathrm{id} \end{cases}$$

verifies

$$\varphi \in C^{r+1,\alpha}([0,\overline{\varepsilon}] \times \overline{\Omega};\overline{\Omega}) \tag{14.21}$$

and that for every $\varepsilon \in [0,\overline{\varepsilon}]$,

$$\varphi_\varepsilon \in \mathrm{Diff}^{r+1,\alpha}(\overline{\Omega};\overline{\Omega}) \quad \text{and} \quad \varphi_\varepsilon = \mathrm{id} \text{ on } \partial\Omega.$$

Finally, inserting (14.18) and (14.19) in (12.3), we immediately deduce (14.8).

Step 3.2. Since (14.17) holds, we deduce, using Theorem 12.7, that for every $0 < \varepsilon_1 \le \varepsilon_2 \le \overline{\varepsilon}$,

$$\varphi_{\varepsilon_2}^*(g_{\varepsilon_2}) = \varphi_{\varepsilon_1}^*(g_{\varepsilon_1}) \quad \text{in } \Omega.$$

Since, using (14.12) and (14.21),

$$\lim_{\varepsilon\to 0} \|g_\varepsilon - g\|_{C^0} = 0 \quad \text{and} \quad \lim_{\varepsilon\to 0} \|\varphi_\varepsilon - \varphi_0\|_{C^1} = 0,$$

we immediately infer that for every $\varepsilon \in (0,\overline{\varepsilon}]$,

$$\varphi_\varepsilon^*(g_\varepsilon) = \varphi_0^*(g) = g.$$

The proof is therefore complete. □

We can now go back to the first proof of Theorem 14.5 using an iteration scheme involving appropriate regularization.

Proof. We split the proof into three steps.

Step 1 (approximation of g and f). Choose $\gamma \in (0,\alpha)$ and $\delta > 0$ with $2\delta \le \alpha - \gamma$ and $\alpha + 2\delta < 1$. We next regularize g and f with the help of Theorem 16.49 (and Remark 16.50(v)) and construct for every $\varepsilon \in (0,1]$, $g_\varepsilon, f_\varepsilon \in C^{r+1,\alpha}(\overline{\Omega};\Lambda^2)$ such that

$$dg_\varepsilon = df_\varepsilon = 0, \quad \nu \wedge g_\varepsilon = \nu \wedge g = \nu \wedge f = \nu \wedge f_\varepsilon \text{ on } \partial\Omega,$$

$$\int_\Omega \langle g_\varepsilon;\psi\rangle = \int_\Omega \langle g;\psi\rangle = \int_\Omega \langle f;\psi\rangle = \int_\Omega \langle f_\varepsilon;\psi\rangle, \ \forall \psi \in \mathscr{H}_T(\Omega;\Lambda^2),$$

$$\|g_\varepsilon - g\|_{C^{0,\gamma}} \le C\varepsilon^{r+\alpha-\gamma}\|g\|_{C^{r,\alpha}},$$

$$\|g_\varepsilon - g\|_{C^{1,\gamma}} \le C\varepsilon^{r-1+\alpha-\gamma}\|g\|_{C^{r,\alpha}},$$

$$\|g_\varepsilon\|_{C^{r+1,\alpha}} \le \frac{C}{\varepsilon}\|g\|_{C^{r,\alpha}} + C\|\nu \wedge g\|_{C^{r+1,\alpha}(\partial\Omega)},$$

$$\|g_\varepsilon\|_{C^{r,\alpha+2\delta}} \leq \frac{C}{\varepsilon^{2\delta}}\|g\|_{C^{r,\alpha}} + C\|\nu \wedge g\|_{C^{r,\alpha+2\delta}(\partial\Omega)},$$

$$\|g_\varepsilon\|_{C^{2,\gamma}} \leq \frac{C}{\varepsilon}\|g\|_{C^{1,\gamma}} + C\|\nu \wedge g\|_{C^{2,\gamma}(\partial\Omega)},$$

where $C = C(r,\alpha,\gamma,\delta,\Omega) > 0$ and similarly for f and f_ε. Note that using the first inequality above, there exists $\overline{\varepsilon}$ such that for every $\varepsilon_1, \varepsilon_2 \leq \overline{\varepsilon}$, we have that

$$\operatorname{rank}\left[t g_{\varepsilon_1} + (1-t) f_{\varepsilon_2}\right] = n \quad \text{in } \overline{\Omega} \text{ and for every } t \in [0,1].$$

Step 2. In this step we show that there exist $\varepsilon_1, \varepsilon_2 \leq \overline{\varepsilon}$ and $\varphi_1, \varphi_3 \in \operatorname{Diff}^{r+1,\alpha}\left(\overline{\Omega};\overline{\Omega}\right)$ such that

$$\begin{cases} \varphi_1^*(g_{\varepsilon_1}) = g & \text{in } \Omega, \\ \varphi_1 = \text{id} & \text{on } \partial\Omega \end{cases} \quad \text{and} \quad \begin{cases} \varphi_3^*(f_{\varepsilon_2}) = f & \text{in } \Omega, \\ \varphi_3 = \text{id} & \text{on } \partial\Omega. \end{cases}$$

For this we will use a combination of Theorem 14.10 and Proposition 14.12. We only show the assertion for g, the one with f being proved exactly in the same way.

Step 2.1. We start with some preliminary estimates. Using the second inequality in Step 1, we deduce that for every ε small enough, recalling that $r \geq 1$ and $\gamma < \alpha$,

$$\frac{1}{2}\|g_\varepsilon\|_{C^{1,\gamma}} \leq \|g\|_{C^{1,\gamma}} \leq 2\|g_\varepsilon\|_{C^{1,\gamma}} \quad \text{and} \quad \left\|\frac{1}{[g_\varepsilon]^{n/2}}\right\|_{C^0} \leq 2\left\|\frac{1}{[g]^{n/2}}\right\|_{C^0}.$$

In what follows, ε will always be assumed small enough. Combining the left-hand side of the previous inequality with the third and fifth inequalities in Step 1, we deduce that there exists $D_1 > 0$, a constant independent of ε, such that, defining

$$\theta(g_\varepsilon) = \frac{1}{\|g_\varepsilon\|_{C^{1,\gamma}}^2} \min\left\{\|g_\varepsilon\|_{C^{1,\gamma}}, \frac{1}{\|g_\varepsilon\|_{C^{2,\gamma}}}, \frac{1}{\|g_\varepsilon\|_{C^{r+1,\alpha}}}\right\},$$

we have

$$\theta(g_\varepsilon) \geq D_1 \varepsilon.$$

Hence, since $\|g_\varepsilon - g\|_{C^{0,\gamma}} \leq C\varepsilon^{r+\alpha-\gamma}$, $r \geq 1$ and $\gamma < \alpha$, we immediately deduce

$$\lim_{\varepsilon \to 0} \frac{\|g_\varepsilon - g\|_{C^{0,\gamma}}}{\theta(g_\varepsilon)} = 0. \tag{14.22}$$

Note also that there exists $D_2 > 0$, a constant independent of ε, such that

$$\|g_\varepsilon\|_{C^0}, \left\|\frac{1}{[g_\varepsilon]^{n/2}}\right\|_{C^0} \leq D_2.$$

Step 2.2. Let $C = C(D_2, r, \alpha, \gamma, \Omega)$ be the constant given in (14.5) of Theorem 14.10. Due to (14.22), the first inequality of (14.5) is satisfied for every $\varepsilon \leq \widetilde{\varepsilon}$ and for some $\widetilde{\varepsilon} \leq \overline{\varepsilon}$. We show the assertion by considering two cases. In the first one, we use Theorem 14.10 to obtain the assertion and in the second one, we use Proposition 14.12.

(i) Suppose that for some $\varepsilon \leq \widetilde{\varepsilon}$, the second inequality of (14.5) is also satisfied, namely

$$\|g_\varepsilon - g\|_{C^{0,\gamma}} \leq C(D_2, r, \alpha, \gamma, \Omega) \frac{\|g_\varepsilon - g\|_{C^{r,\alpha}}}{\|g_\varepsilon\|_{C^{1,\gamma}} \|g_\varepsilon\|_{C^{r+1,\alpha}}}.$$

Hence, we have the claim of Step 2 using Theorem 14.10.

(ii) Suppose that the first case does not hold true. Hence, for all $\varepsilon \leq \widetilde{\varepsilon}$

$$\|g_\varepsilon\|_{C^{1,\gamma}} \|g_\varepsilon\|_{C^{r+1,\alpha}} \|g_\varepsilon - g\|_{C^{0,\gamma}} > C(D_2, r, \alpha, \gamma, \Omega) \|g_\varepsilon - g\|_{C^{r,\alpha}}.$$

Using the first and third inequality of Step 1, the fact that $\|g_\varepsilon\|_{C^{1,\gamma}} \leq 2\|g\|_{C^{1,\gamma}}$, we obtain, recalling that $r \geq 1$ and that $2\delta \leq \alpha - \gamma$,

$$\|g_\varepsilon - g\|_{C^{r,\alpha}} \leq D_3 \varepsilon^{2\delta} \quad \text{for every } 0 < \varepsilon \leq \widetilde{\varepsilon},$$

where D_3 is independent of ε. Combining the above equation with the fact that, by the fourth inequality in Step 1 (where $D_4 > 0$ is independent of ε),

$$\|g_\varepsilon\|_{C^{r,\alpha+2\delta}} \leq \frac{D_4}{\varepsilon^{2\delta}},$$

we immediately deduce from Proposition 16.45 that $g \in C^{r,\alpha+\delta}(\overline{\Omega}; \Lambda^2)$. The assertion then follows directly from Proposition 14.12 once noticed, using Remark 16.50(v), that the g_ε constructed in Proposition 14.12 are the same as the ones defined in Step 1.

Step 3. Since

$$\begin{cases} dg_{\varepsilon_1} = df_{\varepsilon_2} = 0 & \text{in } \Omega, \\ \nu \wedge g_{\varepsilon_1} = \nu \wedge f_{\varepsilon_2} & \text{on } \partial\Omega, \\ \int_\Omega \langle g_{\varepsilon_1}; \psi \rangle = \int_\Omega \langle f_{\varepsilon_2}; \psi \rangle & \text{for every } \psi \in \mathscr{H}_T(\Omega; \Lambda^2), \\ \operatorname{rank}\left[t g_{\varepsilon_1} + (1-t) f_{\varepsilon_2}\right] = n & \text{in } \overline{\Omega} \text{ and for every } t \in [0,1], \end{cases}$$

we can apply Theorem 14.7 to find $\varphi_2 \in \operatorname{Diff}^{r+1,\alpha}(\overline{\Omega}; \overline{\Omega})$ such that

$$\begin{cases} \varphi_2^*(g_{\varepsilon_1}) = f_{\varepsilon_2} & \text{in } \Omega, \\ \varphi_2 = \text{id} & \text{on } \partial\Omega. \end{cases}$$

The claimed solution is then given by

$$\varphi = \varphi_1^{-1} \circ \varphi_2 \circ \varphi_3.$$

This achieves the proof of the theorem. □

14.4.6 A Second Proof of the Main Theorem

We first show Theorem 14.5 for special f and general g with extra regularity only on the boundary and under a smallness assumption.

Proposition 14.13. *Let $\Omega \subset \mathbb{R}^n$ be a bounded open smooth set, $r \geq 1$ and $0 < \delta < \alpha < \alpha + \delta < 1$. Let $g \in C^{r,\alpha}(\overline{\Omega};\Lambda^2)$ with*

$$\nu \wedge g \in C^{r+1,\alpha+\delta}(\partial\Omega;\Lambda^3), \quad dg = 0 \quad \text{and} \quad \operatorname{rank}[g] = n \quad \text{in } \overline{\Omega}.$$

Then for every ε small, there exist

$$g_\varepsilon \in C^{r+1,\alpha+\delta}(\overline{\Omega};\Lambda^2) \quad \text{and} \quad \varphi_\varepsilon \in \operatorname{Diff}^{r+1,\alpha}(\overline{\Omega};\overline{\Omega})$$

such that

$$\begin{cases} \varphi_\varepsilon^*(g_\varepsilon) = g & \text{in } \Omega, \\ \varphi_\varepsilon = \mathrm{id} & \text{on } \partial\Omega, \end{cases}$$

$$dg_\varepsilon = 0, \quad \nu \wedge g_\varepsilon = \nu \wedge g \ \text{ on } \partial\Omega,$$

$$\int_\Omega \langle g_\varepsilon;\psi\rangle = \int_\Omega \langle g;\psi\rangle \ \text{for every } \psi \in \mathcal{H}_T(\Omega;\Lambda^2),$$

$$\lim_{\varepsilon\to 0} \|g_\varepsilon - g\|_{C^{r,\alpha-\delta}(\overline{\Omega})} = 0.$$

Proof. We adopt the same simplification in the notations of the norms as in the proof of Proposition 14.12.

Step 1 (definition of g_ε). Apply Theorem 16.49 and Remark 16.50(v)–16.50(vi). Therefore, for every $\varepsilon \in (0,1]$, there exist $g_\varepsilon \in C^{r+1,\alpha+\delta}(\overline{\Omega};\Lambda^2)$ and a constant $C_1 = C_1(r,\alpha,\delta,\Omega)$ such that for every $\gamma \in [\alpha-\delta,\alpha+\delta]$,

$$dg_\varepsilon = 0 \quad \text{in } \Omega, \quad \nu \wedge g_\varepsilon = \nu \wedge g \quad \text{on } \partial\Omega, \tag{14.23}$$

$$\int_\Omega \langle g_\varepsilon;\psi\rangle = \int_\Omega \langle g;\psi\rangle \quad \text{for every } \psi \in \mathcal{H}_T(\Omega;\Lambda^2), \tag{14.24}$$

$$\|g_\varepsilon\|_{C^{r+1,\gamma}} \leq \frac{C_1}{\varepsilon^{1+\gamma-\alpha}}\|g\|_{C^{r,\alpha}} + C_1\|\nu\wedge g\|_{C^{r+1,\gamma}(\partial\Omega)}, \tag{14.25}$$

$$\|g_\varepsilon - g\|_{C^{r,\alpha-\delta}} \leq C_1\varepsilon^\delta\|g\|_{C^{r,\alpha}}, \tag{14.26}$$

$$\left\|\frac{d}{d\varepsilon}g_\varepsilon\right\|_{C^{0,\alpha}} \leq C_1\|g\|_{C^{1,\alpha}} \quad \text{and} \quad \left\|\frac{d}{d\varepsilon}g_\varepsilon\right\|_{C^{r,\gamma}} \leq \frac{C_1}{\varepsilon^{1+\gamma-\alpha}}\|g\|_{C^{r,\alpha}}. \tag{14.27}$$

Moreover, defining $G:(0,1]\times\overline{\Omega}\to\Lambda^2$ by $G(\varepsilon,x) = g_\varepsilon(x)$, we have

$$G \in C^{r+1,\alpha+\delta}((0;1]\times\overline{\Omega};\Lambda^2) \quad \text{and} \quad \frac{\partial}{\partial\varepsilon}G \in C^\infty((0;1]\times\overline{\Omega};\Lambda^2). \tag{14.28}$$

Since $\operatorname{rank}[g] = n$ in $\overline{\Omega}$ (which is equivalent to $g^{n/2}(x) \neq 0$ for every $x \in \overline{\Omega}$) and since (14.26) holds, there exists $\overline{\varepsilon} \leq 1$ such that for every $\varepsilon \in (0, \overline{\varepsilon}]$,

$$\|g_\varepsilon\|_{C^0} \leq 2\|g\|_{C^0} \quad \text{and} \quad \|1/(g_\varepsilon)^{n/2}\|_{C^0} \leq 2\|1/g^{n/2}\|_{C^0}. \tag{14.29}$$

Hence, combining (14.29) and Notation (v) in Section 14.1, we deduce that for every $\varepsilon \in (0, \overline{\varepsilon}]$ and every $\gamma \in [\alpha - \delta, \alpha + \delta]$,

$$\|(\overline{g}_\varepsilon)^{-1}\|_{C^{r+1,\gamma}} \leq C_2 \|g_\varepsilon\|_{C^{r+1,\gamma}}, \tag{14.30}$$

where $C_2 = C_2(r, \Omega, \|g\|_{C^0}, \|1/g^{n/2}\|_{C^0})$.

Step 2. In this step we will show that for every $\varepsilon \in (0, \overline{\varepsilon}]$, there exist $u_\varepsilon \in C^{r+1,\alpha+\delta}(\overline{\Omega}; \Lambda^1)$ and a constant $C_3 = C_3(r, \alpha, \delta, \Omega, \|g\|_{C^{1,\alpha}}, \|1/g^{n/2}\|_{C^0})$ such that $u_\varepsilon = 0$ on $\partial\Omega$ and

$$d(u_\varepsilon \lrcorner g_\varepsilon) = -\frac{d}{d\varepsilon} g_\varepsilon \quad \text{in } \Omega \tag{14.31}$$

and, for every $\gamma \in [\alpha - \delta, \alpha + \delta]$,

$$\|u_\varepsilon\|_{C^{r+1,\gamma}} \leq \frac{C_3}{\varepsilon^{1+\gamma-\alpha}} \|g\|_{C^{r,\alpha}} + C_3 \|\nu \wedge g\|_{C^{r+1,\alpha+\delta}(\partial\Omega)}. \tag{14.32}$$

Moreover, defining $u : (0, \overline{\varepsilon}] \times \overline{\Omega} \to \Lambda^1$ by $u(\varepsilon, x) = u_\varepsilon(x)$, we will show that $u \in C^{r+1,\alpha+\delta}((0, \overline{\varepsilon}] \times \overline{\Omega}; \Lambda^1)$.

Step 2.1. Since (14.23), (14.24) and (14.28) hold, using Theorem 8.16, we can find for every $\varepsilon \in (0, \overline{\varepsilon}]$, $w_\varepsilon \in C^\infty(\overline{\Omega}; \Lambda^1)$ and a constant $C_4 = C_4(r, \alpha, \delta, \Omega)$ such that

$$dw_\varepsilon = -\frac{d}{d\varepsilon} g_\varepsilon \quad \text{in } \Omega, \quad w_\varepsilon = 0 \quad \text{on } \partial\Omega$$

and, for every integer $q \leq r$ and every $\gamma \in [\alpha - \delta, \alpha + \delta]$,

$$\|w_\varepsilon\|_{C^{q+1,\gamma}} \leq C_4 \left\| \frac{d}{d\varepsilon} g_\varepsilon \right\|_{C^{q,\gamma}}. \tag{14.33}$$

Moreover, defining $w : (0, \overline{\varepsilon}] \times \overline{\Omega} \to \Lambda^1$ by $w(\varepsilon, x) = w_\varepsilon(x)$, we have, using (14.28), $w \in C^\infty((0, \overline{\varepsilon}] \times \overline{\Omega}; \Lambda^1)$.

Step 2.2. Since, by (14.29), we have for every $\varepsilon \in (0, \overline{\varepsilon}]$, $\operatorname{rank}[g_\varepsilon] = n$ in $\overline{\Omega}$, that there exists a unique $u_\varepsilon : \overline{\Omega} \to \Lambda^1$ verifying

$$u_\varepsilon \lrcorner g_\varepsilon = w_\varepsilon.$$

Note that $u_\varepsilon \in C^{r+1,\alpha+\delta}(\overline{\Omega}; \Lambda^1)$ and that $u_\varepsilon = 0$ on $\partial\Omega$. Moreover, defining $u : (0, \overline{\varepsilon}] \times \overline{\Omega} \to \Lambda^1$ by $u(\varepsilon, x) = u_\varepsilon(x)$, we have $u \in C^{r+1,\alpha+\delta}((0, \overline{\varepsilon}] \times \overline{\Omega}; \Lambda^1)$.

Step 2.3. To show Step 2, it only remains to prove (14.32). Using Theorem 16.28, (14.29), (14.30) and (14.33), it follows that

$$\begin{aligned}
\|u_\varepsilon\|_{C^{r+1,\gamma}} &= \|(\overline{g}_\varepsilon)^{-1} w_\varepsilon\|_{C^{r+1,\gamma}} \\
&\le C_5 \|(\overline{g}_\varepsilon)^{-1}\|_{C^{r+1,\gamma}} \|w_\varepsilon\|_{C^0} + C_5 \|(\overline{g}_\varepsilon)^{-1}\|_{C^0} \|w_\varepsilon\|_{C^{r+1,\gamma}} \\
&\le C_6 \|g_\varepsilon\|_{C^{r+1,\gamma}} \|w_\varepsilon\|_{C^{1,\alpha}} + C_6 \|w_\varepsilon\|_{C^{r+1,\gamma}} \\
&\le C_7 \|g_\varepsilon\|_{C^{r+1,\gamma}} \left\| \frac{d}{d\varepsilon} g_\varepsilon \right\|_{C^{0,\alpha}} + C_7 \left\| \frac{d}{d\varepsilon} g_\varepsilon \right\|_{C^{r,\gamma}}
\end{aligned}$$

and hence, appealing to (14.25) and (14.27),

$$\begin{aligned}
&\|u_\varepsilon\|_{C^{r+1,\gamma}} \\
&\le C_8 \left(\frac{1}{\varepsilon^{1+\gamma-\alpha}} \|g\|_{C^{r,\alpha}} + \|\nu \wedge g\|_{C^{r+1,\gamma}(\partial\Omega)} \right) \|g\|_{C^{1,\alpha}} + \frac{C_8}{\varepsilon^{1+\gamma-\alpha}} \|g\|_{C^{r,\alpha}} \\
&\le \frac{C_9}{\varepsilon^{1+\gamma-\alpha}} \|g\|_{C^{r,\alpha}} + C_9 \|\nu \wedge g\|_{C^{r+1,\alpha+\delta}(\partial\Omega)},
\end{aligned}$$

where $C_i = C_i(r, \alpha, \delta, \Omega, \|g\|_{C^{1,\alpha}}, \|1/g^{n/2}\|_{C^0})$. This shows the assertion.

Step 3. We can now conclude the proof.

Step 3.1. Since $u \in C^{r+1,\alpha+\delta}((0,\overline{\varepsilon}] \times \overline{\Omega} : \mathbb{R}^n)$, $u_\varepsilon = 0$ on $\partial\Omega$ and (14.32) holds, we deduce, using Theorem 12.4, that the solution $\varphi : [0,\overline{\varepsilon}] \times \overline{\Omega} \to \overline{\Omega}$, $\varphi(\varepsilon, x) = \varphi_\varepsilon(x)$, of

$$\begin{cases}
\dfrac{d}{d\varepsilon} \varphi_\varepsilon = u_\varepsilon \circ \varphi_\varepsilon, & 0 < \varepsilon \le \overline{\varepsilon}, \\
\varphi_0 = \mathrm{id}
\end{cases}$$

verifies

$$\varphi \in C^{r+1}([0,\overline{\varepsilon}] \times \overline{\Omega}; \overline{\Omega}) \tag{14.34}$$

and that for every $\varepsilon \in [0, \overline{\varepsilon}]$,

$$\varphi_\varepsilon \in \mathrm{Diff}^{r+1,\alpha}(\overline{\Omega}; \overline{\Omega}) \quad \text{and} \quad \varphi_\varepsilon = \mathrm{id} \text{ on } \partial\Omega.$$

Step 3.2. Since (14.31) holds, we deduce, using Theorem 12.7, that for every $0 < \varepsilon_1 \le \varepsilon_2 \le \overline{\varepsilon}$,

$$\varphi_{\varepsilon_2}^*(g_{\varepsilon_2}) = \varphi_{\varepsilon_1}^*(g_{\varepsilon_1}) \quad \text{in } \Omega.$$

Since, using (14.26) and (14.34),

$$\lim_{\varepsilon \to 0} \|g_\varepsilon - g\|_{C^0} = 0 \quad \text{and} \quad \lim_{\varepsilon \to 0} \|\varphi_\varepsilon - \varphi_0\|_{C^1} = 0,$$

we immediately deduce that for every $\varepsilon \in (0, \overline{\varepsilon}]$,

$$\varphi_\varepsilon^*(g_\varepsilon) = \varphi_0^*(g) = g\,.$$

This concludes the proof. □

We now turn to our second proof of Theorem 14.5. We will do it under the stronger assumption that there exists $0 < \alpha < \beta < 1$ so that

$$\nu \wedge f,\ \nu \wedge g \in C^{r+1,\beta}\left(\partial\Omega;\Lambda^3\right).$$

Proof. Step 1. Let $\delta > 0$ small enough so that $[\alpha - \delta, \alpha + \delta] \subset (0,\beta)$. Applying Proposition 14.13 to f and g, there exist for every ε small,

$$f_\varepsilon, g_\varepsilon \in C^{r+1,\alpha+\delta}\left(\overline{\Omega};\Lambda^2\right) \quad \text{and} \quad \varphi_{1,\varepsilon}, \varphi_{2,\varepsilon} \in \operatorname{Diff}^{r+1,\alpha}\left(\overline{\Omega};\overline{\Omega}\right)$$

such that

$$\begin{cases} \varphi_{1,\varepsilon}^*(f_\varepsilon) = f, \quad \varphi_{2,\varepsilon}^*(g_\varepsilon) = g & \text{in } \Omega, \\ \varphi_{1,\varepsilon} = \varphi_{2,\varepsilon} = \text{id} & \text{on } \partial\Omega, \end{cases}$$

$$\lim_{\varepsilon\to 0} \|f_\varepsilon - f\|_{C^{r,\alpha-\delta}} = \lim_{\varepsilon\to 0} \|g_\varepsilon - g\|_{C^{r,\alpha-\delta}} = 0.$$

Using the previous equation, there exists $\varepsilon_0 > 0$ small enough so that for every $t \in [0,1]$,

$$\operatorname{rank}[t g_{\varepsilon_0} + (1-t) f_{\varepsilon_0}] = n \quad \text{in } \overline{\Omega}.$$

Moreover, f_ε and g_ε satisfy

$$dg_\varepsilon = df_\varepsilon = 0, \quad \nu \wedge g_\varepsilon = \nu \wedge f_\varepsilon = \nu \wedge f = \nu \wedge g \text{ on } \partial\Omega,$$

$$\int_\Omega \langle g_\varepsilon;\psi\rangle = \int_\Omega \langle g;\psi\rangle = \int_\Omega \langle f_\varepsilon;\psi\rangle = \int_\Omega \langle f;\psi\rangle,\ \forall \psi \in \mathscr{H}_T\left(\Omega;\Lambda^2\right).$$

Step 2. Using Theorem 14.7, we find $\varphi_3 \in C^{r+1,\alpha+\delta}(\overline{\Omega})$ verifying

$$\begin{cases} \varphi_3^*(g_{\varepsilon_0}) = f_{\varepsilon_0} & \text{in } \Omega, \\ \varphi_3 = \text{id} & \text{on } \partial\Omega. \end{cases}$$

Finally, the diffeomorphism $\varphi = \varphi_{2,\varepsilon_0}^{-1} \circ \varphi_3 \circ \varphi_{1,\varepsilon_0}$ has all of the required properties. $\square$

Chapter 15
The Case $3 \leq k \leq n-1$

The results that will be discussed in this chapter are strongly based on Bandyopadhyay, Dacorogna and Kneuss [9]. For related results see Turiel [97–102].

15.1 A General Theorem for Forms of Rank $= k$

Our first result concerns k-forms of minimal nonzero rank.

Theorem 15.1. *Let $2 \leq k \leq n$, $r \geq 1$ be integers, $0 < \alpha < 1$ and $x_0 \in \mathbb{R}^n$. Let f and g be two $C^{r,\alpha}$ k-forms verifying, in a neighborhood of x_0,*

$$df = dg = 0 \quad \textit{and} \quad \operatorname{rank}[f] = \operatorname{rank}[g] = k.$$

Then there exist a neighborhood U of x_0 and

$$\varphi \in \begin{cases} \operatorname{Diff}^{r,\alpha}(U;\varphi(U)) & \text{if } k < n \\ \operatorname{Diff}^{r+1,\alpha}(U;\varphi(U)) & \text{if } k = n \end{cases}$$

such that $\varphi(x_0) = x_0$ and

$$\varphi^*(g) = f \quad \textit{in } U.$$

In particular, if $g = dx^1 \wedge \cdots \wedge dx^k$, then

$$f = \nabla\varphi^1 \wedge \cdots \wedge \nabla\varphi^k \quad \textit{in } U.$$

Remark 15.2. (i) The case $k = n-1$ is therefore completely solved (cf. Theorem 15.3).

(ii) We recall that the rank of a form is given in Definition 2.28 and Remark 2.31; see also Proposition 2.37.

(iii) Throughout this chapter we will often use the following elementary fact. In order to solve $\varphi^*(g) = f$, it is enough to solve, for some h,

$$\varphi_1^*(h) = g, \quad \varphi_2^*(h) = f$$

and let $\varphi = \varphi_1^{-1} \circ \varphi_2$.

Proof. With no loss of generality, we can assume $x_0 = 0$ and (see Remark 15.2(iii)) $g = dx^1 \wedge \cdots \wedge dx^k$. We split the proof into two steps.

Step 1. We first prove the case $k = n$. Since $f = f_{1\cdots n}\, dx^1 \wedge \cdots \wedge dx^n$ and since $\operatorname{rank}[f] = n > 0$ in a neighborhood of 0, there exists a sufficiently small ball U centered at 0 such that $f_{1\cdots n}(x) \neq 0$ for every $x \in \overline{U}$. Using Theorem 10.1, there exists $\varphi_1 \in \operatorname{Diff}^{r+1,\alpha}(\overline{U}; \overline{U})$ such that $\varphi_1 = \mathrm{id}$ on ∂U and

$$\varphi_1^*(c\, dx^1 \wedge \cdots \wedge dx^n) = f_{1\cdots n}\, dx^1 \wedge \cdots \wedge dx^n \quad \text{in } U,$$

where

$$c = \frac{1}{\operatorname{meas} U} \int_U f_{1\cdots n}.$$

Finally, let

$$\varphi_2(x) = x - \varphi_1(0)$$

and

$$\varphi_3(x) = \varphi_3(x_1, \ldots, x_n) = (x_1, \ldots, x_{n-1}, c x_n).$$

The map $\varphi = \varphi_3 \circ \varphi_2 \circ \varphi_1$ has all of the desired properties.

Step 2. We next suppose that $k < n$. Using Theorem 4.5, there exist a neighborhood V of 0 and $\varphi_1 \in \operatorname{Diff}^{r,\alpha}(V; \varphi_1(V))$ such that $\varphi_1(0) = 0$ and

$$\varphi_1^*(f)(x_1, \ldots, x_n) = a(x_1, \ldots, x_k)\, dx^1 \wedge \cdots \wedge dx^k \quad \text{in } V,$$

where $a \in C^{r-1,\alpha}$ in a neighborhood of 0 in $\mathbb{R}^k$. Using the fact that $\operatorname{rank}[f] = k$ and Proposition 17.1, there exists W, a sufficiently small ball in $\mathbb{R}^k$ centered at 0, such that $a(x) \neq 0$ for every $x \in \overline{W}$. Using Step 1, there exists $\varphi_2 \in \operatorname{Diff}^{r,\alpha}(\overline{W}; \varphi_2(\overline{W}))$ such that $\varphi_2(0) = 0$ and

$$\varphi_2^*(dx^1 \wedge \cdots \wedge dx^k) = a\, dx^1 \wedge \cdots \wedge dx^k.$$

Finally, defining $\widetilde{\varphi}_2 \in \operatorname{Diff}^{r,\alpha}(\overline{W} \times \mathbb{R}^{n-k}; \varphi_2(\overline{W}) \times \mathbb{R}^{n-k})$ by

$$\widetilde{\varphi}_2(x) = (\varphi_2(x_1, \ldots, x_k), x_{k+1}, \ldots, x_n),$$

we get that $\varphi = \widetilde{\varphi}_2 \circ \varphi_1^{-1}$ has all of the desired properties. This concludes the proof. □

15.2 The Case of $(n-1)$-Forms

15.2.1 The Case of Closed $(n-1)$-Forms

The case of closed $(n-1)$-forms is a direct consequence of the results of Section 15.1 (see also Martinet [71]).

Theorem 15.3. *Let $r \geq 1$ be an integer, $0 < \alpha < 1$ and $x_0 \in \mathbb{R}^n$. Let f and g be two closed $C^{r,\alpha}$ $(n-1)$-forms verifying*

$$f(x_0) \neq 0 \quad \text{and} \quad g(x_0) \neq 0.$$

Then there exist a neighborhood U of x_0 and $\varphi \in \operatorname{Diff}^{r,\alpha}(U;\varphi(U))$ such that $\varphi(x_0) = x_0$ and

$$\varphi^*(g) = f \quad \text{in } U.$$

In particular, if $g = dx^1 \wedge \cdots \wedge dx^{n-1}$, then

$$f = \nabla\varphi^1 \wedge \cdots \wedge \nabla\varphi^{n-1} \quad \text{in } U.$$

Proof. Recall first that a nonzero $(n-1)$-form has always (cf. Remark 2.38(i)) its rank equal to $n-1$. Therefore, the hypothesis

$$f(x_0) \neq 0 \quad \text{and} \quad g(x_0) \neq 0$$

is equivalent to

$$\operatorname{rank}[f] = \operatorname{rank}[g] = n-1 \text{ in a neighborhood of } x_0.$$

Applying Theorem 15.1, we have the result. □

Theorem 15.3 reads in a more analytical way when $k = n-1$ (cf. also Barbarosie [11]), since the exterior derivative of an $(n-1)$-form is then essentially the classical divergence operator.

Corollary 15.4. *Let $r \geq 1$ be an integer, $0 < \alpha < 1$ and $x_0 \in \mathbb{R}^n$. Let f be a $C^{r,\alpha}$ vector field satisfying*

$$f(x_0) \neq 0 \quad \text{and} \quad \operatorname{div} f = 0 \text{ in a neighborhood of } x_0.$$

Then there exist a neighborhood U of x_0 and $\varphi \in \operatorname{Diff}^{r,\alpha}(U;\varphi(U))$ such that $\varphi(x_0) = x_0$ and

$$f = *\left(\nabla\varphi^1 \wedge \cdots \wedge \nabla\varphi^{n-1}\right) \quad \text{in } U.$$

15.2.2 The Case of Nonclosed $(n-1)$-Forms

We conclude with the case of nonclosed $(n-1)$-forms.

Theorem 15.5. *Let $x_0 \in \mathbb{R}^n$ and f a C^∞ $(n-1)$-form verifying*

$$f(x_0) \neq 0.$$

Then there exist a neighborhood U of x_0 and

$$\varphi \in C^\infty(U;\varphi(U))$$

such that

$$f = \varphi^n \nabla\varphi^1 \wedge \cdots \wedge \nabla\varphi^{n-1} \quad \text{in } U.$$

If, moreover, $df(x_0) \neq 0$ then, up to restricting U, in addition to the previous equation, φ can be assumed in $\operatorname{Diff}^\infty(U;\varphi(U))$.

Remark 15.6. (i) If $f \in C^r$, then $\varphi^n \in C^{r-1}$ and $\varphi^i \in C^r$, $1 \le i \le n-1$. Moreover, another way to read the conclusion is

$$\varphi^*(x_n\, dx^1 \wedge \cdots \wedge dx^{n-1}) = f.$$

(ii) If $df = 0$ in a neighborhood of x_0, then we have a better result (cf. Theorem 15.3).

(iii) Note that we cannot, in general, ensure that $\varphi(x_0) = x_0$; for a similar result, see Remark 13.11(ii).

Proof. We split the proof into two steps. In the sequel, $*f \in C^\infty(\mathbb{R}^n;\Lambda^1)$ will sometimes be identified with a vector field (see Definition 2.9 for the notation).

Step 1. We prove the main assertion. Since $f(x_0) \neq 0$, using Remark 4.3(ii), there exist a neighborhood $V \subset \mathbb{R}^n$ of x_0 and $\varphi_1 \in \operatorname{Diff}^\infty(V;\varphi_1(V))$ such that $\varphi_1(x_0) = x_0$ and

$$\frac{\partial \varphi_1}{\partial x_n} = (*f)\circ\varphi_1 \quad \text{in } V. \tag{15.1}$$

Using Definition 2.11 and the fact that $(*f)\wedge(*f) = 0$ (since $*f$ is a 1-form), we deduce that $(*f) \lrcorner f = 0$. Thus, using (15.1), Theorem 3.10 and Proposition 3.11, we obtain

$$0 = \varphi_1^*((*f)\lrcorner f) = \varphi_1^\sharp(*f)\lrcorner\varphi_1^*(f) = dx^n \lrcorner \varphi_1^*(f).$$

From the previous equation we immediately deduce

$$\varphi_1^*(f)(x) = a(x_1,\ldots,x_n)dx^1\wedge\cdots\wedge dx^{n-1}, \quad x \in V,$$

where $a \in C^\infty(V)$. Letting $U = \varphi_1(V)$ and

$$\varphi = (\varphi^1, \ldots, \varphi^n) = ((\varphi_1^{-1})^1, \ldots, (\varphi_1^{-1})^{n-1}, a \circ \varphi^{-1}),$$

we have the main assertion, namely

$$f = \varphi^n \nabla \varphi^1 \wedge \cdots \wedge \nabla \varphi^{n-1}.$$

Step 2. We prove the extra assertion. Let φ_1 be the diffeomorphism obtained in Step 1. It verifies, in particular,

$$\varphi_1^*(f)(x) = a(x_1, \ldots, x_n)\, dx^1 \wedge \cdots \wedge dx^{n-1}, \quad x \in V.$$

Since, by hypothesis, $df \neq 0$ in a neighborhood of x_0 and $\varphi_1(x_0) = x_0$, we have

$$d(\varphi_1^*(f)) = \varphi_1^*(df) \neq 0 \text{ in a neighborhood of } x_0$$

and, thus,

$$\frac{\partial a}{\partial x_n}(x_0) \neq 0.$$

Define $\varphi_2 : V \to \mathbb{R}^n$ by

$$\varphi_2(x) = (x_1, \ldots, x_{n-1}, a(x)).$$

Note that

$$\varphi_2^*(x_n\, dx^1 \wedge \cdots \wedge dx^{n-1}) = a(x)\, dx^1 \wedge \cdots \wedge dx^{n-1} \quad \text{in } V$$

and that, taking V smaller if necessary, $\varphi_2 \in \mathrm{Diff}^\infty(V; \varphi_2(V))$. Letting $\varphi = \varphi_2 \circ (\varphi_1)^{-1}$, it follows that $\varphi \in \mathrm{Diff}^\infty(\varphi_1(V); \varphi_2(V))$ and has the desired property. The proof is therefore complete. □

As before, the previous theorem can be seen in a more analytical way (cf. also Barbarosie [11]).

Corollary 15.7. *Let $x_0 \in \mathbb{R}^n$ and let f be a C^∞ vector field satisfying*

$$f(x_0) \neq 0.$$

Then there exist a neighborhood U of x_0 and

$$\varphi \in C^\infty(U; \varphi(U))$$

such that

$$f = *\left(\varphi^n \nabla \varphi^1 \wedge \cdots \wedge \nabla \varphi^{n-1}\right) \quad \text{in } U.$$

If, moreover, $\operatorname{div} f(x_0) \neq 0$, *then, up to restricting U, in addition to the previous equation, φ can be assumed in* $\mathrm{Diff}^\infty(U; \varphi(U))$.

15.3 Simultaneous Resolutions and Applications

15.3.1 Simultaneous Resolution for 1-Forms

We start with a simultaneous resolution of closed 1-forms; see also Cartan [21].

Proposition 15.8. *Let $r \geq 0$, $1 \leq m \leq n$ be integers and $x_0 \in \mathbb{R}^n$. Let $b^1, \ldots, b^m$ and $a^1, \ldots, a^m$ be C^r closed 1-forms verifying*

$$(b^1 \wedge \cdots \wedge b^m)(x_0) \neq 0 \quad \text{and} \quad (a^1 \wedge \cdots \wedge a^m)(x_0) \neq 0.$$

Then there exist a neighborhood U of x_0 and $\varphi \in \operatorname{Diff}^{r+1}(U;\varphi(U))$ such that $\varphi(x_0) = x_0$ and

$$\varphi^*\left(b^i\right) = a^i \quad \text{in } U \text{ and for every } 1 \leq i \leq m.$$

Remark 15.9. (i) When $r = 0$, the fact that a 1-form ω is closed has to be understood in the sense of distributions.

(ii) The result is also valid in Hölder spaces.

(iii) It is interesting to compare the above proposition and Theorem 15.1. In view of Proposition 2.43, we know that any m-form f with $\operatorname{rank}[f] = m$ is a product of 1-forms $a^1, \ldots, a^m$ so that

$$f = a^1 \wedge \cdots \wedge a^m;$$

however, we do not know, in general, that $a^1, \ldots, a^m$ are closed if f is closed (and even that $a^1, \ldots, a^m \in C^r$ if $f \in C^r$). But, Theorem 15.1 shows that there does exist a total decomposition with closed $a^1, \ldots, a^m$; however, we have lost one degree of regularity, namely $a^1, \ldots, a^m \in C^{r-1,\alpha}$ (unless $m = n$). Therefore, if we assume that $a^1, \ldots, a^m$ are closed, then the above proposition is better from the point of view of regularity than Theorem 15.1.

(iv) When $m = n$ and $f \in C^0$, it is, in general, impossible (according to Burago and Kleiner [19] and Mc Mullen [73]) to find closed 1-forms $a^1, \ldots, a^n \in C^0$ so that

$$f = a^1 \wedge \cdots \wedge a^n;$$

although, in view of Theorem 10.1, we can do so if $f \in C^{0,\alpha}$, finding even that $a^1, \ldots, a^n \in C^{0,\alpha}$.

Proof. We split the proof into two steps.

Step 1. With no loss of generality, we can assume $x_0 = 0$. Noticing that if $m < n$, we can choose $1 \leq k_{m+1} < \cdots < k_n \leq n$ and $1 \leq l_{m+1} < \cdots < l_n \leq n$ such that

$$(b^1 \wedge \cdots \wedge b^m \wedge dx^{k_{m+1}} \wedge \cdots \wedge dx^{k_n})(0) \neq 0,$$

$$(a^1 \wedge \cdots \wedge a^m \wedge dx^{l_{m+1}} \wedge \cdots \wedge dx^{l_n})(0) \neq 0.$$

We can therefore assume that $m=n$, letting $b^i = dx^{k_i}$ and $a^i = dx^{l_i}$ for $m+1 \leq i \leq n$. Using Corollary 8.6, we can find a neighborhood V of 0 and, for $1 \leq i \leq n$, $B^i, A^i \in C^{r+1}(V)$ such that

$$dA^i = a^i \quad \text{and} \quad dB^i = b^i \quad \text{in } V \text{ for every } 1 \leq i \leq n.$$

Moreover, adding, if necessary, a constant, we can assume that $A^i(0) = B^i(0) = 0$ for $1 \leq i \leq n$. Finally, define $A, B \in C^{r+1}(U;\mathbb{R}^n)$ by $A = (A^1, \ldots, A^n)$ and $B = (B^1, \ldots, B^n)$. Since $A(0) = B(0) = 0$ and since, identifying n-forms with 0-forms,

$$\det \nabla A(0) = (a^1 \wedge \cdots \wedge a^n)(0) \neq 0 \quad \text{and} \quad \det \nabla B(0) = (b^1 \wedge \cdots \wedge b^n)(0) \neq 0,$$

it follows that $A \in \operatorname{Diff}^{r+1}(U;A(U))$, $B \in \operatorname{Diff}^{r+1}(U;B(U))$ and

$$B^{-1} \circ A \in \operatorname{Diff}^{r+1}(U;(B^{-1} \circ A)(U))$$

for a neighborhood U of 0 small enough. Noticing that for $1 \leq i \leq n$,

$$A^*(dx^i) = a^i \quad \text{and} \quad B^*(dx^i) = b^i \quad \text{in } U,$$

we deduce that

$$(B^{-1} \circ A)^*(b^i) = A^*((B^{-1})^*(b^i)) = A^*(dx^i) = a^i \quad \text{in } U.$$

Therefore, $\varphi = B^{-1} \circ A$ has all of the desired properties and this concludes the proof. □

It is interesting to see that the above proposition can also be global.

Proposition 15.10. *Let $\Omega \subset \mathbb{R}^n$ be a bounded open smooth set with exterior unit normal ν. Let $r \geq 0$ and $1 \leq m \leq n$ be integers. Let $b^1, \ldots, b^m \in C^r(\overline{\Omega};\Lambda^1)$ be closed in Ω and such that*

$$b^1 \wedge \cdots \wedge b^m \wedge dx^{m+1} \wedge \cdots \wedge dx^n \neq 0 \quad \text{in } \overline{\Omega},$$

$$\nu \wedge b^i = \nu \wedge dx^i \quad \text{on } \partial\Omega \text{ for every } 1 \leq i \leq m,$$

$$\int_\Omega \langle b^i;\chi \rangle = \int_\Omega \langle dx^i;\chi \rangle \quad \text{for every } \chi \in \mathcal{H}_T(\Omega;\Lambda^1) \text{ and every } 1 \leq i \leq m.$$

Then there exists $\varphi \in \operatorname{Diff}^{r+1}(\overline{\Omega};\overline{\Omega})$ satisfying $\varphi = \mathrm{id}$ on $\partial\Omega$, and in Ω,

$$\begin{cases} \varphi^*(b^i) = dx^i, & 1 \leq i \leq m, \\ \varphi^*(dx^i) = dx^i, & m+1 \leq i \leq n. \end{cases}$$

Remark 15.11. If Ω is simply connected (cf. Remark 6.6), then $\mathcal{H}_T(\Omega;\Lambda^1) = \{0\}$ and hence the last condition on the b^i is automatically fulfilled.

Proof. Using Theorem 8.16 and the remark following it, we can find, for $1 \leq i \leq m$, $A^i \in C^{r+1}(\overline{\Omega})$ such that

$$\begin{cases} dA^i = b^i - dx^i & \text{in } \Omega, \\ A^i = 0 & \text{on } \partial\Omega. \end{cases}$$

Next, define $B \in C^{r+1}(\overline{\Omega};\mathbb{R}^n)$ by

$$B(x) = (x_1 + A^1(x), \dots, x_m + A^m(x), x_{m+1}, \dots, x_n).$$

Since $B = \mathrm{id}$ on $\partial\Omega$ and since

$$\det \nabla B(x) = (b^1 \wedge \cdots \wedge b^m \wedge dx^{m+1} \wedge \cdots \wedge dx^n)_{1\cdots n}(x) \neq 0$$

for every $x \in \overline{\Omega}$, we immediately deduce from Theorem 19.12 that $B \in \mathrm{Diff}^{r+1}(\overline{\Omega};\overline{\Omega})$. Note that for $1 \le i \le m$, $B^*(dx^i) = dB^i = d(x^i + A^i) = b^i$. Therefore, $\varphi = B^{-1} \in \mathrm{Diff}^{r+1}(\overline{\Omega};\overline{\Omega})$ has all of the required properties. This concludes the proof. □

15.3.2 Applications to k-Forms

We next generalize Proposition 15.8 by mixing 1-forms and 2-forms.

Theorem 15.12. *Let $m, l \ge 0$ be integers and $x_0 \in \mathbb{R}^n$. Let $b^1, \dots, b^m$ and $a^1, \dots, a^m$ be closed C^∞ 1-forms. Let $g_1, \dots, g_l$ and $f_1, \dots, f_l$ be closed C^∞ 2-forms such that, in a neighborhood of x_0,*

$$\operatorname{rank}[g_i] = \operatorname{rank}[f_i] = 2s_i, \; 1 \le i \le l,$$

$$\begin{aligned} \operatorname{rank}\left[g_1 \wedge \cdots \wedge g_l \wedge b^1 \wedge \cdots \wedge b^m\right] &= \operatorname{rank}\left[f_1 \wedge \cdots \wedge f_l \wedge a^1 \wedge \cdots \wedge a^m\right] \\ &= 2(s_1 + \cdots + s_l) + m. \end{aligned}$$

Then there exist a neighborhood U of x_0 and $\varphi \in \mathrm{Diff}^\infty(U;\varphi(U))$ such that $\varphi(x_0) = x_0$ and, in U,

$$\begin{cases} \varphi^*(g_i) = f_i, & 1 \le i \le l, \\ \varphi^*(b^i) = a^i, & 1 \le i \le m. \end{cases}$$

Remark 15.13. (i) When $m = 0$, respectively $l = 0$, the theorem is to be understood as a statement only on 2-forms, respectively only on 1-forms (in this last case, see Proposition 15.8).

(ii) When $0 < \alpha < 1$, $g_i, f_i \in C^{r,\alpha}$ and $b^j, a^j \in C^{r,\alpha}$, the proof will give $\varphi \in \mathrm{Diff}^{r-l+1,\alpha}$.

(iii) Of course, the theorem applies to k-forms, $k = 2l + m$, of the type

$$G = g_1 \wedge \cdots \wedge g_l \wedge b^1 \wedge \cdots \wedge b^m \quad \text{and} \quad F = f_1 \wedge \cdots \wedge f_l \wedge a^1 \wedge \cdots \wedge a^m.$$

We therefore obtain that there exists a diffeomorphism φ such that

$$\varphi^*(G) = F,$$

generalizing a result obtained by Bandyopadhyay and Dacorogna [8].

Proof. We establish the result by induction on l. When $l = 0$, we are in the situation of Proposition 15.8, which has already been proved. Let us suppose that the theorem is true for $l-1$ and prove it for l.

Step 1. Using Remark 15.2(iii), we can assume that

$$f_j = \sum_{i=(s_1+\cdots+s_{j-1})+1}^{s_1+\cdots+s_j} dx^{2i-1} \wedge dx^{2i}, \quad 1 \le j \le l,$$
$$a^i = dx^{2(s_1+\cdots+s_l)+i} \quad \text{for every } 1 \le i \le m.$$

Note that these particular f_j and a^i satisfy all of the hypotheses of the theorem. We find, using Theorem 14.3, a neighborhood U_1 of x_0 and $\varphi_1 \in \text{Diff}^\infty(U_1; \varphi_1(U_1))$ such that $\varphi_1(x_0) = x_0$ and

$$\varphi_1^*(g_1) = f_1 = \sum_{i=1}^{s_1} dx^{2i-1} \wedge dx^{2i} \quad \text{in } U_1 .$$

Step 2. We claim that, in a neighborhood of x_0,

$$\begin{aligned} &\text{rank}[\varphi_1^*(g_2) \wedge \cdots \wedge \varphi_1^*(g_l) \wedge dx^1 \wedge \cdots \wedge dx^{2s_1} \wedge \varphi_1^*(b^1) \wedge \cdots \wedge \varphi_1^*(b^m)] \\ &\quad = 2(s_2 + \cdots + s_l) + (2s_1 + m). \end{aligned} \tag{15.2}$$

Indeed, first note using Proposition 17.1 that, in a neighborhood of x_0,

$$\begin{aligned} &\text{rank}\left[\varphi_1^*(g_1) \wedge \cdots \wedge \varphi_1^*(g_l) \wedge \varphi_1^*(b^1) \wedge \cdots \wedge \varphi_1^*(b^m)\right] \\ &\quad = \text{rank}\left[\varphi_1^*(g_1 \wedge \cdots \wedge g_l \wedge b^1 \wedge \cdots \wedge b^m)\right] \\ &\quad = \text{rank}\left[g_1 \wedge \cdots \wedge g_l \wedge b^1 \wedge \cdots \wedge b^m\right] = 2(s_1 + \cdots + s_l) + m. \end{aligned}$$

Setting

$$h = \varphi_1^*(g_2) \wedge \cdots \wedge \varphi_1^*(g_l) \wedge \varphi_1^*(b^1) \wedge \cdots \wedge \varphi_1^*(b^m)$$

and using Proposition 2.37(iv), we obtain

$$2(s_1 + \cdots + s_l) + m \le 2s_1 + \text{rank}[h] - \dim\left(\Lambda^1_{\varphi_1^*(g_1)} \cap \Lambda^1_h\right).$$

On the other hand, a successive application of the same proposition gives

$$\text{rank}[h] \le 2(s_2 + \cdots + s_l) + m.$$

Combining the two previous inequalities, we get

$$\text{rank}[h] = 2(s_2 + \cdots + s_l) + m \quad \text{and} \quad \Lambda^1_{\varphi_1^*(g_1)} \cap \Lambda^1_h = \{0\}.$$

Finally, noticing that, in a neighborhood of x_0,

$$\Lambda^1_{\varphi_1^*(g_1)} = \operatorname{span}\{dx^1, \ldots, dx^{2s_1}\} = \Lambda^1_{dx^1 \wedge \cdots \wedge dx^{2s_1}},$$

we have the claim (15.2) using again Proposition 2.37(iv). Note also that

$$\begin{aligned} &\operatorname{rank}\left[f_2 \wedge \cdots \wedge f_l \wedge dx^1 \wedge \cdots \wedge dx^{2s_1} \wedge a^1 \wedge \cdots \wedge a^m\right] \\ &= 2(s_2 + \cdots + s_l) + (2s_1 + m). \end{aligned}$$

Step 3. Therefore, using the induction hypothesis, there exist a neighborhood U_2 of x_0 and $\varphi_2 \in \operatorname{Diff}^\infty(U_2; \varphi_2(U_2))$ such that $\varphi_2(x_0) = x_0$ and for every $2 \le i \le l$, $1 \le j \le 2s_1$ and $1 \le k \le m$, the following identities hold in U_2:

$$\varphi_2^*(\varphi_1^*(g_i)) = f_i, \quad \varphi_2^*(dx^j) = dx^j \quad \text{and} \quad \varphi_2^*(\varphi_1^*(b^k)) = a^k.$$

Note, in particular, that $\varphi_2^*(\varphi_1^*(g_1)) = \varphi_2^*(f_1) = f_1$. Setting, choosing if necessary a smaller U_2,

$$\varphi = \varphi_1 \circ \varphi_2,$$

we have $\varphi \in \operatorname{Diff}^\infty(U_2; \varphi(U_2))$ with the claimed properties. □

It is interesting to contrast the algebraic result of Proposition 2.43(iii) with the analytical result of the above theorem, where it is essential to require that the 1-forms and the 2-forms be closed. Although every constant 3-form of rank $= 5$ is a linear pullback (combining Proposition 2.43(iii) and Proposition 2.24(ii)) of

$$(dx^1 \wedge dx^2 + dx^3 \wedge dx^4) \wedge dx^5,$$

we have the following result.

Proposition 15.14. *There exists $F \in C^\infty(\mathbb{R}^5; \Lambda^3)$ with*

$$dF = 0 \quad \text{and} \quad \operatorname{rank}[F] = 5 \quad \text{in } \mathbb{R}^5,$$

which cannot be pulled back locally by a diffeomorphism to the canonical 3-form of rank 5:

$$(dx^1 \wedge dx^2 + dx^3 \wedge dx^4) \wedge dx^5.$$

Proof. We will show that $F = f \wedge a$, where

$$f = \frac{1}{(x_3)^4 + 1} dx^1 \wedge dx^5 + dx^3 \wedge dx^4 \quad \text{and} \quad a = ((x_3)^2 + 1)dx^1 + ((x_3)^4 + 1)dx^2$$

has all of the desired properties. First, note that $dF = 0$ and $\operatorname{rank}[F] = 5$ in $\mathbb{R}^5$. We split the proof of the last assertion of the proposition into three steps.

Step 1. We claim that any 1-divisor c of F must be of the form $c = \lambda a$, where λ is a scalar function. Indeed, if this is not the case, we have that the 1-form $c(x_0)$ is linearly independent of $a(x_0)$ for a certain point $x_0 \in \mathbb{R}^5$. We therefore have

$$F(x_0) \wedge a(x_0) = F(x_0) \wedge c(x_0) = 0 \quad \text{and} \quad c(x_0) \wedge a(x_0) \neq 0.$$

Appealing to Theorem 2.42, we deduce that $F(x_0)$ is totally divisible and, hence (see again Proposition 2.43(ii)), $\operatorname{rank}[F(x_0)] = 3$, a contradiction.

Step 2. We show that if there exist an open set U and $\lambda \in C^1(U)$ such that

$$d(\lambda a) = 0 \quad \text{in } U,$$

then we necessarily have $\lambda \equiv 0$. Indeed, if $d(\lambda a) = 0$ in U, then, in particular,

$$(d(\lambda a))_{13} = (d(\lambda a))_{23} = 0$$

and, hence,

$$\frac{\partial(\lambda(x)(x_3^2+1))}{\partial x_3} = \frac{\partial(\lambda(x)(x_3^4+1))}{\partial x_3} = 0.$$

However, this implies the existence of $u, v \in C^1(U)$ with

$$u(x_1,x_2,x_3,x_4,x_5) = u(x_1,x_2,x_4,x_5),$$

$$v(x_1,x_2,x_3,x_4,x_5) = v(x_1,x_2,x_4,x_5)$$

such that

$$\lambda(x) = \frac{u(x_1,x_2,x_4,x_5)}{x_3^2+1} = \frac{v(x_1,x_2,x_4,x_5)}{x_3^4+1},$$

which is possible only if $u = v = 0$ in U, which proves the claim.

Step 3. We now conclude. If there exists a local diffeomorphism φ satisfying

$$F = \varphi^*((dx^1 \wedge dx^2 + dx^3 \wedge dx^4) \wedge dx^5) = \varphi^*(dx^1 \wedge dx^2 + dx^3 \wedge dx^4) \wedge \varphi^*(dx^5),$$

it follows from Step 1 that

$$\varphi^*(dx^5) = \lambda a.$$

However, this leads to a contradiction, because the form on the left-hand side is closed and nonzero, whereas (cf. Step 2) the form on the right-hand side is either not closed or identically 0. □

We end this chapter with the following result, a particular case of which was proved in Bandyopadhyay and Dacorogna [8].

Theorem 15.15. *Let $4 \leq 2m \leq n$ be integers. Let $x_0 \in \mathbb{R}^n$, f and g be C^∞ closed 2-forms, and a and b be C^∞ closed 1-forms such that, in a neighborhood of x_0,*

$$\operatorname{rank}[f] = \operatorname{rank}[g] = 2m \quad \text{and} \quad \operatorname{rank}[g \wedge b] = \operatorname{rank}[f \wedge a] = 2m-1.$$

Then there exist a neighborhood U of x_0 and $\varphi \in \operatorname{Diff}^\infty(U;\varphi(U))$ such that $\varphi(x_0) = x_0$ and

$$\varphi^*(g) = f \quad \text{and} \quad \varphi^*(b) = a \quad \text{in } U.$$

Remark 15.16. Note that if $\operatorname{rank}[g]=2m=n\geq 4$ and $b\neq 0$, then $g\wedge b\neq 0$; otherwise by Theorem 2.42 there would exist c a 1-form such that

$$g = b \wedge c$$

and, hence, $\operatorname{rank}[g] = 2$, which is a contradiction. We therefore have, by Proposition 2.37(v), that

$$\operatorname{rank}[g\wedge b] = 2m-1.$$

Proof. As usual, we may assume that $x_0 = 0$ and, using Remark 15.2(iii), that

$$f = \omega_m = \sum_{i=1}^{m} dx^{2i-1} \wedge dx^{2i} \quad \text{and} \quad a = dx^1$$

(note that these particular f and a satisfy all of the hypotheses of the theorem, in view of Proposition 2.37(v)). We split the proof into three steps.

Step 1. Let us show that, with no loss of generality, we can assume

$$g = \sum_{i=1}^{m} dx^{2i-1} \wedge dx^{2i} = \omega_m \quad \text{and} \quad b = \sum_{i=1}^{2m} b_i(x_1,\ldots,x_{2m})dx^i$$

and, thus, we can assume that $2m = n$. Since $dg = 0$ and $\operatorname{rank}[g] = 2m$ in a neighborhood of 0, we can apply Theorem 14.3 to find a neighborhood U_1 of 0 and $\varphi_1 \in \operatorname{Diff}^\infty(U_1;\varphi(U_1))$ such that $\varphi_1(0) = 0$ and

$$\varphi_1^*(g) = \sum_{i=1}^{m} dx^{2i-1} \wedge dx^{2i} = \omega_m \quad \text{in } U_1 .$$

We claim that

$$\varphi_1^*(b)(x_1,\ldots,x_n) = c(x_1,\ldots,x_{2m}) = \sum_{i=1}^{2m} c_i(x_1,\ldots,x_{2m})\, dx^i .$$

Once this is shown, we will have the assertion of Step 1. Let us prove the claim. Note that, in a neighborhood of 0,

$$\begin{aligned}
\operatorname{rank}[\omega_m] &= \operatorname{rank}[\varphi_1^*(g)] = \operatorname{rank}[g] = 2m,\\
\operatorname{rank}[c\wedge\omega_m] &= \operatorname{rank}[\varphi_1^*(b\wedge g)] = \operatorname{rank}[b\wedge g] = 2m-1.
\end{aligned}$$

Hence, using Proposition 2.37(v), we get, in a neighborhood of 0,

$$c \in \Lambda^1_{\omega_m} = \operatorname{span}\left\{dx^1,\ldots,dx^{2m}\right\}$$

and, thus,

$$c_i(x) = 0 \quad \text{for } 2m+1 \le i \le n.$$

Finally, combining the previous equation with the fact that $dc = 0$, we immediately deduce that for every $1 \le i \le m$ and every x in a neighborhood of 0,

$$c_i(x_1,\ldots,x_n) = c_i(x_1,\ldots,x_{2m}),$$

which proves the claim.

Step 2. Using Theorem 8.1, we can find a C^∞ function (in a small ball B_ε centered at 0) ρ such that

$$d\rho = b(0) - b.$$

With no loss of generality, we can assume that $\rho(0) = 0$. Let $b_t(x) \in C^\infty\left([0,1] \times B_\varepsilon;\Lambda^1\right)$ be defined by

$$b_t(x) = (1-t)b(0) + tb(x).$$

Since for every $t \in [0,1]$, $b_t(0) = b(0) \neq 0$, there exist $1 \leq i \leq n$ and a neighborhood of 0 in which

$$[b_t \lrcorner\, \omega_m]_i = [\overline{\omega}_m b_t]_i \neq 0 \quad \text{for every } t \in [0,1].$$

Hence, we can apply Remark 8.21 and find a neighborhood U_2 of 0 and $w \in C^\infty\left([0,1] \times U_2;\Lambda^1\right)$, $w(t,x) = w_t(x)$ such that for every $t \in [0,1]$, $w_t(0) = 0$ and

$$dw_t = 0 \quad \text{and} \quad \langle w_t;\overline{\omega}_m b_t\rangle = \rho \quad \text{in } U_2.$$

Finally, define $u \in C^\infty([0,1] \times U_2;\Lambda^1)$, $u = u(t,x) = u_t(x)$, as

$$u_t = \overline{\omega}_m^{-1} w_t \;\Leftrightarrow\; u_t \lrcorner\, \omega_m = w_t\,.$$

Note that for every $t \in [0,1]$, $u_t(0) = 0$ and in U_2, $d\left(u_t \lrcorner\, \omega_m\right) = dw_t = 0$ and since $\overline{\omega}_m \in O(n)$,

$$d(u_t \lrcorner\, b_t) = d(\langle u_t;b_t\rangle) = d\left(\langle w_t;\overline{\omega}_m b_t\rangle\right) = d\rho = -\frac{db_t}{dt}\,.$$

Hence, we deduce from Theorem 12.8 that for every $t \in [0,1]$, the solution ϕ_t of

$$\begin{cases} \dfrac{d}{dt}\phi_t = u_t \circ \phi_t, & 0 \leq t \leq 1, \\ \phi_0 = \text{id} \end{cases}$$

exists in a neighborhood U_3 of 0 and verifies $\phi_t \in \text{Diff}^\infty(U_3;\phi_t(U_3))$ and

$$\phi_t^*(\omega_m) = \omega_m\,, \quad \phi_t^*(b_t) = b(0) \quad \text{in } U_3\,.$$

Step 3. Finally, recalling that $b(0) \in \Lambda^1_{\omega_m}$, there exists, using Proposition 2.24, $A \in \text{GL}(n)$ such that

$$A^*(\omega_m) = \omega_m \quad \text{and} \quad A^*(b(0)) = dx^1.$$

Letting $\psi(x) = Ax$ and $\varphi = \phi_1 \circ \psi$, we get the result and this concludes the proof.

□

Part V
Hölder Spaces

Chapter 16
Hölder Continuous Functions

We recall here the basic definitions of Hölder spaces. We use the following as references in the present chapter: Adams [2], Dacorogna [29], de la Llave and Obaya [36], Edmunds and Evans [40], Fefferman [42], Gilbarg and Trudinger [49] and Hörmander [55].

16.1 Definitions of Continuous and Hölder Continuous Functions

16.1.1 Definitions

In this chapter, for $x = (x_1, \dots, x_N) \in \mathbb{R}^N$, we will write

$$|x| = \max_{1 \le i \le N} \{|x_i|\}.$$

We start by recalling the definition of C^r spaces.

Definition 16.1. Let $r \geq 0$ be an integer and $\Omega \subset \mathbb{R}^n$ be an open set.

(i) $C^0(\Omega)$ is the set of continuous functions $f : \Omega \to \mathbb{R}$.

(ii) $C^r(\Omega)$ is the set of functions $f : \Omega \to \mathbb{R}$ which have all of their partial derivatives of any order up to r continuous; in other words, $D^a f \in C^0(\Omega)$ for every $a \in \mathscr{A}_m$, $0 \leq m \leq r$, where $\mathscr{A}_m$ is the set of all multi-indices of order m. We also set $\nabla^m f = \{D^a f\}_{a \in \mathscr{A}_m}$.

(iii) $C^0(\overline{\Omega})$ is the set of bounded continuous functions $f : \overline{\Omega} \to \mathbb{R}$. We equip this space with the norm

$$\|f\|_{C^0(\overline{\Omega})} = \sup_{x \in \Omega} \{|f(x)|\}.$$

(iv) $C^r\left(\overline{\Omega}\right)$ is the set of $C^r(\Omega)$ bounded functions whose derivatives up to the order r can be extended continuously and in a bounded way to $\overline{\Omega}$. The space $C^r\left(\overline{\Omega}\right)$

is equipped with the following norm:

$$\|f\|_{C^r(\overline{\Omega})} = \sum_{m=0}^{r} \|\nabla^m f\|_{C^0(\overline{\Omega})} .$$

When there is no ambiguity, we drop the dependence on the set $\overline{\Omega}$ and write simply

$$\|f\|_{C^r} = \sum_{m=0}^{r} \|\nabla^m f\|_{C^0} .$$

(v) The set $C_0^r(\Omega)$ denotes the set of functions in $C^r(\Omega)$ with compact support in Ω.

We now give the definitions of Hölder continuous functions.

Definition 16.2. Let $D \subset \mathbb{R}^n$, $f : D \to \mathbb{R}$ and $0 < \alpha \leq 1$. We let

$$[f]_{C^{0,\alpha}(D)} = \sup_{\substack{x,y \in D \\ x \neq y}} \left\{ \frac{|f(x) - f(y)|}{|x-y|^\alpha} \right\} .$$

Let $\Omega \subset \mathbb{R}^n$ be an open set and $r \geq 0$ be an integer. We define the different spaces of *Hölder continuous* functions in the following way:

(i) $C^{0,\alpha}(\Omega)$ is the set of $f \in C^0(\Omega)$ such that

$$[f]_{C^{0,\alpha}(K)} < \infty$$

for every compact set $K \subset \Omega$.

(ii) $C^{0,\alpha}\left(\overline{\Omega}\right)$ is the set of $f \in C^0\left(\overline{\Omega}\right)$ so that

$$\|f\|_{C^{0,\alpha}(\overline{\Omega})} = \|f\|_{C^0(\overline{\Omega})} + [f]_{C^{0,\alpha}(\overline{\Omega})} < \infty.$$

If there is no ambiguity, we drop the dependence on the set $\overline{\Omega}$ and write simply

$$\|f\|_{C^{0,\alpha}} = \|f\|_{C^0} + [f]_{C^{0,\alpha}} .$$

(iii) $C^{r,\alpha}(\Omega)$ is the set of $f \in C^r(\Omega)$ such that

$$[D^a f]_{C^{0,\alpha}(K)} < \infty$$

for every compact set $K \subset \Omega$ and every $a \in \mathscr{A}_r$.

(iv) $C^{r,\alpha}\left(\overline{\Omega}\right)$ is the set of functions $f \in C^r\left(\overline{\Omega}\right)$ so that

$$[D^a f]_{C^{0,\alpha}(\overline{\Omega})} < \infty$$

for every $a \in \mathscr{A}_r$. We equip $C^{r,\alpha}\left(\overline{\Omega}\right)$ with the following norm:

$$\|f\|_{C^{r,\alpha}} = \|f\|_{C^r} + \max_{a \in \mathscr{A}_r} [D^a f]_{C^{0,\alpha}} .$$

Most of the time we will write

$$[\nabla^r f]_{C^{0,\alpha}} = \max_{a\in\mathscr{A}_r} [D^a f]_{C^{0,\alpha}};$$

thus, we adopt the notation

$$\|f\|_{C^{r,\alpha}} = \|f\|_{C^r} + [\nabla^r f]_{C^{0,\alpha}} = \|f\|_{C^{r-1}} + \|\nabla^r f\|_{C^{0,\alpha}}.$$

(v) The set $C_0^{r,\alpha}(\Omega)$ denotes the set of functions in $C^{r,\alpha}(\Omega)$ with compact support in Ω.

Remark 16.3. (i) $C^{r,\alpha}(\overline{\Omega})$ with its norm $\|\cdot\|_{C^{r,\alpha}}$ is a Banach space.

(ii) When the domain has some minimal regularity (say Lipschitz), it will be shown that the norm considered here is equivalent to the following ones (see Corollary 16.13 for the first one and Corollary 16.25 for the second one):

$$\|f\|_{C^{r,\alpha}} = \sum_{m=0}^{r} \|\nabla^m f\|_{C^{0,\alpha}}$$

and

$$\|f\|_{C^{r,\alpha}} = \begin{cases} \|f\|_{C^0} + [\nabla^r f]_{C^{0,\alpha}} & \text{if } 0 < \alpha \leq 1 \\ \|f\|_{C^0} + \|\nabla^r f\|_{C^0} & \text{if } \alpha = 0. \end{cases}$$

We should, however, insist that these norms are, in general, not equivalent for very wild sets.

(iii) When $\alpha = 1$, we note that $C^{0,1}(\overline{\Omega})$ is in fact the set of *Lipschitz continuous* and bounded functions, namely the set of bounded functions f such that there exists a constant $\gamma > 0$ so that

$$|f(x) - f(y)| \leq \gamma |x - y|, \ \forall x, y \in \overline{\Omega}.$$

The best such constant is $\gamma = [f]_{C^{0,1}}$.

(iv) If one wants to include the classical C^r spaces in the context of $C^{r,\alpha}$ spaces, one is led to some inconsistencies. We have decided to write

$$C^r = C^{r,0}.$$

In this case, we set

$$[f]_{C^{0,0}} = 0 \quad \text{and} \quad \|f\|_{C^{0,0}} = \|f\|_{C^0}$$

and similarly for $r \geq 1$,

$$\|f\|_{C^{r,0}} = \|f\|_{C^r}.$$

(v) When $\Omega = \mathbb{R}^n$, in order to remove any ambiguity, we understand $C^{r,\alpha}(\mathbb{R}^n)$ as $C^{r,\alpha}(\overline{\mathbb{R}^n})$ in the sense of point (iv) of the above definition.

(vi) It follows from Theorem 16.11 that if Ω is bounded and Lipschitz, then $f \in C^{r,\alpha}(\overline{\Omega})$ if and only if there exist an open set $O \subset \mathbb{R}^n$ such that $\overline{\Omega} \subset O$ and $g \in C^{r,\alpha}(O)$ with $g = f$ in $\overline{\Omega}$.

16.1.2 Regularity of Boundaries

We used and will use in several places the notion of $C^{r,\alpha}$ sets, in particular, Lipschitz or smooth sets. We now give two classical definitions of such sets.

Definition 16.4. (i) Let $\Omega \subset \mathbb{R}^n$ be an open set, $r \geq 0$ be an integer and $0 \leq \alpha \leq 1$. The set Ω is said to be $C^{r,\alpha}$ if for every $x \in \partial\Omega$, there exist a neighborhood U_x of x and $\varphi_x \in C^{r,\alpha}(\mathbb{R}^{n-1})$ such that, up to a rotation,

$$U_x \cap \Omega = U_x \cap \{y = (y', y_n) \in \mathbb{R}^{n-1} \times \mathbb{R} : y_n > \varphi_x(y')\}.$$

(ii) When Ω is $C^{0,1}$, then Ω will be referred to as *Lipschitz*.

(iii) If Ω is C^{∞}, then Ω is said to be *smooth*.

Definition 16.5. Let $r \geq 1$ be an integer and $0 \leq \alpha \leq 1$. The set Ω is said to be $C^{r,\alpha}$ if for every $x \in \partial\Omega$, there exist a neighborhood U_x of x and $\phi_x \in \mathrm{Diff}^{r,\alpha}(U_x; B)$ (where B denotes the open unit ball in $\mathbb{R}^n$) such that

$$\phi_x(U_x \cap \Omega) = \{y \in B : y_n > 0\}.$$

Remark 16.6. It is easy to see that when $r \geq 1$, both definitions are equivalent.

We now define the meaning of $C^{r,\alpha}(\partial\Omega)$ functions.

Definition 16.7. Let $r \geq 1$ be an integer, $0 \leq \alpha \leq 1$ and $\Omega \subset \mathbb{R}^n$ be a bounded open $C^{r,\alpha}$ set.

(i) The set $C^{r,\alpha}(\partial\Omega)$ is the set of functions $f : \partial\Omega \to \mathbb{R}$ such that for every $x \in \partial\Omega$,

$$(y_1, \ldots, y_{n-1}) \to f \circ \phi_x^{-1}(y_1, \ldots, y_{n-1}, 0)$$

belongs to $C^{r,\alpha}(B')$, where B' is the open unit ball in $\mathbb{R}^{n-1}$ and where ϕ_x is as in Definition 16.5.

(ii) For $f \in C^{r,\alpha}(\partial\Omega)$, we define

$$\|f\|_{C^{r,\alpha}(\partial\Omega)} = \inf_{\substack{F \in C^{r,\alpha}(\overline{\Omega}): \\ F = f \text{ on } \partial\Omega}} \|F\|_{C^{r,\alpha}(\overline{\Omega})}.$$

Remark 16.8. (i) The definition of $C^{r,\alpha}(\partial\Omega)$ is independent of the chosen ϕ_x.

(ii) Note that the set over which the infimum is taken is never empty (see, e.g., Lemma 6.38 in [49]).

(iii) The set $C^{0,\alpha}(\partial\Omega)$ and the associated norm have already been considered in Definition 16.2.

(iv) $C^{r,\alpha}(\partial\Omega)$ with its norm $\|\cdot\|_{C^{r,\alpha}(\partial\Omega)}$ is a Banach space.

16.1.3 Some Elementary Properties

In Section 16.2, it will be more convenient to work with the norm

$$\|f\|_{C_*^{r,\alpha}} = \sum_{m=0}^{r} \|\nabla^m f\|_{C^{0,\alpha}}$$

and the corresponding space

$$C_*^{r,\alpha}\left(\overline{\Omega}\right) = \left\{f \in C^r\left(\overline{\Omega}\right) : \|f\|_{C_*^{r,\alpha}} < \infty\right\}.$$

Note that

$$g, h \in C_*^{r,\alpha} \Rightarrow gh \in C_*^{r,\alpha}$$

and

$$\|gh\|_{C_*^{r,\alpha}} \leq C \|g\|_{C_*^{r,\alpha}} \|h\|_{C_*^{r,\alpha}}.$$

A priori, we have no better result on the product of two Hölder functions. This is why we introduce the space $C_*^{r,\alpha}$ (if, however, Ω is Lipschitz, we have a much better result; see Theorem 16.28). Clearly,

$$\|f\|_{C^{r,\alpha}} \leq \|f\|_{C_*^{r,\alpha}}$$

and, as already said, we will show in Corollary 16.13 the equivalence of the two norms for Lipschitz sets. We now gather some elementary properties.

Proposition 16.9. *Let $\Omega \subset \mathbb{R}^n$ be any bounded open set and $r \geq 0$ be an integer. Then*

$$\left\|\nabla^{r+1} f\right\|_{C^0} \leq [\nabla^r f]_{C^{0,1}} \quad \textit{and} \quad \|f\|_{C^{r+1}} \leq \|f\|_{C^{r,1}}.$$

If $s \geq r$ is an integer and $0 \leq \alpha \leq \beta \leq 1$, then

$$\|f\|_{C^{r,\alpha}} \leq C \|f\|_{C^{r,\beta}} \quad \textit{and} \quad \|f\|_{C_*^{r,\alpha}} \leq C \|f\|_{C_*^{s,\beta}},$$

where $C = \max\{1, \operatorname{diam}\Omega\}$ and

$$\operatorname{diam}\Omega = \sup_{x,y\in\Omega} \{|x-y|\}.$$

Proof. Step 1. Let $1 \leq i \leq n$, e_i be the ith vector of the Euclidean basis and $x \in \Omega$. We have

$$\left|\frac{\partial}{\partial x_i}[\nabla^r f](x)\right| = \left|\lim_{h\to 0}\frac{\nabla^r f(x+he_i)-\nabla^r f(x)}{h}\right| \le [\nabla^r f]_{C^{0,1}}$$

and, thus,

$$\left\|\nabla^{r+1} f\right\|_{C^0} \le [\nabla^r f]_{C^{0,1}}\,,$$

which, in turn, also implies

$$\|f\|_{C^{r+1}} \le \|f\|_{C^{r,1}}\,,$$

as wished.

Step 2. The inequalities

$$\|f\|_{C^{r,\alpha}} \le C\,\|f\|_{C^{r,\beta}} \quad \text{and} \quad \|f\|_{C_*^{r,\alpha}} \le C\,\|f\|_{C_*^{s,\beta}}$$

follow from the observation that for every $0 \le m \le r$,

$$[\nabla^m f]_{C^{0,\alpha}} \le [\nabla^m f]_{C^{0,\beta}} \sup_{x,y\in\overline{\Omega}}\left\{|x-y|^{\beta-\alpha}\right\} \le C\,[\nabla^m f]_{C^{0,\beta}}\,.$$

This concludes the proof of the proposition. □

The above proposition can be strongly improved if one requires some additional regularity on Ω. We discuss here the case of convex sets Ω. The more general case of Lipschitz sets is dealt with in Corollary 16.13.

Proposition 16.10. *Let $\Omega \subset \mathbb{R}^n$ be a bounded convex open set and $r \ge 0$ be an integer. Then*

$$\left\|\nabla^{r+1} f\right\|_{C^0} = [\nabla^r f]_{C^{0,1}} \quad \textit{and} \quad \|f\|_{C^{r+1}} = \|f\|_{C^{r,1}}\,.$$

Let $s \ge r$ be an integer and $0 \le \alpha, \beta \le 1$, with

$$r+\alpha \le s+\beta.$$

Then

$$\|f\|_{C^{r,\alpha}} \le C\,\|f\|_{C^{s,\beta}} \quad \textit{and} \quad \|f\|_{C_*^{r,\alpha}} \le (C+1)\,\|f\|_{C^{r,\alpha}}\,,$$

where $C = \max\{1, \operatorname{diam}\Omega\}$.

Proof. Step 1. For any $x, y \in \Omega$ and $a \in \mathscr{A}_r$, we can write

$$\begin{aligned} D^a f(x) - D^a f(y) &= \int_0^1 \frac{d}{dt}\left[D^a f(y+t(x-y))\right] dt \\ &= \int_0^1 \langle \nabla D^a f(y+t(x-y)); x-y\rangle\, dt. \end{aligned}$$

We hence deduce that

$$[\nabla^r f]_{C^{0,1}} \le \left\|\nabla^{r+1} f\right\|_{C^0}.$$

The above inequality, coupled with Proposition 16.9, leads immediately to the claim, namely

$$\left\|\nabla^{r+1} f\right\|_{C^0} = [\nabla^r f]_{C^{0,1}} \quad \text{and} \quad \|f\|_{C^{r+1}} = \|f\|_{C^{r,1}} .$$

Step 2. We now prove that

$$\|f\|_{C^{r,\alpha}} \leq C\,\|f\|_{C^{s,\beta}} .$$

Observe first that if $s = r$ and thus $0 \leq \alpha \leq \beta \leq 1$, the result is already in Proposition 16.9. So let us assume that $s \geq r+1$ and use Proposition 16.9 and Step 1 to get

$$\|f\|_{C^{r,\alpha}} \leq C\,\|f\|_{C^{r,1}} = C\,\|f\|_{C^{r+1}} \leq C\,\|f\|_{C^s} \leq C\,\|f\|_{C^{s,\beta}} .$$

Step 3. We finally establish that

$$\|f\|_{C_*^{r,\alpha}} \leq (C+1)\,\|f\|_{C^{r,\alpha}} .$$

Assume that $\alpha > 0$; otherwise the result is trivially valid by definition. Let $0 \leq m \leq r-1$ and note first that

$$[\nabla^m f]_{C^{0,\alpha}} \leq C\,[\nabla^m f]_{C^{0,1}} = C\left\|\nabla^{m+1} f\right\|_{C^0} .$$

We therefore deduce that

$$\|\nabla^m f\|_{C^{0,\alpha}} \leq \|\nabla^m f\|_{C^0} + C\left\|\nabla^{m+1} f\right\|_{C^0}$$

and, hence,

$$\sum_{m=0}^{r-1} \|\nabla^m f\|_{C^{0,\alpha}} \leq \|f\|_{C^0} + (C+1)\sum_{m=1}^{r-1} \|\nabla^m f\|_{C^0} + C\,\|\nabla^r f\|_{C^0} .$$

We have therefore obtained that

$$\begin{aligned} \|f\|_{C_*^{r,\alpha}} = \sum_{m=0}^{r} \|\nabla^m f\|_{C^{0,\alpha}} &\leq \|f\|_{C^0} + (C+1)\sum_{m=1}^{r} \|\nabla^m f\|_{C^0} + [\nabla^r f]_{C^{0,\alpha}} \\ &\leq (C+1)\,\|f\|_{C^{r,\alpha}} \end{aligned}$$

and the result is proved. □

16.2 Extension of Continuous and Hölder Continuous Functions

16.2.1 The Main Result and Some Corollaries

The main result of this section is the following extension theorem essentially due to Calderon [20]. We will closely follow the presentation of Stein [92] (sometimes word for word). Although Stein does his extension for Sobolev spaces, exactly the same extension works for Hölder spaces. We will therefore only outline the main points of the proof and refer to Kneuss [60] for details.

Theorem 16.11. *Let $\Omega \subset \mathbb{R}^n$ be a bounded open Lipschitz set. Then there exists a continuous linear extension operator*

$$E : C^{r,\alpha}\left(\overline{\Omega}\right) \to C_0^{r,\alpha}\left(\mathbb{R}^n\right)$$

for any integer $r \geq 0$ and any $0 \leq \alpha \leq 1$. More precisely, there exists a constant $C = C(r,\Omega) > 0$ such that for every $f \in C^{r,\alpha}\left(\overline{\Omega}\right)$,

$$E(f)|_{\overline{\Omega}} = f, \quad \operatorname{supp}[E(f)] \text{ is compact,}$$

$$\|E(f)\|_{C^{r,\alpha}(\mathbb{R}^n)} \leq C\|f\|_{C^{r,\alpha}\left(\overline{\Omega}\right)}.$$

Remark 16.12. We should emphasize that the extension is universal, in the sense that the same extension also leads to

$$\|E(f)\|_{C^{s,\beta}(\mathbb{R}^n)} \leq C\|f\|_{C^{s,\beta}\left(\overline{\Omega}\right)}$$

for any integer s and any $0 \leq \beta \leq 1$, with, of course, $C = C(s,\Omega)$. It is also the very same extension that is valid for Sobolev spaces.

We have, as an immediate consequence of Theorem 16.11, Propositions 16.9 and 16.10, the following result.

Corollary 16.13. *Let $\Omega \subset \mathbb{R}^n$ be a bounded open Lipschitz set, $s \geq r \geq 0$ be integers and $0 \leq \alpha, \beta \leq 1$, with*

$$r + \alpha \leq s + \beta.$$

Then there exists a constant $C = C(s,\Omega) > 0$ such that

$$\|f\|_{C^{r,\alpha}} \leq C\|f\|_{C^{s,\beta}}$$

and

$$\|f\|_{C^{s+1}} \leq \|f\|_{C^{s,1}} \leq C\|f\|_{C^{s+1}}.$$

Moreover (cf. Section 16.1 for the notations), the following inequality holds:

$$\|f\|_{C^{s,\beta}} \leq \|f\|_{C_*^{s,\beta}} \leq C\|f\|_{C^{s,\beta}}$$

and, therefore, the $\|\cdot\|_{C_*^{s,\beta}}$ and the $\|\cdot\|_{C^{s,\beta}}$ norms are equivalent.

Remark 16.14. In particular, if $0 < \alpha < \beta < 1$, we deduce from the corollary that

$$C^r \supset C^{r,\alpha} \supset C^{r,\beta} \supset C^{r,1} \supset C^{r+1}$$

and the imbeddings are continuous. Note, however, that the result is false if the set Ω is not smooth enough (see [29] for an example).

Another immediate corollary is the following. It has been used in the proof of Theorem 11.1.

Corollary 16.15. *Let $r \geq 1$ be an integer and $0 \leq \alpha \leq 1$. Let $\Omega \subset \mathbb{R}^n$ be a bounded open Lipschitz set and $\varphi \in \operatorname{Diff}^{r,\alpha}(\overline{\Omega}; \varphi(\overline{\Omega}))$. Then there exist a neighborhood V of $\overline{\Omega}$ and ψ an extension of φ such that*

$$\psi \in \operatorname{Diff}^{r,\alpha}(V; \psi(V)).$$

Proof. By Theorem 16.11, there exists $\psi \in C^{r,\alpha}(\mathbb{R}^n; \mathbb{R}^n)$, an extension of φ. By continuity, there exists $\varepsilon_1 > 0$ such that

$$\det \nabla \psi(x) \neq 0 \quad \text{in } \Omega + B_{\varepsilon_1}. \tag{16.1}$$

Let us show that there exists $\varepsilon < \varepsilon_1$ such that

$$\psi \in \operatorname{Diff}^{r,\alpha}(\Omega + B_\varepsilon; \psi(\Omega + B_\varepsilon)),$$

which will conclude the proof. It is sufficient, using (16.1), to find $\varepsilon < \varepsilon_1$ such that ψ is one-to-one in $\Omega + B_\varepsilon$. We proceed by contradiction. Then there exist two sequences $x_\nu, y_\nu \in \Omega + B_{\varepsilon_1}$, $\nu \in \mathbb{N}$, such that

$$x_\nu \neq y_\nu, \quad \psi(x_\nu) = \psi(y_\nu) \quad \text{and} \quad x_\nu, y_\nu \in \Omega + B_{1/\nu}, \quad \nu \in \mathbb{N}.$$

Extracting, if necessary, a subsequence, we can assume that $x_\nu \to x$ and $y_\nu \to y$ with $x, y \in \overline{\Omega}$. Therefore, $\psi(x) = \psi(y)$ and, thus, $x = y$, ψ being one-to-one in $\overline{\Omega}$. Since $\det \nabla \psi(x) \neq 0$, we know that ψ is a local diffeomorphism from a neighborhood of x onto a neighborhood of $\psi(x)$. This contradicts the fact that $\psi(x_\nu) = \psi(y_\nu)$ for ν large enough. This concludes the proof. □

We would finally like to mention that when Ω is as regular as the function to be extended, then there is an elementary extension result using rectification of boundary and reflection (see, e.g., Gilbarg and Trudinger [49]). Moreover, when $r = 0$, we have other classical extension theorems. When $r = \alpha = 0$, the Tietze extension theorem (cf., e.g., [87]) is one of them. When $r = 0$ and $0 < \alpha \leq 1$, we have the Mc Shane lemma that we now prove.

Theorem 16.16 (Mc Shane lemma). *Let $D \subset \mathbb{R}^n$ be any set, $0 < \alpha \leq 1$ and $f : D \to \mathbb{R}$, with*

$$\gamma = [f]_{C^{0,\alpha}(D)} < \infty.$$

Part 1. Then the two functions

$$f_+(x) = \inf_{y \in D} \left\{ f(y) + \gamma |x - y|^\alpha \right\},$$
$$f_-(x) = \sup_{y \in D} \left\{ f(y) - \gamma |x - y|^\alpha \right\}$$

are extensions of f satisfying

$$[f_+]_{C^{0,\alpha}(\mathbb{R}^n)} = [f_-]_{C^{0,\alpha}(\mathbb{R}^n)} = [f]_{C^{0,\alpha}(D)} = \gamma.$$

Furthermore, any other extension g of f such that $[g]_{C^{0,\alpha}(\mathbb{R}^n)} = \gamma$ satisfies

$$f_- \leq g \leq f_+.$$

Part 2. If, moreover, D is bounded and $f \in C^{0,\alpha}\left(\overline{D}\right)$, then there exists $g \in C^{0,\alpha}(\mathbb{R}^n)$ such that

$$g|_{\overline{D}} = f, \quad \operatorname{supp} g \text{ is compact,}$$

and

$$[g]_{C^{0,\alpha}(\mathbb{R}^n)} = [f]_{C^{0,\alpha}(\overline{D})} = \gamma.$$

Proof. Proof of Part 1. We divide the proof into two steps.

Step 1. We discuss the case of f_+, the other one being handled similarly.

1) Let us first check that f_+ is indeed an extension of f. Let $x \in D$; we therefore get

$$f(x) \leq f(y) + \gamma|x-y|^\alpha \quad \text{for every } y \in D$$

and, thus,

$$f(x) \leq f_+(x).$$

Now, clearly, choosing $y = x$ in the definition of f_+ leads to $f_+(x) \leq f(x)$. Thus, f_+ is indeed an extension of f.

2) Let $x, z \in \mathbb{R}^n$. Assume, without loss of generality, that $f_+(z) \leq f_+(x)$. For every $\varepsilon > 0$, we can find $y_z \in D$ such that

$$-\varepsilon + f(y_z) + \gamma|z-y_z|^\alpha \leq f_+(z) \leq f(y_z) + \gamma|z-y_z|^\alpha.$$

We hence obtain

$$\begin{aligned} |f_+(x) - f_+(z)| &= f_+(x) - f_+(z) \\ &\leq f(y_z) + \gamma|x-y_z|^\alpha + \varepsilon - f(y_z) - \gamma|z-y_z|^\alpha \\ &\leq \varepsilon + \gamma|x-z|^\alpha. \end{aligned}$$

Letting $\varepsilon \to 0$, we have the claim.

Step 2. Let g be such that $[g]_{C^{0,\alpha}(\mathbb{R}^n)} = \gamma$. We therefore have for $x \in \mathbb{R}^n$ and for every $y \in D$ (and, thus, $g(y) = f(y)$),

$$-\gamma|x-y|^\alpha \leq g(x) - g(y) = g(x) - f(y) \leq \gamma|x-y|^\alpha.$$

This leads to

$$f(y) - \gamma|x-y|^\alpha \leq g(x) \leq f(y) + \gamma|x-y|^\alpha$$

and, hence, $f_-(x) \leq g(x) \leq f_+(x)$, as wished.

Proof of Part 2. We split the discussion into two steps.

Step 1. Let $C > 0$ be such that

$$\|f\|_{C^0(\overline{D})} \leq C.$$

Since D is bounded, we can find $R > 0$ so that

$$\operatorname{dist}\left(\overline{D};\partial B_R\right)=\left(\frac{C}{\gamma}\right)^{1/\alpha}.$$

We then define

$$f_1(x)=\begin{cases} f(x) & \text{if } x\in\overline{D} \\ 0 & \text{if } x\in(B_R)^c=\mathbb{R}^n\setminus B_R. \end{cases}$$

Observe that $f_1\in C^{0,\alpha}\left(\overline{D}\cup(B_R)^c\right)$, with

$$[f_1]_{C^{0,\alpha}\left(\overline{D}\cup(B_R)^c\right)}=\gamma.$$

Indeed, let us prove that for every $x,y\in\overline{D}\cup(B_R)^c$, we have

$$|f_1(x)-f_1(y)|\le\gamma|x-y|^\alpha.$$

This is clearly so if $x,y\in\overline{D}$ or if $x,y\in(B_R)^c$. So let us prove the inequality for $x\in\overline{D}$ and $y\in(B_R)^c$ so that

$$|x-y|^\alpha\ge\left(\operatorname{dist}\left(\overline{D};\partial B_R\right)\right)^\alpha=\frac{C}{\gamma}.$$

We therefore have

$$|f_1(x)-f_1(y)|=|f_1(x)|=|f(x)|\le C=\gamma\frac{C}{\gamma}\le\gamma|x-y|^\alpha,$$

as wished.

Step 2. Use Part 1 to extend f_1 to $\mathbb{R}^n$. We denote this extension g (we can choose, e.g., $g=(f_1)_+$) and we therefore have

$$g|_{\overline{D}}=f,\quad \operatorname{supp} g \text{ is compact} \quad\text{and}\quad [g]_{C^{0,\alpha}(\mathbb{R}^n)}=[f]_{C^{0,\alpha}(\overline{D})}=\gamma.$$

This concludes the proof of the theorem. □

16.2.2 Preliminary Results

The main step in the proof of Theorem 16.11 is the following special case.

Theorem 16.17. *Let* $\varphi\in C^{0,1}(\mathbb{R}^{n-1})$ *and*

$$\Omega=\{x=(x',x_n)\in\mathbb{R}^{n-1}\times\mathbb{R}:x_n>\varphi(x')\}.$$

Then there exists a continuous linear extension operator

$$E:C^{r,\alpha}\left(\overline{\Omega}\right)\to C^{r,\alpha}\left(\mathbb{R}^n\right)$$

for any integer $r \geq 0$ and any $0 \leq \alpha \leq 1$. In particular, there exists a constant $C = C(r,\Omega) > 0$ such that for every $f \in C^{r,\alpha}(\overline{\Omega})$,

$$\|E(f)\|_{C^{r,\alpha}(\mathbb{R}^n)} \leq C\|f\|_{C^{r,\alpha}(\overline{\Omega})}. \tag{16.2}$$

Remark 16.18. (i) Note that (see Section 16.1 for the notations) the theorem immediately gives

$$\|E(f)\|_{C_*^{r,\alpha}(\mathbb{R}^n)} \leq C\|f\|_{C_*^{r,\alpha}(\overline{\Omega})} \quad \text{for every } f \in C_*^{r,\alpha}(\overline{\Omega}). \tag{16.3}$$

(ii) The proof of the previous theorem gives, in fact, a sharper estimate. Indeed, we have that for every $f \in C^{r,\alpha}(\overline{\Omega})$ and every $0 \leq m \leq r$,

$$\begin{aligned} \|\nabla^m E(f)\|_{C^0(\mathbb{R}^n)} &\leq C\|\nabla^m f\|_{C^0(\overline{\Omega})}, \\ [\nabla^m E(f)]_{C^{0,\alpha}(\mathbb{R}^n)} &\leq C[\nabla^m f]_{C^{0,\alpha}(\overline{\Omega})}. \end{aligned}$$

To prove Theorem 16.17, we will use the following three results. The first lemma can be found in Stein [92, p. 182].

Lemma 16.19. *There exists $\psi \in C^0([1,\infty))$ such that for every $N \in \mathbb{N}$, there exists $A_N > 0$ so that*

$$|\psi(\lambda)| \leq \frac{A_N}{\lambda^N} \quad \textit{for every } \lambda \in [1,\infty) \tag{16.4}$$

and, for every $k \geq 1$,

$$\int_1^\infty \psi(\lambda)d\lambda = 1 \quad \textit{and} \quad \int_1^\infty \lambda^k \psi(\lambda)d\lambda = 0. \tag{16.5}$$

The next result is essentially geometrical. In the sequel, we write

$$d(x) = d(x;\overline{\Omega}) = \inf\left\{|x-y| : y \in \overline{\Omega}\right\}.$$

Lemma 16.20. *Let $\varphi \in C^{0,1}(\mathbb{R}^{n-1})$ and*

$$\Omega = \{x = (x',x_n) \in \mathbb{R}^{n-1} \times \mathbb{R} : x_n > \varphi(x')\} \quad \textit{and} \quad \Omega_- = \overline{\Omega}^c.$$

Then for any $x = (x',x_n) \in \Omega_-$,

$$(1 + [\varphi]_{C^{0,1}}) \cdot d(x) \geq \varphi(x') - x_n.$$

Moreover, for every $x,y \in \overline{\Omega_-}$ with $x \neq y$, there exists $z \in \Omega_-$ such that

$$\begin{gathered} (x,z] \cup (y,z] \subset \Omega_-, \\ |x-z| + |z-y| \leq (2 + 4[\varphi]_{C^{0,1}})\,|x-y|. \end{gathered} \tag{16.6}$$

The result is also true for Ω_- replaced by Ω.

Proof (Proof of Lemma 16.20). Step 1. Let $x=(x',x_n)\in\Omega_-$. Note that there exists $z\in\partial\Omega$ such that

$$d(x;\overline{\Omega})=|z-x|=\max_{1\le i\le n}\{|z_i-x_i|\}.$$

We thus have $z=(y',\varphi(y'))$ for some $y'\in\mathbb{R}^{n-1}$ and

$$d(x;\overline{\Omega})=\max\{|y'-x'|;|\varphi(y')-x_n|\}.$$

We then have

$$\begin{aligned}\varphi(x')-x_n=|\varphi(x')-x_n|&\le|\varphi(x')-\varphi(y')|+|\varphi(y')-x_n|\\&\le[\varphi]_{C^{0,1}}|x'-y'|+|\varphi(y')-x_n|\\&\le(1+[\varphi]_{C^{0,1}})\max\{|y'-x'|;|\varphi(y')-x_n|\}\\&=(1+[\varphi]_{C^{0,1}})\cdot d(x;\overline{\Omega}),\end{aligned}$$

which proves the first statement.

Step 2. It therefore remains to prove the second statement. Let $x,y\in\overline{\Omega}_-$ (the case where Ω_- is replaced by Ω is completely analogous). We can assume, without loss of generality, that $x_n\le y_n$. Then

$$z=\left(y',x_n-2[\varphi]_{C^{0,1}}\left|x'-y'\right|\right)$$

has the claimed properties. □

The main ingredient is the construction of a regularized distance, denoted d^*.

Theorem 16.21. *Let $\varphi\in C^{0,1}(\mathbb{R}^{n-1})$, $r\ge0$ be an integer and $0\le\alpha\le1$. Let*

$$\Omega=\{x=(x',x_n)\in\mathbb{R}^{n-1}\times\mathbb{R}:x_n>\varphi(x')\}\quad\text{and}\quad\Omega_-=\overline{\Omega}^c.$$

Then there exist

$$d^*=d^*(\cdot;\overline{\Omega})\in C^\infty(\Omega_-;[0,\infty))$$

and a constant $C=C(r,n,[\varphi]_{C^{0,1}})$ such that for every $x=(x',x_n),y=(y',y_n)\in\Omega_-$,

$$d^*(x)\ge2(\varphi(x')-x_n),\tag{16.7}$$

$$\frac{1}{C}d(x)\le d^*(x)\le Cd(x),\tag{16.8}$$

$$|\nabla^r d^*(x)|\le Cd(x)^{1-r},\tag{16.9}$$

$$|\nabla^r d^*(x)-\nabla^r d^*(y)|\le C|x-y|^\alpha\max\{d(x)^{1-r-\alpha};d(y)^{1-r-\alpha}\}.\tag{16.10}$$

Proof (Proof of Theorem 16.21). Step 1. According to Theorem 2 in Stein [92, p. 171] (the last statement is not explicitly in [92] but is hidden in the proof of the theorem; cf. Kneuss [60] for details), there exist for every closed set $F\subset\mathbb{R}^n$, a constant $C_1=C_1(r,n)$ and a function $\triangle(\cdot,F)\in C^\infty(F^c)$ such that for every $x,y\in F^c$,

$$\frac{1}{C_1}d(x;F) \le \triangle(x;F) \le C_1 d(x;F),$$
$$|\nabla^r(\triangle(x;F))| \le C_1 d(x;F)^{1-r},$$
$$|\nabla^r(\triangle(x;F)) - \nabla^r(\triangle(y;F))| \le C_1|x-y|^\alpha \max\{d(x;F)^{1-r-\alpha}; d(y;F)^{1-r-\alpha}\}.$$

Step 2. From Lemma 16.20, we have for every $x = (x', x_n) \in \Omega_-$,

$$(1 + [\varphi]_{C^{0,1}}) \cdot d(x; \overline{\Omega}) \ge \varphi(x') - x_n.$$

The regularized distance

$$d^*(x) = 2C_1 \, (1 + [\varphi]_{C^{0,1}}) \cdot \Delta(x; \overline{\Omega})$$

has all of the required properties. □

We now return to the proof of Theorem 16.17.

Proof. Step 1. Let $f \in C^{r,\alpha}\left(\overline{\Omega}\right)$. We define the desired extension as

$$E(f)(x', x_n) = \begin{cases} f(x', x_n) & \text{if } (x', x_n) \in \overline{\Omega} \\ \int_1^\infty \psi(\lambda) f(x(\lambda)) d\lambda & \text{if } (x', x_n) \notin \overline{\Omega}, \end{cases}$$

where ψ is as in Lemma 16.19, $d^* = d^*(\cdot; \overline{\Omega})$ is as in Theorem 16.21 and

$$x(\lambda) = (x', x_n + \lambda d^*(x)).$$

Appealing to (16.7), we have for every $\lambda \ge 1$ and every $x \in \Omega_-$,

$$x_n + \lambda d^*(x', x_n) \ge x_n + 2(\varphi(x') - x_n) = \varphi(x') + (\varphi(x') - x_n) > \varphi(x').$$

Combining the above inequality with the fact that f is bounded and (16.4), we get that $E(f)$ is finite and well defined. It remains to show that $E(f) \in C^{r,\alpha}(\mathbb{R}^n)$ and (16.2). We will only prove it for $r \le 2$, the general case being handled in exactly the same way (the key estimate being for $r = 2$). Recall that

$$\Omega_- = \overline{\Omega}^c = \{x = (x', x_n) \in \mathbb{R}^{n-1} \times \mathbb{R} : \varphi(x') > x_n\}$$

and note that

$$\overline{\Omega}_- \cap \overline{\Omega} = \partial\Omega = \partial\Omega_- = \{(x', \varphi(x')) : x' \in \mathbb{R}^{n-1}\}.$$

We will often use the following elementary fact, valid for $g : \mathbb{R}^n \to \mathbb{R}$:

$$[g]_{C^{0,\alpha}(\overline{\Omega})}, [g]_{C^{0,\alpha}(\overline{\Omega}_-)} \le D \quad \Rightarrow \quad [g]_{C^{0,\alpha}(\mathbb{R}^n)} \le 2D. \tag{16.11}$$

Step 2. We prove the theorem first for $r = 0$.

Step 2.1. Let us show that $E(f) \in C^0(\mathbb{R})$ and

$$\|E(f)\|_{C^0(\mathbb{R}^n)} \leq C\|f\|_{C^0(\overline{\Omega})}.$$

Since $f \in C^0\left(\overline{\Omega}\right)$ and (16.4) holds, we easily have $E(f) \in C^0\left(\overline{\Omega}\right) \cap C^0(\Omega_-)$ and for every $x \in \Omega_-$,

$$|E(f)(x)| \leq C\|f\|_{C^0(\overline{\Omega})}.$$

To conclude that $E(f) \in C^0(\mathbb{R}^n)$, it is therefore enough to show that for every $x \in \partial\Omega$,

$$\lim_{\substack{y \to x \\ y \in \Omega_-}} E(f)(y) = E(f)(x).$$

Appealing to (16.4), (16.5), (16.8) and the fact that f is bounded on $\overline{\Omega}$, we can apply dominated convergence theorem and we get the desired convergence.

Step 2.2. We now prove that, in fact, $E(f) \in C^{0,\alpha}(\mathbb{R}^n)$ and

$$[E(f)]_{C^{0,\alpha}(\mathbb{R}^n)} \leq C[f]_{C^{0,\alpha}(\overline{\Omega})}.$$

Using (16.11), it is sufficient to establish the following inequality:

$$[E(f)]_{C^{0,\alpha}(\overline{\Omega_-})} \leq C[f]_{C^{0,\alpha}(\overline{\Omega})}.$$

Let $x, y \in \Omega_-$. Observe that (16.10) (with $\alpha = 1$) implies that

$$|d^*(x) - d^*(y)| \leq C_1|x-y|,$$

which combined with (16.4) leads to the desired inequality; indeed,

$$\begin{aligned}
|E(f)(x) - E(f)(y)| &\leq \int_1^\infty |\psi(\lambda)\left[f(x(\lambda)) - f(y(\lambda))\right]| \, d\lambda \\
&\leq \int_1^\infty |\psi(\lambda)| C_2(1+\lambda^\alpha)[f]_{C^{0,\alpha}(\overline{\Omega})}|x-y|^\alpha \, d\lambda \\
&\leq C_3[f]_{C^{0,\alpha}(\overline{\Omega})}|x-y|^\alpha.
\end{aligned}$$

Step 3. We now consider the case $r = 1$.

Step 3.1. We first prove that $E(f) \in C^1(\mathbb{R}^n)$ and

$$\|E(f)\|_{C^1(\mathbb{R}^n)} \leq C\|f\|_{C^1(\overline{\Omega})}.$$

Since $f \in C^1\left(\overline{\Omega}\right)$ and $E(f) \in C^1\left(\overline{\Omega}\right) \cap C^1(\Omega_-)$ (according to (16.4)), we get for every $x = (x', x_n) \in \Omega_-$,

$$E(f)_{x_i}(x) = \int_1^\infty f_{x_i}(x(\lambda))\psi(\lambda)\, d\lambda + \int_1^\infty f_{x_n}(x(\lambda))d^*_{x_i}(x)\lambda\psi(\lambda)\, d\lambda,$$

where we recall that

$$x(\lambda) = (x', x_n + \lambda d^*(x)).$$

From Lemma 16.19 and the fact (see (16.9)) that $|d^*_{x_i}| \le C$ in Ω_-, we obtain, by the dominated convergence theorem, that for every $x \in \partial\Omega$,

$$\lim_{\substack{y \to x \\ y \in \Omega_-}} E(f)_{x_i}(y) = f_{x_i}(x).$$

Appealing to (16.4) we have for every $x \in \Omega_-$,

$$|E(f)_{x_i}(x)| \le C \|\nabla f\|_{C^0(\overline{\Omega})}.$$

We therefore have $E(f) \in C^1\left(\overline{\Omega}\right) \cap C^1(\overline{\Omega}_-)$ and for every $1 \le i \le n$,

$$\|E(f)_{x_i}\|_{C^0(\overline{\Omega}_-)} \le C \|\nabla f\|_{C^0(\overline{\Omega})}.$$

Clearly, $E(f)$ is differentiable in $\Omega \cup \Omega_-$. Since (16.6) holds, we easily see that $E(f)$ is differentiable on $\partial\Omega$ (see [60] for details) and, thus, $E(f)$ is differentiable in $\mathbb{R}^n$.

Step 3.2. We now show that for every $1 \le i \le n$, $E(f)_{x_i} \in C^{0,\alpha}(\mathbb{R}^n)$ and

$$[E(f)_{x_i}]_{C^{0,\alpha}(\mathbb{R}^n)} \le C [\nabla f]_{C^{0,\alpha}(\overline{\Omega})}.$$

As before, it is enough to prove that

$$[E(f)_{x_i}]_{C^{0,\alpha}(\overline{\Omega}_-)} \le C [\nabla f]_{C^{0,\alpha}(\overline{\Omega})}.$$

Let $x, y \in \Omega_-$ and assume, without loss of generality, that $d(x) \le d(y)$. We have

$$\begin{aligned}
&|E(f)_{x_i}(x) - E(f)_{x_i}(y)| \\
&\le \left| \int_1^\infty \psi(\lambda) \left[f_{x_i}(x(\lambda)) - f_{x_i}(y(\lambda)) \right] d\lambda \right| \\
&\quad + \left| \int_1^\infty \lambda \psi(\lambda) \left[f_{x_n}(x(\lambda)) d^*_{x_i}(x) - f_{x_n}(y(\lambda)) d^*_{x_i}(y) \right] d\lambda \right|
\end{aligned}$$

and thus, as in Step 2.2, the first term is readily estimated by

$$C_1 [f_{x_i}]_{C^{0,\alpha}(\overline{\Omega})} |x - y|^\alpha.$$

The second term is estimated as follows. Since (16.5) holds, we get

$$\begin{aligned}
&\left| \int_1^\infty \lambda \psi(\lambda) \left[f_{x_n}(x(\lambda)) d^*_{x_i}(x) - f_{x_n}(y(\lambda)) d^*_{x_i}(y) \right] d\lambda \right| \\
&\le \left| \int_1^\infty \lambda \psi(\lambda) f_{x_n}(x(\lambda)) \left[d^*_{x_i}(x) - d^*_{x_i}(y) \right] d\lambda \right|
\end{aligned}$$

$$
\begin{aligned}
&+\left|\int_1^\infty \lambda\psi(\lambda)d^*_{x_i}(y)\left[f_{x_n}(x(\lambda))-f_{x_n}(y(\lambda))\right]d\lambda\right|\\
&=\left|\int_1^\infty \lambda\psi(\lambda)[f_{x_n}(x(\lambda))-f_{x_n}(x)]\left[d^*_{x_i}(x)-d^*_{x_i}(y)\right]d\lambda\right|\\
&+\left|\int_1^\infty \lambda\psi(\lambda)d^*_{x_i}(y)\left[f_{x_n}(x(\lambda))-f_{x_n}(y(\lambda))\right]d\lambda\right|
\end{aligned}
$$

and hence the claim, since (16.4) and (16.8)–(16.10) hold (recall that $d(x)\le d(y)$),

$$
\begin{aligned}
&\left|\int_1^\infty \lambda\psi(\lambda)\left[f_{x_n}(x(\lambda))d^*_{x_i}(x)-f_{x_n}(y(\lambda))d^*_{x_i}(y)\right]d\lambda\right|\\
&\le C_1\int_1^\infty \lambda\,|\psi(\lambda)|\,(\lambda d^*(x))^\alpha\,[\nabla f]_{C^{0,\alpha}}\,d(x)^{-\alpha}|x-y|^\alpha d\lambda\\
&\quad+C_2\int_1^\infty \lambda(1+\lambda^\alpha)\,|\psi(\lambda)|\,[\nabla f]_{C^{0,\alpha}}\,|x-y|^\alpha d\lambda\\
&\le C_3\,[\nabla f]_{C^{0,\alpha}}\,|x-y|^\alpha.
\end{aligned}
$$

Step 4. We finally discuss the case $r=2$.

Step 4.1. As before, let us first prove that $E(f)\in C^2(\mathbb{R}^n)$ and

$$\|E(f)\|_{C^2(\mathbb{R}^n)}\le C\|f\|_{C^2(\overline{\Omega})}.$$

Since $f\in C^2\left(\overline{\Omega}\right)$ and $E(f)\in C^2\left(\overline{\Omega}\right)\cap C^2(\Omega_-)$ (according to (16.4)), we have for every $x=(x',x_n)\in\Omega_-$,

$$
\begin{aligned}
E(f)_{x_ix_j}(x)&=\int_1^\infty f_{x_ix_j}(x(\lambda))\psi(\lambda)\,d\lambda+\int_1^\infty f_{x_ix_n}(x(\lambda))\lambda\psi(\lambda)d^*_{x_j}(x)\,d\lambda\\
&+\int_1^\infty f_{x_jx_n}(x(\lambda))\lambda\psi(\lambda)d^*_{x_i}(x)\,d\lambda\\
&+\int_1^\infty f_{x_nx_n}(x(\lambda))\psi(\lambda)\lambda^2d^*_{x_i}(x)d^*_{x_j}(x)\,d\lambda\\
&+\int_1^\infty f_{x_n}(x(\lambda))\lambda\psi(\lambda)d^*_{x_ix_j}(x)\,d\lambda\\
&=A_1(x)+A_2(x)+A_3(x)+A_4(x)+A_5(x).
\end{aligned}
$$

As in Step 3.1, we obtain for every $1\le k\le 4$ and every $x\in\Omega$,

$$|A_k(x)|\le C\|\nabla^2 f\|_{C^0(\overline{\Omega})},$$

whereas for every $x\in\partial\Omega$, we get

$$
\lim_{\substack{y\to x\\ y\in\Omega_-}}A_k(y)=\begin{cases} f_{x_ix_j}(x) & \text{if } k=1\\ 0 & \text{if } k=2,3,4.\end{cases}
$$

It therefore remains to study the behavior of A_5. We have, recalling that $x(\lambda) = (x', x_n + \lambda d^*(x))$,

$$\begin{aligned} f_{x_n}(x(\lambda)) - f_{x_n}(x(1)) &= \int_0^1 \frac{d}{dt} f_{x_n}(x(1+t(\lambda-1)))\,dt \\ &= \int_0^1 d^*(x)(\lambda-1) f_{x_n x_n}(x(1+t(\lambda-1)))\,dt. \end{aligned}$$

We therefore deduce, from (16.5), that

$$\begin{aligned} A_5(x) &= \int_1^\infty f_{x_n}(x(\lambda))\lambda\psi(\lambda) d^*_{x_i x_j}(x)\,d\lambda \\ &= d^*(x)\,d^*_{x_i x_j}(x) \int_1^\infty \lambda(\lambda-1)\psi(\lambda) \int_0^1 f_{x_n x_n}(x(1+t(\lambda-1)))\,dt\,d\lambda. \quad (16.12) \end{aligned}$$

Recalling Lemma 16.19 and using the fact (see (16.8) and (16.9)) that

$$d^*(x) d^*_{x_i x_j}(x) \le C,$$

we get

$$|A_5(x)| \le C \|\nabla^2 f\|_{C^0(\overline{\Omega})} \quad \text{and} \quad \lim_{\substack{y \to x \\ y \in \Omega_-}} A_5(y) = 0.$$

To show that $E(f) \in C^2(\mathbb{R}^n)$, we proceed as in Step 3.1. From the previous estimates we indeed have

$$\|E(f)\|_{C^2(\mathbb{R}^n)} \le C \|f\|_{C^2(\overline{\Omega})}.$$

Step 4.2. We finally have to prove for every $1 \le i \le j$ that $E(f)_{x_i x_j} \in C^{0,\alpha}(\mathbb{R}^n)$ and

$$[E(f)_{x_i x_j}]_{C^{0,\alpha}(\mathbb{R}^n)} \le C \left[\nabla^2 f\right]_{C^{0,\alpha}(\overline{\Omega})}.$$

Using (16.11), it is enough to show that

$$[E(f)_{x_i x_j}]_{C^{0,\alpha}(\overline{\Omega}_-)} \le C \left[\nabla^2 f\right]_{C^{0,\alpha}(\overline{\Omega})}.$$

So let $x, y \in \Omega_-$. With the notations of Step 4.1, we can write

$$|E(f)_{x_i x_j}(x) - E(f)_{x_i x_j}(y)| \le \sum_{i=1}^5 |A_i(x) - A_i(y)|.$$

As in Step 3.2 and using (16.12) for A_5, we get

$$\sum_{i=1}^5 |A_i(x) - A_i(y)| \le C \left[\nabla^2 f\right]_{C^{0,\alpha}} |x-y|^\alpha.$$

This finishes the proof of the theorem. □

16.2.3 Proof of the Main Theorem

We finally turn to the proof of our main theorem.

Proof (Proof of Theorem 16.11). Step 1. We start by appropriately covering the boundary of Ω.

Step 1.1. Since Ω is Lipschitz and bounded, we can find an integer N, $x_i \in \partial\Omega$, $\varepsilon_i > 0$ and $\varphi_i \in C^{0,1}(\mathbb{R}^{n-1})$, $1 \leq i \leq N$, such that

$$\partial\Omega \subset \bigcup_{i=1}^{N} B_{\varepsilon_i}(x_i)$$

and, up to a rotation,

$$\overline{\Omega} \cap \overline{B}_{\varepsilon_i}(x_i) = \overline{\Omega}_i \cap \overline{B}_{\varepsilon_i}(x_i), \tag{16.13}$$

where

$$\Omega_i = \{x = (x', x_n) \in \mathbb{R}^{n-1} \times \mathbb{R} : x_n > \varphi_i(x')\}.$$

Choose $0 < \varepsilon < \min_{1 \leq i \leq N} \varepsilon_i$ such that

$$\partial\Omega \subset \bigcup_{i=1}^{N} B_{\varepsilon_i - \varepsilon}(x_i) \tag{16.14}$$

and define

$$c = \max_{1 \leq i \leq N} \{\varepsilon_i / \varepsilon\} > 1.$$

Step 1.2. We then define some auxiliary functions. Let $\lambda_i \in C^\infty(\mathbb{R}^n; [0,1])$, $1 \leq i \leq N$, be such that

$$\lambda_i = 1 \text{ in } B_{\varepsilon_i - \varepsilon/2}(x_i) \quad \text{and} \quad \operatorname{supp}(\lambda_i) \subset B_{\varepsilon_i - \varepsilon/4}(x_i).$$

Let $\lambda_0, \lambda_+, \lambda_- \in C^\infty(\mathbb{R}^n; [0,1])$ verify

$$\lambda_0 = 1 \text{ in } \Omega \quad \text{and} \quad \operatorname{supp}(\lambda_0) \subset \Omega + B_{\varepsilon/2},$$

$$\lambda_+ = 1 \text{ in } \partial\Omega + B_{\varepsilon/2} \quad \text{and} \quad \operatorname{supp}(\lambda_+) \subset \partial\Omega + B_\varepsilon,$$

$$\lambda_- = 1 \text{ in } \Omega \cap \left(\partial\Omega + B_{\varepsilon/2}\right)^c \quad \text{and} \quad \operatorname{supp}(\lambda_-) \subset \Omega.$$

Then let

$$\Lambda_+ = \lambda_0 \left(\frac{\lambda_+}{\lambda_+ + \lambda_-}\right) \quad \text{and} \quad \Lambda_- = \lambda_0 \left(\frac{\lambda_-}{\lambda_+ + \lambda_-}\right).$$

Since

$$\operatorname{supp}(\lambda_0) \subset \{x \in \mathbb{R}^n : \lambda_+ + \lambda_- \geq 1\},$$

the functions Λ_+ and Λ_- belong to $C_0^\infty(\mathbb{R}^n)$. Note also that

$$\Lambda_+ + \Lambda_- = \lambda_0.$$

Finally, since (using (16.14))

$$\operatorname{supp}(\Lambda_+) \subset \partial\Omega + B_{\varepsilon/2} \subset \bigcup_{i=1}^N B_{\varepsilon_i - \frac{\varepsilon}{2}}(x_i)$$

and noticing that

$$\sum_{i=1}^N \lambda_i^2 \geq 1 \quad \text{in} \bigcup_{i=1}^N B_{\varepsilon_i - \frac{\varepsilon}{2}}(x_i),$$

we obtain

$$\frac{\Lambda_+}{\sum_{i=1}^N \lambda_i^2} \in C_0^\infty(\mathbb{R}^n). \tag{16.15}$$

Step 2. We now make explicit the extension E.

Step 2.1 (Simplification). We show that to prove the theorem, we can restrict ourselves to the space $C_*^{r,\alpha}(\overline{\Omega})$ (see Section 16.1 for the notations). More precisely, we prove that it is enough to show the existence of

$$E : C_*^{r,\alpha}(\overline{\Omega}) \to C_0^{r,\alpha}(\mathbb{R}^n) \tag{16.16}$$

and

$$\|E(f)\|_{C^{r,\alpha}(\mathbb{R}^n)} \leq \|f\|_{C_*^{r,\alpha}(\overline{\Omega})} \quad \text{for every } f \in C_*^{r,\alpha}(\overline{\Omega}). \tag{16.17}$$

Thus, suppose that (16.16) and (16.17) hold true and let $f \in C^{r,\alpha}(\overline{\Omega})$. It is enough to show that

$$f \in C_*^{r,\alpha}(\overline{\Omega}) \quad \text{and} \quad \|f\|_{C_*^{r,\alpha}(\overline{\Omega})} \leq C\|f\|_{C^{r,\alpha}(\overline{\Omega})} \tag{16.18}$$

to have the claim. Since, in particular, $f \in C^r(\overline{\Omega}) = C_*^r(\overline{\Omega})$, we have

$$E(f) \in C_0^r(\mathbb{R}^n) \quad \text{and} \quad \|E(f)\|_{C^r(\mathbb{R}^n)} \leq C\|f\|_{C^r(\overline{\Omega})}.$$

Let $0 \leq m \leq r-1$ and $R \geq 1/2$ large enough so that $\overline{\Omega} \subset B_R$. Using Proposition 16.10, we hence obtain

$$[\nabla^m f]_{C^{0,\alpha}(\overline{\Omega})} \leq [\nabla^m E(f)]_{C^{0,\alpha}(\overline{B_R})} \leq 2R\|\nabla^{m+1}E(f)\|_{C^0(\overline{B_R})} \leq 2RC\|f\|_{C^r(\overline{\Omega})},$$

which directly implies (16.18) and shows the assertion.

Step 2.2 (Conclusion). For $f \in C_*^{r,\alpha}(\overline{\Omega})$ and $x \in \mathbb{R}^n$, the desired extension is given by

$$E(f)(x) = \Lambda_+(x)\left\{\frac{\sum_{i=1}^N \lambda_i(x) E_i(f_i)(x)}{\sum_{i=1}^N \lambda_i^2(x)}\right\} + \Lambda_-(x) f(x),$$

where $E_i : C_*^{r,\alpha}(\overline{\Omega}_i) \to C^{r,\alpha}(\mathbb{R}^n)$ is the extension operator of Theorem 16.17 (see also Remark 16.18(i)) applied to Ω_i and $f_i : \overline{\Omega}_i \to \mathbb{R}$ is defined by

$$f_i = \begin{cases} \lambda_i f & \text{in } \overline{\Omega}_i \cap \overline{B}_{\varepsilon_i}(x_i), \\ 0 & \text{in } \overline{\Omega}_i \backslash \overline{B}_{\varepsilon_i}(x_i). \end{cases}$$

Let us show that $E(f)$ is well defined and has all of the desired properties. We recall that

$$g,h \in C_*^{r,\alpha} \Rightarrow gh \in C^{r,\alpha}$$

and

$$\|gh\|_{C^{r,\alpha}} \leq C\|g\|_{C_*^{r,\alpha}} \|h\|_{C_*^{r,\alpha}}.$$

(i) First, we show that $f_i \in C_*^{r,\alpha}(\overline{\Omega}_i)$ and, for an appropriate $C = C(r,\Omega)$,

$$\|f_i\|_{C_*^{r,\alpha}(\overline{\Omega}_i)} \leq C\|f\|_{C_*^{r,\alpha}(\overline{\Omega})}. \tag{16.19}$$

— Since $f \in C_*^{r,\alpha}(\overline{\Omega})$ and $\lambda_i \in C^\infty(\mathbb{R}^n)$, we immediately obtain, using (16.13), that $f_i \in C_*^{r,\alpha}(\overline{\Omega}_i \cap \overline{B}_{\varepsilon_i}(x_i))$ and that

$$\|f_i\|_{C_*^{r,\alpha}(\overline{\Omega}_i \cap \overline{B}_{\varepsilon_i}(x_i))} \leq C\|f\|_{C_*^{r,\alpha}(\overline{\Omega})}. \tag{16.20}$$

— Then, recalling that $\operatorname{supp}(\lambda_i) \subset B_{\varepsilon_i - \varepsilon/4}$, we have that

$$\operatorname{supp}(f_i) \subset \overline{\Omega}_i \cap B_{\varepsilon_i - \varepsilon/4} \tag{16.21}$$

and, therefore, $f_i \in C^r(\overline{\Omega}_i)$ and

$$\|f_i\|_{C^r(\overline{\Omega}_i)} \leq C\|f\|_{C^r(\overline{\Omega})}.$$

— Finally, we show that $f_i \in C_*^{r,\alpha}(\overline{\Omega}_i)$ and (16.19). For this, it is enough to show that for every integer $0 \leq m \leq r$ and every $x,y \in \overline{\Omega}_i$,

$$|\nabla^m f_i(x) - \nabla^m f_i(y)| \leq C\|f\|_{C_*^{r,\alpha}(\overline{\Omega})}|x-y|^\alpha.$$

We only prove the assertion for $x \in \overline{\Omega}_i \cap B_{\varepsilon_i - \varepsilon/4}(x_i)$ and $y \in \overline{\Omega}_i \cap (B_{\varepsilon_i}(x_i))^c$, the other cases being trivial since (16.20) and (16.21) hold. First, note that

$$|y-x| \geq \frac{\varepsilon}{4}.$$

Note also that any $z \in \overline{\Omega}_i \cap \partial B_{\varepsilon_i - \varepsilon/4}(x_i)$ verifies (cf. Step 1.1 for the definition of c)

$$|x-z| \leq 2(\varepsilon_i - \varepsilon/4) \leq 2c\,\varepsilon.$$

Therefore, using (16.20) and (16.21), we find

$$\begin{aligned}|\nabla^m f_i(x)-\nabla^m f_i(y)| &\le |\nabla^m f_i(x)-\nabla^m f_i(z)|+|\nabla^m f_i(z)-\nabla^m f_i(y)|\\ &= |\nabla^m f_i(x)-\nabla^m f_i(z)| \le C\|f\|_{C_*^{r,\alpha}(\overline{\Omega})}|x-z|^\alpha\\ &\le C\|f\|_{C_*^{r,\alpha}(\overline{\Omega})}(2c\varepsilon)^\alpha \le 8cC\|f\|_{C_*^{r,\alpha}(\overline{\Omega})}(\varepsilon/4)^\alpha\\ &\le 8cC\|f\|_{C_*^{r,\alpha}(\overline{\Omega})}|x-y|^\alpha,\end{aligned}$$

which proves the assertion.

(ii) E is well defined in view of (16.15) and the fact that $\operatorname{supp}(\Lambda_-)\subset\Omega$.

(iii) Since $\Lambda_+ + \Lambda_- = \lambda_0 = 1$ in Ω and, for every $1\le i\le N$, $\lambda_i E_i(f_i)=\lambda_i^2 f$ in Ω, we obtain that

$$E(f)=f \quad \text{in } \Omega.$$

(iv) By construction, $\operatorname{supp}(\Lambda_+)\subset\Omega+B_{\varepsilon/2}$ and $\operatorname{supp}(\Lambda_-)\subset\Omega$; we hence deduce that

$$\operatorname{supp} E(f)\subset\Omega+B_{\varepsilon/2}.$$

(v) Since $\Lambda_-\in C_0^\infty(\mathbb{R}^n)$ with $\operatorname{supp}(\Lambda_-)\subset\Omega$, we get that

$$\Lambda_- f\in C_0^\infty(\mathbb{R}^n) \quad \text{and} \quad \|\Lambda_- f\|_{C^{r,\alpha}(\mathbb{R}^n)}\le C\|f\|_{C_*^{r,\alpha}(\overline{\Omega})}.$$

(vi) Since, for $1\le i\le n$,

$$f_i\in C_*^{r,\alpha}(\overline{\Omega}_i) \quad \text{and} \quad \frac{\Lambda_+}{(\sum_{i=1}^N \lambda_i^2)},\lambda_i\in C_0^\infty(\mathbb{R}^n)$$

and (cf. (16.3) and (16.19))

$$\|E_i(f_i)\|_{C_*^{r,\alpha}(\mathbb{R}^n)}\le C\|f\|_{C_*^{r,\alpha}(\overline{\Omega})},$$

we easily have (using point (iv)) that $E(f)\in C_0^{r,\alpha}(\mathbb{R}^n)$ and

$$\|E(f)\|_{C^{r,\alpha}(\mathbb{R}^n)}\le C\|f\|_{C_*^{r,\alpha}(\overline{\Omega})}.$$

This concludes the proof of the theorem. □

16.3 Compact Imbeddings

We now turn to the compactness of the imbeddings.

Theorem 16.22. *Let $\Omega\subset\mathbb{R}^n$ be a bounded open Lipschitz set. Let $s\ge r\ge 0$ be integers and $0\le\alpha,\beta\le 1$, with*

$$r+\alpha<s+\beta.$$

Then the imbedding

$$C^{s,\beta}\left(\overline{\Omega}\right) \hookrightarrow C^{r,\alpha}\left(\overline{\Omega}\right)$$

is compact.

Proof. Let $\{f_\nu\}_{\nu\in\mathbb{N}} \subset C^{s,\beta}\left(\overline{\Omega}\right)$ with $\|f_\nu\|_{C^{s,\beta}} \leq C$ for all ν. We have to show that we can extract a convergent subsequence in $C^{r,\alpha}\left(\overline{\Omega}\right)$.

Step 1. We first assume $r = s$ and thus $\alpha < \beta$.

Step 1.1. We deal with the case $r = s = 0$. From the Ascoli–Arzela theorem we find a subsequence, still denoted by f_ν, which converges to f in $C^0\left(\overline{\Omega}\right)$. We now show that $[f - f_\nu]_{C^{0,\alpha}}$ also converges to 0. Since the convergence is trivial, with our convention, when $\alpha = 0$, we assume below that $\alpha > 0$. Let $\delta > 0$ and $x \neq y \in \overline{\Omega}$ with $|x-y| \leq \delta$. Then

$$\begin{aligned}\frac{|(f-f_\nu)(x)-(f-f_\nu)(y)|}{|x-y|^\alpha} &= \lim_{\mu\to\infty}\frac{|(f_\mu-f_\nu)(x)-(f_\mu-f_\nu)(y)|}{|x-y|^\alpha}\\ &\leq \sup_\mu [f_\mu - f_\nu]_{C^{0,\beta}}|x-y|^{\beta-\alpha} \leq 2C\delta^{\beta-\alpha}.\end{aligned}$$

We, moreover, have

$$\frac{|(f-f_\nu)(x)-(f-f_\nu)(y)|}{|x-y|^\alpha} \leq 2\|f-f_\nu\|_{C^0}\delta^{-\alpha} \qquad \text{if } |x-y| \geq \delta.$$

For any given $\varepsilon > 0$, we can take $\delta > 0$ so small that $2C\delta^{\beta-\alpha} \leq \varepsilon$. Using the convergence in $C^0\left(\overline{\Omega}\right)$, we can then take $m \in \mathbb{N}$ such that

$$2\|f-f_\nu\|_{C^0}\delta^{-\alpha} \leq \varepsilon \quad \text{for every } \nu \geq m.$$

We therefore obtain that

$$[f-f_\nu]_{C^{0,\alpha}} \leq \varepsilon \quad \text{for every } \nu \geq m,$$

which concludes Step 1.1.

Step 1.2. We deal with the case $r = s \geq 1$. Due to Corollary 16.13, the $\|\cdot\|_{C^{s,\beta}}$ and the $\|\cdot\|_{C_*^{s,\beta}}$ norms are equivalent. Therefore, $\|\nabla^t f_\nu\|_{C^{0,\beta}}$ is bounded for $0 \leq t \leq s$. Appealing to Step 1.1 and extracting iteratively subsequences, we obtain $g_t \in C^{0,\alpha}\left(\overline{\Omega};\mathbb{R}^{n^t}\right)$ such that

$$\nabla^t f_\nu \to g_t \quad \text{in } C^{0,\alpha}\left(\overline{\Omega}\right) \quad \text{as } \nu\to\infty \quad \text{for every } 0\leq t\leq s.$$

Since we have uniform convergence in all derivatives, we find that $g_t = \nabla^t g_0$ and, thus, f_ν converge to g_0 in $C^{s,\alpha}$.

Step 2. Consider the case $r < s$.

Step 2.1. We suppose $r < s$ and $\beta > 0$. In view of Step 1, the imbedding

$$C^{s,\beta}\left(\overline{\Omega}\right) \hookrightarrow C^{s,0}\left(\overline{\Omega}\right)$$

is compact. From Corollary 16.13, we have that the imbedding

$$C^{s,0}\left(\overline{\Omega}\right) \hookrightarrow C^{r,\alpha}\left(\overline{\Omega}\right)$$

is continuous. So the composition of these two imbeddings is compact.

Step 2.2. Let now $r < s$ and $\beta = 0$.

(i) We first assume $s = r+1$ and thus $\alpha < 1$. Step 1 gives the compactness of the imbedding

$$C^{r,1}\left(\overline{\Omega}\right) \hookrightarrow C^{r,\alpha}\left(\overline{\Omega}\right).$$

Combining this with the continuity of the imbedding

$$C^{r+1,0}\left(\overline{\Omega}\right) \hookrightarrow C^{r,1}\left(\overline{\Omega}\right)$$

(cf. Corollary 16.13), we have the desired result.

(ii) It remains to deal with the case $s > r+1$ and $\beta = 0$, and thus $s-1 \geq r+\alpha$. We use Step 1 to obtain the compactness of the imbedding

$$C^{s-1,1}\left(\overline{\Omega}\right) \hookrightarrow C^{s-1,0}\left(\overline{\Omega}\right)$$

and combine it with the continuous imbeddings (cf. Remark 16.14)

$$C^{s,0}\left(\overline{\Omega}\right) \hookrightarrow C^{s-1,1}\left(\overline{\Omega}\right) \quad \text{and} \quad C^{s-1,0}\left(\overline{\Omega}\right) \hookrightarrow C^{r,\alpha}\left(\overline{\Omega}\right).$$

This concludes the proof of the theorem. □

16.4 A Lower Semicontinuity Result

The following lower semicontinuity result (cf. Dacorogna [28]) has been used on several occasions.

Proposition 16.23. *Let $r \geq 0$ be an integer and $0 < \alpha \leq 1$. Let $\Omega \subset \mathbb{R}^n$ be a bounded open Lipschitz set. Let $R > 0$ and*

$$C_R = \left\{f \in C^{r,\alpha}\left(\overline{\Omega}\right) : \|f\|_{C^{r,\alpha}} \leq R\right\}.$$

Let $\{f_\nu\} \subset C_R$ be a sequence such that

$$f_\nu \to f \text{ in } C^0\left(\overline{\Omega}\right) \text{ as } \nu \to \infty;$$

then $f \in C_R$ and

$$\|f\|_{C^{r,\alpha}} \leq \liminf_{\nu\to\infty} \|f_\nu\|_{C^{r,\alpha}}.$$

Proof. First, define

$$L = \liminf_{\nu\to\infty} \|f_\nu\|_{C^{r,\alpha}}$$

and then choose a subsequence such that

$$L = \lim_{i \to \infty} \|f_{v_i}\|_{C^{r,\alpha}} . \tag{16.22}$$

From Theorem 16.22 we deduce that there is a further subsequence such that

$$f_{v_{i_j}} \to f \text{ in } C^r\left(\overline{\Omega}\right) \text{ as } j \to \infty. \tag{16.23}$$

Finally, we let $x, y \in \overline{\Omega}$ and observe that

$$|\nabla^r f(x) - \nabla^r f(y)| \leq 2\left\|\nabla^r f_{v_{i_j}} - \nabla^r f\right\|_{C^0} + \left[\nabla^r f_{v_{i_j}}\right]_{C^{0,\alpha}} |x-y|^\alpha .$$

Letting $j \to \infty$ and using (16.23), we get

$$[\nabla^r f]_{C^{0,\alpha}} \leq \liminf_{j \to \infty} \left[\nabla^r f_{v_{i_j}}\right]_{C^{0,\alpha}} .$$

Combining the above inequality with (16.22), we have indeed obtained that

$$\|f\|_{C^{r,\alpha}} \leq L,$$

which is our claim. □

16.5 Interpolation and Product

Throughout this section, we follow Hörmander [55].

16.5.1 Interpolation

We start with a preliminary result.

Proposition 16.24. *Let $\Omega \subset \mathbb{R}^n$ be a bounded open Lipschitz set and $r \geq 0$ be an integer. Then*

$$\|f\|_{C^r(\overline{\Omega})} \leq C\left(\|f\|_{C^0(\overline{\Omega})} + \sup_{x,y \in \Omega}\{|\nabla^r f(x) - \nabla^r f(y)|\}\right)$$

for some constant $C = C(r, \Omega)$.

As a immediate consequence of the previous proposition, we have the following result.

Corollary 16.25. *Let $\Omega \subset \mathbb{R}^n$ be a bounded open Lipschitz set. Let $r \geq 0$ be an integer and $0 \leq \alpha \leq 1$. Then there exists a constant $C = C(r, \Omega)$ such that for every $f \in C^{r,\alpha}(\overline{\Omega})$,*

$$\|f\|_{C^{r,\alpha}(\overline{\Omega})} \leq \begin{cases} C\left(\|f\|_{C^0(\overline{\Omega})} + [\nabla^r f]_{C^{0,\alpha}(\overline{\Omega})}\right) & \text{if } 0 < \alpha \leq 1 \\ C\left(\|f\|_{C^0(\overline{\Omega})} + \|\nabla^r f\|_{C^0(\overline{\Omega})}\right) & \text{if } \alpha = 0. \end{cases}$$

In other words, the norms $\|\cdot\|_{C^{r,\alpha}}$ *and*

$$\begin{cases} \|\cdot\|_{C^0} + [\nabla^r \cdot]_{C^{0,\alpha}} & \text{if } 0 < \alpha \leq 1 \\ \|\cdot\|_{C^0} + \|\nabla^r \cdot\|_{C^0} & \text{if } \alpha = 0 \end{cases}$$

are equivalent.

We now proceed with the proof of Proposition 16.24.

Proof. Step 1. Since Ω is bounded and Lipschitz, we easily deduce the existence of $\varepsilon > 0$ with the following property: For every $x \in \overline{\Omega}$ there exists $\overline{x} \in \Omega$ such that

$$[x, y] \subset \overline{\Omega}, \quad \text{for every } y \in B_\varepsilon(\overline{x}). \tag{16.24}$$

Note that the previous equation implies, in particular, that $B_\varepsilon(\overline{x}) \subset \overline{\Omega}$.

Step 2. Fix $x \in \overline{\Omega}$. For every $y \in B_\varepsilon(\overline{x})$ we consider the Taylor polynomial of f of degree r at x, denoted by $T_x^r f(y)$; it is defined through the function (this is justified by (16.24))

$$F(u) = f(x + u(y - x)), \quad u \in [0, 1],$$

and it is given by

$$T_x^r f(y) = \sum_{k=0}^{r} \frac{1}{k!} F^{(k)}(0).$$

Recalling that

$$F(1) = \sum_{k=0}^{r} \frac{F^{(k)}(0)}{k!} + \int_0^1 \frac{(1-u)^{r-1}}{(r-1)!} \left[F^{(r)}(u) - F^{(r)}(0)\right] du,$$

we obtain

$$f(y) - T_x^r f(y) = \int_0^1 \frac{(1-u)^{r-1}}{(r-1)!} \left[F^{(r)}(u) - F^{(r)}(0)\right] du.$$

We hence have

$$|f(y) - T_x^r f(y)| \leq C_1 \sup_{z \in \Omega} \{|\nabla^r f(z) - \nabla^r f(x)|\},$$

which implies

$$|T_x^r f(y)| \leq C_1 (\|f\|_{C^0} + \sup_{z \in \Omega} \{|\nabla^r f(z) - \nabla^r f(x)|\}).$$

We assert that the previous inequality, holding true for every $y \in B_{\varepsilon}(\overline{x})$, implies a bound on all the coefficients (i.e., the derivatives of f at x up the order r) appearing in $T_x^r f$ and thus proves the theorem. We will have the claim once the following assertion is proved (cf. Step 3). If

$$\left| \sum_{0 \le i_1, \dots, i_n \le r} a_{i_1 \cdots i_n} v_1^{i_1} \cdots v_n^{i_n} \right| \le c \quad \text{for every } v_i \in [b_i - \varepsilon, b_i + \varepsilon],\ 1 \le i \le n,$$

where $(v_1, \dots, v_n) = x - y$ and $b_i = x_i - \overline{x}_i$, then all the $a_{i_1 \cdots i_n}$ verify

$$|a_{i_1 \cdots i_n}| \le Cc$$

for an appropriate constant $C = C(r, \varepsilon, \operatorname{diam} \Omega) = C(r, \Omega)$, recalling that ε only depends on Ω. This will prove Step 2 and thus the proposition.

Step 3. We prove the above assertion. By induction it is easily seen that we can restrict ourselves to the case $n = 1$. So we have to show that if

$$\left| \sum_{i=0}^{r} a_i v^i \right| \le c \quad \text{for every } v \in [b - \varepsilon, b + \varepsilon],$$

then we have

$$|a_i| \le Cc$$

for an appropriate constant $C = C(r, \varepsilon, b)$. This is easily achieved as follows. Define, for $0 \le i \le r$,

$$t_i = b - \varepsilon + \frac{2\varepsilon i}{r}, \quad a = (a_0, \dots, a_r) \in \mathbb{R}^{r+1},$$

and $B \in \mathbb{R}^{(r+1)\times(r+1)}$ by

$$B = \begin{pmatrix} 1 & t_0 & t_0^2 & \cdots & t_0^r \\ 1 & t_1 & t_1^2 & \cdots & t_1^r \\ \vdots & \vdots & \vdots & \ddots & \vdots \\ 1 & t_r & t_r^2 & \cdots & t_r^r \end{pmatrix}.$$

By hypothesis we have that $|Ba| \le c$. Moreover, using well-known properties of Vandermonde matrices, we know that

$$\det B = \prod_{0 \le i < j \le r} (t_j - t_i) > 0.$$

Therefore, since

$$a = B^{-1} Ba = \frac{(\operatorname{adj} B)^t}{\det B} Ba,$$

we immediately have the result. □

We next state and prove the main interpolation theorem.

Theorem 16.26. *Let $\Omega \subset \mathbb{R}^n$ be a bounded open Lipschitz set, $s \geq r \geq t \geq 0$ be integers and $0 \leq \alpha, \beta, \gamma \leq 1$ with*

$$t+\gamma \leq r+\alpha \leq s+\beta.$$

Let $\lambda \in [0,1]$ be such that

$$r+\alpha = \lambda(s+\beta)+(1-\lambda)(t+\gamma).$$

Then there exists a constant $C = C(s,\Omega) > 0$ such that

$$\|f\|_{C^{r,\alpha}} \leq C \|f\|_{C^{s,\beta}}^{\lambda} \|f\|_{C^{t,\gamma}}^{1-\lambda}.$$

Proof. In view of Theorem 16.11, there is no loss of generality in assuming that Ω is convex. We will also make in the proof of the theorem, and only here, an exception to the convention

$$[f]_{C^{0,0}} = 0.$$

Here, we will adopt that

$$[f]_{C^{0,0}} = \|f\|_{C^0}.$$

Step 1. We first prove that if k is an integer such that

$$t+\gamma \leq k \leq s \leq s+\beta,$$

then there exists a constant $C = C(s,\Omega) > 0$ such that

$$\left\|\nabla^k f\right\|_{C^0} \leq C\left(\left[\nabla^t f\right]_{C^{0,\gamma}} + \left[\nabla^s f\right]_{C^{0,\beta}}\right). \tag{16.25}$$

Replacing k by $k-t$, it is enough to prove the result when $t=0$ (and, thus, $\gamma \leq k \leq s \leq s+\beta$), namely

$$\left\|\nabla^k f\right\|_{C^0} \leq C\left([f]_{C^{0,\gamma}} + \left[\nabla^s f\right]_{C^{0,\beta}}\right). \tag{16.26}$$

If $\gamma = 0$, then (16.26) is an immediate consequence of Proposition 16.24. It remains to prove (16.26) when $\gamma > 0$. Then since in particular $k > 0$, we remark that all of the terms in (16.26) remain unchanged if we add a constant to f. Therefore, it is enough to establish (16.26) only for those f such that there exists $x_0 \in \overline{\Omega}$ with $f(x_0) = 0$. We hence deduce for every $x \in \overline{\Omega}$,

$$|f(x)| \leq [f]_{C^{0,\gamma}} |x-x_0|^{\gamma},$$

and, thus, there exists a constant $C_2 = C_2(\Omega) > 0$ such that

$$\|f\|_{C^0} \leq C_2 [f]_{C^{0,\gamma}}.$$

Combining the previous equation with Proposition 16.24, we immediately obtain the claim. This finishes Step 1.

Step 2. We next prove that

$$[\nabla^r f]_{C^{0,\alpha}} \leq C\left([\nabla^s f]_{C^{0,\beta}} + [\nabla^t f]_{C^{0,\gamma}}\right)^{\lambda} [\nabla^t f]_{C^{0,\gamma}}^{1-\lambda}. \tag{16.27}$$

Step 2.1. We first establish the inequality when

$$r \leq t+\gamma \leq r+\alpha \leq s+\beta \leq r+1$$

and in fact we will establish a sharper form of the inequality, namely

$$[\nabla^r f]_{C^{0,\alpha}} \leq C\,[\nabla^s f]_{C^{0,\beta}}^{\lambda} [\nabla^t f]_{C^{0,\gamma}}^{1-\lambda}.$$

Since the seminorms of $C^{0,1}$ and C^1 are the same (cf. Proposition 16.10), we can infer from the above inequalities that $t=r=s$ and $0 \leq \gamma \leq \alpha \leq \beta \leq 1$, with

$$\alpha = \lambda\beta + (1-\lambda)\gamma.$$

If $\gamma=\alpha$ (and hence $\beta=\alpha$), (16.27) is then trivial, so we assume that $\gamma<\alpha$. Since

$$\frac{|\nabla^t f(x) - \nabla^t f(y)|}{|x-y|^{\alpha}} = \left(\frac{|\nabla^t f(x) - \nabla^t f(y)|}{|x-y|^{\beta}}\right)^{\lambda} \left(\frac{|\nabla^t f(x) - \nabla^t f(y)|}{|x-y|^{\gamma}}\right)^{1-\lambda},$$

we deduce, recalling that $t=s=r$, that

$$[\nabla^r f]_{C^{0,\alpha}} \leq [\nabla^s f]_{C^{0,\beta}}^{\lambda} [\nabla^t f]_{C^{0,\gamma}}^{1-\lambda}$$

if $\gamma>0$ and if $\gamma=0$ that

$$[\nabla^r f]_{C^{0,\alpha}} \leq 2\,[\nabla^s f]_{C^{0,\beta}}^{\lambda} \|\nabla^t f\|_{C^0}^{1-\lambda} = 2\,[\nabla^s f]_{C^{0,\beta}}^{\lambda} [\nabla^t f]_{C^{0,0}}^{1-\lambda}.$$

This establishes the sharper version of (16.27) under the assumptions of Step 2.1.

Step 2.2. We next prove (16.27) when $\alpha=0$ and thus

$$r = \lambda(s+\beta) + (1-\lambda)(t+\gamma)$$

and we have to show that

$$[\nabla^r f]_{C^{0,0}} = \|\nabla^r f\|_{C^0} \leq C\left([\nabla^s f]_{C^{0,\beta}} + [\nabla^t f]_{C^{0,\gamma}}\right)^{\lambda} [\nabla^t f]_{C^{0,\gamma}}^{1-\lambda}. \tag{16.28}$$

If $[\nabla^s f]_{C^{0,\beta}} \leq [\nabla^t f]_{C^{0,\gamma}}$, the result follows from (16.25), since then

$$\begin{aligned}\|\nabla^r f\|_{C^0} &\leq C_1\left([\nabla^t f]_{C^{0,\gamma}} + [\nabla^s f]_{C^{0,\beta}}\right) \leq 2C_1\,[\nabla^t f]_{C^{0,\gamma}} \\ &= 2C_1\,[\nabla^t f]_{C^{0,\gamma}}^{\lambda} [\nabla^t f]_{C^{0,\gamma}}^{1-\lambda} \\ &\leq 2C_1\left([\nabla^s f]_{C^{0,\beta}} + [\nabla^t f]_{C^{0,\gamma}}\right)^{\lambda} [\nabla^t f]_{C^{0,\gamma}}^{1-\lambda}.\end{aligned}$$

So we may assume that $[\nabla^t f]_{C^{0,\gamma}} < [\nabla^s f]_{C^{0,\beta}}$. Note that the strict inequality implies that $[\nabla^t f]_{C^{0,\gamma}} \neq 0$, otherwise $[\nabla^s f]_{C^{0,\beta}} = 0$ also. We fix $x \in \Omega$ and define for $\theta \in (0,1)$ and for $y \in \Omega$ (recall that Ω is convex),

$$f_\theta(y) = f((1-\theta)x + \theta y).$$

We immediately find that

$$\|\nabla^r f_\theta\|_{C^0} = \theta^r \|\nabla^r f\|_{C^0},$$

$$[\nabla^t f_\theta]_{C^{0,\gamma}} \leq \theta^{t+\gamma} [\nabla^t f]_{C^{0,\gamma}} \quad \text{and} \quad [\nabla^s f_\theta]_{C^{0,\beta}} \leq \theta^{s+\beta} [\nabla^s f]_{C^{0,\beta}}.$$

We choose $\theta \in (0,1)$ such that

$$\theta^{(s+\beta)-(t+\gamma)} = \frac{[\nabla^t f]_{C^{0,\gamma}}}{[\nabla^s f]_{C^{0,\beta}}} < 1.$$

Invoking (16.25) applied to f_θ, we get

$$\begin{aligned}
\theta^r \|\nabla^r f\|_{C^0} &\leq C_1 \left(\theta^{t+\gamma} [\nabla^t f]_{C^{0,\gamma}} + \theta^{s+\beta} [\nabla^s f]_{C^{0,\beta}}\right) \\
&= 2C_1 \theta^{t+\gamma} [\nabla^t f]_{C^{0,\gamma}} = 2C_1 \theta^{s+\beta} [\nabla^s f]_{C^{0,\beta}} \\
&= 2C_1 \left(\theta^{s+\beta} [\nabla^s f]_{C^{0,\beta}}\right)^\lambda \left(\theta^{t+\gamma} [\nabla^t f]_{C^{0,\gamma}}\right)^{1-\lambda} \\
&\leq 2C_1 \theta^r \left([\nabla^s f]_{C^{0,\beta}} + [\nabla^t f]_{C^{0,\gamma}}\right)^\lambda [\nabla^t f]_{C^{0,\gamma}}^{1-\lambda}.
\end{aligned}$$

This achieves the proof of Step 2.2.

Step 2.3. We now consider the general case

$$r + \alpha = \lambda (s+\beta) + (1-\lambda)(t+\gamma).$$

Four cases can happen.

Case 1: $r \leq t+\gamma \leq r+\alpha \leq s+\beta \leq r+1$, which has already been dealt in Step 2.1.

Case 2: $t+\gamma \leq r \leq r+\alpha \leq r+1 \leq s+\beta$. We combine three interpolations, namely for r and $r+1$ between $[t+\gamma, s+\beta]$ and for $r+\alpha$ between $[r, r+1]$. More precisely, we let

$$\begin{aligned}
r+\alpha &= (1-\alpha) r + \alpha (r+1), \\
r &= \mu (s+\beta) + (1-\mu)(t+\gamma), \\
r+1 &= \nu (s+\beta) + (1-\nu)(t+\gamma).
\end{aligned}$$

From Step 2.1, we have

$$[\nabla^r f]_{C^{0,\alpha}} \leq C_1 [\nabla^r f]_{C^{0,0}}^{1-\alpha} [\nabla^{r+1} f]_{C^{0,0}}^{\alpha} = C_1 \|\nabla^r f\|_{C^0}^{1-\alpha} \|\nabla^{r+1} f\|_{C^0}^{\alpha},$$

whereas from Step 2.2, we get

$$\|\nabla^r f\|_{C^0} = [\nabla^r f]_{C^{0,0}} \leq C_2 \left([\nabla^s f]_{C^{0,\beta}} + [\nabla^t f]_{C^{0,\gamma}}\right)^{\mu} [\nabla^t f]_{C^{0,\gamma}}^{1-\mu},$$
$$\left\|\nabla^{r+1} f\right\|_{C^0} = \left[\nabla^{r+1} f\right]_{C^{0,0}} \leq C_3 \left([\nabla^s f]_{C^{0,\beta}} + [\nabla^t f]_{C^{0,\gamma}}\right)^{\nu} [\nabla^t f]_{C^{0,\gamma}}^{1-\nu}.$$

Combining the three inequalities, we have that (16.27) is valid under the hypotheses of Case 2.

Case 3: $t+\gamma \leq r \leq r+\alpha \leq s+\beta \leq r+1$. This is dealt with as in Case 2; it is enough to interpolate r between $[t+\gamma, s+\beta]$, $r+\alpha$ between $[r, s+\beta]$ and combine the results.

Case 4: $r \leq t+\gamma \leq r+\alpha \leq r+1 \leq s+\beta$. We also proceed as in Case 2 and interpolate $r+1$ between $[t+\gamma, s+\beta]$ and $r+\alpha$ between $[t+\gamma, r+1]$.

Thus, Step 2 is established.

Step 3. We are now in a position to conclude. We trivially have

$$\|f\|_{C^0} = \|f\|_{C^0}^{\lambda} \|f\|_{C^0}^{1-\lambda} \leq \|f\|_{C^{s,\beta}}^{\lambda} \|f\|_{C^{t,\gamma}}^{1-\lambda}. \tag{16.29}$$

We immediately deduce from (16.27) that

$$[\nabla^r f]_{C^{0,\alpha}} \leq C_1 \left(\|f\|_{C^{s,\beta}} + \|f\|_{C^{t,\gamma}}\right)^{\lambda} \|f\|_{C^{t,\gamma}}^{1-\lambda},$$

which combined with Proposition 16.10 (or Corollary 16.13) leads to

$$[\nabla^r f]_{C^{0,\alpha}} \leq C_2 \|f\|_{C^{s,\beta}}^{\lambda} \|f\|_{C^{t,\gamma}}^{1-\lambda}.$$

Finally, combining the above inequality with Corollary 16.25 and (16.29), we have the result, namely

$$\|f\|_{C^{r,\alpha}} \leq C_3 \left(\|f\|_{C^0} + [\nabla^r f]_{C^{0,\alpha}}\right) \leq C \|f\|_{C^{s,\beta}}^{\lambda} \|f\|_{C^{t,\gamma}}^{1-\lambda}.$$

This concludes the proof of the theorem. □

We have as an immediate corollary the following.

Corollary 16.27. *Let $\Omega \subset \mathbb{R}^n$ be a bounded open Lipschitz set, $s \geq r \geq k \geq 1$ be integers and $0 \leq \alpha, \beta \leq 1$. Then there exists a constant $C = C(s, \Omega) > 0$ such that*

$$\|f\|_{C^{r,\alpha}} \|g\|_{C^{s,\beta}} \leq C \left[\|f\|_{C^{r-k,\alpha}} \|g\|_{C^{s+k,\beta}} + \|g\|_{C^{r-k,\alpha}} \|f\|_{C^{s+k,\beta}}\right].$$

Proof. Set

$$\lambda = \frac{(s+\beta) - (r-k+\alpha)}{(s+k+\beta) - (r-k+\alpha)} \in [0,1].$$

Note that

$$\lambda(r-k+\alpha)+(1-\lambda)(s+k+\beta)=r+\alpha,$$
$$\lambda(s+k+\beta)+(1-\lambda)(r-k+\alpha)=s+\beta.$$

Appealing to Theorem 16.26, we get

$$\begin{aligned}\|f\|_{C^{r,\alpha}}\|g\|_{C^{s,\beta}} &\leq C_1\left[\|f\|_{C^{r-k,\alpha}}^{\lambda}\|f\|_{C^{s+k,\beta}}^{1-\lambda}\right]\cdot C_1\left[\|g\|_{C^{s+k,\beta}}^{\lambda}\|g\|_{C^{r-k,\alpha}}^{1-\lambda}\right]\\ &\leq C_2\left[\|f\|_{C^{r-k,\alpha}}\|g\|_{C^{s+k,\beta}}\right]^{\lambda}\cdot\left[\|f\|_{C^{s+k,\beta}}\|g\|_{C^{r-k,\alpha}}\right]^{1-\lambda}\\ &\leq C_2\left[\|f\|_{C^{r-k,\alpha}}\|g\|_{C^{s+k,\beta}}+\|g\|_{C^{r-k,\alpha}}\|f\|_{C^{s+k,\beta}}\right]\end{aligned}$$

and thus the corollary. □

16.5.2 Product and Quotient

From the previous results we deduce the following inequality.

Theorem 16.28. *Let $\Omega\subset\mathbb{R}^n$ be a bounded open Lipschitz set, $r\geq 0$ an integer and $0\leq\alpha\leq 1$. Then there exists a constant $C=C(r,\Omega)>0$ such that*

$$\|fg\|_{C^{r,\alpha}}\leq C\left(\|f\|_{C^{r,\alpha}}\|g\|_{C^0}+\|f\|_{C^0}\|g\|_{C^{r,\alpha}}\right).$$

Proof. Step 1. We start with the case $r=0$. If $\alpha=0$, the result is trivial, so we assume that $\alpha>0$. We first observe that

$$\begin{aligned}[fg]_{C^{0,\alpha}} &= \sup_{\substack{x,y\in\overline{\Omega}\\ x\neq y}}\left\{\frac{|(f(x)-f(y))g(x)+f(y)(g(x)-g(y))|}{|x-y|^{\alpha}}\right\}\\ &\leq \|g\|_{C^0}[f]_{C^{0,\alpha}}+\|f\|_{C^0}[g]_{C^{0,\alpha}}\end{aligned}$$

and hence the claim follows, since

$$\|fg\|_{C^{0,\alpha}}=\|fg\|_{C^0}+[fg]_{C^{0,\alpha}}.$$

Step 2. We then proceed by induction on r. Observe that from Corollary 16.27, we have

$$\|f\|_{C^1}\|g\|_{C^{r-1,\alpha}}\leq C_1\left(\|f\|_{C^0}\|g\|_{C^{r,\alpha}}+\|f\|_{C^{r,\alpha}}\|g\|_{C^0}\right),$$
$$\|f\|_{C^{r-1,\alpha}}\|g\|_{C^1}\leq C_2\left(\|f\|_{C^0}\|g\|_{C^{r,\alpha}}+\|f\|_{C^{r,\alpha}}\|g\|_{C^0}\right).$$

We now use the hypothesis of induction to find

$$\begin{aligned}\|\nabla(fg)\|_{C^{r-1,\alpha}} &\leq C_3\left(\|\nabla f\|_{C^{r-1,\alpha}}\|g\|_{C^0}+\|\nabla f\|_{C^0}\|g\|_{C^{r-1,\alpha}}\right)\\ &\quad+C_3\left(\|f\|_{C^{r-1,\alpha}}\|\nabla g\|_{C^0}+\|f\|_{C^0}\|\nabla g\|_{C^{r-1,\alpha}}\right),\end{aligned}$$

which, combined with the above estimates, leads to

$$\|\nabla(fg)\|_{C^{r-1,\alpha}} \le C_4 \left(\|f\|_{C^0} \|g\|_{C^{r,\alpha}} + \|f\|_{C^{r,\alpha}} \|g\|_{C^0}\right).$$

Since

$$\|fg\|_{C^{r,\alpha}} = \|fg\|_{C^0} + \|\nabla(fg)\|_{C^{r-1,\alpha}},$$

we have indeed established the theorem. □

As a corollary, we have the following proposition.

Proposition 16.29. *Let $\Omega \subset \mathbb{R}^n$ be a bounded open Lipschitz set, $r \ge 0$ an integer and $0 \le \alpha \le 1$. Let $f \in C^{r,\alpha}\left(\overline{\Omega}\right)$ be such that*

$$f(x) \ge f_0 > 0, \quad \forall x \in \overline{\Omega}.$$

Then there exists a constant $C = C(r, \Omega) > 0$ such that

$$\left\|\frac{1}{f}\right\|_{C^{r,\alpha}} \le C \frac{\|f\|_{C^0}^r}{f_0^{r+2}} \|f\|_{C^{r,\alpha}}.$$

In particular, if there exists a constant $c > 0$ so that

$$\left\|\frac{1}{f}\right\|_{C^0}, \|f\|_{C^0} \le c,$$

then there exists a constant $C = C(c, r, \Omega) > 0$ such that

$$\left\|\frac{1}{f}\right\|_{C^{r,\alpha}} \le C \|f\|_{C^{r,\alpha}}.$$

Proof. Step 1. We start with the case $r = 0$. If $\alpha = 0$, the result below is trivial, so we assume that $\alpha > 0$. We have

$$\begin{aligned}\left\|\frac{1}{f}\right\|_{C^{0,\alpha}} &= \left\|\frac{1}{f}\right\|_{C^0} + \left[\frac{1}{f}\right]_{C^{0,\alpha}} \\ &\le \frac{1}{f_0} + \sup_{\substack{x,y\in\overline{\Omega} \\ x\neq y}} \left\{ \frac{\left|\frac{1}{f(x)} - \frac{1}{f(y)}\right|}{|x-y|^\alpha} \right\}\end{aligned}$$

and thus

$$\begin{aligned}\left\|\frac{1}{f}\right\|_{C^{0,\alpha}} &\le \frac{1}{f_0} + \frac{1}{f_0^2} \sup_{\substack{x,y\in\overline{\Omega} \\ x\neq y}} \left\{ \frac{|f(x) - f(y)|}{|x-y|^\alpha} \right\} = \frac{1}{f_0^2}\left(f_0 + [f]_{C^{0,\alpha}}\right) \\ &\le \frac{1}{f_0^2}\left(\|f\|_{C^0} + [f]_{C^{0,\alpha}}\right) = \frac{1}{f_0^2} \|f\|_{C^{0,\alpha}},\end{aligned}$$

as wished.

Step 2. We then proceed by induction and assume the result for $(r-1)$ and prove it for r. We have, appealing to Theorem 16.28,

$$\begin{aligned}\left\|\frac{1}{f}\right\|_{C^{r,\alpha}} &= \left\|\frac{1}{f}\right\|_{C^0} + \left\|\frac{\nabla f}{f^2}\right\|_{C^{r-1,\alpha}} \\ &\leq \frac{1}{f_0} + C_1 \|\nabla f\|_{C^{r-1,\alpha}} \frac{1}{f_0^2} + C_1 \|\nabla f\|_{C^0} \frac{1}{f_0} \left\|\frac{1}{f}\right\|_{C^{r-1,\alpha}}.\end{aligned}$$

Using the hypothesis of induction, we find

$$\begin{aligned}\left\|\frac{1}{f}\right\|_{C^{r,\alpha}} &\leq \frac{1}{f_0} + C_1 \|f\|_{C^{r,\alpha}} \frac{1}{f_0^2} + C_2 \|f\|_{C^1} \frac{1}{f_0} \frac{\|f\|_{C^0}^{r-1}}{f_0^{r+1}} \|f\|_{C^{r-1,\alpha}} \\ &\leq \frac{C_3}{f_0^{r+2}} \left(f_0^{r+1} + f_0^r \|f\|_{C^{r,\alpha}} + \|f\|_{C^0}^{r-1} \|f\|_{C^1} \|f\|_{C^{r-1,\alpha}}\right).\end{aligned}$$

We next invoke Corollary 16.27 to get that

$$\|f\|_{C^1} \|f\|_{C^{r-1,\alpha}} \leq C_4 \|f\|_{C^0} \|f\|_{C^{r,\alpha}}$$

and we hence obtain

$$\begin{aligned}\left\|\frac{1}{f}\right\|_{C^{r,\alpha}} &\leq \frac{C_5}{f_0^{r+2}} \left(f_0^{r+1} + f_0^r \|f\|_{C^{r,\alpha}} + \|f\|_{C^0}^r \|f\|_{C^{r,\alpha}}\right) \\ &\leq C \frac{\|f\|_{C^0}^r}{f_0^{r+2}} \|f\|_{C^{r,\alpha}}.\end{aligned}$$

Thus, the proposition is proved. □

Finally, the above proposition, combined with Theorem 16.28, leads to the following corollary for the inverse of matrices.

Corollary 16.30. *Let $\Omega \subset \mathbb{R}^n$ be a bounded open Lipschitz set, $r \geq 0$ an integer and $0 \leq \alpha \leq 1$. Let $A \in C^{r,\alpha}\left(\overline{\Omega};\mathbb{R}^{n\times n}\right)$ and $c > 0$ be such that*

$$\left\|\frac{1}{\det A}\right\|_{C^0}, \|A\|_{C^0} \leq c.$$

Then there exists a constant $C = C(c,r,\Omega) > 0$ such that

$$\left\|A^{-1}\right\|_{C^{r,\alpha}} \leq C \|A\|_{C^{r,\alpha}}.$$

In particular, if there exists a constant $c > 0$ so that

$$\|A\|_{C^0}, \left\|A^{-1}\right\|_{C^0} \leq c,$$

then there exists a constant $C = C(c,r,\Omega) > 0$ such that

$$\left\|A^{-1}\right\|_{C^{r,\alpha}} \leq C \|A\|_{C^{r,\alpha}}.$$

16.6 Composition and Inverse

16.6.1 Composition

We follow here Hörmander [55]. We start with the composition of two Hölder continuous functions.

Theorem 16.31. *Let $\Omega \subset \mathbb{R}^n$, $O \subset \mathbb{R}^m$ be bounded open Lipschitz sets, $r \geq 0$ an integer and $0 \leq \alpha, \beta \leq 1$. Let $g \in C^{r,\alpha}(\overline{O})$. If $r = 0$ and $f \in C^{0,\beta}(\overline{\Omega}; \overline{O})$, then*

$$\|g \circ f\|_{C^{0,\alpha\beta}(\overline{\Omega})} \leq \|g\|_{C^{0,\alpha}(\overline{O})} \|f\|^{\alpha}_{C^{0,\beta}(\overline{\Omega})} + \|g\|_{C^0(\overline{O})}.$$

If $r \geq 1$ and $f \in C^{r,\alpha}(\overline{\Omega}; \overline{O})$, then there exists a constant $C = C(r, \Omega, O) > 0$ such that

$$\|g \circ f\|_{C^{r,\alpha}(\overline{\Omega})} \leq C \left[\|g\|_{C^{r,\alpha}(\overline{O})} \|f\|^{r+\alpha}_{C^1(\overline{\Omega})} + \|g\|_{C^1(\overline{O})} \|f\|_{C^{r,\alpha}(\overline{\Omega})} + \|g\|_{C^0(\overline{O})} \right].$$

Proof. In the sequel, we remove the dependence on the sets Ω and O, since this dependence will be clear from the context. We will use several times that the C^1 norm is equivalent to the $C^{0,1}$ norm (cf. Corollary 16.13).

Step 1. We start with the case $r = 0$. If $\alpha = 0$ or if $\beta = 0$, then the result is trivial since

$$\|g \circ f\|_{C^{0,0}} = \|g \circ f\|_{C^0} \leq \|g\|_{C^0}.$$

So we assume that $r = 0 < \alpha, \beta \leq 1$. We therefore have, since $g \in C^{0,\alpha}(\overline{O})$,

$$|g(f(x)) - g(f(y))| \leq [g]_{C^{0,\alpha}} |f(x) - f(y)|^{\alpha} \leq [g]_{C^{0,\alpha}} \left[[f]_{C^{0,\beta}} |x - y|^{\beta} \right]^{\alpha}$$

and hence the claim

$$\|g \circ f\|_{C^{0,\alpha\beta}} \leq \|g\|_{C^{0,\alpha}} \|f\|^{\alpha}_{C^{0,\beta}} + \|g\|_{C^0}.$$

Step 2. We next discuss the case $r = 1$ and $0 < \alpha \leq 1$, the case $\alpha = 0$ being already settled in Step 1 (just take $\alpha = \beta = 1$). We have, by definition,

$$\|g \circ f\|_{C^{1,\alpha}} = \|g \circ f\|_{C^0} + \|\nabla (g \circ f)\|_{C^{0,\alpha}}.$$

Since

$$\|\nabla (g \circ f)\|_{C^{0,\alpha}} = \|(\nabla g \circ f) \nabla f\|_{C^{0,\alpha}},$$

we have, by Step 1 and by Theorem 16.28,

$$\begin{aligned} \|\nabla (g \circ f)\|_{C^{0,\alpha}} &\leq C_1 \|\nabla g \circ f\|_{C^{0,\alpha}} \|\nabla f\|_{C^0} + C_1 \|\nabla g \circ f\|_{C^0} \|\nabla f\|_{C^{0,\alpha}} \\ &\leq C_2 \left[\|g\|_{C^{1,\alpha}} \|f\|^{\alpha}_{C^1} + \|g\|_{C^1} \right] \|f\|_{C^1} + C_2 \|g\|_{C^1} \|f\|_{C^{1,\alpha}} \end{aligned}$$

and hence our claim,

$$\|g\circ f\|_{C^{1,\alpha}} \le C\left[\|g\|_{C^{1,\alpha}}\|f\|_{C^1}^{1+\alpha}+\|g\|_{C^1}\|f\|_{C^{1,\alpha}}+\|g\|_{C^0}\right].$$

Step 3. We proceed by induction on $r \ge 2$. We write

$$\begin{aligned}\|g\circ f\|_{C^{r,\alpha}} &= \|g\circ f\|_{C^0}+\|\nabla(g\circ f)\|_{C^{r-1,\alpha}}\\ &= \|g\circ f\|_{C^0}+\|(\nabla g\circ f)\nabla f\|_{C^{r-1,\alpha}}.\end{aligned}$$

We use the hypothesis of induction and Theorem 16.28 to get

$$\begin{aligned}&\|(\nabla g\circ f)\nabla f\|_{C^{r-1,\alpha}}\\ &\quad\le C_1\|\nabla g\circ f\|_{C^{r-1,\alpha}}\|\nabla f\|_{C^0}+C_1\|\nabla g\circ f\|_{C^0}\|\nabla f\|_{C^{r-1,\alpha}}\\ &\quad\le C_2\left[\|\nabla g\|_{C^{r-1,\alpha}}\|f\|_{C^1}^{(r-1)+\alpha}+\|\nabla g\|_{C^1}\|f\|_{C^{r-1,\alpha}}+\|\nabla g\|_{C^0}\right]\|f\|_{C^1}\\ &\qquad+C_2\|g\|_{C^1}\|f\|_{C^{r,\alpha}}.\end{aligned}$$

We therefore find

$$\begin{aligned}\|g\circ f\|_{C^{r,\alpha}} \le \|g\|_{C^0}+C_2\|g\|_{C^{r,\alpha}}\|f\|_{C^1}^{r+\alpha}+C_2\|g\|_{C^2}\|f\|_{C^{r-1,\alpha}}\|f\|_{C^1}&\\ +C_3\|g\|_{C^1}\|f\|_{C^{r,\alpha}}.&\end{aligned} \tag{16.30}$$

In order to conclude, it is enough to estimate

$$\|g\|_{C^2}\|f\|_{C^{r-1,\alpha}}\|f\|_{C^1}.$$

We use Theorem 16.26 to get

$$\|g\|_{C^2}\le C_4\left(\|g\|_{C^1}\right)^{\frac{r-2+\alpha}{r-1+\alpha}}\left(\|g\|_{C^{r,\alpha}}\right)^{\frac{1}{r-1+\alpha}},$$

$$\|f\|_{C^{r-1,\alpha}}\le C_4\left(\|f\|_{C^{r,\alpha}}\right)^{\frac{r-2+\alpha}{r-1+\alpha}}\left(\|f\|_{C^1}\right)^{\frac{1}{r-1+\alpha}}$$

and thus

$$\begin{aligned}\|g\|_{C^2}\|f\|_{C^{r-1,\alpha}}\|f\|_{C^1} &\le C_5\left(\|g\|_{C^1}\|f\|_{C^{r,\alpha}}\right)^{\frac{r-2+\alpha}{r-1+\alpha}}\left(\|g\|_{C^{r,\alpha}}\|f\|_{C^1}^{r+\alpha}\right)^{\frac{1}{r-1+\alpha}}\\ &\le C_6\left[\|g\|_{C^1}\|f\|_{C^{r,\alpha}}+\|g\|_{C^{r,\alpha}}\|f\|_{C^1}^{r+\alpha}\right].\end{aligned} \tag{16.31}$$

Combining (16.30) and (16.31), we have the claim. □

16.6.2 Inverse

We easily deduce, from the previous results, an estimate on the inverse.

Theorem 16.32. *Let $\Omega, O \subset \mathbb{R}^n$ be bounded open Lipschitz sets, $r \geq 1$ be an integer and $0 \leq \alpha \leq 1$. Let $c > 0$. Let $f \in C^{r,\alpha}\left(\overline{\Omega};\overline{O}\right)$ and $g \in C^{r,\alpha}\left(\overline{O};\overline{\Omega}\right)$ be such that*

$$g \circ f = \mathrm{id} \quad \textit{and} \quad \|g\|_{C^1(\overline{O})}, \|f\|_{C^1(\overline{\Omega})} \leq c.$$

Then there exists a constant $C = C(c, r, \Omega, O) > 0$ such that

$$\|f\|_{C^{r,\alpha}(\overline{\Omega})} \leq C \|g\|_{C^{r,\alpha}(\overline{O})}.$$

Proof. *Step 1.* Since $g \circ f = \mathrm{id}$, we obtain

$$\|\mathrm{id}\|_{C^0} = \|g \circ f\|_{C^0} \leq \|g\|_{C^0}$$

and, hence, combining with the fact that $\|f\|_{C^1} \leq c$, we find that there exists a constant $C_1 = C_1(c, \Omega) > 0$ such that

$$\|f\|_{C^0} \leq \|f\|_{C^1} \leq c = \frac{c}{\|\mathrm{id}\|_{C^0}} \|\mathrm{id}\|_{C^0} \leq \frac{c}{\|\mathrm{id}\|_{C^0}} \|g\|_{C^0} = C_1 \|g\|_{C^0}.$$

Step 2. Appealing to Corollary 16.30, we get that there exists a constant $C_2 = C_2(c, r, \Omega) > 0$ such that

$$\|\nabla f\|_{C^{r-1,\alpha}} \leq C_2 \|\nabla g \circ f\|_{C^{r-1,\alpha}}. \tag{16.32}$$

Therefore, the case $r = 1$ is immediate, invoking Theorem 16.31, since we then have

$$\|\nabla f\|_{C^{0,\alpha}} \leq C_2 \|\nabla g \circ f\|_{C^{0,\alpha}} \leq C_3 \left[\|\nabla g\|_{C^{0,\alpha}} + \|\nabla g\|_{C^0}\right] \leq C \|g\|_{C^{1,\alpha}},$$

which, combined with Step 1, gives the claim.

Step 3. We now proceed by induction and apply Theorem 16.31 to (16.32). We find

$$\begin{aligned} \|\nabla f\|_{C^{r-1,\alpha}} &\leq C_2 \|\nabla g \circ f\|_{C^{r-1,\alpha}} \\ &\leq C_3 \left[\|\nabla g\|_{C^{r-1,\alpha}} + \|\nabla g\|_{C^1} \|f\|_{C^{r-1,\alpha}} + \|\nabla g\|_{C^0}\right] \end{aligned}$$

and therefore, appealing to Step 1 and to the above inequality,

$$\|f\|_{C^{r,\alpha}} \leq C_4 \left[\|g\|_{C^{r,\alpha}} + \|g\|_{C^2} \|f\|_{C^{r-1,\alpha}}\right].$$

Applying the hypothesis of induction, we deduce that

$$\|f\|_{C^{r,\alpha}} \leq C_5 \left[\|g\|_{C^{r,\alpha}} + \|g\|_{C^2} \|g\|_{C^{r-1,\alpha}}\right]. \tag{16.33}$$

From Corollary 16.27, we get that

$$\|g\|_{C^2} \|g\|_{C^{r-1,\alpha}} \leq C_6 \|g\|_{C^{r,\alpha}}$$

and, thus, combining with (16.33), we have the claim

$$\|f\|_{C^{r,\alpha}} \leq C\,\|g\|_{C^{r,\alpha}}\,.$$

This concludes the proof of the theorem. □

16.6.3 A Further Result

Finally, the next result (cf. Rivière and Ye [85]) has been explicitly used in Theorem 12.4.

Lemma 16.33. *Let $r \geq 1$ be an integer, $0 \leq \alpha \leq 1$, $\Omega \subset \mathbb{R}^n$ be a bounded open Lipschitz set, $g \in C^{r,\alpha}(\overline{\Omega})$ and $u \in C^{r,\alpha}(\overline{\Omega};\overline{\Omega})$. Then for every $x, y \in \overline{\Omega}$, there exists a constant $C = C(r,\Omega)$ such that*

$$\begin{aligned} |\nabla^r(g\circ u)(x) - \nabla^r(g\circ u)(y)| &\leq C\|\nabla g\|_{C^0}\|u - \operatorname{id}\|_{C^{r,\alpha}}|x-y|^\alpha \\ &\quad + C\|g\|_{C^{r,\alpha}}\|u-\operatorname{id}\|_{C^1}(1+\|u\|_{C^1}^{r+\alpha})|x-y|^\alpha \\ &\quad + C|\nabla^r g(u(x)) - \nabla^r g(u(y))|\,. \end{aligned}$$

Remark 16.34. The estimate implies the following inequality:

$$\begin{aligned} [\nabla^r(g\circ u)]_{C^{0,\alpha}} &\leq C\,[(\nabla^r g)\circ u]_{C^{0,\alpha}} + C\|g\|_{C^1}\|u-\operatorname{id}\|_{C^{r,\alpha}} \\ &\quad + C\|g\|_{C^{r,\alpha}}(1+\|u\|_{C^1}^{r+\alpha})\|u-\operatorname{id}\|_{C^1}\,. \end{aligned}$$

Note, however, that the lemma is more precise.

Proof. We prove the lemma by induction. We split the proof into two steps. In the sequel, $C_1, C_2, \ldots$ will denote generic constants depending on r and Ω. We will use several times that the C^1 norm is equivalent to the $C^{0,1}$ norm (cf. Corollary 16.13).

Step 1 (the case $r = 1$). We show the result when $r = 1$. Let $x, y \in \overline{\Omega}$. We obtain

$$\begin{aligned} |\nabla(g\circ u)(x) - \nabla(g\circ u)(y)| &= |\nabla g(u(x))\cdot\nabla u(x) - \nabla g(u(y))\cdot\nabla u(y)| \\ &= |\nabla g(u(x))\cdot[\nabla u(x) - \nabla u(y)] \\ &\quad - [\nabla g(u(y)) - \nabla g(u(x))]\cdot\nabla u(y)| \end{aligned}$$

and thus

$$\begin{aligned} &|\nabla(g\circ u)(x) - \nabla(g\circ u)(y)| \\ &\leq \|\nabla g\|_{C^0}|\nabla u(x) - \nabla u(y)| + \|\nabla u\|_{C^0}|\nabla g(u(y)) - \nabla g(u(x))| \\ &= \|\nabla g\|_{C^0}|\nabla(u-\operatorname{id})(x) - \nabla(u-\operatorname{id})(y)| + \|\nabla u\|_{C^0}|\nabla g(u(y)) - \nabla g(u(x))| \\ &\leq \|\nabla g\|_{C^0}\|u-\operatorname{id}\|_{C^{1,\alpha}}|x-y|^\alpha + \|\nabla u\|_{C^0}|\nabla g(u(y)) - \nabla g(u(x))|\,. \end{aligned}$$

Since the first term of the previous sum has the desired form, it is enough to estimate the second one. We obtain

$$\begin{aligned}&\|\nabla u\|_{C^0}\,|\nabla g(u(x))-\nabla g(u(y))|\\&\quad\le(\|u-\mathrm{id}\|_{C^1}+1)\,|\nabla g(u(x))-\nabla g(u(y))|\\&\quad\le C_1\|u-\mathrm{id}\|_{C^1}\|g\|_{C^{1,\alpha}}\|u\|_{C^1}^{\alpha}|x-y|^{\alpha}+|\nabla g(u(x))-\nabla g(u(y))|\\&\quad\le C_1\|u-\mathrm{id}\|_{C^1}\|g\|_{C^{1,\alpha}}(1+\|u\|_{C^1}^{1+\alpha})|x-y|^{\alpha}+|\nabla g(u(x))-\nabla g(u(y))|\,.\end{aligned}$$

Since both terms on the right-hand side of the previous inequality have the desired form, Step 1 is shown.

Step 2 (induction). Assume that the result holds true for $r-1$ and let us show it for $r\ge 2$. We have to establish that for every $x,y\in\overline{\Omega}$,

$$\begin{aligned}&|\nabla^r(g\circ u)(x)-\nabla^r(g\circ u)(y)|\\&\quad\le C\|\nabla g\|_{C^0}\|u-\mathrm{id}\|_{C^{r,\alpha}}|x-y|^{\alpha}\\&\qquad+C\|g\|_{C^{r,\alpha}}\|u-\mathrm{id}\|_{C^1}(1+\|u\|_{C^1}^{r+\alpha})|x-y|^{\alpha}\\&\qquad+C|\nabla^r g(u(x))-\nabla^r g(u(y))|\,.\end{aligned}\tag{16.34}$$

Let $x,y\in\overline{\Omega}$. We write

$$\begin{aligned}|\nabla^r(g\circ u)(x)-\nabla^r(g\circ u)(y)|&=\left|\nabla^{r-1}[\nabla(g\circ u)](x)-\nabla^{r-1}[\nabla(g\circ u)](y)\right|\\&=\left|\nabla^{r-1}[\nabla g\circ u\cdot\nabla u](x)-\nabla^{r-1}[\nabla g\circ u\cdot\nabla u](y)\right|\end{aligned}$$

and therefore get

$$\begin{aligned}&|\nabla^r(g\circ u)(x)-\nabla^r(g\circ u)(y)|\\&\quad\le\left|\nabla^{r-1}[\nabla g\circ u\cdot\nabla(u-\mathrm{id})](x)-\nabla^{r-1}[\nabla g\circ u\cdot\nabla(u-\mathrm{id})](y)\right|\\&\qquad+\left|\nabla^{r-1}[\nabla g\circ u](x)-\nabla^{r-1}[\nabla g\circ u](y)\right|.\end{aligned}$$

Hence, using the induction hypothesis, we obtain

$$\begin{aligned}&|\nabla^r(g\circ u)(x)-\nabla^r(g\circ u)(y)|\\&\quad\le\|\nabla g\circ u\cdot\nabla(u-\mathrm{id})\|_{C^{r-1,\alpha}}|x-y|^{\alpha}+C_3\|\nabla\nabla g\|_{C^0}\|u-\mathrm{id}\|_{C^{r-1,\alpha}}|x-y|^{\alpha}\\&\qquad+C_3\|\nabla g\|_{C^{r-1,\alpha}}\|u-\mathrm{id}\|_{C^1}(1+\|u\|_{C^1}^{r-1+\alpha})|x-y|^{\alpha}\\&\qquad+C_3|\nabla^r g(u(x))-\nabla^r g(u(y))|\,.\end{aligned}$$

We now estimate separately each of the four terms of the previous inequality and show that they have the same form as the right-hand side of (16.34). Noticing first that the fourth term has already the desired form, it is enough to estimate the first three terms.

1) Estimate of $\|\nabla g\circ u\cdot\nabla(u-\mathrm{id})\|_{C^{r-1,\alpha}}|x-y|^{\alpha}$. Using Theorems 16.28 and 16.31, we obtain

$$\begin{aligned}
&\|\nabla g\circ u\cdot\nabla(u-\mathrm{id})\|_{C^{r-1,\alpha}}|x-y|^\alpha\\
&\quad\le C_4\left(\|\nabla g\circ u\|_{C^0}\|\nabla(u-\mathrm{id})\|_{C^{r-1,\alpha}}+\|\nabla g\circ u\|_{C^{r-1,\alpha}}\|\nabla(u-\mathrm{id})\|_{C^0}\right)|x-y|^\alpha\\
&\quad\le C_5\|\nabla g\|_{C^0}\|u-\mathrm{id}\|_{C^{r,\alpha}}|x-y|^\alpha\\
&\qquad+C_6\|u-\mathrm{id}\|_{C^1}\left[\|\nabla g\|_{C^{r-1,\alpha}}\|u\|_{C^1}^{r-1+\alpha}+\|\nabla g\|_{C^1}\|u\|_{C^{r-1,\alpha}}+\|\nabla g\|_{C^0}\right]|x-y|^\alpha
\end{aligned}$$

and thus

$$\begin{aligned}
&\|\nabla g\circ u\cdot\nabla(u-\mathrm{id})\|_{C^{r-1,\alpha}}|x-y|^\alpha\\
&\quad\le C_5\|\nabla g\|_{C^0}\|u-\mathrm{id}\|_{C^{r,\alpha}}|x-y|^\alpha\\
&\qquad+C_7\|u-\mathrm{id}\|_{C^1}\|g\|_{C^{r,\alpha}}(1+\|u\|_{C^1}^{r+\alpha})|x-y|^\alpha\\
&\qquad+C_6\|u-\mathrm{id}\|_{C^1}\|\nabla g\|_{C^1}\|u\|_{C^{r-1,\alpha}}|x-y|^\alpha.
\end{aligned}$$

Since the first two terms of the last sum have the desired form, it is enough to estimate the last term. We have

$$\begin{aligned}
&\|u-\mathrm{id}\|_{C^1}\|\nabla g\|_{C^1}\|u\|_{C^{r-1,\alpha}}|x-y|^\alpha\\
&\quad\le\|u-\mathrm{id}\|_{C^1}\|\nabla g\|_{C^1}\|u-\mathrm{id}\|_{C^{r-1,\alpha}}|x-y|^\alpha+C_8\|u-\mathrm{id}\|_{C^1}\|\nabla g\|_{C^1}|x-y|^\alpha\\
&\quad\le\|u-\mathrm{id}\|_{C^1}\|\nabla g\|_{C^1}\|u-\mathrm{id}\|_{C^{r-1,\alpha}}|x-y|^\alpha+C_8\|u-\mathrm{id}\|_{C^1}\|g\|_{C^{r,\alpha}}|x-y|^\alpha.
\end{aligned}$$

The last term of the previous sum having the desired form, it remains to estimate the first term. Using Theorem 16.26, we get, letting $\lambda=1/(r-1+\alpha)$,

$$\|u-\mathrm{id}\|_{C^{r-1,\alpha}}\le C_9(\|u-\mathrm{id}\|_{C^{r,\alpha}})^{1-\lambda}(\|u-\mathrm{id}\|_{C^1})^\lambda,\tag{16.35}$$

$$\|\nabla g\|_{C^1}\le C_9(\|\nabla g\|_{C^{r-1,\alpha}})^\lambda(\|\nabla g\|_{C^0})^{1-\lambda}.\tag{16.36}$$

Therefore, using (16.35) and (16.36), we obtain

$$\begin{aligned}
&\|u-\mathrm{id}\|_{C^1}\|u-\mathrm{id}\|_{C^{r-1,\alpha}}\|\nabla g\|_{C^1}|x-y|^\alpha\\
&\quad\le C_9\|u-\mathrm{id}\|_{C^1}\left(\|u-\mathrm{id}\|_{C^{r,\alpha}}\|\nabla g\|_{C^0}\right)^{1-\lambda}\left(\|u-\mathrm{id}\|_{C^1}\|\nabla g\|_{C^{r-1,\alpha}}\right)^\lambda|x-y|^\alpha\\
&\quad=C_9\left(\|u-\mathrm{id}\|_{C^{r,\alpha}}\|\nabla g\|_{C^0}\right)^{1-\lambda}\left(\|u-\mathrm{id}\|_{C^1}^{\frac{1}{\lambda}+1}\|\nabla g\|_{C^{r-1,\alpha}}\right)^\lambda|x-y|^\alpha\\
&\quad\le C_9\left(\|u-\mathrm{id}\|_{C^{r,\alpha}}\|\nabla g\|_{C^0}+\|u-\mathrm{id}\|_{C^1}^{r+\alpha}\|g\|_{C^{r,\alpha}}\right)|x-y|^\alpha,
\end{aligned}$$

where we have used the fact that $1/\lambda+1=r+\alpha$. The assertion is proved since the two terms of the last inequality have the desired form.

2) Estimate of $\|\nabla\nabla g\|_{C^0}\|u-\mathrm{id}\|_{C^{r-1,\alpha}}|x-y|^\alpha$. Combining (16.35) and (16.36), we obtain

$$\|u-\mathrm{id}\|_{C^{r-1,\alpha}}\|\nabla g\|_{C^1}\le C_9\left(\|u-\mathrm{id}\|_{C^{r,\alpha}}\|\nabla g\|_{C^0}+\|u-\mathrm{id}\|_{C^1}\|\nabla g\|_{C^{r-1,\alpha}}\right)$$

and thus

$$\|\nabla\nabla g\|_{C^0}\|u-\operatorname{id}\|_{C^{r-1,\alpha}}|x-y|^{\alpha}\leq\|\nabla g\|_{C^1}\|u-\operatorname{id}\|_{C^{r-1,\alpha}}|x-y|^{\alpha}$$
$$\leq C_9\left[\|\nabla g\|_{C^0}\|u-\operatorname{id}\|_{C^{r,\alpha}}+\|\nabla g\|_{C^{r-1,\alpha}}\|u-\operatorname{id}\|_{C^1}\right]|x-y|^{\alpha}.$$

The assertion is proved since the two terms of the last inequality have the desired form.

3) Estimate of $\|\nabla g\|_{C^{r-1,\alpha}}\|u-\operatorname{id}\|_{C^1}(1+\|u\|_{C^1}^{r-1+\alpha})|x-y|^{\alpha}$. We get

$$\|\nabla g\|_{C^{r-1,\alpha}}\|u-\operatorname{id}\|_{C^1}(1+\|u\|_{C^1}^{r-1+\alpha})|x-y|^{\alpha}$$
$$\leq C_{10}\|g\|_{C^{r,\alpha}}\|u-\operatorname{id}\|_{C^1}(1+\|u\|_{C^1}^{r+\alpha})|x-y|^{\alpha}.$$

This concludes the proof. □

16.7 Difference of Composition

We often need to estimate the difference of two functions that are obtained through composition, namely

$$\|g\circ u-g\circ v\|_{C^{r,\alpha}}.$$

This has been used in our study of the pullback equation. We will give several theorems, following Bandyopadhyay and Dacorogna [8]. At first glance, one would think that the above quantity can be estimated, in a continuous way, in terms of $\|g\|_{C^{r,\alpha}}$ and $\|u-v\|_{C^{r,\alpha}}$. This is, in general, impossible if $0<\alpha<1$, as will be seen in the example below; one needs to consider higher norms $\|g\|_{C^{s,\beta}}$ with

$$r+\alpha<s+\beta.$$

We will consider three theorems. The first and easiest one is when $s=r+1$ and $\beta=\alpha$, the second one is when $s=r$ and $0\leq\alpha<\beta\leq1$ and the third one is when $s=r+1$ and $0\leq\beta<\alpha\leq1$.

Example 16.35. Let $r\geq0$ be an integer and $0<\alpha<1$. Let $g\in C^{r,\alpha}([0,2])$ be given by

$$g(x)=x^{r+\alpha}.$$

Let, for $\varepsilon,x\in\overline{\Omega}=[0,1]$,

$$u(x)=x+\varepsilon\quad\text{and}\quad v(x)=x.$$

Note that for every integer $t\geq0$ and every $0\leq\gamma\leq1$,

$$\|u-v\|_{C^{t,\gamma}}=\varepsilon.$$

It is easily proved that

$$\|g\circ u-g\circ v\|_{C^{r,\alpha}}\not\to0\quad\text{as }\varepsilon\to0.$$

We show this for $r=0$ (the general case is proved similarly). It is enough to observe that

$$\lambda_\varepsilon(x)=\frac{[g(v(x))-g(u(x))]-[g(v(0))-g(u(0))]}{|x|^\alpha}=\frac{|x|^\alpha+\varepsilon^\alpha-|x+\varepsilon|^\alpha}{|x|^\alpha}$$

does not tend uniformly to 0 as $\varepsilon\to 0$. Indeed, choosing $x=t\varepsilon$ for any $t>0$, we have the claim.

Before starting our analysis, we recall, from Theorem 16.31, that if $0\le\alpha\le 1$ and $\|u\|_{C^1}\le c$, then

$$\|g\circ u\|_{C^{0,\alpha}}\le C\|g\|_{C^{0,\alpha}} \tag{16.37}$$

if $r=0$, whereas if $r\ge 1$, then

$$\|g\circ u\|_{C^{r,\alpha}}\le C\|g\|_{C^{r,\alpha}}+C\|g\|_{C^1}\|u\|_{C^{r,\alpha}}. \tag{16.38}$$

16.7.1 A First Result

We start with the easiest case.

Theorem 16.36. *Let $\Omega\subset\mathbb{R}^n$ and $O\subset\mathbb{R}^m$ be bounded open Lipschitz sets. Let $r\ge 0$ be an integer and $0\le\alpha\le 1$. Let $c>0$, $g\in C^{r+1,\alpha}(\overline{O})$ and $u,v\in C^{r,\alpha}(\overline{\Omega};\overline{O})\cap C^1(\overline{\Omega};\overline{O})$, with*

$$\|u\|_{C^1(\overline{\Omega})},\|v\|_{C^1(\overline{\Omega})}\le c.$$

Then there exists $C=C(c,r,\Omega,O)>0$ so that if $r=0$,

$$\|g\circ u-g\circ v\|_{C^{0,\alpha}(\overline{\Omega})}\le C\|g\|_{C^{1,\alpha}(\overline{O})}\|u-v\|_{C^0(\overline{\Omega})}+C\|g\|_{C^1(\overline{O})}\|u-v\|_{C^{0,\alpha}(\overline{\Omega})},$$

whereas, when $r\ge 1$,

$$\begin{aligned}&\|g\circ u-g\circ v\|_{C^{r,\alpha}(\overline{\Omega})}\\&\quad\le C\|g\|_{C^{r+1,\alpha}(\overline{O})}\|u-v\|_{C^0(\overline{\Omega})}\\&\qquad+C\|g\|_{C^2(\overline{O})}\left[\|u\|_{C^{r,\alpha}(\overline{\Omega})}+\|v\|_{C^{r,\alpha}(\overline{\Omega})}\right]\|u-v\|_{C^0(\overline{\Omega})}\\&\qquad+C\|g\|_{C^1(\overline{O})}\|u-v\|_{C^{r,\alpha}(\overline{\Omega})}.\end{aligned}$$

Proof. In view of Theorem 16.11, we can assume that Ω is convex. We write

$$\begin{aligned}g\circ u-g\circ v&=\int_0^1\frac{d}{dt}[g(v+t(u-v))]\,dt\\&=\int_0^1\langle\nabla g(v+t(u-v));u-v\rangle\,dt\end{aligned}$$

and deduce, from Theorem 16.28, that

$$\begin{aligned}\|g\circ u-g\circ v\|_{C^{r,\alpha}}\leq C_1\|u-v\|_{C^0}\int_0^1\|\nabla g(v+t(u-v))\|_{C^{r,\alpha}}dt\\+C_1\|u-v\|_{C^{r,\alpha}}\int_0^1\|\nabla g(v+t(u-v))\|_{C^0}dt.\end{aligned}$$

We first deal with the case $r=0$. We have from (16.37) and since $\|u\|_{C^1},\|v\|_{C^1}\leq c$ that

$$\|g\circ u-g\circ v\|_{C^{0,\alpha}}\leq C_2\|u-v\|_{C^0}\|\nabla g\|_{C^{0,\alpha}}+C_2\|u-v\|_{C^{0,\alpha}}\|\nabla g\|_{C^0},$$

which is our claim. We then discuss the case $r\geq 1$. Appealing to (16.38) and to the fact that $\|u\|_{C^1},\|v\|_{C^1}\leq c$, we get

$$\begin{aligned}&\|g\circ u-g\circ v\|_{C^{r,\alpha}}\\&\quad\leq C_2\|u-v\|_{C^0}\left[\|\nabla g\|_{C^{r,\alpha}}+\|\nabla g\|_{C^1}\left(\|u\|_{C^{r,\alpha}}+\|v\|_{C^{r,\alpha}}\right)\right]\\&\qquad+C_2\|u-v\|_{C^{r,\alpha}}\|\nabla g\|_{C^0}\end{aligned}$$

and thus the result. □

16.7.2 A Second Result

Before going to our next results, we need the following lemma.

Lemma 16.37. *Let $\Omega\subset\mathbb{R}^n$ and $O\subset\mathbb{R}^m$ be bounded open Lipschitz sets. Let $s\geq r\geq 1$ be integers and $0\leq\alpha,\beta\leq 1$ with*

$$r+\alpha<s+\beta\leq r+1+\alpha.$$

Let $c>0$, $g\in C^{s,\beta}\left(\overline{O}\right)$ and $u,v\in C^{r,\alpha}\left(\overline{\Omega};\overline{O}\right)$, with

$$\|u\|_{C^1(\overline{\Omega})},\quad\|v\|_{C^1(\overline{\Omega})}\leq c.$$

If $r=1$, then there exists $C=C(c,\Omega,O)>0$ such that

$$\begin{aligned}&\|\nabla(g\circ u)-\nabla(g\circ v)\|_{C^{0,\alpha}(\overline{\Omega})}\\&\quad\leq C\|\nabla g\circ u-\nabla g\circ v\|_{C^{0,\alpha}(\overline{\Omega})}+C\|g\|_{C^1(\overline{O})}\|u-v\|_{C^{1,\alpha}(\overline{\Omega})}\\&\qquad+C\|g\|_{C^{s,\beta}(\overline{O})}\left(1+\min\left\{\|u\|_{C^{1,\alpha}(\overline{\Omega})},\|v\|_{C^{1,\alpha}(\overline{\Omega})}\right\}\right)\|u-v\|_{C^0(\overline{\Omega})}^{(s+\beta)-(1+\alpha)}.\end{aligned}$$

If $r \geq 2$, then there exists $C = C(c, r, \Omega, O) > 0$ so that

$$
\begin{aligned}
&\|\nabla(g \circ u) - \nabla(g \circ v)\|_{C^{r-1,\alpha}(\overline{\Omega})} \\
&\leq C\|\nabla g \circ u - \nabla g \circ v\|_{C^{r-1,\alpha}(\overline{\Omega})} \\
&\quad + C\|g\|_{C^1(\overline{O})}\|u-v\|_{C^{r,\alpha}(\overline{\Omega})} + C\|g\|_{C^{s,\beta}(\overline{O})}\|u-v\|_{C^0(\overline{\Omega})}^{(s+\beta)-(r+\alpha)} \\
&\quad + C\|g\|_{C^2(\overline{O})}\left[\|u\|_{C^{r,\alpha}(\overline{\Omega})} + \|v\|_{C^{r,\alpha}(\overline{\Omega})}\right]\|u-v\|_{C^0(\overline{\Omega})}.
\end{aligned}
$$

Remark 16.38. Note that the proof gives that the lemma is also valid in the limit case when $s = r+1$, $\alpha = 1$ and $\beta = 0$.

Proof. Step 1. We start with a preliminary computation. We prove that for every $r \geq 1$ and $0 \leq \alpha, \beta \leq 1$, with

$$r + \alpha < s + \beta \leq r + 1 + \alpha,$$

we can find a constant $C = C(c, r, \Omega) > 0$ such that

$$\|g\|_{C^{r,\alpha}}\|u-v\|_{C^1} \leq C\|g\|_{C^{s,\beta}}\|u-v\|_{C^0}^{(s+\beta)-(r+\alpha)} + C\|g\|_{C^1}\|u-v\|_{C^{r,\alpha}}.$$

We divide the proof into two cases.

Case 1: $s = r$ and, therefore, $0 \leq \alpha < \beta \leq 1$. Observe that since

$$0 < \theta = \frac{(\beta-\alpha)(r+\alpha)}{r+\beta-1} \leq 1$$

and $\|u\|_{C^1}, \|v\|_{C^1} \leq c$, we have

$$\|u-v\|_{C^1} \leq C_1\|u-v\|_{C^1}^{\theta},$$

and thus, by interpolation,

$$
\begin{aligned}
\|u-v\|_{C^1} &\leq C_2\left(\|u-v\|_{C^{r,\alpha}}^{\frac{1}{r+\alpha}}\|u-v\|_{C^0}^{\frac{r+\alpha-1}{r+\alpha}}\right)^{\theta} \\
&= C_2\|u-v\|_{C^{r,\alpha}}^{\frac{\beta-\alpha}{r+\beta-1}}\|u-v\|_{C^0}^{\frac{(\beta-\alpha)(r+\alpha-1)}{r+\beta-1}}.
\end{aligned}
$$

Using again an interpolation, we have

$$\|g\|_{C^{r,\alpha}} \leq C_3\|g\|_{C^{r,\beta}}^{\frac{r+\alpha-1}{r+\beta-1}}\|g\|_{C^1}^{\frac{\beta-\alpha}{r+\beta-1}}.$$

Combining the two estimates, we get

$$
\begin{aligned}
\|g\|_{C^{r,\alpha}}\|u-v\|_{C^1} &\leq C_4\left(\|g\|_{C^{r,\beta}}\|u-v\|_{C^0}^{\beta-\alpha}\right)^{\frac{r+\alpha-1}{r+\beta-1}}\left(\|g\|_{C^1}\|u-v\|_{C^{r,\alpha}}\right)^{\frac{\beta-\alpha}{r+\beta-1}} \\
&\leq C_5\|g\|_{C^{r,\beta}}\|u-v\|_{C^0}^{\beta-\alpha} + C_5\|g\|_{C^1}\|u-v\|_{C^{r,\alpha}},
\end{aligned}
$$

as wished.

Case 2: $s = r+1$ and, therefore, $0 \le \beta \le \alpha \le 1$. We proceed in a very similar way. We first note that since

$$0 \le \theta = \frac{(1+\beta-\alpha)(r+\alpha)}{r+\beta} \le 1$$

and $\|u\|_{C^1}, \|v\|_{C^1} \le c$, we obtain

$$\|u-v\|_{C^1} \le C_1 \|u-v\|_{C^1}^{\theta}$$

and hence, by interpolation,

$$\begin{aligned}\|u-v\|_{C^1} &\le C_2 \left(\|u-v\|_{C^{r,\alpha}}^{\frac{1}{r+\alpha}} \|u-v\|_{C^0}^{\frac{r+\alpha-1}{r+\alpha}} \right)^{\theta} \\ &= C_2 \|u-v\|_{C^{r,\alpha}}^{\frac{1+\beta-\alpha}{r+\beta}} \|u-v\|_{C^0}^{\frac{(1+\beta-\alpha)(r+\alpha-1)}{r+\beta}} .\end{aligned}$$

Interpolating once more, we have

$$\|g\|_{C^{r,\alpha}} \le C_3 \|g\|_{C^{r+1,\beta}}^{\frac{r+\alpha-1}{r+\beta}} \|g\|_{C^1}^{\frac{1+\beta-\alpha}{r+\beta}} .$$

Combining the two estimates, we deduce

$$\begin{aligned}\|g\|_{C^{r,\alpha}} \|u-v\|_{C^1} &\le C_4 \left(\|g\|_{C^{r+1,\beta}} \|u-v\|_{C^0}^{1+\beta-\alpha} \right)^{\frac{r+\alpha-1}{r+\beta}} \left(\|g\|_{C^1} \|u-v\|_{C^{r,\alpha}} \right)^{\frac{1+\beta-\alpha}{r+\beta}} \\ &\le C_5 \|g\|_{C^{r+1,\beta}} \|u-v\|_{C^0}^{1+\beta-\alpha} + C_5 \|g\|_{C^1} \|u-v\|_{C^{r,\alpha}} ,\end{aligned}$$

as claimed.

Step 2. It easily follows from the observation

$$\begin{aligned}&\|\nabla(g\circ u) - \nabla(g\circ v)\|_{C^{r-1,\alpha}} = \|(\nabla g\circ u)\cdot \nabla u - (\nabla g\circ v)\cdot \nabla v\|_{C^{r-1,\alpha}} \\ &\quad \le C_1 \|\nabla g\circ u - \nabla g\circ v\|_{C^{r-1,\alpha}} \|\nabla u\|_{C^0} + C_1 \|\nabla g\circ u - \nabla g\circ v\|_{C^0} \|\nabla u\|_{C^{r-1,\alpha}} \\ &\quad\quad + C_1 \|\nabla g\circ v\|_{C^{r-1,\alpha}} \|\nabla u - \nabla v\|_{C^0} + C_1 \|\nabla g\circ v\|_{C^0} \|\nabla u - \nabla v\|_{C^{r-1,\alpha}} ,\end{aligned}$$

using the fact that $\|u\|_{C^1} \le c$, that

$$\begin{aligned}&\|\nabla(g\circ u) - \nabla(g\circ v)\|_{C^{r-1,\alpha}} \\ &\quad \le C \|\nabla g\circ u - \nabla g\circ v\|_{C^{r-1,\alpha}} + C \|\nabla g\circ u - \nabla g\circ v\|_{C^0} \|u\|_{C^{r,\alpha}} \\ &\quad\quad + C \|\nabla g\circ v\|_{C^{r-1,\alpha}} \|u-v\|_{C^1} + C \|g\|_{C^1} \|u-v\|_{C^{r,\alpha}} .\end{aligned}$$

Step 3. We now discuss the case $r = 1$.

Step 3.1. Let us first prove that

$$\|\nabla g\circ u - \nabla g\circ v\|_{C^0} \le C \|g\|_{C^{s,\beta}} \|u-v\|_{C^0}^{(s+\beta)-(r+\alpha)} .$$

We have to consider two cases.

Case 1: $s = r = 1$ and, therefore, $0 \leq \alpha < \beta \leq 1$. Observe that we immediately have the claim, since $\|u\|_{C^1}, \|v\|_{C^1} \leq c$ implies

$$\begin{aligned}\|\nabla g \circ u - \nabla g \circ v\|_{C^0} &\leq C_1 \|\nabla g\|_{C^{0,\beta}} \|u - v\|_{C^0}^{\beta} \leq C_2 \|g\|_{C^{1,\beta}} \|u - v\|_{C^0}^{\beta-\alpha} \\ &= C_2 \|g\|_{C^{s,\beta}} \|u - v\|_{C^0}^{(s+\beta)-(r+\alpha)}.\end{aligned}$$

Case 2: $s = r + 1 = 2$ and, therefore, $0 \leq \beta \leq \alpha \leq 1$. Since $\|u\|_{C^1}, \|v\|_{C^1} \leq c$, we get (recalling that the C^1 norm is equivalent to the $C^{0,1}$ norm, in view of Corollary 16.13)

$$\begin{aligned}\|\nabla g \circ u - \nabla g \circ v\|_{C^0} &\leq C_1 \|\nabla g\|_{C^1} \|u - v\|_{C^0} \leq C_1 \|g\|_{C^2} \|u - v\|_{C^0} \\ &\leq C_1 \|g\|_{C^{2,\beta}} \|u - v\|_{C^0}^{1+\beta-\alpha} \|u - v\|_{C^0}^{\alpha-\beta}.\end{aligned}$$

Using once more that $\|u\|_{C^1}, \|v\|_{C^1} \leq c$, we have indeed established Step 3.1.

Step 3.2. We are now in a position to conclude. We have (recall that $r = 1$), from Steps 1 and 2 and (16.37),

$$\begin{aligned}&\|\nabla (g \circ u) - \nabla (g \circ v)\|_{C^{r-1,\alpha}} \\ &\quad \leq C_1 \|\nabla g \circ u - \nabla g \circ v\|_{C^{r-1,\alpha}} + C_1 \|\nabla g \circ u - \nabla g \circ v\|_{C^0} \|u\|_{C^{r,\alpha}} \\ &\quad\quad + C_1 \|g\|_{C^{s,\beta}} \|u - v\|_{C^0}^{(s+\beta)-(r+\alpha)} + C_1 \|g\|_{C^1} \|u - v\|_{C^{r,\alpha}}\end{aligned}$$

and, therefore, from Step 3.1,

$$\begin{aligned}\|\nabla (g \circ u) - \nabla (g \circ v)\|_{C^{r-1,\alpha}} &\leq C \|\nabla g \circ u - \nabla g \circ v\|_{C^{r-1,\alpha}} + C \|g\|_{C^1} \|u - v\|_{C^{r,\alpha}} \\ &\quad + C \|g\|_{C^{s,\beta}} (1 + \|u\|_{C^{1,\alpha}}) \|u - v\|_{C^0}^{(s+\beta)-(r+\alpha)}.\end{aligned}$$

Since in the previous process we have privileged u with respect to v, the result is also valid with the interchange of u and v and thus we get the lemma for the case $r = 1$.

Step 4. We finally prove the result for $r \geq 2$.

Step 4.1. Let us first show that

$$\begin{aligned}\|g\|_{C^2} \|v\|_{C^{r-1,\alpha}} \|u - v\|_{C^1} &\leq C \|g\|_{C^2} \|v\|_{C^{r,\alpha}} \|u - v\|_{C^0} \\ &\quad + C \|g\|_{C^1} \|u - v\|_{C^{r,\alpha}} + C \|g\|_{C^{s,\beta}} \|u - v\|_{C^0}^{(s+\beta)-(r+\alpha)}.\end{aligned}$$

This is easily obtained by the usual interpolation argument. Indeed, recalling that $\|v\|_{C^1} \leq c$,

$$\begin{aligned}&\|v\|_{C^{r-1,\alpha}} \|u - v\|_{C^1} \\ &\quad \leq C_1 \left(\|v\|_{C^{r,\alpha}} \|u - v\|_{C^0}\right)^{\frac{r+\alpha-2}{r+\alpha-1}} \left(\|v\|_{C^1} \|u - v\|_{C^{r-1,\alpha}}\right)^{\frac{1}{r+\alpha-1}} \\ &\quad \leq C_2 \|v\|_{C^{r,\alpha}} \|u - v\|_{C^0} + C_2 \|u - v\|_{C^{r-1,\alpha}}.\end{aligned}$$

Moreover, from Corollary 16.27,

$$\|g\|_{C^2}\|u-v\|_{C^{r-1,\alpha}} \le C_3\|g\|_{C^{r,\alpha}}\|u-v\|_{C^1} + C_3\|g\|_{C^1}\|u-v\|_{C^{r,\alpha}},$$

which combined with Step 1 and the previous inequality leads to the desired estimate.

Step 4.2. Returning to Step 2 and appealing to (16.38), we get

$$\begin{aligned}
&\|\nabla(g\circ u)-\nabla(g\circ v)\|_{C^{r-1,\alpha}}\\
&\quad\le C_1\|\nabla g\circ u-\nabla g\circ v\|_{C^{r-1,\alpha}} + C_1\|\nabla g\circ u-\nabla g\circ v\|_{C^0}\|u\|_{C^{r,\alpha}}\\
&\qquad + C_1\left[\|g\|_{C^{r,\alpha}}+\|g\|_{C^2}\|v\|_{C^{r-1,\alpha}}\right]\|u-v\|_{C^1} + C_1\|g\|_{C^1}\|u-v\|_{C^{r,\alpha}}.
\end{aligned}$$

Since

$$\|\nabla g\circ u-\nabla g\circ v\|_{C^0} \le C_2\|\nabla g\|_{C^1}\|u-v\|_{C^0} \le C_2\|g\|_{C^2}\|u-v\|_{C^0},$$

we get, combining with Step 1 and Step 4.1,

$$\begin{aligned}
&\|\nabla(g\circ u)-\nabla(g\circ v)\|_{C^{r-1,\alpha}}\\
&\quad\le C\|\nabla g\circ u-\nabla g\circ v\|_{C^{r-1,\alpha}}\\
&\qquad + C\|g\|_{C^1}\|u-v\|_{C^{r,\alpha}} + C\|g\|_{C^{s,\beta}}\|u-v\|_{C^0}^{(s+\beta)-(r+\alpha)}\\
&\qquad + C\|g\|_{C^2}\left[\|u\|_{C^{r,\alpha}}+\|v\|_{C^{r,\alpha}}\right]\|u-v\|_{C^0},
\end{aligned}$$

which is exactly the claim of the lemma for $r\ge 2$. □

We now turn to our second result.

Theorem 16.39. *Let $\Omega\subset\mathbb{R}^n$ and $O\subset\mathbb{R}^m$ be bounded open Lipschitz sets. Let $r\ge 0$ be an integer and $0\le\alpha<\beta\le 1$. Let $c>0$, $g\in C^{r,\beta}(\overline{O})$ and $u,v\in C^{r,\alpha}(\overline{\Omega};\overline{O})\cap C^1(\overline{\Omega};\overline{O})$, with*

$$\|u\|_{C^1(\overline{\Omega})},\quad \|v\|_{C^1(\overline{\Omega})}\le c.$$

If $r=0$, then there exists $C=C(c,\Omega,O)>0$ so that

$$\|g\circ u-g\circ v\|_{C^{0,\alpha}(\overline{\Omega})} \le C\|g\|_{C^{0,\beta}(\overline{O})}\|u-v\|_{C^0(\overline{\Omega})}^{\beta-\alpha}.$$

If $r\ge 1$, then there exists $C=C(c,r,\Omega,O)>0$ so that

$$\begin{aligned}
&\|g\circ u-g\circ v\|_{C^{r,\alpha}(\overline{\Omega})}\\
&\quad\le C\|g\|_{C^{r,\beta}(\overline{O})}\left(1+\min\left\{\|u\|_{C^{1,\alpha}(\overline{\Omega})},\|v\|_{C^{1,\alpha}(\overline{\Omega})}\right\}\right)\|u-v\|_{C^0(\overline{\Omega})}^{\beta-\alpha}\\
&\qquad + C\|g\|_{C^r(\overline{O})}\left[\|u\|_{C^{r,\alpha}(\overline{\Omega})}+\|v\|_{C^{r,\alpha}(\overline{\Omega})}\right]\|u-v\|_{C^0(\overline{\Omega})}\\
&\qquad + C\|g\|_{C^1(\overline{O})}\|u-v\|_{C^{r,\alpha}(\overline{\Omega})}.
\end{aligned}$$

Proof. The proof is divided into two steps.

Step 1. We first prove the theorem when $r=0$. We get, by interpolation,

$$\begin{aligned}\|g\circ u-g\circ v\|_{C^{0,\alpha}} &\leq C_1\,\|g\circ u-g\circ v\|_{C^0}^{1-\frac{\alpha}{\beta}}\,\|g\circ u-g\circ v\|_{C^{0,\beta}}^{\frac{\alpha}{\beta}}\\ &\leq C_1\left[\|g\|_{C^{0,\beta}}\,\|u-v\|_{C^0}^{\beta}\right]^{1-\frac{\alpha}{\beta}}\left[\|g\circ u\|_{C^{0,\beta}}+\|g\circ v\|_{C^{0,\beta}}\right]^{\frac{\alpha}{\beta}}.\end{aligned}$$

Combining the above estimate with (16.37) and the fact that $\|u\|_{C^1},\|v\|_{C^1}\leq c$, we find

$$\begin{aligned}\|g\circ u-g\circ v\|_{C^{0,\alpha}} &\leq C\left[\|g\|_{C^{0,\beta}}\,\|u-v\|_{C^0}^{\beta}\right]^{1-\frac{\alpha}{\beta}}\|g\|_{C^{0,\beta}}^{\frac{\alpha}{\beta}}\\ &\leq C\,\|g\|_{C^{0,\beta}}\,\|u-v\|_{C^0}^{\beta-\alpha},\end{aligned}$$

as wished.

Step 2. We now deal with the case $r\geq 1$ and we will proceed by induction.

Step 2.1. Consider the case $r=1$. According to Lemma 16.37, we have

$$\begin{aligned}&\|\nabla(g\circ u)-\nabla(g\circ v)\|_{C^{0,\alpha}}\\ &\quad\leq C_1\,\|\nabla g\circ u-\nabla g\circ v\|_{C^{0,\alpha}}+C_1\,\|g\|_{C^1}\,\|u-v\|_{C^{1,\alpha}}\\ &\qquad+C_1\,\|g\|_{C^{1,\beta}}\left(1+\min\left\{\|u\|_{C^{1,\alpha}},\|v\|_{C^{1,\alpha}}\right\}\right)\|u-v\|_{C^0}^{\beta-\alpha}.\end{aligned}$$

Since

$$\|g\circ u-g\circ v\|_{C^0}\leq\|g\|_{C^1}\,\|u-v\|_{C^0}$$

and, appealing to Step 1,

$$\|\nabla g\circ u-\nabla g\circ v\|_{C^{0,\alpha}}\leq C_2\,\|g\|_{C^{1,\beta}}\,\|u-v\|_{C^0}^{\beta-\alpha},$$

we get, from the above three inequalities,

$$\begin{aligned}&\|g\circ u-g\circ v\|_{C^{1,\alpha}}\\ &\quad\leq C\,\|g\|_{C^{1,\beta}}\left(1+\min\left\{\|u\|_{C^{1,\alpha}},\|v\|_{C^{1,\alpha}}\right\}\right)\|u-v\|_{C^0}^{\beta-\alpha}+C\,\|g\|_{C^1}\,\|u-v\|_{C^{1,\alpha}}.\end{aligned}$$

The theorem for $r=1$ thus follows.

Step 2.2. We finally prove the claim by induction and start with assuming the result for $(r-1)$ and we prove it for $r\geq 2$. The hypothesis of induction gives

$$\begin{aligned}&\|\nabla g\circ u-\nabla g\circ v\|_{C^{r-1,\alpha}}\\ &\quad\leq C_1\,\|g\|_{C^{r,\beta}}\left(1+\min\left\{\|u\|_{C^{1,\alpha}},\|v\|_{C^{1,\alpha}}\right\}\right)\|u-v\|_{C^0}^{\beta-\alpha}\\ &\qquad+C_1\,\|g\|_{C^r}\left[\|u\|_{C^{r-1,\alpha}}+\|v\|_{C^{r-1,\alpha}}\right]\|u-v\|_{C^0}\\ &\qquad+C_1\,\|g\|_{C^2}\,\|u-v\|_{C^{r-1,\alpha}}.\end{aligned}$$

The only term in the above inequality that does not have the proper form is the last one, but we readily estimate it by Corollary 16.27:

$$\|g\|_{C^2}\|u-v\|_{C^{r-1,\alpha}} \leq C_2\|g\|_{C^1}\|u-v\|_{C^{r,\alpha}} + C_2\|g\|_{C^{r,\alpha}}\|u-v\|_{C^1}.$$

Finally, appealing to Step 1 of Lemma 16.37, we obtain

$$\|g\|_{C^2}\|u-v\|_{C^{r-1,\alpha}} \leq C_3\|g\|_{C^1}\|u-v\|_{C^{r,\alpha}} + C_3\|g\|_{C^{r,\beta}}\|u-v\|_{C^0}^{\beta-\alpha}.$$

We therefore have

$$\begin{aligned}\|\nabla g\circ u-\nabla g\circ v\|_{C^{r-1,\alpha}} &\leq C_4\|g\|_{C^{r,\beta}}\left(1+\min\left\{\|u\|_{C^{1,\alpha}},\|v\|_{C^{1,\alpha}}\right\}\right)\|u-v\|_{C^0}^{\beta-\alpha}\\ &\quad + C_4\|g\|_{C^r}\left[\|u\|_{C^{r-1,\alpha}}+\|v\|_{C^{r-1,\alpha}}\right]\|u-v\|_{C^0}\\ &\quad + C_4\|g\|_{C^1}\|u-v\|_{C^{r,\alpha}}. \end{aligned} \tag{16.39}$$

Next, observe that in view of Lemma 16.37, we have

$$\begin{aligned}\|\nabla(g\circ u)-\nabla(g\circ v)\|_{C^{r-1,\alpha}} &\leq C_5\|\nabla g\circ u-\nabla g\circ v\|_{C^{r-1,\alpha}}\\ &\quad + C_5\|g\|_{C^1}\|u-v\|_{C^{r,\alpha}} + C_5\|g\|_{C^{r,\beta}}\|u-v\|_{C^0}^{\beta-\alpha}\\ &\quad + C_5\|g\|_{C^2}\left[\|u\|_{C^{r,\alpha}}+\|v\|_{C^{r,\alpha}}\right]\|u-v\|_{C^0}. \end{aligned} \tag{16.40}$$

Since

$$\|g\circ u-g\circ v\|_{C^{r,\alpha}} = \|g\circ u-g\circ v\|_{C^0} + \|\nabla(g\circ u)-\nabla(g\circ v)\|_{C^{r-1,\alpha}},$$

we have, using (16.40), that

$$\begin{aligned}\|g\circ u-g\circ v\|_{C^{r,\alpha}} &\leq C_6\|\nabla g\circ u-\nabla g\circ v\|_{C^{r-1,\alpha}} + C_6\|g\|_{C^{r,\beta}}\|u-v\|_{C^0}^{\beta-\alpha}\\ &\quad + C_6\|g\|_{C^2}\left[\|u\|_{C^{r,\alpha}}+\|v\|_{C^{r,\alpha}}\right]\|u-v\|_{C^0} + C_6\|g\|_{C^1}\|u-v\|_{C^{r,\alpha}}. \end{aligned}$$

Combining the above inequality with (16.39), we have indeed established the theorem. □

Setting

$$u = \mathrm{id}+w \quad \text{and} \quad v = \mathrm{id}$$

in Theorem 16.39, we have the following corollary.

Corollary 16.40. *Let $\Omega \subset \mathbb{R}^n$ be a bounded open Lipschitz set. Let $r \geq 0$ be an integer and $0 \leq \alpha < \beta \leq 1$. Let $c > 0$, $g \in C^{r,\beta}(\overline{\Omega})$ and $w \in C^{r,\alpha}(\overline{\Omega};\overline{\Omega}) \cap C^1(\overline{\Omega};\overline{\Omega})$, with*

$$\|w\|_{C^1} \leq c.$$

Then there exists $C = C(c,r,\Omega) > 0$ so that

$$\|g\circ(\mathrm{id}+w)-g\circ\mathrm{id}\|_{C^{r,\alpha}} \leq C\|g\|_{C^{r,\beta}}\|w\|_{C^0}^{\beta-\alpha} + C\|g\|_{C^{r,\alpha}}\|w\|_{C^{r,\alpha}}.$$

16.7.3 A Third Result

We now discuss our third result.

Theorem 16.41. *Let $\Omega \subset \mathbb{R}^n$ and $O \subset \mathbb{R}^m$ be bounded open Lipschitz sets. Let $r \geq 0$ be an integer and $0 \leq \beta \leq \alpha \leq 1$. Let $c > 0$, $g \in C^{r+1,\beta}\left(\overline{O}\right)$ and $u, v \in C^{r,\alpha}\left(\overline{\Omega};\overline{O}\right) \cap C^1\left(\overline{\Omega};\overline{O}\right)$, with*

$$\|u\|_{C^1(\overline{\Omega})}\,, \quad \|v\|_{C^1(\overline{\Omega})} \leq c.$$

If $r = 0$, then there exists $C = C(c, \Omega, O) > 0$ so that

$$\|g \circ u - g \circ v\|_{C^{0,\alpha}(\overline{\Omega})} \leq C \|g\|_{C^{1,\beta}(\overline{O})} \|u - v\|_{C^1(\overline{\Omega})}^{1+\beta-\alpha}.$$

If $r \geq 1$, then there exists $C = C(c, r, \Omega, O) > 0$ so that

$$\begin{aligned}
&\|g \circ u - g \circ v\|_{C^{r,\alpha}(\overline{\Omega})} \\
&\leq C \|g\|_{C^{r+1,\beta}(\overline{O})} \left(1 + \min\left\{\|u\|_{C^{1,\alpha}(\overline{\Omega})}, \|v\|_{C^{1,\alpha}(\overline{\Omega})}\right\}\right) \|u - v\|_{C^1(\overline{\Omega})}^{1+\beta-\alpha} \\
&\quad + C \|g\|_{C^r(\overline{O})} \left[\|u\|_{C^{r,\alpha}(\overline{\Omega})} + \|v\|_{C^{r,\alpha}(\overline{\Omega})}\right] \|u - v\|_{C^0(\overline{\Omega})} \\
&\quad + C \|g\|_{C^1(\overline{O})} \|u - v\|_{C^{r,\alpha}(\overline{\Omega})}.
\end{aligned}$$

Proof. The proof will be divided into two steps.

Step 1. We prove the theorem for $r = 0$. We start by observing that from Theorem 16.36, we have

$$\begin{aligned}
\|g \circ u - g \circ v\|_{C^{0,\beta}} &\leq C_1 \|g\|_{C^{1,\beta}} \|u - v\|_{C^0} + C_1 \|g\|_{C^1} \|u - v\|_{C^{0,\beta}} \\
&\leq C_2 \|g\|_{C^{1,\beta}} \|u - v\|_{C^1}\,,
\end{aligned}$$

whereas from Theorem 16.39 and from the fact that $\|u\|_{C^1}, \|v\|_{C^1} \leq c$, we get

$$\begin{aligned}
\|g \circ u - g \circ v\|_{C^1} &\leq C_3 \|g\|_{C^{1,\beta}} \|u - v\|_{C^0}^{\beta} + C_3 \|g\|_{C^1} \|u - v\|_{C^1} \\
&\leq C_4 \|g\|_{C^{1,\beta}} \|u - v\|_{C^1}^{\beta}.
\end{aligned}$$

We now combine interpolation with the two above inequalities to get the claim; more precisely,

$$\begin{aligned}
\|g \circ u - g \circ v\|_{C^{0,\alpha}} &\leq C_5 \left(\|g \circ u - g \circ v\|_{C^{0,\beta}}\right)^{\frac{1-\alpha}{1-\beta}} \left(\|g \circ u - g \circ v\|_{C^1}\right)^{\frac{\alpha-\beta}{1-\beta}} \\
&\leq C_6 \|g\|_{C^{1,\beta}} \|u - v\|_{C^1}^{1+\beta-\alpha}.
\end{aligned}$$

Step 2. We then proceed by induction on r.

Step 2.1. When $r = 1$, we have, from Lemma 16.37 (and the remark following it) and Step 1, that

$$
\begin{aligned}
&\|\nabla(g\circ u)-\nabla(g\circ v)\|_{C^{0,\alpha}}\\
&\quad\le C\,\|g\|_{C^{2,\beta}}\,\|u-v\|_{C^1}^{1+\beta-\alpha}+C\,\|g\|_{C^1}\,\|u-v\|_{C^{1,\alpha}}\\
&\qquad+C\,\|g\|_{C^{2,\beta}}\left(1+\min\left\{\|u\|_{C^{1,\alpha}},\|v\|_{C^{1,\alpha}}\right\}\right)\|u-v\|_{C^0}^{1+\beta-\alpha}.
\end{aligned}
$$

The claim then follows at once.

Step 2.2. We now consider the case $r\ge 2$. Lemma 16.37 gives

$$
\begin{aligned}
\|\nabla(g\circ u)-\nabla(g\circ v)\|_{C^{r-1,\alpha}}\le{}& C_1\,\|\nabla g\circ u-\nabla g\circ v\|_{C^{r-1,\alpha}}\\
&+C_1\,\|g\|_{C^1}\,\|u-v\|_{C^{r,\alpha}}+C_1\,\|g\|_{C^{r+1,\beta}}\,\|u-v\|_{C^0}^{1+\beta-\alpha}\\
&+C_1\,\|g\|_{C^2}\left[\|u\|_{C^{r,\alpha}}+\|v\|_{C^{r,\alpha}}\right]\|u-v\|_{C^0}.
\end{aligned}
$$

From the hypothesis of induction, we obtain

$$
\begin{aligned}
&\|\nabla g\circ u-\nabla g\circ v\|_{C^{r-1,\alpha}}\\
&\quad\le C_2\,\|g\|_{C^{r+1,\beta}}\left(1+\min\left\{\|u\|_{C^{1,\alpha}},\|v\|_{C^{1,\alpha}}\right\}\right)\|u-v\|_{C^1}^{1+\beta-\alpha}\\
&\qquad+C_2\,\|g\|_{C^r}\left(\|u\|_{C^{r-1,\alpha}}+\|v\|_{C^{r-1,\alpha}}\right)\|u-v\|_{C^0}+C_2\,\|g\|_{C^2}\,\|u-v\|_{C^{r-1,\alpha}}.
\end{aligned}
$$

Combining the two inequalities, we get

$$
\begin{aligned}
\|g\circ u-g\circ v\|_{C^{r,\alpha}}\le{}& C_3\,\|g\|_{C^{r+1,\beta}}\left(1+\min\left\{\|u\|_{C^{1,\alpha}},\|v\|_{C^{1,\alpha}}\right\}\right)\|u-v\|_{C^1}^{1+\beta-\alpha}\\
&+C_3\,\|g\|_{C^r}\left(\|u\|_{C^{r,\alpha}}+\|v\|_{C^{r,\alpha}}\right)\|u-v\|_{C^0}+C_3\,\|g\|_{C^1}\,\|u-v\|_{C^{r,\alpha}}\\
&+C_3\,\|g\|_{C^2}\,\|u-v\|_{C^{r-1,\alpha}}.
\end{aligned}
$$

The only term in the above inequality that does not have the proper form is the last one. To write it appropriately, we combine Corollary 16.27, namely

$$
\|g\|_{C^2}\,\|u-v\|_{C^{r-1,\alpha}}\le C_4\,\|g\|_{C^1}\,\|u-v\|_{C^{r,\alpha}}+C_4\,\|g\|_{C^{r,\alpha}}\,\|u-v\|_{C^1},
$$

with Step 1 of Lemma 16.37, namely

$$
\|g\|_{C^{r,\alpha}}\,\|u-v\|_{C^1}\le C_5\,\|g\|_{C^{r+1,\beta}}\,\|u-v\|_{C^0}^{1+\beta-\alpha}+C_5\,\|g\|_{C^1}\,\|u-v\|_{C^{r,\alpha}},
$$

to obtain

$$
\|g\|_{C^2}\,\|u-v\|_{C^{r-1,\alpha}}\le C_6\,\|g\|_{C^1}\,\|u-v\|_{C^{r,\alpha}}+C_6\,\|g\|_{C^{r+1,\beta}}\,\|u-v\|_{C^0}^{1+\beta-\alpha}.
$$

This concludes the proof of the theorem. □

Setting

$$
u=\mathrm{id}+w\quad\text{and}\quad v=\mathrm{id}
$$

in Theorem 16.41, we have the following corollary.

Corollary 16.42. *Let $\Omega \subset \mathbb{R}^n$ be a bounded open Lipschitz set. Let $r \geq 0$ be an integer and $0 \leq \beta \leq \alpha \leq 1$. Let $c > 0$, $g \in C^{r+1,\beta}\left(\overline{\Omega}\right)$ and $w \in C^{r,\alpha}\left(\overline{\Omega};\overline{\Omega}\right) \cap C^1\left(\overline{\Omega};\overline{\Omega}\right)$, with*

$$\|w\|_{C^1} \leq c.$$

Then there exists $C = C(c, r, \Omega) > 0$ so that

$$\|g \circ (\mathrm{id} + w) - g \circ \mathrm{id}\|_{C^{r,\alpha}} \leq C \|g\|_{C^{r+1,\beta}} \|w\|_{C^1}^{1+\beta-\alpha} + C \|g\|_{C^{r,\alpha}} \|w\|_{C^{r,\alpha}}.$$

16.8 The Smoothing Operator

16.8.1 The Main Theorem

The result of this section is an essential tool in the Nash–Moser theorem, see Krantz and Parks [62]. We have also used it in a significant way in Theorems 10.1 and 14.5. Our presentation follows Hörmander [55].

Theorem 16.43. *Let $\Omega \subset \mathbb{R}^n$ be a bounded open Lipschitz set. Let $s \geq r \geq t \geq 0$ be integers and $0 \leq \alpha, \beta, \gamma \leq 1$ be such that*

$$t + \gamma \leq r + \alpha \leq s + \beta.$$

Let $f \in C^{r,\alpha}\left(\overline{\Omega}\right)$. Then for every $0 < \varepsilon \leq 1$, there exist a constant $C = C(s, \Omega) > 0$ and $f_\varepsilon \in C^\infty\left(\overline{\Omega}\right)$ such that

$$\|f_\varepsilon\|_{C^{s,\beta}} \leq \frac{C}{\varepsilon^{(s+\beta)-(r+\alpha)}} \|f\|_{C^{r,\alpha}},$$

$$\|f - f_\varepsilon\|_{C^{t,\gamma}} \leq C \varepsilon^{(r+\alpha)-(t+\gamma)} \|f\|_{C^{r,\alpha}},$$

$$\left\|\frac{d}{d\varepsilon} f_\varepsilon\right\|_{C^{s,\beta}} \leq \frac{C}{\varepsilon^{(s+\beta)-(r+\alpha)+1}} \|f\|_{C^{r,\alpha}},$$

$$\left\|\frac{d}{d\varepsilon} f_\varepsilon\right\|_{C^{t,\gamma}} \leq C \varepsilon^{(r+\alpha)-(t+\gamma)-1} \|f\|_{C^{r,\alpha}}.$$

Moreover, defining $F : (0,1] \times \overline{\Omega} \to \mathbb{R}$ by $F(\varepsilon, x) = f_\varepsilon(x)$,

$$F \in C^\infty((0,1] \times \overline{\Omega}; \Lambda^k).$$

Remark 16.44. (i) The theorem coupled with Corollary 16.13 gives that

$$\|f_\varepsilon\|_{C^{t,\gamma}} \leq C \|f\|_{C^{r,\alpha}}.$$

(ii) The construction of f_ε is universal, in the sense that the four inequalities remain true replacing $(r,s,t,\alpha,\beta,\gamma)$ by $(r',s',t',\alpha',\beta',\gamma')$ as far as $f \in C^{r',\alpha'}(\overline{\Omega})$ and with a constant $C' = C'(s',\Omega) > 0$.

Proof. Step 1. Choose any $\Phi \in C_0^\infty(\mathbb{R}^n)$ with $\Phi \equiv 1$ near the origin. Define next the smoothing kernel as

$$\varphi(x) = \int_{\mathbb{R}^n} e^{2\pi i \langle \xi; x \rangle} \Phi(\xi)\, d\xi.$$

Note that since φ is the inverse of the Fourier transform of Φ, we have $\varphi \in \mathscr{S}(\mathbb{R}^n)$, the Schwartz space. Therefore, for any a and b multi-indices, we can find $c = c(a,b)$ such that

$$|D^a \varphi(x)| \leq c(1+|x|)^{-|b|}.$$

Moreover, since $\Phi \equiv 1$ near the origin, we find, using the formula for the inverse Fourier transform,

$$\int_{\mathbb{R}^n} \varphi(x)\, dx = \Phi(0) = 1,$$

and for any multi-index a with $|a| \geq 1$ and any multi-index b with $|a| > |b|$, since $D^a\left(\xi^b \Phi(\xi)\right)\big|_{\xi=0} = 0$, we have

$$\int_{\mathbb{R}^n} x^a D^b \varphi(x)\, dx = 0. \tag{16.41}$$

The desired f_ε is then given by

$$f_\varepsilon = \varphi_\varepsilon * f,$$

where

$$\varphi_\varepsilon(x) = \frac{1}{\varepsilon^n} \varphi\left(\frac{x}{\varepsilon}\right).$$

Step 2. From now on, we assume, without loss of generality, that f has been extended to $\mathbb{R}^n$ as in Theorem 16.11. For any integer $r \geq 1$, we know from Proposition 16.10 or Corollary 16.13 that the norms $\|\cdot\|_{C^{r,0}}$ and $\|\cdot\|_{C^{r-1,1}}$ are equivalent. We therefore adopt in the present proof the following convention: If $r + \alpha = k \in \mathbb{N}$ with $0 \leq \alpha \leq 1$, then $r = k$ and $\alpha = 0$.

We then prove the first inequality. Observe that

$$\|f_\varepsilon\|_{C^{r,\alpha}} \leq \|\varphi\|_{L^1} \|f\|_{C^{r,\alpha}} \leq C_1 \|f\|_{C^{r,\alpha}} \tag{16.42}$$

and let us first prove the result when $(s+\beta) - (r+\alpha) = k \in \mathbb{N}$, which, with our convention, implies $\alpha = \beta$. Note that

$$\nabla^k f_\varepsilon = \frac{1}{\varepsilon^k} \psi_\varepsilon * f,$$

where

$$\psi_\varepsilon(x) = \frac{1}{\varepsilon^n} \nabla^k \varphi\left(\frac{x}{\varepsilon}\right).$$

We then deduce, as in (16.42), that

$$\left\|\nabla^k f_\varepsilon\right\|_{C^{r,\alpha}} \le \frac{C_2}{\varepsilon^k}\|f\|_{C^{r,\alpha}},$$

and thus, combining with (16.42) and the fact that $\varepsilon \le 1$, we get the claim, namely

$$\|f_\varepsilon\|_{C^{s,\beta}} = \|f_\varepsilon\|_{C^{r+k,\alpha}} \le \frac{C}{\varepsilon^{(s+\beta)-(r+\alpha)}}\|f\|_{C^{r,\alpha}}. \tag{16.43}$$

The case $(s+\beta)-(r+\alpha) \notin \mathbb{N}$, is obtained, from the above one, with the usual interpolation argument. More precisely, we let $k \in \mathbb{N}$ be such that

$$r+k+\alpha < s+\beta < r+k+1+\alpha.$$

We then let $\lambda \in [0,1]$ be such that

$$s+\beta = \lambda(r+k+1+\alpha)+(1-\lambda)(r+k+\alpha) = r+k+\alpha+\lambda$$

and we apply Theorem 16.26 and (16.43) to get

$$\|f_\varepsilon\|_{C^{s,\beta}} \le C_1\|f_\varepsilon\|^{\lambda}_{C^{r+k+1,\alpha}}\|f_\varepsilon\|^{1-\lambda}_{C^{r+k,\alpha}} \le \frac{C}{\varepsilon^{(s+\beta)-(r+\alpha)}}\|f\|_{C^{r,\alpha}}.$$

Step 3. We next establish the second inequality.

Step 3.1. We first prove it for $t+\gamma=0$, namely

$$\|f-f_\varepsilon\|_{C^0} \le C\varepsilon^{r+\alpha}\|f\|_{C^{r,\alpha}}. \tag{16.44}$$

We use the definition of the smoothing operator to get

$$\begin{aligned} f_\varepsilon(x)-f(x) &= \int_{\mathbb{R}^n}\frac{1}{\varepsilon^n}\varphi\left(\frac{y}{\varepsilon}\right)[f(x-y)-f(x)]\,dy \\ &= \int_{\mathbb{R}^n}\varphi(z)[f(x-\varepsilon z)-f(x)]\,dz. \end{aligned}$$

Note that from the above identity, we immediately have (16.44) when $r=0$. So from now on, we assume that $r \ge 1$. We next set

$$F(u) = f(x-u\varepsilon z), \quad u \in [0,1],$$

so that

$$\begin{aligned} f(x-\varepsilon z)-f(x) &= F(1)-F(0) \\ &= \sum_{k=1}^{r}\frac{F^{(k)}(0)}{k!}+\int_0^1\frac{(1-u)^{r-1}}{(r-1)!}\left[F^{(r)}(u)-F^{(r)}(0)\right]du. \end{aligned}$$

Combining the two identities with (16.41), we deduce that

$$f_\varepsilon(x) - f(x) = \int_0^1 \int_{\mathbb{R}^n} \frac{(1-u)^{r-1}}{(r-1)!} \varphi(z) \left[F^{(r)}(u) - F^{(r)}(0) \right] dz\, du.$$

We have therefore proved (16.44), since

$$\|f - f_\varepsilon\|_{C^0} \le C\varepsilon^{r+\alpha} \|f\|_{C^{r,\alpha}}.$$

Step 3.2. We now discuss the general case $t+\gamma \ge 0$. From Theorem 16.26, we have

$$\|f - f_\varepsilon\|_{C^{t,\gamma}} \le C_1 \|f - f_\varepsilon\|_{C^{r,\alpha}}^{\frac{t+\gamma}{r+\alpha}} \|f - f_\varepsilon\|_{C^0}^{1-\frac{t+\gamma}{r+\alpha}},$$

and since

$$\|f - f_\varepsilon\|_{C^{r,\alpha}} \le C_2 \|f\|_{C^{r,\alpha}}$$

we have, appealing to (16.44), the claim

$$\|f - f_\varepsilon\|_{C^{t,\gamma}} \le C\varepsilon^{(r+\alpha)-(t+\gamma)} \|f\|_{C^{r,\alpha}}.$$

Step 4. We now prove the third inequality. We start by noting that

$$\begin{aligned} \frac{d}{d\varepsilon} f_\varepsilon &= \frac{-n}{\varepsilon^{n+1}} \int_{\mathbb{R}^n} \varphi\left(\frac{x-y}{\varepsilon}\right) f(y)\, dy \\ &\quad - \frac{1}{\varepsilon^{n+1}} \int_{\mathbb{R}^n} \left\langle \frac{x-y}{\varepsilon}; \nabla\varphi\left(\frac{x-y}{\varepsilon}\right)\right\rangle f(y)\, dy. \end{aligned}$$

Writing

$$\psi(z) = \langle z; \nabla\varphi(z)\rangle \quad \text{and} \quad \psi_\varepsilon(z) = \frac{1}{\varepsilon^n}\psi\left(\frac{z}{\varepsilon}\right),$$

we deduce that

$$\frac{d}{d\varepsilon} f_\varepsilon = \frac{-n}{\varepsilon}\varphi_\varepsilon * f - \frac{1}{\varepsilon}\psi_\varepsilon * f = \frac{-1}{\varepsilon}(n\varphi_\varepsilon + \psi_\varepsilon) * f.$$

Observe that the kernel

$$\chi = n\varphi + \psi$$

has essentially the same properties as the kernel φ. The third inequality therefore follows by the same argument as the one of Step 2.

Step 5. The fourth inequality follows as in Steps 3 and 4. Indeed, write

$$f_\varepsilon(x) - f(x) = \int_{\mathbb{R}^n} \frac{1}{\varepsilon^n}\varphi\left(\frac{y}{\varepsilon}\right) [f(x-y) - f(x)]\, dy$$

and then, with the same notations as in Step 4, we obtain

$$\begin{aligned}\frac{d}{d\varepsilon}f_\varepsilon &= \frac{-1}{\varepsilon}\int_{\mathbb{R}^n}\frac{1}{\varepsilon^n}\chi\left(\frac{y}{\varepsilon}\right)[f(x-y)-f(x)]\,dy\\ &= \frac{-1}{\varepsilon}\int_{\mathbb{R}^n}\chi(z)\,[f(x-\varepsilon z)-f(x)]\,dz.\end{aligned}$$

Since χ has essentially the same properties as φ (in particular, it satisfies (16.41)), the argument of Step 3 then applies and we get the last inequality of the theorem.

Step 6. The last statement follows from the fact that the regularization is obtained through convolution. □

In the same spirit, we have the following elementary result.

Proposition 16.45. *Let $r \geq 0$ be an integer, $0 \leq \alpha < \beta \leq 1$, $\varepsilon_0 > 0$, $C > 0$ and $\Omega \subset \mathbb{R}^n$ be a bounded open set. Let $f \in C^{r,\alpha}(\overline{\Omega})$ and, for every $\varepsilon \in (0,\varepsilon_0]$, $f_\varepsilon \in C^{r,\beta}(\overline{\Omega})$ verifying*

$$\|f_\varepsilon\|_{C^{r,\beta}} \leq \frac{C}{\varepsilon^{\beta-\alpha}} \quad \text{and} \quad \|f_\varepsilon - f\|_{C^{r,\alpha}} \leq C\varepsilon^{\beta-\alpha}. \tag{16.45}$$

Then $f \in C^{r,(\alpha+\beta)/2}(\overline{\Omega})$ and

$$\|f\|_{C^{r,(\alpha+\beta)/2}} \leq \|f\|_{C^r} + 2C + \frac{2\|f\|_{C^r}}{\varepsilon_0^{\alpha+\beta}}.$$

Proof. First, notice that we trivially have, for every $x,y \in \overline{\Omega}$ with $|x-y| \geq \varepsilon_0^2$, that

$$|\nabla^r f(x) - \nabla^r f(y)| \leq 2\|f\|_{C^r} \leq \frac{2\|f\|_{C^r}}{\varepsilon_0^{\alpha+\beta}}|x-y|^{(\alpha+\beta)/2}.$$

Let $x,y \in \overline{\Omega}$ with $|x-y| \in (0,\varepsilon_0^2)$ and define $\varepsilon_1 = |x-y|^{1/2}$. Noticing that $\varepsilon_1 \in (0,\varepsilon_0)$ and using (16.45), we deduce

$$\begin{aligned}&|\nabla^r f(x) - \nabla^r f(y)|\\ &\leq |\nabla^r f_{\varepsilon_1}(x) - \nabla^r f_{\varepsilon_1}(y)| + |\nabla^r f(x) - \nabla^r f_{\varepsilon_1}(x) + \nabla^r f_{\varepsilon_1}(y) - \nabla^r f(y)|\\ &\leq \frac{C}{\varepsilon_1^{\beta-\alpha}}|x-y|^\beta + C\varepsilon_1^{\beta-\alpha}|x-y|^\alpha = 2C|x-y|^{(\alpha+\beta)/2},\end{aligned}$$

which concludes the proof. □

16.8.2 A First Application

We now give a direct consequence of the above theorem. It has been used in Chapter 10.

Proposition 16.46. *Let $\Omega \subset \mathbb{R}^n$ be a bounded open Lipschitz set, $r \geq 0$ be an integer and $0 \leq \alpha \leq 1$. Let $f \in C^{r,\alpha}\left(\overline{\Omega}\right)$ and $c > 0$ be such that*

$$\left\|\frac{1}{f}\right\|_{C^0}, \quad \|f\|_{C^{0,\alpha}} \leq c.$$

Then for every $\varepsilon > 0$ small, there exists $f_\varepsilon \in C^\infty\left(\overline{\Omega}\right)$ with

$$\int_\Omega \frac{f}{f_\varepsilon} = \operatorname{meas}\Omega.$$

Moreover, there exists a constant $C = C(c, r, \Omega) > 0$ such that for every integer $0 \leq t \leq r$ and every $0 \leq \gamma \leq \alpha$,

$$\|f_\varepsilon\|_{C^{t,\gamma}} \leq C\|f\|_{C^{t,\gamma}}, \quad \|f_\varepsilon\|_{C^{t+1,\gamma}} \leq \frac{C}{\varepsilon}\|f\|_{C^{t,\gamma}},$$

$$\|f_\varepsilon - 1\|_{C^{t,\gamma}} \leq C\|f-1\|_{C^{t,\gamma}}, \quad \|f_\varepsilon - 1\|_{C^{t+1,\gamma}} \leq \frac{C}{\varepsilon}\|f-1\|_{C^{t,\gamma}},$$

$$\left\|\frac{f}{f_\varepsilon} - 1\right\|_{C^{t,\gamma}} \leq C\|f-1\|_{C^{t,\gamma}}, \quad \left\|\frac{f}{f_\varepsilon} - 1\right\|_{C^{0,\gamma}} \leq C\varepsilon^{\alpha-\gamma}.$$

Proof. We first find, by Theorem 16.43, a constant $C_1 = C_1(r, \Omega) > 0$ and $g_\varepsilon \in C^\infty\left(\overline{\Omega}\right)$ such that

$$\|g_\varepsilon\|_{C^{t,\gamma}} \leq C_1\|f\|_{C^{t,\gamma}} \quad \text{and} \quad \|g_\varepsilon\|_{C^{t+1,\gamma}} \leq \frac{C_1}{\varepsilon}\|f\|_{C^{t,\gamma}},$$

$$\|g_\varepsilon - f\|_{C^{t,\gamma}} \leq C_1\varepsilon^{(r+\alpha)-(t+\gamma)}\|f\|_{C^{r,\alpha}} \quad \text{and} \quad \|g_\varepsilon - f\|_{C^0} \leq C_1\varepsilon^\alpha\|f\|_{C^{0,\alpha}}.$$

Since in the construction of g_ε in Step 1 of Theorem 16.43 we defined

$$g_\varepsilon = \varphi_\varepsilon * f$$

and since $\varphi_\varepsilon * 1 = 1$, we also have

$$\|g_\varepsilon - 1\|_{C^{r,\alpha}} = \|\varphi_\varepsilon * (f-1)\|_{C^{r,\alpha}} \leq \frac{C_1}{\varepsilon^{(r+\alpha)-(t+\gamma)}}\|f-1\|_{C^{t,\gamma}},$$

$$\|g_\varepsilon - f\|_{C^{t,\gamma}} = \|\varphi_\varepsilon * (f-1) - (f-1)\|_{C^{t,\gamma}} \leq C_1\varepsilon^{(r+\alpha)-(t+\gamma)}\|f-1\|_{C^{r,\alpha}}.$$

Set, for ε small enough,

$$\lambda_\varepsilon = \frac{1}{\operatorname{meas}\Omega}\int_\Omega \frac{f}{g_\varepsilon} \quad \text{and} \quad f_\varepsilon = \lambda_\varepsilon g_\varepsilon.$$

We claim that f_ε verifies all of the conclusions of the proposition. This will be checked in Step 2, but before that, we need an estimate on λ_ε.

Step 1 (estimate on λ_ε). First, note that since

$$\|g_\varepsilon - f\|_{C^0} \le C_1 \varepsilon^\alpha \|f\|_{C^{0,\alpha}},$$

we have, for ε small enough,

$$\left\|\frac{1}{g_\varepsilon}\right\|_{C^0} \le 2c.$$

Thus, we immediately deduce that

$$|\lambda_\varepsilon - 1| \le 2c\|g_\varepsilon - f\|_{C^0},$$

which implies

$$|\lambda_\varepsilon - 1| \le C_2\|f-1\|_{C^0} \quad \text{and} \quad |\lambda_\varepsilon - 1| \le C_2 \varepsilon^\alpha \|f-1\|_{C^{0,\alpha}}.$$

Step 2. Let us check all of the properties. We assume that ε is small enough.

(i) We clearly have

$$\int_\Omega \frac{f}{f_\varepsilon} = \operatorname{meas}\Omega.$$

(ii) We find

$$\|f_\varepsilon\|_{C^{t,\gamma}} = \lambda_\varepsilon \|g_\varepsilon\|_{C^{t,\gamma}} \le C\|f\|_{C^{t,\gamma}}.$$

(iii) Observe that

$$\|f_\varepsilon\|_{C^{t+1,\gamma}} = \lambda_\varepsilon \|g_\varepsilon\|_{C^{t+1,\gamma}} \le \frac{C}{\varepsilon}\|f\|_{C^{t,\gamma}}.$$

(iv) Note that

$$\begin{aligned}
\|f_\varepsilon - 1\|_{C^{t,\gamma}} &= \|f_\varepsilon - g_\varepsilon + g_\varepsilon - 1\|_{C^{t,\gamma}} \le |\lambda_\varepsilon - 1|\,\|g_\varepsilon\|_{C^{t,\gamma}} + \|g_\varepsilon - 1\|_{C^{t,\gamma}} \\
&\le C_1C_2\|f-1\|_{C^0}\|f\|_{C^{t,\gamma}} + C_1\|f-1\|_{C^{t,\gamma}} \\
&\le C_1C_2\|f-1\|_{C^0}[1+\|f-1\|_{C^{t,\gamma}}] + C_1\|f-1\|_{C^{t,\gamma}}
\end{aligned}$$

and hence

$$\|f_\varepsilon - 1\|_{C^{t,\gamma}} \le C\|f-1\|_{C^{t,\gamma}}.$$

(v) Similarly, we have

$$\begin{aligned}
\|f_\varepsilon - 1\|_{C^{t+1,\gamma}} &= \|f_\varepsilon - g_\varepsilon + g_\varepsilon - 1\|_{C^{t+1,\gamma}} \\
&\le |\lambda_\varepsilon - 1|\,\|g_\varepsilon\|_{C^{t+1,\gamma}} + \|g_\varepsilon - 1\|_{C^{t+1,\gamma}} \\
&\le \frac{C_1}{\varepsilon}C_2\|f-1\|_{C^0}\|f\|_{C^{t,\gamma}} + \frac{C_1}{\varepsilon}\|f-1\|_{C^{t,\gamma}} \\
&\le \frac{C_1}{\varepsilon}C_2\|f-1\|_{C^0}[1+\|f-1\|_{C^{t,\gamma}}] + \frac{C_1}{\varepsilon}\|f-1\|_{C^{t,\gamma}}
\end{aligned}$$

and hence

$$\|f_\varepsilon - 1\|_{C^{t+1,\gamma}} \le \frac{C}{\varepsilon}\,\|f-1\|_{C^{t,\gamma}}.$$

(vi) We next obtain

$$\begin{aligned}\left\|\frac{f}{f_\varepsilon}-1\right\|_{C^{t,\gamma}} &= \left\|\frac{f-f_\varepsilon}{f_\varepsilon}\right\|_{C^{t,\gamma}} \\ &\le C_3\left\|\frac{1}{f_\varepsilon}\right\|_{C^{t,\gamma}}\|f-f_\varepsilon\|_{C^0}+C_3\left\|\frac{1}{f_\varepsilon}\right\|_{C^0}\|f-f_\varepsilon\|_{C^{t,\gamma}} \\ &\le C_4\|f_\varepsilon\|_{C^{t,\gamma}}\|f-f_\varepsilon\|_{C^0}+C_4\|f-1\|_{C^{t,\gamma}}.\end{aligned}$$

Since, using (iv),

$$\begin{aligned}\|f_\varepsilon\|_{C^{t,\gamma}}\|f-f_\varepsilon\|_{C^0} &\le C_5(\|f_\varepsilon-1\|_{C^{t,\gamma}}+1)(\|f-1\|_{C^0}+\|f_\varepsilon-1\|_{C^0}) \\ &\le C_6\|f-1\|_{C^{t,\gamma}},\end{aligned}$$

we get

$$\left\|\frac{f}{f_\varepsilon}-1\right\|_{C^{t,\gamma}} \le C\|f-1\|_{C^{t,\gamma}}.$$

(vii) Finally, note that

$$\begin{aligned}\|f_\varepsilon - f\|_{C^{0,\gamma}} = \|f_\varepsilon - g_\varepsilon + g_\varepsilon - f\|_{C^{0,\gamma}} &\le |\lambda_\varepsilon - 1|\,\|g_\varepsilon\|_{C^{0,\gamma}} + \|g_\varepsilon - f\|_{C^{0,\gamma}} \\ &\le C_1C_2\varepsilon^\alpha\,\|f\|_{C^{0,\gamma}} + C_1\varepsilon^{\alpha-\gamma}\|f\|_{C^{0,\alpha}} \le C_7\varepsilon^{\alpha-\gamma}\|f\|_{C^{0,\alpha}}\end{aligned}$$

and thus

$$\left\|\frac{f}{f_\varepsilon}-1\right\|_{C^{0,\gamma}} \le C\varepsilon^{\alpha-\gamma}.$$

This concludes the proof of the proposition. □

16.8.3 A Second Application

In Chapter 10 we used the following proposition.

Proposition 16.47. *Let $\Omega \subset \mathbb{R}^n$ be a bounded open Lipschitz set. Let $s \ge r \ge t \ge 0$ be integers. Let $0 \le \alpha, \beta, \gamma \le 1$ be such that*

$$t+\gamma \le r+\alpha \le s+\beta.$$

Let $f \in C^{r,\alpha}\left(\overline{\Omega}\right)$ with $f>0$ and $c>0$ be such that

$$\left\|\frac{1}{f}\right\|_{C^0},\quad \|f\|_{C^0} \le c.$$

Then for every $\varepsilon > 0$ small, there exist $f_\varepsilon \in C^\infty\left(\overline{\Omega}\right)$ and a constant $C = C(c,s,\Omega)$ such that

$$\int_\Omega f_\varepsilon = \int_\Omega f,$$

$$\|f_\varepsilon\|_{C^{s,\beta}} \leq \frac{C}{\varepsilon^{(s+\beta)-(r+\alpha)}}\|f\|_{C^{r,\alpha}},$$

$$\|f_\varepsilon - f\|_{C^{t,\gamma}} \leq C\varepsilon^{(r+\alpha)-(t+\gamma)}[\|f\|_{C^{r,\alpha}} + \|f\|^2_{C^{r,\alpha}}],$$

$$\left\|\frac{d}{d\varepsilon}f_\varepsilon\right\|_{C^{s,\beta}} \leq \frac{C}{\varepsilon^{(s+\beta)-(r+\alpha)+1}}[\|f\|_{C^{r,\alpha}} + \|f\|^2_{C^{r,\alpha}}],$$

$$\left\|\frac{d}{d\varepsilon}f_\varepsilon\right\|_{C^{t,\gamma}} \leq C\varepsilon^{(r+\alpha)-(t+\gamma)-1}[\|f\|_{C^{r,\alpha}} + \|f\|^2_{C^{r,\alpha}}].$$

Moreover, defining, for some $\overline{\varepsilon}$ small enough, $F : (0,\overline{\varepsilon}] \times \overline{\Omega} \to \mathbb{R}$ by $F(\varepsilon,x) = f_\varepsilon(x)$, we have $F \in C^\infty((0,\overline{\varepsilon}] \times \overline{\Omega})$.

Remark 16.48. The construction is universal in the sense of Remark 16.44.

Proof. We split the proof into three steps.

Step 1. Applying Theorem 16.43 to f, we get a constant $C_1 = C_1(s,\Omega)$ and $h_\varepsilon \in C^\infty(\overline{\Omega})$, $\varepsilon \in (0,1]$, such that

$$\|h_\varepsilon\|_{C^{s,\beta}} \leq \frac{C_1}{\varepsilon^{(s+\beta)-(r+\alpha)}}\|f\|_{C^{r,\alpha}}, \tag{16.46}$$

$$\|h_\varepsilon - f\|_{C^{t,\gamma}} \leq C_1\varepsilon^{(r+\alpha)-(t+\gamma)}\|f\|_{C^{r,\alpha}}, \tag{16.47}$$

$$\left\|\frac{d}{d\varepsilon}h_\varepsilon\right\|_{C^{s,\beta}} \leq \frac{C_1}{\varepsilon^{(s+\beta)-(r+\alpha)+1}}\|f\|_{C^{r,\alpha}}, \tag{16.48}$$

$$\left\|\frac{d}{d\varepsilon}h_\varepsilon\right\|_{C^{t,\gamma}} \leq C_1\varepsilon^{(r+\alpha)-(t+\gamma)-1}\|f\|_{C^{r,\alpha}}. \tag{16.49}$$

Moreover, defining $H : (0,1] \times \overline{\Omega} \to \mathbb{R}$ by $H(\varepsilon,x) = h_\varepsilon(x)$, we have

$$H \in C^\infty((0,1] \times \overline{\Omega}).$$

Using (16.47), there exists $\overline{\varepsilon} \leq 1$ such that for every $\varepsilon \in (0,\overline{\varepsilon}]$,

$$\|h_\varepsilon\|_{C^0} \leq 2\|f\|_{C^0} \quad \text{and} \quad \inf_\Omega h_\varepsilon \geq \frac{\inf_\Omega f}{2}. \tag{16.50}$$

From now on, $C_2, C_3, \ldots$ will be generic constants depending only on c, s and Ω.

Step 2. For every $\varepsilon \in (0,\overline{\varepsilon}]$, define

$$\lambda(\varepsilon) = \frac{\int_\Omega f}{\int_\Omega h_\varepsilon}.$$

We trivially have $\lambda \in C^\infty((0,\overline{\varepsilon}])$. Moreover, using Step 1, we obtain the following properties:

(i) Using (16.50), we obtain

$$0 < \lambda(\varepsilon) \le \frac{2\|f\|_{C^0}}{\inf_\Omega f}. \tag{16.51}$$

(ii) Appealing to (16.47) and (16.50), we get

$$|1-\lambda(\varepsilon)| = \left| \frac{\int_\Omega (h_\varepsilon - f)}{\int_\Omega h_\varepsilon} \right| \le \frac{2\,\mathrm{meas}\,\Omega \, \|h_\varepsilon - f\|_{C^0}}{\mathrm{meas}\,\Omega \cdot \inf_\Omega f} \le C_2 \varepsilon^{r+\alpha} \|f\|_{C^{r,\alpha}}. \tag{16.52}$$

(iii) Since (16.49) and (16.50) hold, we find

$$\begin{aligned} |\lambda'(\varepsilon)| &= \left| \int_\Omega f \right| \left| \int_\Omega \frac{d}{d\varepsilon} h_\varepsilon \right| \left| \left(\int_\Omega h_\varepsilon \right)^{-2} \right| \\ &\le \|f\|_{C^0}\, \mathrm{meas}\,\Omega \left\| \frac{d}{d\varepsilon} h_\varepsilon \right\|_{C^0} \frac{4\,\mathrm{meas}\,\Omega}{(\mathrm{meas}\,\Omega)^2 (\inf_\Omega f)^2} \\ &\le C_3 \varepsilon^{r+\alpha-1} \|f\|_{C^{r,\alpha}}. \end{aligned} \tag{16.53}$$

Step 3. Let us show that

$$f_\varepsilon = \lambda(\varepsilon) h_\varepsilon, \quad \varepsilon \in (0,\overline{\varepsilon}],$$

has all of the required properties. First, we obviously have

$$\int_\Omega f_\varepsilon = \int_\Omega f.$$

We now show the inequalities.

(i) Appealing to (16.46) and (16.51), we have

$$\|f_\varepsilon\|_{C^{s,\beta}} \le \frac{C_4}{\varepsilon^{(s+\beta)-(r+\alpha)}} \|f\|_{C^{r,\alpha}},$$

which shows the first inequality.

(ii) Using (16.47), (16.51) and (16.52) and recalling that $\varepsilon \le 1$, we get

$$\begin{aligned} \|f_\varepsilon - f\|_{C^{t,\gamma}} &\le \lambda(\varepsilon) \|h_\varepsilon - f\|_{C^{t,\gamma}} + |1-\lambda(\varepsilon)| \|f\|_{C^{t,\gamma}} \\ &\le C_5 \varepsilon^{(r+\alpha)-(t+\gamma)} \|f\|_{C^{r,\alpha}} + C_5 \varepsilon^{r+\alpha} \|f\|^2_{C^{r,\alpha}} \\ &\le C_5 \varepsilon^{(r+\alpha)-(t+\gamma)} [\|f\|_{C^{r,\alpha}} + \|f\|^2_{C^{r,\alpha}}]. \end{aligned}$$

Therefore, the second inequality is shown.

(iii) Since (16.46), (16.48), (16.51) and (16.53) hold and recalling that $\varepsilon \leq 1$, we have

$$\left\|\frac{d}{d\varepsilon} f_\varepsilon\right\|_{C^{s,\beta}} \leq \lambda(\varepsilon)\left\|\frac{d}{d\varepsilon} h_\varepsilon\right\|_{C^{s,\beta}} + |\lambda'(\varepsilon)| \|h_\varepsilon\|_{C^{s,\beta}}$$
$$\leq \frac{C_6}{\varepsilon^{(s+\beta)-(r+\alpha)+1}} \|f\|_{C^{r,\alpha}} + \frac{C_6 \varepsilon^{r+\alpha-1}}{\varepsilon^{(s+\beta)-(r+\alpha)}} \|f\|^2_{C^{r,\alpha}}$$
$$\leq \frac{C_6}{\varepsilon^{(s+\beta)-(r+\alpha)+1}} [\|f\|_{C^{r,\alpha}} + \|f\|^2_{C^{r,\alpha}}],$$

which proves the third inequality.

(iv) Using (16.46), (16.49), (16.51) and (16.53) and recalling that $\varepsilon \leq 1$, we obtain

$$\left\|\frac{d}{d\varepsilon} f_\varepsilon\right\|_{C^{t,\gamma}} \leq \lambda(\varepsilon)\left\|\frac{d}{d\varepsilon} h_\varepsilon\right\|_{C^{t,\gamma}} + |\lambda'(\varepsilon)| \|h_\varepsilon\|_{C^{t,\gamma}}$$
$$\leq C_7\, \varepsilon^{(r+\alpha)-(t+\gamma)-1} \|f\|_{C^{r,\alpha}} + C_7 \varepsilon^{r+\alpha-1} \|f\|_{C^{r,\alpha}} \|f\|_{C^{t,\gamma}}$$
$$\leq C_8\, \varepsilon^{(r+\alpha)-(t+\gamma)-1} [\|f\|_{C^{r,\alpha}} + \|f\|^2_{C^{r,\alpha}}],$$

which establishes the last inequality.

(v) Finally, we have that $F : (0, \overline{\varepsilon}] \times \overline{\Omega} \to \mathbb{R}$ defined by $F(\varepsilon, x) = f_\varepsilon(x)$ verifies $F \in C^\infty((0, \overline{\varepsilon}] \times \overline{\Omega})$, which concludes the proof. □

16.9 Smoothing Operator for Differential Forms

The results presented here are in Bandyopadhyay and Dacorogna [8] and Dacorogna and Kneuss [32] (cf. also [7]). We will use the following functional notations. Let $\Omega \subset \mathbb{R}^n$ be an open smooth set, $r \geq 0$ an integer and $0 \leq \alpha \leq 1$.

(i) We denote by $C^{r,\alpha}\left(\overline{\Omega}; \Lambda^k\right)$ the set of k-forms

$$g = \sum_{1 \leq i_1 < \cdots < i_k \leq n} g_{i_1 \cdots i_k} dx^{i_1} \wedge \cdots \wedge dx^{i_k},$$

so that $g_{i_1 \cdots i_k} \in C^{r,\alpha}\left(\overline{\Omega}\right)$.

(ii) For $x \in \partial\Omega$, we denote by $\nu = \nu(x)$ the exterior unit normal to Ω. By

$$\nu \wedge g \in C^{r,\alpha}\left(\partial\Omega; \Lambda^{k+1}\right)$$

we mean that the tangential part of g is in $C^{r,\alpha}$; more precisely, the $(k+1)$-form Φ defined by

$$\Phi(x) = \nu(x) \wedge g(x)$$

is such that

$$\Phi \in C^{r,\alpha}\left(\partial\Omega;\Lambda^{k+1}\right).$$

We now approximate closed forms in $C^{r,\alpha}\left(\overline{\Omega};\Lambda^k\right)$ by smooth closed forms in a precise way.

Theorem 16.49. *Let $\Omega \subset \mathbb{R}^n$ be a bounded open smooth set and ν be the exterior unit normal. Let $s \geq r \geq t \geq 0$ with $s \geq 1$ and $1 \leq k \leq n-1$ be integers. Let $0 < \alpha, \beta, \gamma < 1$ be such that*

$$t + \gamma \leq r + \alpha \leq s + \beta.$$

Let $g \in C^{r,\alpha}\left(\overline{\Omega};\Lambda^k\right)$ with

$$dg = 0 \text{ in } \Omega \quad \text{and} \quad \nu \wedge g \in C^{s,\beta}\left(\partial\Omega;\Lambda^{k+1}\right).$$

Then for every $\varepsilon \in (0,1]$, there exist $g_\varepsilon \in C^\infty\left(\Omega;\Lambda^k\right) \cap C^{s,\beta}\left(\overline{\Omega};\Lambda^k\right)$ and a constant $C = C(s,\alpha,\beta,\gamma,\Omega) > 0$ such that

$$dg_\varepsilon = 0 \text{ in } \Omega, \quad \nu \wedge g_\varepsilon = \nu \wedge g \text{ on } \partial\Omega,$$

$$\int_\Omega \langle g_\varepsilon;\psi\rangle = \int_\Omega \langle g;\psi\rangle \quad \text{for every } \psi \in \mathscr{H}_T\left(\Omega;\Lambda^k\right),$$

$$\|g_\varepsilon - g\|_{C^{t,\gamma}(\overline{\Omega})} \leq C\varepsilon^{(r+\alpha)-(t+\gamma)}\|g\|_{C^{r,\alpha}(\overline{\Omega})},$$

$$\|g_\varepsilon\|_{C^{s,\beta}(\overline{\Omega})} \leq \frac{C}{\varepsilon^{(s+\beta)-(r+\alpha)}}\|g\|_{C^{r,\alpha}(\overline{\Omega})} + C\|\nu \wedge g\|_{C^{s,\beta}(\partial\Omega)},$$

$$\left\|\frac{d}{d\varepsilon}g_\varepsilon\right\|_{C^{s,\beta}(\overline{\Omega})} \leq \frac{C}{\varepsilon^{(s+\beta)-(r+\alpha)+1}}\|g\|_{C^{r,\alpha}(\overline{\Omega})},$$

$$\left\|\frac{d}{d\varepsilon}g_\varepsilon\right\|_{C^{t,\gamma}(\overline{\Omega})} \leq C\varepsilon^{(r+\alpha)-(t+\gamma)-1}\|g\|_{C^{r,\alpha}(\overline{\Omega})}.$$

Moreover, defining $\Gamma : (0,1] \times \overline{\Omega} \to \Lambda^k$ by $\Gamma(\varepsilon,x) = g_\varepsilon(x)$, then

$$\Gamma \in C^{s,\beta}\left((0,1] \times \overline{\Omega};\Lambda^k\right) \quad \text{and} \quad \frac{\partial\Gamma}{\partial\varepsilon} \in C^\infty\left((0,1] \times \overline{\Omega};\Lambda^k\right).$$

Remark 16.50. (i) The result is valid for $k = n$, as a direct consequence of Theorem 16.43 (cf. Proposition 16.47 and Theorem 6.5). It holds for any Lipschitz set and, moreover, it gives $g_\varepsilon \in C^\infty\left(\overline{\Omega};\Lambda^n\right)$.

(ii) The result is, of course, trivially true for $k = 0$.

(iii) We recall that $\mathscr{H}_T\left(\Omega;\Lambda^k\right)$ and $\mathscr{H}_N\left(\Omega;\Lambda^k\right)$ are defined in Definition 6.1 and if Ω is contractible and since $1 \leq k \leq n-1$, then

$$\mathscr{H}_T\left(\Omega;\Lambda^k\right) = \mathscr{H}_N\left(\Omega;\Lambda^k\right) = \{0\}.$$

(iv) We will prove not only that

$$\int_{\Omega} \langle g_{\varepsilon}; \psi \rangle = \int_{\Omega} \langle g; \psi \rangle \quad \text{for every } \psi \in \mathscr{H}_T(\Omega; \Lambda^k)$$

but also that there exist G and G_{ε} such that $g_{\varepsilon} - g = d(G_{\varepsilon} - G)$ with $G_{\varepsilon} - G = 0$ on $\partial\Omega$.

(v) The construction is universal in the sense of Remark 16.44.

(vi) The constant $C = C(s, \alpha, \beta, \gamma, \Omega)$ is uniform in (α, β, γ) in the sense that if

$$0 < a \leq \alpha, \beta, \gamma \leq b < 1,$$

then $C = C(s, a, b, \Omega)$.

(vii) If $r = 0$, the condition $dg = 0$ is understood in the sense of distributions.

Before starting the proof of the theorem, we need the equivalent of the theorem, but for functions.

Lemma 16.51. *Let $\Omega \subset \mathbb{R}^n$ be a bounded open smooth set. Let $s \geq r \geq t \geq 0$ be integers. Let $0 < \alpha, \beta, \gamma < 1$ be such that*

$$t + \gamma \leq r + \alpha \leq s + \beta.$$

Let $f \in C^{r,\alpha}(\overline{\Omega}) \cap C^{s,\beta}(\partial\Omega)$. Then for every $\varepsilon \in (0,1]$, there exist $f_{\varepsilon} \in C^{\infty}(\Omega) \cap C^{s,\beta}(\overline{\Omega})$ and a constant $C = C(s, \alpha, \beta, \gamma, \Omega) > 0$ such that

$$f_{\varepsilon} = f \quad \text{on } \partial\Omega,$$

$$\|f_{\varepsilon} - f\|_{C^{t,\gamma}(\overline{\Omega})} \leq C\varepsilon^{(r+\alpha)-(t+\gamma)} \|f\|_{C^{r,\alpha}(\overline{\Omega})},$$

$$\|f_{\varepsilon}\|_{C^{s,\beta}(\overline{\Omega})} \leq \frac{C}{\varepsilon^{(s+\beta)-(r+\alpha)}} \|f\|_{C^{r,\alpha}(\overline{\Omega})} + C\|f\|_{C^{s,\beta}(\partial\Omega)},$$

$$\left\|\frac{d}{d\varepsilon} f_{\varepsilon}\right\|_{C^{s,\beta}(\overline{\Omega})} \leq \frac{C}{\varepsilon^{(s+\beta)-(r+\alpha)+1}} \|f\|_{C^{r,\alpha}(\overline{\Omega})},$$

$$\left\|\frac{d}{d\varepsilon} f_{\varepsilon}\right\|_{C^{t,\gamma}(\overline{\Omega})} \leq C\varepsilon^{(r+\alpha)-(t+\gamma)-1} \|f\|_{C^{r,\alpha}(\overline{\Omega})}.$$

Moreover, defining $F : (0,1] \times \overline{\Omega} \to \mathbb{R}$ by $F(\varepsilon, x) = f_{\varepsilon}(x)$, then

$$F \in C^{s,\beta}((0,1] \times \overline{\Omega}) \quad \text{and} \quad \frac{\partial F}{\partial \varepsilon} \in C^{\infty}((0,1] \times \overline{\Omega}).$$

Proof. For the sake of alleviating the notations, we will write in the present proof, for example, $\|g\|_{C^{r,\beta}}$ instead of $\|g\|_{C^{r,\beta}(\overline{\Omega})}$. When we will be considering norms on the boundary of Ω, we will keep the notation $\|g\|_{C^{r,\beta}(\partial\Omega)}$.

We first find (see Theorem 16.43) $g_\varepsilon \in C^\infty(\overline{\Omega})$ and a constant $C_1 = C_1(s,\Omega)$ such that

$$\|g_\varepsilon - f\|_{C^{t,\gamma}} \leq C_1 \varepsilon^{(r+\alpha)-(t+\gamma)} \|f\|_{C^{r,\alpha}},$$
$$\|g_\varepsilon\|_{C^{s,\beta}} \leq \frac{C_1}{\varepsilon^{(s+\beta)-(r+\alpha)}} \|f\|_{C^{r,\alpha}},$$
$$\left\|\frac{d}{d\varepsilon} g_\varepsilon\right\|_{C^{s,\beta}} \leq \frac{C_1}{\varepsilon^{(s+\beta)-(r+\alpha)+1}} \|f\|_{C^{r,\alpha}},$$
$$\left\|\frac{d}{d\varepsilon} g_\varepsilon\right\|_{C^{t,\gamma}} \leq C_1 \varepsilon^{(r+\alpha)-(t+\gamma)-1} \|f\|_{C^{r,\alpha}}.$$

Moreover, defining $G : (0,1] \times \overline{\Omega} \to \mathbb{R}$ by $G(\varepsilon,x) = g_\varepsilon(x)$, we have $G \in C^\infty((0,1] \times \overline{\Omega})$. We then fix the boundary data as follows. Let $f_\varepsilon \in C^\infty(\Omega) \cap C^{s,\beta}(\overline{\Omega})$ be the solution of

$$\begin{cases} \Delta f_\varepsilon = \Delta g_\varepsilon & \text{in } \Omega, \\ f_\varepsilon = f & \text{on } \partial\Omega \end{cases} \Leftrightarrow \begin{cases} \Delta [f_\varepsilon - g_\varepsilon] = 0 & \text{in } \Omega, \\ f_\varepsilon - g_\varepsilon = f - g_\varepsilon & \text{on } \partial\Omega. \end{cases}$$

Using Schauder estimates (the classical estimates assume $s \geq 2$; however, they are also valid when $s = 0$ or $s = 1$; see Agmon, Douglis and Nirenberg [3], Gilbarg and Hörmander [48], Lieberman [69, 70] and Widman [105]), there exists $C_2 = C_2(s,\beta,\gamma,\Omega)$ such that

$$\|f_\varepsilon\|_{C^{s,\beta}} \leq C_2 \left[\|g_\varepsilon\|_{C^{s,\beta}} + \|f\|_{C^{s,\beta}(\partial\Omega)}\right],$$
$$\|f_\varepsilon - g_\varepsilon\|_{C^{t,\gamma}} \leq C_2 \|f - g_\varepsilon\|_{C^{t,\gamma}(\partial\Omega)} \leq C_2 \|f - g_\varepsilon\|_{C^{t,\gamma}}.$$

Moreover, defining $F : (0,1] \times \overline{\Omega} \to \mathbb{R}$ by $F(\varepsilon,x) = f_\varepsilon(x)$, we have $F \in C^{s,\beta}((0,1] \times \overline{\Omega})$. Finally, noticing that $\frac{d}{d\varepsilon} f_\varepsilon$ verifies

$$\begin{cases} \Delta \frac{d}{d\varepsilon} f_\varepsilon = \Delta \frac{d}{d\varepsilon} g_\varepsilon & \text{in } \Omega, \\ \frac{d}{d\varepsilon} f_\varepsilon = 0 & \text{on } \partial\Omega \end{cases} \Leftrightarrow \begin{cases} \Delta \left[\frac{d}{d\varepsilon}(f_\varepsilon - g_\varepsilon)\right] = 0 & \text{in } \Omega, \\ \frac{d}{d\varepsilon}(f_\varepsilon - g_\varepsilon) = -\frac{d}{d\varepsilon} g_\varepsilon & \text{on } \partial\Omega, \end{cases}$$

there exists $C_3 = C_3(s,\beta,\gamma,\Omega)$ such that

$$\left\|\frac{d}{d\varepsilon} f_\varepsilon\right\|_{C^{s,\beta}} \leq C_3 \left\|\frac{d}{d\varepsilon} g_\varepsilon\right\|_{C^{s,\beta}}, \quad \left\|\frac{d}{d\varepsilon} f_\varepsilon\right\|_{C^{t,\gamma}} \leq C_3 \left\|\frac{d}{d\varepsilon} g_\varepsilon\right\|_{C^{t,\gamma}},$$

and

$$\frac{\partial F}{\partial \varepsilon} \in C^\infty((0,1] \times \overline{\Omega}).$$

The combination of the properties of f_ε and g_ε gives the result. □

We can now go back to the proof of Theorem 16.49.

Proof. We adopt the same simplification in the notations of the norms as in Lemma 16.51.

Step 1. We first show that we can find

$$G \in C^{r+1,\alpha}(\overline{\Omega};\Lambda^{k-1}) \cap C^{s+1,\beta}(\partial\Omega;\Lambda^{k-1}), \quad h \in C^{\infty}(\overline{\Omega};\Lambda^{k}) \cap \mathscr{H}(\Omega;\Lambda^{k}),$$

and a constant $C = C(s,\alpha,\beta,\Omega)$ such that

$$g = dG + h,$$
$$\|h\|_{C^{r,\alpha}} + \|G\|_{C^{r+1,\alpha}} \leq C\,\|g\|_{C^{r,\alpha}}\,,$$
$$\|h\|_{C^{s,\beta}} + \|G\|_{C^{s+1,\beta}(\partial\Omega)} \leq C\left[\|g\|_{C^{r,\alpha}} + \|\nu \wedge g\|_{C^{s,\beta}(\partial\Omega)}\right].$$

We proceed as follows. In the sequel, $C_1, C_2, \ldots$ will denote generic constants depending only on s, α, β and Ω.

(i) We first find (solving the Dirichlet problem component by component) $g^{(1)} \in C^{s,\beta}\left(\overline{\Omega};\Lambda^k\right) \cap C^{\infty}\left(\Omega;\Lambda^k\right)$ verifying

$$\begin{cases} \Delta g^{(1)} = 0 & \text{in } \Omega, \\ g^{(1)} = \nu \lrcorner (\nu \wedge g) & \text{on } \partial\Omega. \end{cases}$$

We, moreover, have

$$\left\|g^{(1)}\right\|_{C^{r,\alpha}} \leq C_1\,\|g\|_{C^{r,\alpha}} \quad \text{and} \quad \left\|g^{(1)}\right\|_{C^{s,\beta}} \leq C_1\,\|\nu \wedge g\|_{C^{s,\beta}(\partial\Omega)}\,.$$

Since $|\nu| = 1$, we have, using (2.7),

$$\nu \wedge g^{(1)} = \nu \wedge g.$$

Observe also that since $dg = 0$, then, using Theorem 3.23, we have

$$\nu \wedge dg^{(1)} = 0.$$

The above argument is valid only if $r \geq 1$; however, it still holds if $r = 0$ (recall that $s \geq 1$) and if we use Proposition 7.6(iv). More precisely, we have for every $\varphi \in C^{\infty}\left(\overline{\Omega};\Lambda^{k+1}\right)$,

$$\begin{aligned} \int_{\partial\Omega} \langle \nu \wedge dg^{(1)};\varphi\rangle &= \int_{\Omega} \langle dg^{(1)};\delta\varphi\rangle = \int_{\partial\Omega} \langle \nu \wedge g^{(1)};\delta\varphi\rangle \\ &= \int_{\partial\Omega} \langle \nu \wedge g;\delta\varphi\rangle = \int_{\Omega} \langle g;\delta\delta\varphi\rangle = 0. \end{aligned}$$

Since φ is arbitrary, we have indeed proved that

$$\nu \wedge dg^{(1)} = 0.$$

Note that we also have

$$\int_{\Omega} \left\langle dg^{(1)}; \psi \right\rangle = 0 \text{ for every } \psi \in \mathscr{H}_T\left(\Omega; \Lambda^{k+1}\right).$$

This last identity follows from the fact that, using Theorem 3.28,

$$\int_{\Omega} \left\langle dg^{(1)}; \psi \right\rangle = \int_{\partial\Omega} \left\langle \nu \wedge g^{(1)}; \psi \right\rangle = \int_{\partial\Omega} \langle \nu \wedge g; \psi \rangle = \int_{\Omega} \langle dg; \psi \rangle = 0.$$

If $r = 0$, we proceed similarly but use Proposition 7.6(iv). More precisely, we obtain

$$\int_{\Omega} \left\langle dg^{(1)}; \psi \right\rangle = \int_{\partial\Omega} \left\langle \nu \wedge g^{(1)}; \psi \right\rangle = \int_{\partial\Omega} \langle \nu \wedge g; \psi \rangle = \int_{\Omega} \langle g; \delta\psi \rangle = 0.$$

(ii) We next solve

$$\begin{cases} dg^{(2)} = -dg^{(1)} & \text{in } \Omega, \\ g^{(2)} = 0 & \text{on } \partial\Omega \end{cases}$$

and we have $g^{(2)} \in C^{s,\beta}\left(\overline{\Omega}; \Lambda^k\right)$. This is possible (according to Theorem 8.16) since

$$\nu \wedge dg^{(1)} = 0 \quad \text{and} \quad \int_{\Omega} \left\langle dg^{(1)}; \psi \right\rangle = 0 \text{ for every } \psi \in \mathscr{H}_T\left(\Omega; \Lambda^{k+1}\right).$$

We also have

$$\left\|g^{(2)}\right\|_{C^{r,\alpha}} \le C_2 \|g\|_{C^{r,\alpha}} \quad \text{and} \quad \left\|g^{(2)}\right\|_{C^{s,\beta}} \le C_2 \|\nu \wedge g\|_{C^{s,\beta}(\partial\Omega)}.$$

(iii) We then set

$$g^{(3)} = g^{(2)} + g^{(1)}$$

and observe that $g^{(3)} \in C^{s,\beta}\left(\overline{\Omega}; \Lambda^k\right)$,

$$dg^{(3)} = 0 \quad \text{and} \quad \nu \wedge g^{(3)} = \nu \wedge g^{(1)} = \nu \wedge g.$$

Apply the Hodge–Morrey decomposition and find (in view of Theorem 6.12)

$$G^{(3)} \in C_N^{s+1,\beta}\left(\Omega; \Lambda^{k-1}\right), \quad \beta^{(3)} \in C_N^{s+1,\beta}\left(\overline{\Omega}; \Lambda^{k+1}\right)$$

and

$$h^{(3)} \in \mathscr{H}_N\left(\Omega; \Lambda^k\right) \subset C^{\infty}\left(\overline{\Omega}; \Lambda^k\right)$$

such that

$$g^{(3)} = dG^{(3)} + \delta\beta^{(3)} + h^{(3)}.$$

Note that $\delta\beta^{(3)} = 0$, due to the orthogonality of the decomposition and since $dg^{(3)} = 0$ and $\beta^{(3)} \in C_N^{s+1,\beta}$. Observe that

$$\left\|h^{(3)}\right\|_{C^{r,\alpha}} + \left\|G^{(3)}\right\|_{C^{r+1,\alpha}} \leq C_3 \left\|g^{(3)}\right\|_{C^{r,\alpha}} \leq C_4 \|g\|_{C^{r,\alpha}},$$
$$\left\|h^{(3)}\right\|_{C^{s,\beta}} + \left\|G^{(3)}\right\|_{C^{s+1,\beta}} \leq C_3 \left\|g^{(3)}\right\|_{C^{s,\beta}} \leq C_4 \|\nu \wedge g\|_{C^{s,\beta}(\partial\Omega)}.$$

(iv) We now apply again the Hodge–Morrey decomposition (cf. Theorem 6.12) to get

$$g - g^{(3)} = dG^{(4)} + \delta\beta^{(4)} + h^{(4)},$$

where

$$G^{(4)} \in C_T^{r+1,\alpha}\left(\overline{\Omega};\Lambda^{k-1}\right), \quad \beta^{(4)} \in C_T^{r+1,\alpha}\left(\overline{\Omega};\Lambda^{k+1}\right)$$

and

$$h^{(4)} \in \mathscr{H}_T\left(\Omega;\Lambda^k\right) \subset C^\infty\left(\overline{\Omega};\Lambda^k\right).$$

Note that $\delta\beta^{(4)} = 0$ since, using Theorem 3.28, the orthogonality of the decomposition and the fact that $dg = dg^{(3)} = 0$, we have

$$\int_\Omega \left|\delta\beta^{(4)}\right|^2 = \int_\Omega \left\langle g - g^{(3)}; \delta\beta^{(4)}\right\rangle$$
$$= \int_{\partial\Omega} \left\langle \nu \wedge \left[g - g^{(3)}\right]; \beta^{(4)}\right\rangle = 0.$$

The above argument is valid only if $r \geq 1$; however, it still holds if $r = 0$ by density and by Proposition 7.6(iv). We also have, using Remark 6.4,

$$\left\|h^{(4)}\right\|_{C^{s,\beta}} \leq C_5 \left\|h^{(4)}\right\|_{C^{r,\alpha}} \leq C_6 \left\|g - g^{(3)}\right\|_{C^{r,\alpha}} \leq C_7 \|g\|_{C^{r,\alpha}}.$$

Similarly, as above, we get

$$\left\|G^{(4)}\right\|_{C^{r+1,\alpha}} \leq C_7 \|g\|_{C^{r,\alpha}}.$$

(v) Finally, we adjust $G^{(4)} \in C_T^{r+1,\alpha}(\overline{\Omega};\Lambda^{k-1})$ so as to have $G^{(5)} \in C^{r+1,\alpha}(\overline{\Omega};\Lambda^{k-1})$ with $G^{(5)} = 0$ on $\partial\Omega$. If $k = 1$, then nothing is to be done; just choose $G^{(5)} = G^{(4)}$. So we may assume that $k \geq 2$. The claim then follows from Lemma 8.11, since we can find $A \in C^{r+2,\alpha}\left(\overline{\Omega};\Lambda^{k-2}\right)$ such that

$$dA = -G^{(4)} \quad \text{on } \partial\Omega.$$

It then suffices to set

$$G^{(5)} = G^{(4)} + dA.$$

We also have
$$\left\|G^{(5)}\right\|_{C^{r+1,\alpha}} \leq C_8 \left\|G^{(4)}\right\|_{C^{r+1,\alpha}} \leq C_9 \|g\|_{C^{r,\alpha}} .$$

(vi) We conclude that
$$G = G^{(3)} + G^{(5)} \quad \text{and} \quad h = h^{(3)} + h^{(4)}$$
have all of the desired properties. Indeed,
$$G \in C^{r+1,\alpha}(\overline{\Omega};\Lambda^{k-1}) \cap C^{s+1,\beta}(\partial\Omega;\Lambda^{k-1}),$$
$h \in C^{\infty}(\overline{\Omega};\Lambda^k) \cap \mathscr{H}(\Omega;\Lambda^k)$ and
$$dG = dG^{(3)} + dG^{(5)} = g^{(3)} - h^{(3)} + g - g^{(3)} - h^{(4)} = g - h^{(3)} - h^{(4)}.$$

By construction, we deduce the estimates
$$\|h\|_{C^{r,\alpha}} + \|G\|_{C^{r+1,\alpha}} \leq C_{10} \|g\|_{C^{r,\alpha}} ,$$
$$\|h\|_{C^{s,\beta}} + \|G\|_{C^{s+1,\beta}(\partial\Omega)} \leq C_{11} \left[\|g\|_{C^{r,\alpha}} + \|\nu \wedge g\|_{C^{s,\beta}(\partial\Omega)}\right]$$
since $G = G^{(3)}$ on $\partial\Omega$.

Step 2. Applying Lemma 16.51 on each component of G, we get
$$G_\varepsilon \in C^{\infty}(\Omega;\Lambda^{k-1}) \cap C^{s+1,\beta}(\overline{\Omega};\Lambda^{k-1}),$$
as in the lemma (in particular, $G_\varepsilon = G$ on $\partial\Omega$). Setting $g_\varepsilon = dG_\varepsilon + h$, we have the claim. Before checking the inequalities, observe that, by construction, using Theorem 3.23,
$$dg_\varepsilon = 0 \text{ in } \Omega, \quad \nu \wedge g_\varepsilon = \nu \wedge g \text{ on } \partial\Omega,$$
and since $g_\varepsilon - g = d(G_\varepsilon - G)$ with $G_\varepsilon - G = 0$ on $\partial\Omega$, we have, using Theorem 3.28,
$$\int_\Omega \langle g_\varepsilon - g; \psi\rangle \, dx = 0 \quad \text{for every } \psi \in \mathscr{H}_T(\Omega;\Lambda^k).$$

In the sequel, $C_1, C_2, \ldots$ will denote generic constants depending only on s, α, β, γ and Ω. The first inequality follows from
$$\begin{aligned}\|g_\varepsilon - g\|_{C^{t,\gamma}} &\leq C_1 \|G_\varepsilon - G\|_{C^{t+1,\gamma}} \leq C_2 \varepsilon^{(r+1+\alpha)-(t+1+\gamma)} \|G\|_{C^{r+1,\alpha}} \\ &\leq C_3\, \varepsilon^{(r+\alpha)-(t+\gamma)} \|g\|_{C^{r,\alpha}} .\end{aligned}$$

To obtain the second inequality, first remark that
$$\begin{aligned}\|g_\varepsilon\|_{C^{s,\beta}} &\leq C_4 \left[\|h\|_{C^{s,\beta}} + \|G_\varepsilon\|_{C^{s+1,\beta}}\right] \\ &\leq \frac{C_5}{\varepsilon^{(s+1+\beta)-(r+1+\alpha)}} \|G\|_{C^{r+1,\alpha}} + C_5 \left(\|h\|_{C^{s,\beta}} + \|G\|_{C^{s+1,\beta}(\partial\Omega)}\right)\end{aligned}$$

and hence, since $0 < \varepsilon \leq 1$,

$$\|g_\varepsilon\|_{C^{s,\beta}} \leq \frac{C_6}{\varepsilon^{(s+\beta)-(r+\alpha)}} \|g\|_{C^{r,\alpha}} + C_6 \|\nu \wedge g\|_{C^{s,\beta}(\partial\Omega)}.$$

The third one comes from

$$\begin{aligned} \left\| \frac{d}{d\varepsilon} g_\varepsilon \right\|_{C^{s,\beta}} &\leq C_7 \left\| \frac{d}{d\varepsilon} G_\varepsilon \right\|_{C^{s+1,\beta}} \leq \frac{C_8}{\varepsilon^{(s+1+\beta)-(r+1+\alpha)+1}} \|G\|_{C^{r+1,\alpha}} \\ &\leq \frac{C_9}{\varepsilon^{(s+\beta)-(r+\alpha)+1}} \|g\|_{C^{r,\alpha}} \end{aligned}$$

and the last one follows from

$$\begin{aligned} \left\| \frac{d}{d\varepsilon} g_\varepsilon \right\|_{C^{t,\gamma}} &\leq C_{10} \left\| \frac{d}{d\varepsilon} G_\varepsilon \right\|_{C^{t+1,\gamma}} \\ &\leq C_{11} \varepsilon^{(r+1+\alpha)-(t+\gamma+1)-1} \|G\|_{C^{r+1,\alpha}} \\ &\leq C_{12} \varepsilon^{(r+\alpha)-(t+\gamma)-1} \|g\|_{C^{r,\alpha}}. \end{aligned}$$

Finally, defining $\Gamma : (0,1] \times \overline{\Omega} \to \Lambda^k$ by $\Gamma(\varepsilon, x) = g_\varepsilon(x)$, we have that

$$\Gamma \in C^{s,\beta}\left((0,1] \times \overline{\Omega}; \Lambda^k\right) \quad \text{and} \quad \frac{\partial \Gamma}{\partial \varepsilon} \in C^\infty\left((0,1] \times \overline{\Omega}; \Lambda^k\right),$$

which concludes the proof. ☐

Part VI
Appendix

Chapter 17
Necessary Conditions

In the following proposition we gather some elementary necessary conditions (cf. [8], [9] and [31]).

Proposition 17.1. *Let* $\Omega \subset \mathbb{R}^n$ *be a bounded open smooth set and* $\varphi \in \operatorname{Diff}^1\left(\overline{\Omega}; \varphi\left(\overline{\Omega}\right)\right)$. *Let* $1 \leq k \leq n$, $f \in C^1\left(\overline{\Omega}; \Lambda^k\right)$ *and* $g \in C^1\left(\varphi\left(\overline{\Omega}\right); \Lambda^k\right)$ *be such that*

$$\varphi^*(g) = f \quad in\ \Omega.$$

(i) Then for every $x \in \Omega$,

$$\operatorname{rank}\left[g(\varphi(x))\right] = \operatorname{rank}\left[f(x)\right],$$

$$\operatorname{rank}\left[dg(\varphi(x))\right] = \operatorname{rank}\left[df(x)\right].$$

In particular,

$$dg = 0 \ \ in\ \varphi(\Omega) \ \Leftrightarrow\ df = 0 \ \ in\ \Omega.$$

(ii) If $\det \nabla \varphi > 0$ *and* $n = mk$, *with* m *an integer, then*

$$\int_{\Omega} f^m = \int_{\varphi(\Omega)} g^m,$$

where $f^m = \underbrace{f \wedge \cdots \wedge f}_{m\ times}$.

(iii) If $\varphi(x) = x$ *for* $x \in \partial\Omega$, *then*

$$\nu \wedge f = \nu \wedge g \quad on\ \partial\Omega,$$

where ν *is the exterior unit normal to* Ω.

Remark 17.2. (i) Recall that we have denoted by rank what is denoted by rank_1 in Chapter 2. In fact, statement (i) remains true for all ranks and coranks of any order. More precisely, for every $x \in \Omega$, $0 \leq s \leq k$ and $0 \leq t \leq n-k$, we have

$$\operatorname{rank}_s[g(\varphi(x))] = \operatorname{rank}_s[f(x)],$$

$$\operatorname{corank}_t[g(\varphi(x))] = \operatorname{corank}_t[f(x)],$$

and for every $0 \le s \le k+1$ and $0 \le t \le n-k-1$,

$$\operatorname{rank}_s[dg(\varphi(x))] = \operatorname{rank}_s[df(x)],$$

$$\operatorname{corank}_t[dg(\varphi(x))] = \operatorname{corank}_t[df(x)].$$

We will establish these facts in the proof of the proposition.

(ii) Note that the equation

$$\nu \wedge g = \nu \wedge f \quad \text{on } \partial\Omega$$

is equivalent to

$$i^*(g) = i^*(f),$$

where $i : \partial\Omega \to \mathbb{R}^n$ is the inclusion map (cf. Remark 3.22).

Proof. (i) The statements

$$\operatorname{rank}_s[g(\varphi(x))] = \operatorname{rank}_s[f(x)],$$

$$\operatorname{corank}_t[g(\varphi(x))] = \operatorname{corank}_t[f(x)]$$

come directly from Proposition 2.33(vi), whereas the claims

$$\operatorname{rank}_s[dg(\varphi(x))] = \operatorname{rank}_s[df(x)],$$

$$\operatorname{corank}_t[dg(\varphi(x))] = \operatorname{corank}_t[df(x)]$$

follow as above and from the observation that

$$\varphi^*(g) = f \Rightarrow \varphi^*(dg) = df.$$

(ii) Since f^m and g^m are volume forms and

$$\varphi^*(g^m) = f^m,$$

we have the claim using the change of variables formula.

(iii) Let $1 \le i \le n$. Since $\varphi(x) = x$ for $x \in \partial\Omega$, we have, appealing to Theorem 3.23, that

$$\nu \wedge \left(d\varphi^i - dx^i\right) = 0 \quad \text{on } \partial\Omega.$$

Thus, invoking Theorem 2.42, there exist

$$\phi_i : \partial\Omega \to \mathbb{R}$$

such that

$$d\varphi^i = dx^i + \phi_i \nu \quad \text{on } \partial\Omega.$$

Taking any index $I \in \mathscr{T}_k$, where

$$\mathscr{T}_k = \{I = (i_1, \dots, i_k) \in \mathbb{N}^k,\ 1 \le i_1 < \cdots < i_k \le n\},$$

we find

$$\begin{aligned} d\varphi^I &= \left(dx^{i_1} + \phi_{i_1}\nu\right) \wedge \cdots \wedge \left(dx^{i_k} + \phi_{i_k}\nu\right) \\ &= dx^I + \nu \wedge \Phi \end{aligned}$$

for some $(k-1)$-form Φ and hence

$$\nu \wedge d\varphi^I = \nu \wedge dx^I.$$

We thus find, for $x \in \partial\Omega$, that

$$\begin{aligned} \nu \wedge f = \nu \wedge \varphi^*(g) &= \nu \wedge \sum_I g_I(\varphi(x))\, d\varphi^I = \sum_I g_I(x)\, \nu \wedge d\varphi^I \\ &= \sum_I g_I(x)\, \nu \wedge dx^I = \nu \wedge \sum_I g_I(x)\, dx^I = \nu \wedge g. \end{aligned}$$

This concludes the proof. □

For results on the pullback equation when the rank is not conserved, see Golubitsky and Tischler [53], Martinet [71], Pelletier [81], Roussarie [86] and Zhitomirskiĭ [108]; see also Chapter 11 for the case $k = n$.

Before stating another necessary condition, we need the next theorem (cf. Bandyopadhyay [7]). The proof closely follows the standard proofs of the classical Poincaré lemma for contractible sets.

Theorem 17.3. *Let $1 \le k \le n$ be integers, $\Omega \subset \mathbb{R}^n$ be an open set and $F \in C^2([0,1] \times \Omega; \Omega)$, $F = F(t,x) = F_t(x)$. Let $g \in C^1(\Omega; \Lambda^k)$ be closed. Then there exists $G \in C^1(\Omega; \Lambda^{k-1})$ such that*

$$F_1^*(g) - F_0^*(g) = dG.$$

Moreover, if there exists a set $N \subset \Omega$ such that

$$F(t,x) = x \quad \text{for every } (t,x) \in [0,1] \times N,$$

then G can be chosen such that it additionally satisfies

$$G(x) = 0 \quad \text{for every } x \in N.$$

Remark 17.4. (i) Noticing that with exactly the same proof, the theorem is still valid with Ω replaced by $\overline{\Omega}$, for Ω bounded, we have the following special case (used in

Theorem 14.10). Let $F \in C^2\left([0,1] \times \overline{\Omega};\overline{\Omega}\right)$ such that $F(t,x) = x$ for every $x \in \partial\Omega$ and every $t \in [0,1]$. Then there exists $G \in C^1\left(\overline{\Omega};\Lambda^{k-1}\right)$ verifying

$$\begin{cases} dG = F_1^*(g) - F_0^*(g) & \text{in } \Omega, \\ G = 0 & \text{on } \partial\Omega. \end{cases}$$

(ii) The same proof shows that if $r \geq 1$ is an integer, $0 \leq \alpha \leq 1$, $F \in C^{r+1,\alpha}$ and $g \in C^{r,\alpha}$, then G is in $C^{r,\alpha}$.

Proof. Step 1. We start with a preliminary step. Let

$$i_0, i_1 : \Omega \to \mathbb{R} \times \Omega$$

be defined by

$$i_0(x) = (0,x) \quad \text{and} \quad i_1(x) = (1,x).$$

Hence, we have, in Ω,

$$F_0 = F \circ i_0 \quad \text{and} \quad F_1 = F \circ i_1 .$$

Note that any $\omega \in C^1\left([0,1] \times \Omega;\Lambda^k\left(\mathbb{R}^{n+1}\right)\right)$ can be written as

$$\omega(t,x) = \sum_{I \in \mathcal{T}_k} a_I(t,x)\, dx^I + \sum_{J \in \mathcal{T}_{k-1}} b_J(t,x) dt \wedge dx^J, \tag{17.1}$$

where the $a_I, b_J \in C^1([0,1] \times \Omega)$ are uniquely determined. We now define for $1 \leq k \leq n+1$ the maps

$$L_k : C^0\left([0,1] \times \Omega;\Lambda^k\left(\mathbb{R}^{n+1}\right)\right) \to C^0\left(\Omega;\Lambda^{k-1}\left(\mathbb{R}^n\right)\right)$$

by

$$L_k(\omega)(x) = \sum_{J \in \mathcal{T}_{k-1}} \left(\int_0^1 b_J(t,x) dt\right) dx^J.$$

We claim that for every $\omega \in C^1\left([0,1] \times \Omega;\Lambda^k\left(\mathbb{R}^{n+1}\right)\right)$, the following identity holds:

$$i_1^*(\omega) - i_0^*(\omega) = L_{k+1}(d_{t,x}\omega) + d_x L_k(\omega) \quad \text{in } \Omega. \tag{17.2}$$

To prove this, note that

$$\begin{aligned} d_{t,x}\omega &= \sum_{I \in \mathcal{T}_k} \left(d_x a_I + \frac{\partial a_I}{\partial t} dt\right) \wedge dx^I + \sum_{J \in \mathcal{T}_{k-1}} d_x b_J \wedge dt \wedge dx^J \\ &= \sum_{I \in \mathcal{T}_k} \left(d_x a_I + \frac{\partial a_I}{\partial t} dt\right) \wedge dx^I - \sum_{J \in \mathcal{T}_{k-1}} dt \wedge d_x b_J \wedge dx^J . \end{aligned}$$

Therefore,

$$
\begin{aligned}
L_{k+1}(d_{t,x}\omega) &= \sum_{I\in\mathscr{T}_k}\left(\int_0^1 \frac{\partial a_I}{\partial t}dt\right)dx^I - \sum_{J\in\mathscr{T}_{k-1}}\left(\int_0^1 d_x b_J(t,x)dt\right)\wedge dx^J \\
&= \sum_{I\in\mathscr{T}_k}(a_I\circ i_1 - a_I\circ i_0)\,dx^I - \sum_{J\in\mathscr{T}_{k-1}} d_x\left(\int_0^1 b_J(t,x)dt\right)\wedge dx^J \\
&= \sum_{I\in\mathscr{T}_k}(a_I\circ i_1 - a_I\circ i_0)\,dx^I - d_x\left(\sum_{J\in\mathscr{T}_{k-1}}\left(\int_0^1 b_J(t,x)dt\right)\wedge dx^J\right) \\
&= i_1^*(\omega) - i_0^*(\omega) - d_x L_k(\omega),
\end{aligned}
$$

which proves (17.2).

Step 2. We claim that

$$G = L_k\left(F^*\left(g\right)\right)$$

has all of the desired properties. Let $\omega = F^*\left(g\right)$. Since g is closed in Ω, we get

$$d_{t,x}\omega = d_{t,x}(F^*(g)) = F^*(dg) = 0.$$

We therefore obtain

$$
\begin{aligned}
F_1^*(g) - F_0^*(g) &= i_1^*\left(F^*(g)\right) - i_0^*\left(F^*(g)\right) = i_1^*(\omega) - i_0^*(\omega) \\
&= L_{k+1}(d_{t,x}\omega) + d_x L_k(\omega) = d_x L_k(\omega) = dG.
\end{aligned}
$$

The first part of the theorem is therefore proved. It remains to show that if $F\left(t,x_0\right) = x_0$ for every $t\in[0,1]$, then the above G satisfies $G(x_0)=0$. First, note that for every $t\in[0,1]$ and every $1\le i\le n$, we have

$$\frac{\partial F^i}{\partial t}\left(t,x_0\right) = 0$$

and hence

$$dF^i\left(t,x_0\right) = \frac{\partial F^i}{\partial t}\left(t,x_0\right)dt + \sum_{j=1}^n \frac{\partial F^i}{\partial x_j}\left(t,x_0\right)dx^j = \sum_{j=1}^n \frac{\partial F^i}{\partial x_j}\left(t,x_0\right)dx^j.$$

We therefore deduce that

$$F^*\left(g\right)\left(t,x_0\right) = \sum_{I\in\mathscr{T}_k} g_I\left(F\left(t,x_0\right)\right)dF^I = \sum_{I\in\mathscr{T}_k} g_I\left(x_0\right)dF^I = \sum_{I\in\mathscr{T}_k} a_I\left(t,x_0\right)dx^I$$

for some appropriate a_I and, hence, the corresponding $b_J\left(t,x_0\right)$ in (17.1) is zero. We thus obtain, by definition of L_k, that

$$G(x_0) = L_k(F^*\left(g\right))(x_0) = 0.$$

This concludes the proof of the theorem. □

As a corollary, we obtain the following necessary condition, whenever a solution of the pullback equation is achieved by the flow method.

Corollary 17.5. *Let* $1 \leq k \leq n$ *be integers,* $\Omega \subset \mathbb{R}^n$ *be an open set and*

$$\varphi \in C^2\left([0,1] \times \overline{\Omega};\overline{\Omega}\right), \ \varphi = \varphi(t,x) = \varphi_t(x).$$

Let $f, g \in C^1\left(\overline{\Omega};\Lambda^k\right)$ *be closed. Assume that*

$$\varphi_1^*(g) = f \quad \text{and} \quad \varphi_0 = \mathrm{id} \quad \text{in } \Omega,$$
$$\varphi_t = \mathrm{id} \quad \text{on } \partial\Omega \quad \text{for every } 0 \leq t \leq 1.$$

Then for every $\chi \in C^\infty\left(\overline{\Omega};\Lambda^k\right)$ *with* $\delta\chi = 0$ *in* Ω*, the following equality holds true:*

$$\int_\Omega \langle g;\chi\rangle = \int_\Omega \langle f;\chi\rangle.$$

Proof. In view of Remark 17.4, there exists $G \in C^1\left(\overline{\Omega};\Lambda^{k-1}\right)$ verifying

$$\begin{cases} dG = \varphi_1^*(g) - \varphi_0^*(g) = f - g & \text{in } \Omega, \\ G = 0 & \text{on } \partial\Omega. \end{cases}$$

Therefore, by partial integration, we obtain

$$\int_\Omega \langle f - g;\chi\rangle = \int_\Omega \langle dG;\chi\rangle = \int_\Omega \langle G;\delta\chi\rangle = 0.$$

This concludes the proof of the corollary. □

Chapter 18
An Abstract Fixed Point Theorem

The following theorem is particularly useful when dealing with nonlinear problems, once good estimates are known for the linearized problem. We give it under a general form, because we have used it this way in Theorems 14.1 and 14.10. However, in many instances, Corollary 18.2 is amply sufficient. Our theorem will lean on the following hypotheses.

(H_{XY}) Let $X_1 \supset X_2$ be Banach spaces and $Y_1 \supset Y_2$ be normed spaces such that the following property holds: If

$$u_\nu \xrightarrow{X_1} u \quad \text{and} \quad \|u_\nu\|_{X_2} \le r,$$

then $u \in X_2$ and

$$\|u\|_{X_2} \le r.$$

(H_L) Let $L : X_2 \to Y_2$ be such that there exist a linear right inverse operator $L^{-1} : Y_2 \to X_2$ (namely $LL^{-1} = \text{id}$ on Y_2) and $k_1, k_2 > 0$ such that for every $f \in Y_2$,

$$\|L^{-1} f\|_{X_i} \le k_i \|f\|_{Y_i}, \quad i = 1, 2.$$

(H_Q) There exists $\rho > 0$ such that

$$Q : B_\rho = \{u \in X_2 : \|u\|_{X_1} \le \rho\} \to Y_2$$

$Q(0) = 0$ and for every $u, v \in B_\rho$, the following two inequalities hold:

$$\|Q(u) - Q(v)\|_{Y_1} \le c_1(\|u\|_{X_1}, \|v\|_{X_1}) \|u - v\|_{X_1}, \tag{18.1}$$

$$\|Q(v)\|_{Y_2} \le c_2(\|v\|_{X_1}, \|v\|_{X_2}), \tag{18.2}$$

where $c_1, c_2 : [0, +\infty) \times [0, +\infty) \to [0, +\infty)$ are separately increasing.

Theorem 18.1 (Fixed point theorem). *Let X_1, X_2, Y_1, Y_2, L and Q satisfy the hypotheses (H_{XY}), (H_L) and (H_Q). Then for every $f \in Y_2$ verifying*

$$2k_1 \|f\|_{Y_1} \le \rho \quad \textit{and} \quad 2k_1 c_1(2k_1 \|f\|_{Y_1}, 2k_1 \|f\|_{Y_1}) \le 1, \tag{18.3}$$

$$c_2(2k_1\|f\|_{Y_1},2k_2\|f\|_{Y_2}) \le \|f\|_{Y_2}, \tag{18.4}$$

there exists $u \in B_\rho \subset X_2$ such that

$$Lu = Q(u)+f \quad \text{and} \quad \|u\|_{X_i} \le 2k_i\|f\|_{Y_i},\ i=1,2. \tag{18.5}$$

We have as an immediate consequence of the theorem the following result.

Corollary 18.2. *Let X be a Banach space and Y a normed space. Let $L : X \to Y$ be such that there exist a linear right inverse operator $L^{-1} : Y \to X$ (namely $LL^{-1} = \mathrm{id}$ on Y) and $k > 0$ such that*

$$\|L^{-1}f\|_X \le k\|f\|_Y.$$

Let $\rho > 0$ and

$$Q : B_\rho = \{u \in X : \|u\|_X \le \rho\} \to Y,$$

with $Q(0) = 0$ and, for every $u,v \in B_\rho$,

$$\|Q(u)-Q(v)\|_Y \le c(\|u\|_X,\|v\|_X)\|u-v\|_X$$

and where $c : [0,+\infty) \times [0,+\infty) \to [0,+\infty)$ is separately increasing. If

$$2k\|f\|_Y \le \rho \quad \text{and} \quad 2kc(2k\|f\|_Y,2k\|f\|_Y) \le 1,$$

then there exists $u \in B_\rho \subset X$ such that

$$Lu = Q(u)+f \quad \text{and} \quad \|u\|_X \le 2k\|f\|_Y.$$

We now turn to the proof of Theorem 18.1.

Proof. We set

$$N(u) = Q(u)+f.$$

We next define

$$B = \{u \in X_2 : \|u\|_{X_i} \le 2k_i\|f\|_{Y_i}\ i=1,2\}.$$

We endow B with $\|\cdot\|_{X_1}$ norm; the property (H_{XY}) ensures that B is closed. We now want to show that $L^{-1}N : B \to B$ is a contraction mapping (cf. Claims 1 and 2 below). Applying the Banach fixed point theorem we will have indeed found a solution verifying (18.5), since $LL^{-1} = \mathrm{id}$.

Claim 1. Let us first show that $L^{-1}N$ is a contraction on B. To show this, let $u,v \in B$ and use (18.1) and (18.3) to get that

$$\begin{aligned}\|L^{-1}N(u)-L^{-1}N(v)\|_{X_1} &\le k_1\|N(u)-N(v)\|_{Y_1} = k_1\|Q(u)-Q(v)\|_{Y_1}\\ &\le k_1c_1(\|u\|_{X_1},\|v\|_{X_1})\|u-v\|_{X_1}\\ &\le k_1c_1(2k_1\|f\|_{Y_1},2k_1\|f\|_{Y_1})\|u-v\|_{X_1}\\ &\le \frac{1}{2}\|u-v\|_{X_1}.\end{aligned}$$

Claim 2. We next show $L^{-1}N : B \to B$ is well defined. First, note that

$$\|L^{-1}N(0)\|_{X_1} \le k_1 \|N(0)\|_{Y_1} = k_1 \|f\|_{Y_1}.$$

Therefore, using Claim 1, we obtain

$$\begin{aligned}\|L^{-1}N(u)\|_{X_1} &\le \|L^{-1}N(u) - L^{-1}N(0)\|_{X_1} + \|L^{-1}N(0)\|_{X_1} \\ &\le \frac{1}{2}\|u\|_{X_1} + k_1\|f\|_{Y_1} \le 2k_1\|f\|_{Y_1}.\end{aligned}$$

It remains to show that

$$\|L^{-1}N(u)\|_{X_2} \le 2k_2\|f\|_{Y_2}.$$

Using (18.2), we have

$$\begin{aligned}\|L^{-1}N(u)\|_{X_2} &\le k_2\|N(u)\|_{Y_2} \le k_2\|Q(u)\|_{Y_2} + k_2\|f\|_{Y_2} \\ &\le k_2 c_2\left(\|u\|_{X_1}, \|u\|_{X_2}\right) + k_2\|f\|_{Y_2} \\ &\le k_2\left[c_2(2k_1\|f\|_{Y_1}, 2k_2\|f\|_{Y_2}) + \|f\|_{Y_2}\right]\end{aligned}$$

and hence, appealing to (18.4),

$$\|L^{-1}N(u)\|_{X_2} \le k_2\left[\|f\|_{Y_2} + \|f\|_{Y_2}\right] = 2k_2\|f\|_{Y_2}.$$

This concludes the proof of Claim 2 and thus of the theorem. □

For the sake of illustration, we give here an academic example loosely related to our problem.

Example 18.3. Let $\Omega \subset \mathbb{R}^n$ be a bounded contractible smooth set and $0 < \alpha < 1$. Let $r \ge 1$ and $1 \le k \le n-2$ be integers. Consider the form $w : \mathbb{R}^n \to \Lambda^k$, where

$$w = \sum_{I \in \mathscr{T}_k} w_I dx^I,$$

where $\mathscr{T}_k$ is the set of ordered k-indices. Let $I_1, \ldots, I_{k+1} \in \mathscr{T}_k$; then there exists $\varepsilon > 0$ such that for every $f \in C^{r,\alpha}\left(\overline{\Omega}; \Lambda^{k+1}\right)$ with

$$\|f\|_{C^{r,\alpha}} \le \varepsilon, \quad df = 0 \quad \text{and} \quad \nu \wedge f = 0 \text{ on } \partial\Omega,$$

there exists $w \in C^{r+1,\alpha}\left(\overline{\Omega}; \Lambda^k\right)$ satisfying

$$\begin{cases} dw - \bigwedge_{r=1}^{k+1} dw_{I_r} = f & \text{in } \Omega, \\ w = 0 & \text{on } \partial\Omega. \end{cases}$$

The proof immediately follows from Corollary 18.2 if we set

$$X = \{w \in C^{r+1,\alpha}(\overline{\Omega}; \Lambda^k) : w = 0 \text{ on } \partial\Omega\},$$

$$Y = \{f \in C^{r,\alpha}(\overline{\Omega}; \Lambda^{k+1}) : df = 0 \text{ in } \Omega \text{ and } \nu \wedge f = 0 \text{ on } \partial\Omega\},$$

L to be the operator constructed in Theorem 8.16 ($Lw = f$ being equivalent to $dw = f$) and

$$Q(a) = \bigwedge_{r=1}^{k+1} da_{I_r}\,.$$

Since $k \geq 1$, the following estimate holds:

$$\|Q(u) - Q(v)\|_Y \leq c(\|u\|_X, \|v\|_X)\|u - v\|_X,$$

with

$$c(s,t) = C\,(s+t)\,,$$

where C is a constant given using Theorem 16.28.

Chapter 19
Degree Theory

19.1 Definition and Main Properties

We begin recalling some results on the topological degree (see, e.g., [43] or [88] for further details). We start by defining the degree for C^1 maps.

Definition 19.1. Let Ω be a bounded open set in $\mathbb{R}^n$, $\varphi \in C^1(\overline{\Omega};\mathbb{R}^n)$ and

$$Z_\varphi = \{x \in \overline{\Omega} : \det \nabla\varphi(x) = 0\}.$$

For every $p \in \mathbb{R}^n$ such that

$$p \notin \varphi(\partial\Omega) \cup \varphi(Z_\varphi),$$

we define the integer $\deg(\varphi, \Omega, p)$ as

$$\deg(\varphi, \Omega, p) = \sum_{x \in \Omega : \varphi(x) = p} \operatorname{sign}(\det \nabla\varphi(x)),$$

with the convention $\deg(\varphi, \Omega, p) = 0$ if $\{x \in \Omega : \varphi(x) = p\} = \emptyset$.

Remark 19.2. The fact that $p \notin \varphi(Z_\varphi)$ ensures that the set $\{\varphi^{-1}(p)\}$ is finite and therefore, $\deg(\varphi, \Omega, p)$ is well defined.

It is possible to extend the definition of $\deg(\varphi, \Omega, p)$ to $\varphi \in C^0(\overline{\Omega};\mathbb{R}^n)$ and $p \notin \varphi(\partial\Omega)$. Keeping the same notation, this extension of the degree has the following properties.

Theorem 19.3. *Let Ω be a bounded open set in $\mathbb{R}^n$, $\varphi, \psi \in C^0(\overline{\Omega};\mathbb{R}^n)$ and $H \in C^0([0,1] \times \overline{\Omega};\mathbb{R}^n)$. The following properties are then verified:*

(i) If $p \notin H(t,x)$ for every $0 \le t \le 1$ and $x \in \partial\Omega$, then

$$\deg(H(0,\cdot), \Omega, p) = \deg(H(1,\cdot), \Omega, p).$$

In particular, if $\varphi = \psi$ *on* $\partial\Omega$ *and* $p \notin \varphi(\partial\Omega)$*, then (choosing* $H(t,x) = (1-t)\varphi(x) + t\psi(x)$*)*

$$\deg(\varphi,\Omega,p) = \deg(\psi,\Omega,p).$$

(ii) If $p \notin \varphi(\partial\Omega)$ *is such that* $\deg(\varphi,\Omega,p) \neq 0$*, then there exists* $x \in \Omega$ *such that* $\varphi(x) = p$.

(iii) Let $p \notin \varphi(\partial\Omega)$*. Then for every* p' *in the connected component of* $(\varphi(\partial\Omega))^c$ *containing* p*,*

$$\deg(\varphi,\Omega,p) = \deg(\varphi,\Omega,p').$$

We have the following immediate corollary.

Corollary 19.4. *Let* Ω *be a bounded open set in* $\mathbb{R}^n$ *and* $\varphi \in C^0(\overline{\Omega};\mathbb{R}^n)$ *such that* $\varphi = \mathrm{id}$ *on* $\partial\Omega$*. Then*

$$\deg(\varphi,\Omega,p) = \begin{cases} 1 & \text{if } p \in \Omega \\ 0 & \text{if } p \notin \overline{\Omega} \end{cases} \tag{19.1}$$

and

$$\varphi(\Omega) \supset \Omega \quad \text{and} \quad \varphi(\overline{\Omega}) \supset \overline{\Omega}. \tag{19.2}$$

Proof. Noticing that

$$\deg(\mathrm{id},\Omega,p) = \begin{cases} 1 & \text{if } p \in \Omega \\ 0 & \text{if } p \notin \overline{\Omega}, \end{cases}$$

we immediately obtain (19.1) and (19.2) using Theorems 19.3(i) and 19.3(ii). □

Finally we recall the Sard theorem and the invariance of domain theorem (see, e.g., [43]).

Theorem 19.5 (Sard theorem). *Let* Ω *be a bounded open Lipschitz set in* $\mathbb{R}^n$ *and* $\varphi \in C^1(\overline{\Omega};\mathbb{R}^n)$*. Then*

$$\operatorname{meas}(\varphi(Z_\varphi)) = 0.$$

Theorem 19.6 (Invariance of domain theorem). *Let* Ω *be a bounded open set in* $\mathbb{R}^n$ *and* $\varphi \in C^0(\overline{\Omega};\mathbb{R}^n)$ *be a one-to-one map. Then*

$$\varphi(\Omega) \text{ is open} \quad \text{and} \quad \varphi(\partial\Omega) = \partial(\varphi(\Omega)).$$

19.2 General Change of Variables Formula

First, we recall the classic change of variables formula.

Theorem 19.7. *Let Ω and U be two bounded open sets in $\mathbb{R}^n$, $\varphi \in \operatorname{Diff}^1(\overline{\Omega};\overline{U})$ and $g \in C^0(\overline{\Omega})$. Then the following formula holds:*

$$\int_U g(y)\,dy = \int_{\varphi(\Omega)} g(y)\,dy = \int_\Omega g(\varphi(x))|\det\nabla\varphi(x)|\,dx.$$

We now give a generalization of the above theorem whose proof can be found, for example, in [43, Theorem 5.27].

Theorem 19.8. *Let Ω be a bounded open Lipschitz set in $\mathbb{R}^n$. Let $\varphi \in C^1(\overline{\Omega};\mathbb{R}^n)$ and $g \in C^0(\mathbb{R}^n)$. Then the following formula holds:*

$$\int_{\mathbb{R}^n} g(y)\deg(\varphi,\Omega,y)\,dy = \int_\Omega g(\varphi(x))\det\nabla\varphi(x)\,dx.$$

Remark 19.9. (i) Although $\deg(\varphi,\Omega,y)$ is not defined for $y \in \varphi(\partial\Omega)$, the left-hand side makes sense since $\operatorname{meas}(\varphi(\partial\Omega)) = 0$. This last equality holds due to the fact that φ is Lipschitz and $\operatorname{meas}(\partial\Omega) = 0$.

(ii) If $\varphi \in \operatorname{Diff}^1(\overline{\Omega};\overline{\varphi(\Omega)})$ then, by definition of deg,

$$\deg(\varphi,\Omega,y) = \begin{cases} \operatorname{sign}(\det\nabla\varphi) & \text{if } y \in \varphi(\Omega) \\ 0 & \text{if } y \notin \overline{\varphi(\Omega)}. \end{cases}$$

Therefore, Theorem 19.8 is indeed a generalization of Theorem 19.7.

(iii) If $\varphi = \mathrm{id}$ on $\partial\Omega$, then, by Corollary 19.4, we have that

$$\deg(\varphi,\Omega,y) = \begin{cases} 1 & \text{if } y \in \Omega \\ 0 & \text{if } y \notin \overline{\Omega} \end{cases}$$

and, thus,

$$\int_\Omega g(y) = \int_\Omega g(\varphi(x))\det\nabla\varphi(x)\,dx. \tag{19.3}$$

As direct consequence we have the following result (cf. also Theorem 8.35 in [28]).

Corollary 19.10. *Let Ω be a bounded open Lipschitz set in $\mathbb{R}^n$ and $u \in C^1(\overline{\Omega};\mathbb{R}^n)$, with $u = 0$ on $\partial\Omega$. Then*

$$\int_\Omega \det(I + \nabla u) = \operatorname{meas}\Omega,$$

where I stands for the identity matrix in $\mathbb{R}^{n\times n}$. As a consequence, if $\varphi \in \operatorname{Diff}^1(\overline{\Omega};\overline{\Omega})$, with $\varphi = \mathrm{id}$ on $\partial\Omega$, then

$$\det\nabla\varphi > 0 \quad \text{in } \overline{\Omega}.$$

Proof. Letting $g = 1$ and $\varphi = \mathrm{id} + u$, we have the result applying the identity (19.3). □

19.3 Local and Global Invertibility

As an application of these above properties we have the following results. We first give a sufficient condition for a map to be a homeomorphism.

Lemma 19.11. *Let Ω be a bounded open set in $\mathbb{R}^n$ and $\varphi \in C^0(\overline{\Omega};\mathbb{R}^n)$ be one-to-one such that $\varphi = \mathrm{id}$ on $\partial\Omega$. Then $\varphi \in \mathrm{Hom}(\overline{\Omega};\overline{\Omega})$.*

Proof. It can be easily seen, working on each connected component of Ω, that we can assume Ω to be connected. By the boundedness of Ω and the continuity of φ, we have that $\varphi(F)$ is closed for every closed set F in $\mathbb{R}^n$ such that $F \subset \Omega$. Since φ is one-to-one, we obtain

$$\varphi \in \mathrm{Hom}(\overline{\Omega};\varphi(\overline{\Omega})).$$

Let us prove that

$$\varphi(\overline{\Omega}) = \overline{\Omega},$$

which will end the proof. Due to (19.2), it is enough to prove that

$$\varphi(\overline{\Omega}) \subset \overline{\Omega}. \tag{19.4}$$

By Theorem 19.6, we have that $\varphi(\partial\Omega) = \partial(\varphi(\Omega))$. Thus, since $\varphi = \mathrm{id}$ on $\partial\Omega$ and φ is one-to-one, we get

$$\partial\Omega = \partial(\varphi(\Omega)) \quad \text{and} \quad \varphi(\Omega) \cap \partial\Omega = \emptyset. \tag{19.5}$$

Suppose by contradiction that there exists $x \in \overline{\Omega}$ such that $\varphi(x) \in (\overline{\Omega})^c$. Since φ is the identity map on $\partial\Omega$, we have that $x \in \Omega$. Let us now consider $y \in \Omega$ such that $\varphi(y) \in \Omega$ (such a y surely exists by (19.2)) and let $c \in C^0([0,1];\Omega)$ be a path connecting x and y. Then, by continuity, there exists $0 < t < 1$ such that $\varphi(c(t)) \in \partial\Omega$, contradicting (19.5), which concludes the proof. □

We now provide a sufficient condition for the invertibility of functions in $C^1(\overline{\Omega};\mathbb{R}^n)$. A similar result can be found in Meisters and Olech [75].

Theorem 19.12. *Let Ω be a bounded open set in $\mathbb{R}^n$ and let $\varphi \in C^1(\overline{\Omega};\mathbb{R}^n)$ be such that*

$$\begin{cases} \det\nabla\varphi > 0 & \text{in } \overline{\Omega}, \\ \varphi = \mathrm{id} & \text{on } \partial\Omega. \end{cases}$$

Then $\varphi \in \mathrm{Diff}^1(\overline{\Omega};\overline{\Omega})$. Moreover, if $c > 0$ is such that

$$\left\| \frac{1}{\det\nabla\varphi} \right\|_{C^0}, \|\nabla\varphi\|_{C^0} \le c,$$

then there exists a constant $C = C(c,\Omega) > 0$ such that

$$\|\varphi^{-1}\|_{C^1} \le C\,\|\varphi\|_{C^1}.$$

Remark 19.13. Under the weaker hypotheses, $\det \nabla \varphi \geq 0$ in $\overline{\Omega}$, $\varphi = \mathrm{id}$ on $\partial\Omega$ and $Z_\varphi \cap \Omega$ does not have accumulation points, it can be proved that $\varphi \in C^1(\overline{\Omega};\overline{\Omega}) \cap \mathrm{Hom}(\overline{\Omega};\overline{\Omega})$; see [60].

Proof. We divide the proof into three steps.

Step 1. We first prove that $\varphi(\Omega) = \Omega$. Using (19.2), we know that

$$\varphi(\Omega) \supset \Omega.$$

Let us show the reverse inclusion (i.e., $\varphi(\Omega) \subset \Omega$). We first prove that $\varphi(\Omega) \subset \overline{\Omega}$ and then conclude. By contradiction, let $x \in \Omega$ be such that $\varphi(x) \notin \overline{\Omega}$. By definition of the degree, we obtain

$$0 < \sum_{y \in \Omega : \varphi(y) = \varphi(x)} \operatorname{sign}(\det \nabla \varphi(y)) = \deg(\varphi, \Omega, \varphi(x)),$$

whereas, using (19.1), we have

$$\deg(\varphi, \Omega, \varphi(x)) = 0,$$

which is absurd.

To conclude, suppose that $x \in \Omega$ and $\varphi(x) \in \overline{\Omega} \setminus \Omega = \partial\Omega$. By the inverse function theorem, which can be applied since $\det \nabla \varphi(x) > 0$, there exists a neighborhood of x such that the restriction of φ on this set is one-to-one and onto a neighborhood of $\varphi(x) \in \partial\Omega$. In particular, this implies the existence of $y \in \Omega$ such that $\varphi(y) \notin \overline{\Omega}$, which contradicts what has just been proved.

Step 2. Since $\varphi(\Omega) = \Omega$ and $\varphi = \mathrm{id}$ on $\partial\Omega$, we have that

$$\varphi(\overline{\Omega}) = \overline{\Omega}.$$

Moreover, $\varphi(\partial\Omega) \cap \varphi(\Omega) = \partial\Omega \cap \Omega = \emptyset$. Thus, it suffices to show that the restriction of φ to Ω is one-to-one to conclude. We reason by contradiction. We assume that there exists $p \in \Omega$ which is the image of at least two elements in Ω. By definition of the degree, we obtain

$$2 < \sum_{x \in \Omega : \varphi(x) = p} \operatorname{sign}(\det \nabla \varphi(x)) = \deg(\varphi, \Omega, p),$$

whereas, using (19.1), we have

$$\deg(\varphi, \Omega, p) = 1,$$

which is the desired contradiction.

Step 3. We finally establish the estimate. We clearly have

$$\left\|\varphi^{-1}\right\|_{C^0} = \|\varphi\|_{C^0} = \|\mathrm{id}\|_{C^0}.$$

We also have

$$\left\|\nabla\varphi^{-1}\right\|_{C^0} = \left\|(\nabla\varphi)^{-1} \circ \varphi\right\|_{C^0} = \left\|(\nabla\varphi)^{-1}\right\|_{C^0}.$$

Since

$$(\nabla\varphi)^{-1} = \frac{(\operatorname{adj}\nabla\varphi)^t}{\det\nabla\varphi},$$

we immediately get that

$$\|\nabla\varphi^{-1}\|_{C^0} \leq C\|\nabla\varphi\|_{C^0}.$$

This finishes the proof. □

We also have a necessary condition for φ to be a C^1 homeomorphism.

Proposition 19.14. *Let $\Omega \subset \mathbb{R}^n$ be a bounded open set in $\mathbb{R}^n$ and $\varphi \in C^1(\overline{\Omega};\overline{\Omega}) \cap \operatorname{Hom}(\overline{\Omega};\overline{\Omega})$, with $\varphi = \operatorname{id}$ on $\partial\Omega$. Then*

$$\det\nabla\varphi(x) \geq 0 \quad \text{in } \overline{\Omega} \quad \text{and} \quad \operatorname{int}(Z_\varphi) = \emptyset.$$

Proof. We split the proof into two steps.

Step 1. We show that $\det\nabla\varphi \geq 0$. By contradiction, suppose that there exists $y \in \overline{\Omega}$ such that $\det\nabla\varphi(y) < 0$. By continuity, without loss of generality, we can assume that $y \in \Omega$. In particular, $y \notin Z_\varphi$ and since φ is one-to-one, we obtain

$$\varphi(y) \notin \varphi(Z_\varphi) \cup \varphi(\partial\Omega) = \varphi(Z_\varphi) \cup \partial\Omega.$$

By definition of $\deg(\varphi,\Omega,\varphi(y))$ and by (19.1), we have

$$1 = \deg(\varphi,\Omega,\varphi(y)) = \sum_{z:\varphi(z)=\varphi(y)} \operatorname{sign}(\det\nabla\varphi(z)).$$

Since $\operatorname{sign}(\det\nabla\varphi(y)) = -1$, the above equality implies that $\varphi^{-1}(\varphi(y))$ is not a singleton, which is absurd.

Step 2. We prove that $\operatorname{int}(Z_\varphi) = \emptyset$. By contradiction, suppose that $\operatorname{int}(Z_\varphi) \neq \emptyset$. By continuity of φ^{-1}, we have that

$$\varphi\left(\operatorname{int}\left(Z_\varphi\right)\right) = (\varphi^{-1})^{-1}(\operatorname{int}(Z_\varphi)) \neq \emptyset$$

and $(\varphi^{-1})^{-1}\left(\operatorname{int}(Z_\varphi)\right)$ is open. Therefore,

$$\operatorname{int}\left(\varphi\left(Z_\varphi\right)\right) \neq \emptyset,$$

contradicting the Sard theorem. □

We conclude with some other necessary conditions.

Proposition 19.15. *Let Ω be a bounded open set in $\mathbb{R}^n$ and let $\varphi \in C^1(\overline{\Omega};\mathbb{R}^n)$ be such that*

$$\begin{cases} \det\nabla\varphi \geq 0 & \text{in } \Omega, \\ \varphi = \operatorname{id} & \text{on } \partial\Omega. \end{cases} \tag{19.6}$$

Then

$$\operatorname{int}(\varphi(\Omega)) = \Omega. \tag{19.7}$$

Moreover, if

$$\operatorname{int}(Z_\varphi) = \emptyset, \tag{19.8}$$

then

$$\varphi(\overline{\Omega}) = \overline{\Omega}. \tag{19.9}$$

Finally, if (19.8) does not hold, then (19.9) may fail for some $\varphi \in C^1(\overline{\Omega};\mathbb{R}^n)$.

Proof. We divide the proof into three steps.

Step 1. We already know that $\varphi(\Omega) \supset \Omega$ and thus

$$\operatorname{int}(\varphi(\Omega)) \supset \Omega.$$

Let us show the reverse inclusion. We proceed by contradiction and assume that

$$\operatorname{int}(\varphi(\Omega)) \cap \Omega^c \neq \emptyset;$$

thus, since $\operatorname{int}(\varphi(\Omega))$ is open,

$$\operatorname{int}(\varphi(\Omega)) \cap \left(\overline{\Omega}\right)^c \neq \emptyset.$$

Therefore, there exist y and $\varepsilon > 0$ such that

$$B_\varepsilon(y) \subset \operatorname{int}(\varphi(\Omega)) \cap \left(\overline{\Omega}\right)^c \subset \varphi(\Omega) \cap \left(\overline{\Omega}\right)^c. \tag{19.10}$$

We claim that

$$\varphi\left(\Omega \setminus Z_\varphi\right) \subset \Omega. \tag{19.11}$$

Indeed, let $y \in \varphi\left(\Omega \setminus Z_\varphi\right)$ and let us show that $y \in \Omega$. Since $y \in \varphi\left(\Omega \setminus Z_\varphi\right)$, we can find $x \in \Omega \setminus Z_\varphi$ such that $y = \varphi(x)$. Since $x \notin Z_\varphi$ and (19.6) holds, we deduce that

$$\det \nabla\varphi(x) > 0.$$

We then proceed exactly as in Step 1 of the proof of Theorem 19.12 to get that

$$y = \varphi(x) \in \Omega.$$

We next combine (19.10) and (19.11) to find

$$\begin{aligned} B_\varepsilon(y) \subset \varphi(\Omega) \cap \left(\overline{\Omega}\right)^c &\subset \left[\varphi\left(\Omega \setminus Z_\varphi\right) \cap \left(\overline{\Omega}\right)^c\right] \cup \left[\varphi\left(Z_\varphi\right) \cap \left(\overline{\Omega}\right)^c\right] \\ &\subset \left[\Omega \cap \left(\overline{\Omega}\right)^c\right] \cup \left[\varphi\left(Z_\varphi\right) \cap \left(\overline{\Omega}\right)^c\right] = \varphi\left(Z_\varphi\right) \cap \left(\overline{\Omega}\right)^c \end{aligned}$$

and, thus,

$$B_\varepsilon(y) \subset \varphi\left(Z_\varphi\right),$$

contradicting the Sard theorem.

Step 2. Let us next prove that (19.8) implies (19.9). Since $\varphi = \mathrm{id}$ on $\partial\Omega$ and (19.2) holds, it is enough to prove that

$$\varphi(\Omega) \subset \overline{\Omega}.$$

By Step 1, we already know that $\varphi\left(\Omega \setminus Z_\varphi\right) \subset \Omega$. Therefore, it remains to establish that

$$\varphi\left(Z_\varphi \cap \Omega\right) \subset \overline{\Omega}.$$

So let $x \in Z_\varphi \cap \Omega$; then (using (19.8)) there exists a sequence $x_\nu \in \Omega \setminus Z_\varphi$ such that $x_\nu \to x$. Since (19.11) holds, we deduce that $\varphi(x_\nu) \in \Omega$ and, hence, $\varphi(x) \in \overline{\Omega}$. We therefore have the claim.

Step 3. We show that (19.9) may fail if (19.8) does not hold. Set $\Omega = B_1$ be the unit ball in $\mathbb{R}^2$ and consider

$$\varphi(x_1,x_2) = \rho(x_1^2+x_2^2)(x_1,x_2) + \eta(x_1^2+x_2^2)(x_1,0),$$

where

$$\begin{cases} \rho \in C^\infty([0,1];[0,\infty)), \\ \rho \equiv 0 \text{ in } [0,1/2] \quad \rho \equiv 1 \text{ in } [3/4,1], \\ \rho' \geq 0 \text{ in } [0,1] \end{cases} \quad \text{and} \quad \begin{cases} \eta \in C^\infty([0,1];[0,\infty)), \\ \operatorname{supp}\eta \subset (0,1/2), \\ \eta(1/4) = 4. \end{cases}$$

Let us verify the hypotheses of the proposition. Obviously, $\varphi \in C^1(\overline{B}_1;\mathbb{R}^n)$ and $\operatorname{supp}(\varphi - \mathrm{id}) \subset B_1$. Let us now check that $\det\nabla\varphi \geq 0$. We separately consider two cases.

Case 1 $(1/2 \leq |x|^2 \leq 1)$. A straightforward computation implies that

$$\begin{aligned} \det\nabla\varphi(x) &= (2x_1^2\rho' + \rho)(2x_2^2\rho' + \rho) - 4x_1^2x_2^2\rho'^2 \\ &= 4x_1^2x_2^2\rho'^2 + 2|x|^2\rho\rho' + \rho^2 - 4x_1^2x_2^2\rho'^2 \\ &= 2|x|^2\rho\rho' + \rho^2 \geq 0. \end{aligned}$$

Case 2 $(0 \leq |x|^2 \leq 1/2)$. By definition of φ, it immediately follows that $\det\nabla\varphi = 0$. Thus, $\det\nabla\varphi \geq 0$.

Since

$$\varphi(1/2,0) = \eta(1/4)(1/2,0) = (2,0) \notin \overline{B}_1,$$

we have the claim and this concludes the proof of the proposition. □

References

1. Abraham R., Marsden J.E. and Ratiu T., *Manifolds, tensor analysis, and applications*, second edition, Springer-Verlag, New York, 1988.
2. Adams R.A., *Sobolev spaces*, Academic Press, New York, 1975; second edition with Fournier J.J.F., 2003.
3. Agmon S., Douglis A. and Nirenberg L., Estimates near the boundary for solutions of elliptic partial differential equations satisfying general boundary conditions I, *Commun. Pure Appl. Math.,* **12** (1959), 623–727.
4. Agmon S., Douglis A. and Nirenberg L., Estimates near the boundary for solutions of elliptic partial differential equations satisfying general boundary conditions II, *Commun. Pure Appl. Math.,* **17** (1964), 35–92.
5. Alt H.W., *Lineare Funktionalanalysis*, fourth edition, Springer-Verlag, Berlin, 2002.
6. Amrouche C., Bernardi C., Dauge M., Girault V., Vector potentials in three-dimensional non-smooth domains, *Math. Methods Appl. Sci.,* **21** (1998), 823–864.
7. Bandyopadhyay S., private communication, 2010.
8. Bandyopadhyay S. and Dacorogna B., On the pullback equation $\varphi^*(g) = f$, *Ann. Inst. H. Poincaré Anal. Non Linéaire,* **26** (2009), 1717–1741.
9. Bandyopadhyay S., Dacorogna B. and Kneuss O., The pullback equation for degenerate forms, *Discrete Continuous Dynami. Syst. A,* **27** (2010), 657–691.
10. Banyaga A., Formes-volume sur les variétés à bord, *Enseignement Math.,* **20** (1974), 127–131.
11. Barbarosie C., Representation of divergence-free vector fields, *Q. Appl. Math.*, to appear.
12. Bogovski M.E., Solution of the first boundary value problem for the equation of continuity of an incompressible medium, *Soviet Math. Dokl.,* **20** (1979), 1094–1098.
13. Bolik J., H Weyl's boundary value problems for differential forms, *Differ. Integ. Eqs.,* **14** (2001), 937–952.
14. Borchers W. and Sohr H., On the equations $\operatorname{rot} v = g$ and $\operatorname{div} u = f$ with zero boundary conditions, *Hokkaido Math. J.,* **19** (1990), 67–87.
15. Bourbaki N., *Éléments de mathématique. Algèbre. Chapitres 1 à 3*, Hermann, Paris, 1970.
16. Bourgain J. and Brézis H., Sur l'équation $\operatorname{div} u = f$, *C. R. Acad. Sci. Paris Sér. I Math.,* **334** (2002), 973–976.
17. Brézis H., *Analyse fonctionelle, théorie et applications*, Dunod, Paris, 1999.
18. Bryant R.L., Chern S.S., Gardner R.B., Goldschmidt H.L. and Griffiths P.A., *Exterior differential systems*, Mathematical Sciences Research Institute Publications 18, Springer-Verlag, New York, 1991.
19. Burago D. and Kleiner B., Separated nets in Euclidean space and Jacobian of biLipschitz maps, *Geom. Funct. Anal.,* **8** (1998), 273–282.
20. Calderon A.P., Lebesgue spaces of differentiable functions and distributions, *Proc. Symp. Pure Math.*, Vol. IV, American Mathematical Society, Providence, RI, 1961, pp. 33–49.
21. Cartan E., *Selecta*, Gauthier-Villars, Paris, 1939.

22. Coddington E.A. and Levinson N., *Theory of ordinary differential equations*, McGraw-Hill Book Company Inc., New York, 1955.
23. Csató G., PhD thesis, EPFL Lausanne (2012).
24. Csató G. and Dacorogna B., An identity involving exterior derivatives and applications to Gaffney inequality, *Discrete Continuous Dynam. Syst.*, to appear.
25. Cupini G., Dacorogna B. and Kneuss O., On the equation $\det \nabla u = f$ with no sign hypothesis, *Calc. Var. Partial Differ. Equ.*, **36** (2009), 251–283.
26. Dacorogna B., A relaxation theorem and its applications to the equilibrium of gases, *Arch. Rational Mech. Anal.,* **77** (1981), 359–386.
27. Dacorogna B., Existence and regularity of solutions of $dw = f$ with Dirichlet boundary conditions, *Nonlinear Problems in Mathematical Physics and Related Topics*, Int. Math. Ser. (N. Y.), **1**, Kluwer/Plenum, New York (2002), 67–82.
28. Dacorogna B., *Direct methods in the calculus of variations*, second edition, Springer-Verlag, New York, 2007.
29. Dacorogna B., *Introduction to the calculus of variations*, second edition, Imperial College Press, London, 2009.
30. Dacorogna B., Fusco N. and Tartar L., On the solvability of the equation $\operatorname{div} u = f$ in L^1 and in C^0, *Atti Accad. Naz. Lincei Cl. Sci. Fis. Mat. Natur. Rend. Lincei* (9) *Mat. Appl.*, **14** (2003), 239–245.
31. Dacorogna B. and Kneuss O., Divisibility in Grassmann algebra, *Linear Multilinear Alg.*, to appear.
32. Dacorogna B. and Kneuss O., A global version of Darboux theorem with optimal regularity and Dirichlet condition, *Advan. Differ. Eq.*, **16** (2011), 325–360.
33. Dacorogna B. and Moser J., On a partial differential equation involving the Jacobian determinant, *Ann. Inst. H. Poincaré Anal. Non Linéaire,* **7** (1990), 1–26.
34. Darboux G., Sur le problème de Pfaff, *Bull Sci.Math.* **6** (1882), 14–36, 49–68.
35. Dautray R. and Lions J.L., *Analyse mathématique et calcul numérique*, Masson, Paris, 1988.
36. de la Llave R. and Obaya R., Regularity of the composition operator in spaces of Hölder functions, *Discrete Continuous Dynami. Syst. A,* **5** (1999), 157–184.
37. do Carmo M. P., *Differential forms and applications*, Universitext, Springer-Verlag, Berlin, 1994.
38. Duff G.F. and Spencer D.C., Harmonic tensors on Riemannian manifolds with boundary, *Ann. Math.,* **56** (1952), 128–156.
39. Duvaut G. and Lions J.L., *Inequalities in mechanics and physics*, Springer-Verlag, Berlin, 1976.
40. Edmunds D.E. and Evans W.D., *Spectral theory and differential operators*, Oxford Science Publications, Oxford, 1987.
41. Evans L.C., *Partial differential equations*, Graduate Studies in Mathematics, **19**, American Mathematical Society, Providence, RI, 1998.
42. Fefferman C., Whitney's extension problems and interpolation of data, *Bull. Amer. Math. Soc.,* **46** (2009), 207–220.
43. Fonseca I. and Gangbo W., *Degree theory in analysis and applications*, Oxford University Press, New York, 1995.
44. Gaffney M.P., The harmonic operator for exterior differential forms, *Proc. Natl. Acad. Sci. U. S. A.,* **37** (1951), 48–50.
45. Gaffney M.P., Hilbert space methods in the theory of harmonic integrals, *Trans. Amer. Math. Soc,* **78** (1955), 426–444.
46. Galdi G.P., *An introduction to the mathematical theory of the Navier-Stokes equations*, Springer-Verlag, New York, 1994.
47. Georgescu V., Some boundary value problems for differential forms on compact Riemannian manifolds, *Ann. Mat. Pura Appl.* **122** (1979), 159–198.
48. Gilbarg D. and Hörmander L., Intermediate Schauder estimates, *Arch. Rational Mech. Anal.,* **74** (1980), 297–318.
49. Gilbarg D. and Trudinger N.S., *Elliptic partial differential equations of second order*, Springer-Verlag, Berlin, 1977.

50. Girault V. and Raviart P.A., *Finite element approximation of the Navier-Stokes equations*, Lecture Notes in Math. **749**, Springer-Verlag, Berlin, 1979.
51. Godbillon C., *Géométrie différentielle et mécanique analytique*, Hermann, Paris, 1969.
52. Godement R., *Cours d'algèbre*, Hermann, Paris, 1963.
53. Golubitsky M. and Tischler D., A survey on the singularities and stability of differential forms, Astérisque, **59-60**, Société Mathématique de France, Paris, 1978.
54. Greub W. H., *Mutilinear algebra*, Springer-Verlag, New York, 1967.
55. Hörmander L., The boundary problems of physical geodesy, *Arch. Rational Mech. Anal.*, **62** (1976), 1–52.
56. Horn R.A. and Johnson C.A., *Matrix analysis*, Cambridge University Press, Cambridge, 1985.
57. Iwaniec T. and Martin G., *Geometric function theory and non-linear analysis*, Oxford University Press, Oxford, 2001.
58. Iwaniec T., Scott C. and Stroffolini B., Nonlinear Hodge theory on manifolds with boundary, *Annali Mat. Pura Appl.*, **177** (1999), 37–115.
59. Kapitanskii L.V. and Pileckas K., Certain problems of vector analysis, *J. Soviet Math.*, **32** (1986), 469–483.
60. Kneuss O., PhD thesis, EPFL Lausanne (2011).
61. Krantz S.G. and Parks H.R., *The geometry of domains in space*, Birkhäuser, Boston, 1999.
62. Krantz S.G. and Parks H.R., *The implicit function theorem. History, theory and applications*, Birkhäuser, Boston, 2002.
63. Kress R., Potentialtheoretische Randwertprobleme bei Tensorfelderrn beliebiger Dimensions und beliebigen Ranges, *Arch. Rational Mech. Anal.*, **47** (1972), 59–80.
64. Ladyzhenskaya O.A., *The mathematical theory of viscous incompressible flow*, Gordon and Breach, New York, 1969.
65. Ladyzhenskaya O.A. and Solonnikov V.A., Some problems of vector analysis and generalized formulations of boundary value problems for the Navier-Stokes equations, *J. Soviet Math.*, **10** (1978), 257–286.
66. Ladyzhenskaya O.A. and Uraltseva N.N., *Linear and quasilinear elliptic equations*, Academic Press, New York, 1968.
67. Lang S., *Algebra*, Springer-Verlag, New York, 2002.
68. Lee J.M., *Introduction to smooth manifolds*, Springer-Verlag, New York, 2003.
69. Lieberman G.M., The quasilinear Dirichlet problem with decreased regularity at the boundary, *Commun. Partial Differ. Equ.*, **6** (1981), 437–497.
70. Lieberman G.M., The Dirichlet problem for quasilinear elliptic equations with Hölder continuous boundary values, *Arch. Rational Mech. Anal.*, **79** (1982), 305–323.
71. Martinet J., Sur les singularités des formes différentielles, *Ann. Inst. Fourier*, **20** (1970), 95–178.
72. Mc Duff D. and Salamon D., *Introduction to symplectic topology*, second edition, Oxford Science Publications, Oxford, 1998.
73. Mc Mullen C.T., Lipschitz maps and nets in Euclidean space, *Geom. Funct. Anal.*, **8** (1998), 304–314.
74. Marcus M., *Finite dimensional multilinear algebra. Part II*, Marcel Dekker, New York, 1975.
75. Meisters G.H. and Olech C., Locally one-to-one mappings and a classical theorem on schlicht functions, *Duke Math. J.*, **30** (1970), 63–80.
76. Morrey C.B., A variational method in the theory of harmonic integrals II, *Amer. J. Math.*, **78** (1956), 137–170.
77. Morrey C.B., *Multiple integrals in the calculus of variations*, Springer-Verlag, Berlin, 1966.
78. Moser J., On the volume elements on a manifold, *Trans. Amer. Math. Soc.*, **120** (1965), 286–294.
79. Necas J., *Les méthodes directes en théorie des équations elliptiques*, Masson, Paris, 1967.
80. Olver P.J., *Equivalence, invariants, and symmetry*, Cambridge University Press, Cambridge, 1995.

81. Pelletier F., Singularités d'ordre supérieur de 1-formes, 2-formes et équations de Pfaff, *Inst. Hautes Études Sci. Publ. Math.*, **61** (1985), 129–169.
82. Postnikov M., *Leçons de géométrie: algèbre linéaire et géométrie différentielle,* Mir, Moscow, 1988.
83. Preiss D., Additional regularity for Lipschitz solutions of pde, *J. Reine Angew. Math.,* **485** (1997), 197–207.
84. Reimann H.M., Harmonische Funktionen und Jacobi-Determinanten von Diffeomorphismen, *Comment. Math. Helv.,* **47** (1972), 397–408.
85. Rivière T. and Ye D., Resolutions of the prescribed volume form equation, *Nonlinear Differ. Eq. Appl.,* **3** (1996), 323–369.
86. Roussarie R., Modèles locaux de champs et de formes, Astérisque, **30**, Société Mathématique de France, Paris, 1975.
87. Rudin W., *Real and complex analysis,* Mc Graw-Hill, New York, 1966.
88. Schwartz J.T., *Nonlinear functional analysis,* Gordon and Breach, New York, 1969.
89. Schwarz G., *Hodge decomposition—A method for solving boundary value problems*, Lecture Notes in Math. **1607**, Springer-Verlag, Berlin, 1995.
90. Serre, D., *Matrices. Theory and applications,* Springer-Verlag, New York, 2002.
91. Spivak M., *A comprehensive introduction to differential geometry*, third edition, Publish or Perish, Houston, TX, 1999.
92. Stein E.M., *Singular integrals and differentiability properties of functions*, Princeton University Press, Princeton, NJ, 1970.
93. Sternberg S., *Lectures on differential geometry,* Prentice-Hall, Englewood Cliffs, NJ, 1964.
94. Tartar L., *Topics in nonlinear analysis,* Preprint, University of Wisconsin, Madison, 1975.
95. Tartar L., unpublished, 1978.
96. Taylor M.E., *Partial differential equations*, Vol. 1, Springer-Verlag, New York, 1996.
97. Turiel F.J., Some remarks concerning the transitive geometric structures defined by exterior forms, *Seminar on Differential Geometry*, 1981–1982, *Exp. No. 7*, Université des Sciences et Techniques du Languedoc, Montpellier, 1982.
98. Turiel F.J., Classification locale des 3-formes fermées infinitésimalement transitives à cinq variables, *Cahiers Mathématiques,* **30**, Université des Sciences et Techniques du Languedoc, Montpellier, 1984.
99. Turiel F.J., Un théorème de Darboux pour un couple de formes symplectiques. Feuilletages riemanniens, quantification géométrique et mécanique (Lyon, 1986), Hermann, Paris, 1988. pp. 103–121.
100. Turiel F.J., Classification locale simultanée de deux formes symplectiques compatibles, *Manuscr. Math.* **82** (1994), 349–362.
101. Turiel F.J., Classification globale des formes différentielles transitives sur la sphère S^5, Feuilletages et systèmes intégrables (Montpellier, 1995), Progr. Math., **145**, Birkhäuser Boston, 1997, pp. 157–168.
102. Turiel F.J., Sur certaines n-formes en dimension $2n$, *C. R. Acad. Sci. Paris Sér. I Math.* **333** (2001), 471–474.
103. Von Wahl W., On necessary and sufficient conditions for the solvability of the equations $\operatorname{rot} u = \gamma$ and $\operatorname{div} u = \varepsilon$ with u vanishing on the boundary, Lecture Notes in Math. **1431**, Springer-Verlag, Berlin, 1990, pp. 152–157.
104. Von Wahl W., *Vorlesung über das Aussenraumproblem für die instationären Gleichungen von Navier-Stokes*; Rudolph-Lipschitz-Vorlesung. Sonderforschungsbereich 256. Nichtlineare Partielle Differentialgleichungen, Bonn, 1989.
105. Widman K.O., Inequalities for the Green function and boundary continuity of the gradient of solutions of elliptic differential equations, *Math. Scand.,* **21** (1967), 17–37.
106. Ye D., Prescribing the Jacobian determinant in Sobolev spaces, *Ann. Inst. H. Poincaré Anal. Non Linéaire,* **11** (1994), 275–296.
107. Zehnder E., Note on smoothing symplectic and volume preserving diffeomorphisms, Lecture Notes in Mathematics **597**, Springer-Verlag, Berlin, 1976, pp. 828–855.
108. Zhitomirskiĭ M.Y., Degeneration of differential 1−forms and Pfaffian structures, *Usp. Mat. Nauk* **46** (1991), 47–78; translation in *Russian Math. Surveys* **46** (1991), 53–90.

Further Reading

1. Abraham R. and Marsden J. E., *Foundations of mechanics*, second edition, Advanced Book Program, Benjamin/Cummings Publishing Co., Inc., Reading, MA, 1978.
2. Bachman D., *A Geometric approach to differential forms*, Birkhäuser, Boston, 2006.
3. Bolik J., Boundary value problems for differential forms on compact Riemannian manifolds, Part II, *Analysis (Munich)* **27** (2007), 477–493.
4. Bourgain J. and Brézis H., New estimates for elliptic equations and Hodge type systems, *J. Eur. Math. Soc.*, **9** (2007), 277–315.
5. Brézis H. and Nguyen H.M., The Jacobian determinant revisited, preprint (2010).
6. Brézis H. and Nguyen H.M., On the distributional Jacobian of maps from $\mathbb{S}^n$ into $\mathbb{S}^n$ in fractional Sobolev and Hölder spaces, preprint (2010).
7. Cartan E., *Les systèmes différentiels extérieurs et leurs applications géométriques,* Hermann, Paris, 1945.
8. Choquet-Bruhat Y., De Witt-Morette C. and Dillard-Bleick M., *Analysis, manifolds and physics*, North-Holland, Amsterdam, 1977.
9. Debever R., Quelques problèmes d'équivalence de formes différentielles alternées, *Acad. Roy. Belgique Bull. Cl. Sci.*, **31** (1946), 262–277.
10. Deheuvels R., *Tenseurs et spineurs*, Presses Universitaires de France, Paris, 1993.
11. De Rham G., *Differentiable manifolds*, Springer-Verlag, Berlin, 1984.
12. Flanders H., *Differential forms with applications to the physical sciences*, Dover, New York, 1989.
13. Folland G.B., *Introduction to partial differential equations*, Princeton University Press, Princeton, NJ, 1976.
14. Gardner R.B. and Shadwick W.F., An equivalence problem for a two-form and a vector field on $\mathbb{R}^3$. Differential geometry, global analysis, and topology (Halifax, NS, 1990), CMS Conf. Proc. Amer. Math. Soc. 12, American Mathematical Society, Providence, RI, 1991, pp. 41–50.
15. Giaquinta M., Modica G. and Soucek J., *Cartesian currents in the calculus of variations 1*, Springer-Verlag, Berlin, 1998.
16. Hörmander L., *Linear partial differential operators,* Springer-Verlag, Berlin, 1963.
17. Ivey T.A. and Landsberg J.M., *Cartan for beginners: differential geometry via moving frames and exterior differential systems*, Graduate Studies in Mathematics, **61**, American Mathematical Society, Providence, RI, 2003.
18. John F., *Partial differential equations*, Springer-Verlag, Berlin, 1982.
19. Kobayashi S. and Nomizu K., *Foundations of differential geometry. Vol. I*, John Wiley & Sons Inc., New York, 1996.
20. Lions J.L. and Magenes E., *Non-homogeneous boundary value problems and applications I,II,III*, Springer-Verlag, Berlin, 1972.
21. Mitrea D. and Mitrea M., Boundary integral methods for harmonic differential forms in Lipschitz domains, *Electron. Res. Announce. Ameri. Math. Soc.*, **2** (1996), 92–97.

22. Morrey C.B. and Eells J., A variational method in the theory of harmonic integrals, *Ann. Math.*, **63** (1956), 91–128.
23. Papy G., Sur l'arithmétique dans les algèbres de Grassman, *Acad. Roy. Belgique. Cl. Sci. Mém. Coll.*, **26** (1952), 1–108.
24. Picard R., An elementary proof for a compact imbedding result in generalized electromagnetic theory, *Math. Zeitschr.*, **187** (1984), 151–164.
25. Triebel H., *Theory of function spaces II*, Monographs in Mathematics **84**, Birkhäuser, Basel, 1992.
26. Von Wahl W., Estimating ∇u by $\operatorname{div} u$ and $\operatorname{curl} u$, *Math. Methods Appl. Sci.*, **15** (1992), 123–143.

Notations

General Notations

- $|\cdot|$ and $\langle\cdot\,;\cdot\rangle$ denote the usual norm and scalar product in $\mathbb{R}^n$.
- For $E\subset\mathbb{R}^n$, $\overline{E}$, respectively ∂E, $\operatorname{int}E$, $\operatorname{meas}E$, $\operatorname{diam}E$ and E^c stands for the closure, respectively the boundary, the interior, the Lebesgue measure, the diameter and the complement of E.
- $B_\varepsilon(x)=\{y\in\mathbb{R}^n:|y-x|<\varepsilon\}$ and $B_\varepsilon=B_\varepsilon(0)$.
- For $E\subset\mathbb{R}^n$ we denote

$$1_E(x)=\begin{cases}1 & \text{if } x\in E\\ 0 & \text{otherwise.}\end{cases}$$

- For $x=(x_1,\ldots,x_n)\in\mathbb{R}^n$, we let $x'=(x_1,\ldots,x_{n-1})\in\mathbb{R}^{n-1}$.
- For $A,B\subset\mathbb{R}^n$, we let

$$A+B=\{x=a+b:a\in A\quad\text{and}\quad b\in B\}.$$

- The vector space spanned by E is denoted by $\operatorname{span}E$.
- The support of a function f is denoted by $\operatorname{supp}f$.
- $\mathbb{R}^{m\times n}$ is the set of matrices with m rows and n columns. For $A\in\mathbb{R}^{m\times n}$, we write its entries as

$$A=\left(A^i_j\right)^{1\leq i\leq m}_{1\leq j\leq n}.$$

- A^t is the transpose of A.
- $\mathrm{GL}(n)$, respectively $O(n)$, stands for set of invertible, respectively orthogonal, $n\times n$ matrices.
- For a $n\times n$ matrix A, we denote by $\det A$ the determinant of A and $\operatorname{adj}A$ stands for the adjugate matrix of A.
- The dimension of a vector space X is denoted $\dim X$.
- For integers $0\leq s\leq n$, we let

$$\binom{n}{s}=\frac{n!(n-s)!}{s!}.$$

- For two integers i and j, the Kronecker symbol is denoted by δ_{ij} abbreviating

$$\delta_{ij}=\begin{cases}0 & \text{if}\quad i\neq j\\ 1 & \text{if}\quad i=j.\end{cases}$$

Exterior Algebra

- For an integer $k \geq 1$, $\operatorname{Sym}(k)$ stands for the set of permutations of $\{1,\ldots,k\}$. For $\sigma \in \operatorname{Sym}(k)$, we denote by $\operatorname{sign}(\sigma) \in \{-1,1\}$ the sign of σ.
- The set of exterior k-forms over $\mathbb{R}^n$ is denoted by $\Lambda^k(\mathbb{R}^n)$.
- $\mathscr{T}_k$ stands for $\{(i_1,\ldots,i_k) \in \mathbb{N}^k : 1 \leq i_1 < \cdots < i_k \leq n\}$.
- The $\wedge$, respectively $\lrcorner$ and $*$ operator stands for the exterior product, respectively the interior product and Hodge star operator; see Section 2.1.
- The $\widehat{\ }$ sign above a term in a sequence of indices, forms, or coefficients means that the corresponding term is omitted; for example,

$$e^1 \wedge \widehat{e^2} \wedge e^3 = e^1 \wedge e^2.$$

- For an integer m and an exterior form f, we write

$$f^m = \underbrace{f \wedge \cdots \wedge f}_{m \text{ times}}.$$

- For a matrix A and an exterior form f, $A^*(f)$ stands for the pullback of f by A; see Section 2.1.

- For $f \in \Lambda^k(\mathbb{R}^n)$, we let

$$\Lambda_f^1 = \{u \in \Lambda^1(\mathbb{R}^n) : \exists g \in \Lambda^{k-1}(\mathbb{R}^n) \quad \text{with} \quad g \lrcorner f = u\}.$$

- For an integer s, $\operatorname{Anh}_\wedge(f,s)$, respectively $\operatorname{Anh}_\lrcorner(f,s)$, stands for the space of exterior, respectively interior, annihilators of order s of f (see Section 2.2).
- The rank, respectively the corank, of order s of f is denoted by $\operatorname{rank}_s[f]$, respectively $\operatorname{corank}_s[f]$.
- The exterior, respectively interior, matrix of order s associated to f is given by $\overline{f}_{\wedge,s}$, respectively $\overline{f}_{\lrcorner,s}$ (see Notation 2.30).
- Except in Chapter 2, we write, in order not to burden the notations, $\operatorname{rank}[f]$ for $\operatorname{rank}_1[f]$ and $\overline{f}$ for $\overline{f}_{\lrcorner,1}$.

Differential Forms

- For a function $\varphi : \mathbb{R}^m \to \mathbb{R}^n$ and $I \in \mathscr{T}_k$, we denote (cf. Definition 3.8),

$$d\varphi^I = d\varphi^{i_1} \wedge \cdots \wedge d\varphi^{i_k}.$$

- For the notations $L^p(\Omega;\Lambda^k)$, respectively $W^{r,p}(\Omega;\Lambda^k)$, $C^{r,\alpha}(\Omega;\Lambda^k)$; see Chapter 3.
- For the spaces with vanishing tangential component $C_T^{r,\alpha}(\overline{\Omega};\Lambda^k)$ and $W_T^{r,p}(\Omega;\Lambda^k)$ and the spaces with vanishing normal component $C_N^{r,\alpha}(\overline{\Omega};\Lambda^k)$ and $W_N^{r,p}(\Omega;\Lambda^k)$, see Definition 3.24.
- For the set of harmonic fields $\mathscr{H}(\Omega;\Lambda^k)$, $\mathscr{H}_T(\Omega;\Lambda^k)$ and $\mathscr{H}_N(\Omega;\Lambda^k)$, see Definition 6.1.
- For a sufficiently smooth open set Ω, ν stands for the exterior unit normal of Ω.
- The operator d, respectively δ and Δ, stands for the exterior derivative, respectively the interior derivative and the Laplacian operator for forms; see Chapter 3.

- The 2-form ω_m denotes the standard symplectic 2-form of rank $2m$; that is,

$$\omega_m = \sum_{i=1}^{m} dx^{2i-1} \wedge dx^{2i}.$$

- For a map φ and a differential form f, we denote $\varphi^*(f)$ the pullback of f by φ. For the notation $\varphi^\sharp(f)$, see Theorem 3.10.
- For a map u and a differential form f, $\mathscr{L}_u f$ stands for the Lie derivative of f with respect to u; see Notation 4.1. For the notations L^u and K^u, see Definition 5.1.

Functions Spaces

- For $1 \leq p \leq \infty$ and $r \geq 0$, $W^{r,p}$ denote Sobolev spaces, namely the spaces of L^p functions, whose weak partial derivatives of order up to r exist and are all in L^p. We make the convention $W^{0,p} = L^p$.
- For the definition of the spaces C^r, C_0^r, $C^{r,\alpha}$, $C_0^{r,\alpha}$, $C_*^{r,\alpha}$, C^∞ and C_0^∞ and the norms $\|\cdot\|_{C^r}$, $\|\cdot\|_{C^{r,\alpha}}$, and $\|\cdot\|_{C_*^{r,\alpha}}$, see Section 16.1.
- The set of diffeomorphisms of class $C^{r,\alpha}$ from U onto V is denoted by $\mathrm{Diff}^{r,\alpha}(U;V)$. The set of homeomorphisms of class $C^{0,\alpha}$ from U onto V is denoted by $\mathrm{Hom}^{0,\alpha}(U;V)$.
- For a bounded open set Ω and $f \in C^0(\overline{\Omega})$, we write

$$F^+ = f^{-1}((0,\infty)) \quad \text{and} \quad F^- = f^{-1}((-\infty,0)).$$

For $x \in F^\pm$,

$$F_x^\pm \text{ denotes the connected component of } F^\pm \text{ containing } x.$$

- For a function f, $D^a f$, respectively $\nabla^r f$ and $\partial f/\partial v$, stands for the derivative of f with respect to the multi-index a, respectively the set of derivatives of order r and the derivative of f in the direction v. $\mathscr{A}_r$ stands for the set of multi-indices of order r. Sometimes the gradient ∇f of a function f is also denoted by $\operatorname{grad} f$.
- For a vector field u, $\operatorname{div} u$ denotes the divergence of u. For the notation $\operatorname{curl}^* u$; see Notation 9.1.
- For a closed set F, the function $d^*(x) = d^*(x;F)$ is a regularization of the usual distance function $d(x) = d(x;F)$ from x to F (see Theorem 16.21).

Index

MIX
Papier aus verantwortungsvollen Quellen
Paper from responsible sources
FSC® C105338

If you have any concerns about our products,
you can contact us on
ProductSafety@springernature.com

In case Publisher is established outside the EU,
the EU authorized representative is:
Springer Nature Customer Service Center GmbH
Europaplatz 3, 69115 Heidelberg, Germany

Printed by Libri Plureos GmbH
in Hamburg, Germany